Arabidopsis Protocols

METHODS IN MOLECULAR BIOLOGY™

John M. Walker, SERIES EDITOR

83. **Receptor Signal Transduction Protocols**, edited by *R. A. J. Challiss, 1997*
82. ***Arabidopsis* Protocols**, edited by *José M Martinez-Zapater and Julio Salinas, 1998*
81. **Plant Virology Protocols**, edited by *Gary D. Foster, 1998*
80. **Immunochemical Protocols (2nd. ed.)**, edited by *John Pound, 1998*
79. **Polyamine Protocols**, edited by *David M. L. Morgan, 1998*
78. **Antibacterial Peptide Protocols**, edited by *William M. Shafer, 1997*
77. **Protein Synthesis:** *Methods and Protocols*, edited by *Robin Martin, 1998*
76. **Glycoanalysis Protocols**, edited by *Elizabeth F. Hounsel, 1998*
75. **Basic Cell Culture Protocols**, edited by *Jeffrey W. Pollard and John M. Walker, 1997*
74. **Ribozyme Protocols**, edited by *Philip C. Turner, 1997*
73. **Neuropeptide Protocols**, edited by *G. Brent Irvine and Carvell H. Williams, 1997*
72. **Neurotransmitter Methods**, edited by *Richard C. Rayne, 1997*
71. **PRINS and *In Situ* PCR Protocols**, edited by *John R. Gosden, 1997*
70. **Sequence Data Analysis Guidebook**, edited by *Simon R. Swindell, 1997*
69. **cDNA Library Protocols**, edited by *Ian G. Cowell and Caroline A. Austin, 1997*
68. **Gene Isolation and Mapping Protocols**, edited by *Jacqueline Boultwood, 1997*
67. **PCR Cloning Protocols:** *From Molecular Cloning to Genetic Engineering*, edited by *Bruce A. White, 1996*
66. **Epitope Mapping Protocols**, edited by *Glenn E. Morris, 1996*
65. **PCR Sequencing Protocols**, edited by *Ralph Rapley, 1996*
64. **Protein Sequencing Protocols**, edited by *Bryan J. Smith, 1996*
63. **Recombinant Proteins:** *Detection and Isolation Protocols*, edited by *Rocky S. Tuan, 1996*
62. **Recombinant Gene Expression Protocols**, edited by *Rocky S. Tuan, 1996*
61. **Protein and Peptide Analysis by Mass Spectrometry**, edited by *John R. Chapman, 1996*
60. **Protein NMR Protocols**, edited by *David G. Reid, 1996*
59. **Protein Purification Protocols**, edited by *Shawn Doonan, 1996*
58. **Basic DNA and RNA Protocols**, edited by *Adrian J. Harwood, 1996*
57. **In Vitro Mutagenesis Protocols**, edited by *Michael K. Trower, 1996*
56. **Crystallographic Methods and Protocols**, edited by *Christopher Jones, Barbara Mulloy, and Mark Sanderson, 1996*
55. **Plant Cell Electroporation and Electrofusion Protocols**, edited by *Jac A. Nickoloff, 1995*
54. **YAC Protocols**, edited by *David Markie, 1995*
53. **Yeast Protocols:** *Methods in Cell and Molecular Biology*, edited by *Ivor H. Evans, 1996*
52. **Capillary Electrophoresis:** *Principles, Instrumentation, and Applications*, edited by *Kevin D. Altria, 1996*
51. **Antibody Engineering Protocols**, edited by *Sudhir Paul, 1995*
50. **Species Diagnostics Protocols:** *PCR and Other Nucleic Acid Methods*, edited by *Justin P. Clapp, 1996*
49. **Plant Gene Transfer and Expression Protocols**, edited by *Heddwyn Jones, 1995*
48. **Animal Cell Electroporation and Electrofusion Protocols**, edited by *Jac A. Nickoloff, 1995*
47. **Electroporation Protocols for Microorganisms**, edited by *Jac A. Nickoloff, 1995*
46. **Diagnostic Bacteriology Protocols**, edited by *Jenny Howard and David M. Whitcombe, 1995*
45. **Monoclonal Antibody Protocols**, edited by *William C. Davis, 1995*
44. ***Agrobacterium* Protocols**, edited by *Kevan M. A. Gartland and Michael R. Davey, 1995*
43. **In Vitro Toxicity Testing Protocols**, edited by *Sheila O'Hare and Chris K. Atterwill, 1995*
42. **ELISA:** *Theory and Practice*, by *John R. Crowther, 1995*
41. **Signal Transduction Protocols**, edited by *David A. Kendall and Stephen J. Hill, 1995*
40. **Protein Stability and Folding:** *Theory and Practice*, edited by *Bret A. Shirley, 1995*
39. **Baculovirus Expression Protocols**, edited by *Christopher D. Richardson, 1995*
38. **Cryopreservation and Freeze-Drying Protocols**, edited by *John G. Day and Mark R. McLellan, 1995*
37. **In Vitro Transcription and Translation Protocols**, edited by *Martin J. Tymms, 1995*
36. **Peptide Analysis Protocols**, edited by *Ben M. Dunn and Michael W. Pennington, 1994*
35. **Peptide Synthesis Protocols**, edited by *Michael W. Pennington and Ben M. Dunn, 1994*
34. **Immunocytochemical Methods and Protocols**, edited by *Lorette C. Javois, 1994*
33. ***In Situ* Hybridization Protocols**, edited by *K. H. Andy Choo, 1994*
32. **Basic Protein and Peptide Protocols**, edited by *John M. Walker, 1994*
31. **Protocols for Gene Analysis**, edited by *Adrian J. Harwood, 1994*
30. **DNA–Protein Interactions**, edited by *G. Geoff Kneale, 1994*
29. **Chromosome Analysis Protocols**, edited by *John R. Gosden, 1994*
28. **Protocols for Nucleic Acid Analysis by Nonradioactive Probes**, edited by *Peter G. Isaac, 1994*
27. **Biomembrane Protocols:** *II. Architecture and Function*, edited by *John M. Graham and Joan A. Higgins, 1994*

METHODS IN MOLECULAR BIOLOGY™

Arabidopsis Protocols

Edited by

José M. Martínez-Zapater

and

Julio Salinas

Centro de Investigación y Tecnología, Instituto Nacional de Investigación y Tecnología Agraria y Alimentaria, Madrid, Spain

Humana Press ✺ Totowa, New Jersey

© 1998 Humana Press Inc.
999 Riverview Drive, Suite 208
Totowa, New Jersey 07512

All rights reserved. No part of this book may be reproduced, stored in a retrieval system, or transmitted in any form or by any means, electronic, mechanical, photocopying, microfilming, recording, or otherwise without written permission from the Publisher. Methods in Molecular Biology™ is a trademark of The Humana Press Inc.

All authored papers, comments, opinions, conclusions, or recommendations are those of the author(s), and do not necessarily reflect the views of the publisher.

This publication is printed on acid-free paper. ∞
ANSI Z39.48-1984 (American Standards Institute) Permanence of Paper for Printed Library Materials.

Cover illustration: From Fig. 3 in Chapter 38, "β–Glucuronidase Enzyme Histochemistry on Semithin Sections of Plastic-Embedded *Arabidopsis* Explants," by Marc De Block and Mieke Van Lijsebettens.

For additional copies, pricing for bulk purchases, and/or information about other Humana titles, contact Humana at the above address or at any of the following numbers: Tel: 973-256-1699; Fax: 973-256-8341; E-mail: humana@mindspring.com, or visit our Website at www.humanapress.com

Photocopy Authorization Policy:
Authorization to photocopy items for internal or personal use, or the internal or personal use of specific clients, is granted by Humana Press Inc., provided that the base fee of US $8.00 per copy, plus US $00.25 per page, is paid directly to the Copyright Clearance Center at 222 Rosewood Drive, Danvers, MA 01923. For those organizations that have been granted a photocopy license from the CCC, a separate system of payment has been arranged and is acceptable to Humana Press Inc. The fee code for users of the Transactional Reporting Service is: [0-89603-391-0/98 $8.00 + $00.25].

Printed in the United States of America. 10 9 8 7 6 5 4 3 2

Library of Congress Cataloging in Publication Data

Main entry under title:

Methods in molecular biology™.

Arabidopsis Protocols/edited by José M. Martínez Zapater and Julio Salinas
    p. cm.—(Methods in molecular biology™; vol. 82)
    Includes index.
    ISBN 0-89603-391-0 (alk. paper)
    1. Arabidopsis—Laboratory manuals. 2. Arabidopsis—Molecular aspects—Laboratory manuals.
    I. Martínez-Zapater, José M. II. Salinas, Julio. III. Series: Methods in molecular biology (Totowa, NJ); 82.
QK495.C9A735 1998
583'.64—dc21
DNLM/DLC
for Library of Congress
    97-44454
    CIP

Preface

The use of Arabidopsis in genetics and biology laboratories can be traced back to the beginning of the century, when the small size of this plant species, its short generation time, and the thousands of seeds produced by each plant attracted the interest of geneticists. In the middle of the century, this plant had already gained a place in several genetics laboratories both in Europe and the United States, being widely used in mutagenesis experiments. However, exponential growth in the use of Arabidopsis as a model system did not get started until the 1980s, when the small size of its genome was realized, thus triggering the interest of geneticists and molecular biologists, not only from the plant field, but from other research fields needing a novel model system.

In recent years, high expectations for Plant Biotechnology in the next century, together with the need for basic information in every area of plant biology, have served to mobilize resources and orient much new research towards the plant field, where the special biological features of Arabidopsis have further focused attention on this useful species. The high level of interest in Arabidopsis has worked as a catalyst and, today, the system has been developed to the point where any gene identified on the basis of a phenotype can be cloned with a reasonable amount of time and effort. In little more than 10 years from the beginning of this exponential growth phase, the system has reached a self-sustaining stage that enhances its everyday utility as a result of the accumulating flow of powerful information and newly available tools. In short, it is now widely accepted that Arabidopsis has become the system of choice for approaching fundamental questions in plant biology.

Arabidopsis Protocols provides to both experienced researchers and beginners in the field a comprehensive set of up-to-date protocols covering the many methods developed during work with this species, from growing plants, to explants, to gene cloning strategies. The book also includes sections on genetic, transformation, and gene expression analyses that will be especially helpful for scientists involved in mutation analysis or in producing and analyzing transgenic plants. In many cases, the protocols can also be applied to other plant species with minor adjustments. Although a few chapters go beyond the scope of a protocol, providing useful information for the arabidopsologist, the majority of them contain step-by-step protocols, in line with other volumes in this series. Altogether, these *Protocols* form a useful reference for laboratories working not only with Arabidopsis, but also with other plant species

The editors want to thank C. R. Somerville, S. Somerville, and M. Koornneef for their helpful discussions and advice at the beginning of this project, and all colleagues who have contributed their protocols to this latest Humana Press publication.

José M. Martínez-Zapater
Julio Salinas

Contents

Preface v
Contributors xi
List of Color Plates xv

PART I. ARABIDOPSIS CULTURE

1　Growth of Plants and Preservation of Seeds
　　Randy Scholl, Luz Rivero-Lepinckas, and Deborah Crist 1
2　Sterile Techniques in *Arabidopsis*
　　Peter McCourt and Kallie Keith 13
3　Control of Pests and Diseases of *Arabidopsis*
　　Mary Anderson 19
4　Establishment and Maintenance of Cell Suspension Cultures
　　Jaideep Mathur and Csaba Koncz 27
5　Callus Culture and Regeneration
　　Jaideep Mathur and Csaba Koncz 31
6　Protoplast Isolation, Culture, and Regeneration
　　Jaideep Mathur and Csaba Koncz 35

PART II. PURIFICATION OF SUBCELLULAR ORGANELLES AND MACROMOLECULES

7　Preparation of Physiologically Active Chloroplasts from *Arabidopsis*
　　Ljerka Kunst 43
8　Purification of Mitochondria from *Arabidopsis*
　　Mathieu Klein, Stefan Binder, and Axel Brennicke 49
9　Preparation of DNA from *Arabidopsis*
　　Jianming Li and Joanne Chory 55
10　High Molecular Weight DNA Extraction from *Arabidopsis*
　　David Bouchez and Christine Camilleri 61
11　Chloroplast DNA Isolation
　　George S. Mourad 71
12　Purification of Mitochondrial DNA from Green Tissues of *Arabidopsis*
　　**Mathieu Klein, Rudolf Hiesel, Charles André,
　　and Axel Brennicke** 79

13 Preparation of RNA
   Clifford D. Carpenter and Anne E. Simon 85

PART III. MUTAGENESIS AND GENETIC ANALYSIS

14 Seed Mutagenesis of *Arabidopsis*
   Jonathan Lightner and Timothy Caspar 91

15 Genetic Analysis
   Maarten Koornneef, Carlos Alonso-Blanco, and Piet Stam 105

16 Cytogenetic Analysis of *Arabidopsis*
   John S. Heslop-Harrison 119

17 PCR-Based Identification of T-DNA Insertion Mutants
   Rodney G. Winkler and Kenneth A. Feldmann 129

PART IV. GENE MAPPING IN *ARABIDOPSIS*

18 The Use of Recombinant Inbred Lines (RILs) for Genetic Mapping
   Carlos Alonso-Blanco, Maarten Koornneef, and Piet Stam 137

19 AFLP™ Fingerprinting of *Arabidopsis*
   Pieter Vos 147

20 Building a High-Density Genetic Map Using the AFLP™ Technology
   Martin T. R. Kuiper 157

21 Use of Cleaved Amplified Polymorphic Sequences (CAPS)
   as Genetic Markers in *Arabidopsis thaliana*
   *Jane Glazebrook, Eliana Drenkard, Daphne Preuss,
   and Frederick Ausubel* 173

22 Mapping Mutations with ARMS
   Anton R. Schäffner 183

23 Mapping Cloned Sequences on YACs
   *Francis D. Agyare, Gus Lagos, Deval Lashkari, Ronald W. Davis,
   and Bertrand Lemieux* 199

PART V. TRANSIENT AND STABLE TRANSFORMATION

24 Transient Gene Expression in Protoplasts of *Arabidopsis thaliana*
   Steffen Abel and Athanasios Theologis 209

25 Transient Expression of Foreign Genes in Tissues of *Arabidopsis
   thaliana* by Bombardment-Mediated Transformation
   Motoaki Seki, Asako Iida, and Hiromichi Morikawa 219

26 Root Transformation by *Agrobacterium tumefaciens*
   *Annette C. Vergunst, Ellen C. de Waal,
   and Paul J. J. Hooykaas* 227

Contents

27 Transformation of *Arabidopsis thaliana* C24 Leaf Discs
   by *Agrobacterium tumefaciens*
   Eric van der Graaf and Paul J. J. Hooykas 245

28 *In Planta Agrobacterium*-Mediated Transformation
   of Adult *Arabidopsis thaliana* Plants by Vacuum Infiltration
   Nicole Bechtold and Georges Pelletier 259

29 PEG-Mediated Protoplast Transformation with Naked DNA
   Jaideep Mathur and Csaba Koncz 267

PART VI. GENE CLONING STRATEGIES

30 Cloning Genes of *Arabidopsis thaliana* by Chromosome Walking
   Jeffrey Leung and Jérôme Giraudat 277

31 Chromosome Landing Using an AFLP™-Based Strategy
   **Ann Van Gysel, Gerda Cnops, Peter Breyne, Marc Van Montagu,
   and Maria Teresa Cervera** 305

32 Transposon Tagging with *Ac/Ds* in *Arabidopsis*
   Deborah Long and George Coupland 315

33 Transposon Tagging with the *EN-I* System
   Andy Pereira and Mark G. M. Aarts 329

34 Cloning Genes from T-DNA Tagged Mutants
   Brian P. Dilkes and Kenneth A. Feldmann 339

PART VII. GENE EXPRESSION ANALYSES

35 *In Situ* Hybridization
   Gary N. Drews 353

36 Whole-Mount *In Situ* Hybridization in Plants
   **Janice de Almeida Engler, Marc Van Montagu,
   and Gilbert Engler** 373

37 Bacterial and Coelenterate Luciferases as Reporter Genes
   in Plant Cells
   William H. R. Langridge and Aladar A. Szalay 385

38 β-Glucoronidase Enzyme Histochemistry on Semithin Sections
   of Plastic-Embedded *Arabidopsis* Explants
   Marc De Block and Mieke Van Lijsebettens 397

39 *In Situ* Hybridization to RNA in Whole *Arabidopsis* Plants
   Qingzhong Kong and Anne E. Simon 409

40 In Vivo Footprinting in *Arabidopsis*
   Anna-Lisa Paul and Robert J. Ferl 417

Index 431

Contributors

MARK G. M. AARTS • *Centrum voor Plantenveredelings en Reproduktieonderzoek (CPRO-DLO), Wageningen, The Netherlands*
STEFFEN ABEL • *Department of Plant Biology, Plant Gene Expression Center, Albany, CA*
FRANCIS D. AGYARE • *Department of Biology, Faculty of Pure and Applied Sciences, York University, Ontario, Canada*
CARLOS ALONSO-BLANCO • *Department of Genetics, Agricultural University of Wageningen, The Netherlands*
MARY ANDERSON • *Nottinghum Arabidopsis Stock Centre, Department of Life Science, Nottingham University, UK*
CHARLES ANDRÉ • *Universitat Ulm, Allgemeine Botanik, Germany*
FREDERICK M. AUSUBEL • *Department of Genetics, Harvard Medical School; and Department of Molecular Biology, Massachusetts General Hospital, Boston, MA*
NICOLE BECHTOLD • *Laboratoire de Génétique et Amélioration des Plantes, INRA-Centre de Versailles, France*
STEFAN BINDER • *Allgemeine Botanik, Universitat Ulm, Germany*
AXEL BRENNICKE • *Allgemeine Botanik, Universitat Ulm, Germany*
PETER BREYNE • *Laboratorium voor Genetica, Universiteit Gent, Belgium*
DAVID BOUCHEZ • *Laboratoire de Biologie Cellulaire, INRA-Centre de Versailles, France*
CHRISTINE CAMILLERI • *Laboratoire de Biologie Cellulaire, INRA-Centre de Versailles, France*
CLIFFORD D. CARPENTER • *Department of Biochemistry and Molecular Biology, University of Massachusettes, Amherst, MA*
TIM CASPAR • *Dupont Central Research and Development, Experimental Station, Wilmington, DE*
MARÍA TERESA CERVERA • *Laboratorium voor Genetica, Universiteit Gent, Belgium*
JOANNE CHORY • *Plant Biology Laboratory, Salk Institute, La Jolla, CA*
GERDA CNOPS • *Laboratorium voor Genetica, Universiteit Gent, Belgium*
GEORGE COUPLAND • *Department of Molecular Genetics, John Innes Centre, Norwich, UK*

DEBORAH CRIST • *Arabidopsis Biological Resource Center, Ohio State University, Columbus, OH*

RONALD W. DAVIS • *Stanford DNA Sequence and Technology Center, Department of Biochemistry, School of Medicine, Stanford University, CA*

JANICE DE ALMEIDA ENGLER • *Laboratorium voor Genetica, Universiteit Gent, Belgium*

MARC DE BLOCK • *Plant Genetic Systems NV, Gent, Belgium*

ELLEN C. DE WAAL • *Clusius Laboratory, Institute of Molecular Plant Sciences, Faculty of Mathematics and Natural Sciences, Leiden, The Netherlands*

BRIAN P. DILKES • *Department of Plant Sciences, University of Arizona, Tucson, AZ*

ELIANA DRENKARD • *Harvard Medical School and Department of Molecular Biology, Massachusetts General Hospital, Boston, MA*

GARY N. DREWS • *Department of Biology, University of Utah, Salt Lake City, UT*

GILBERT ENGLER • *Laboratorium voor Genetica, Universiteit Gent, Belgium*

KENNETH A. FELDMANN • *Department of Plant Sciences, University of Arizona, Tucson, AZ*

ROBERT FERL • *Horticultural Sciences Department, Institute of Food and Agricultural Sciences, Gainesville, FL*

JÉRÔME GIRAUDAT • *Institut des Sciences Végétales, France*

JANE GLAZEBROOK • *Center for Agricultural Biotechnology, University of Maryland, Biotechnology Institute, College Park, MD*

RUDOLF HIESEL • *Allgemeine Botanik, Universitat Ulm, Germany*

JOHN S. HESLOP-HARRISON • *Department of Cell Biology, John Innes Centre, Norwich, UK*

PAUL J. J. HOOYKAAS • *Clusius Laboratory, Institute of Molecular Plant Sciences, Faculty of Mathematics and Natural Sciences, Leiden, The Netherlands*

ASAKO IISA • *Biotechnology Laboratory, Sumitomo Chemical Co., Hyogo, Japan*

KALLIE KEITH • *Department of Botany, University of Toronto, Canada*

MATHIEU KLEIN • *Allgemeine Botanik, Universitat Ulm, Germany*

CSABA KONCZ • *Max-Planck-Institut für Zuchtungsforschung, Cologne, Germany*

QINGZHONG KONG • *Department of Biochemistry and Molecular Biology, University of Massachusetts, Amherst, MA*

Contributors

MAARTEN KOORNNEEF • *Department of Genetics, Agricultural University of Wageningen, The Netherlands*

MARTIN T. R. KUIPER • *Keygene, Wageningen, The Netherlands*

LJERKA KUNST • *Department of Botany, University of British Columbia, Vancouver, BC, Canada*

GUS LAGOS • *Department of Biology, Faculty of Pure and Applied Sciences, York University, Ontario, Canada*

WILLIAM H. R. LANGRIDGE • *Center for Molecular Biology and Gene Therapy, Loma Linda University, CA*

DEVAL LASHKARI • *Stanford DNA Sequence and Technology Center, Department of Biochemistry, School of Medicine, Stanford University, CA*

BERTRAND LEMIEUX • *Department of Biology, Faculty of Pure and Applied Sciences, York University, Ontario, Canada*

JEFFREY LEUNG • *Institut des Sciences Végétales, France*

JIANMING LI • *Plant Biology Laboratory, Salk Institute, La Jolla, CA*

JONATHAN LIGHTNER • *Dupont Central Research and Development, Experimental Station, Wilmington, DE*

DEBORAH LONG • *Department of Molecular Genetics, John Innes Centre, Norwich, UK*

JAIDEEP MATHUR • *Max-Planck-Institut für Zuchtungsforschung, Cologne, Germany*

PETER MCCOURT • *Department of Botany, University of Toronto, Canada*

HIROMICHI MORIKAWA • *Laboratory of Plant Molecular Biology, The Institute of Physical and Chemical Research (RIKEN), Tsukuba Life Science Center, Ibaraki, Japan*

GEORGE MOURAD • *Department of Biology, Indiana University-Purdue University Fort Wayne, IN*

ANNA-LISA PAUL • *Horticultural Sciences Department, Institute of Food and Agricultural Sciences, University of Florida, Gainesville, FL*

GEORGES PELLETIER • *Laboratoire de Génétique et Amélioration des Plantes, INRA-Centre de Versailles, France*

ANDY PEREIRA • *Centrum voor Plantenveredelings-en Reproduktieonderzoek (CPRO-DLO), Wageningen, The Netherlands*

DAPHNE PREUSS • *Department of Molecular Genetics and Cell Biology, University of Chicago, IL*

LUZ RIVERO-LEPINCKAS • *Arabidopsis Biological Resource Center, Ohio State University, Columbus, OH*

ANTON R. SCHÄFFNER • *GSF Research Center, Institute of Biochemical Plant Pathology, Oberschleissheim, Germany*

RANDY SCHOLL • *Arabidopsis Biological Resource Center, Ohio State University, Columbus, OH*
MOTOAKI SEKI • *Laboratory of Plant Molecular Biology, The Institute of Physical and Chemical Research (RIKEN), Tsukuba Life Science Center, Ibaraki, Japan*
ANNE E. SIMON • *Department of Biochemistry and Molecular Biology, University of Massachusetts, Amherst, MA*
PIET STAM • *Department of Genetics, Agricultural University of Wageningen, The Netherlands*
ALADAR A. SZALAY • *Center for Molecular Biology and Gene Therapy, Loma Linda University, CA*
ATHANASIOS THEOLOGIS • *Department of Plant Biology, Plant Gene Expression Center, Albany, CA*
ERIC VAN DER GRAAF • *Clusius Laboratory, Institute of Molecular Plant Sciences, Faculty of Mathematics and Natural Sciences, Leiden, The Netherlands*
ANNE VAN GYSEL • *Laboratorium voor Genetica, Universiteit Gent, Belgium*
MIEKE VAN LIJSEBETTENS • *Laboratorium voor Genetica, Universiteit Gent, Belgium*
MARC VAN MONTAGU • *Laboratorium voor Genetica, Universiteit Gent, Belgium*
ANNETTE C. VERGUNST • *Clusius Laboratory, Institute of Molecular Plant Sciences, Faculty of Mathematics and Natural Sciences, Leiden, The Netherlands*
PIETER VOS • *Keygene, Wageningen, The Netherlands*
RODNEY G. WINKLER • *Department of Plant Sciences, University of Arizona, Tucson, AZ*

List of Color Plates

Color plates appear as an insert following p. 230

Plate 1 (Fig. 1 from Chapter 24). Histochemical localization of GUS activity in *Arabidopsis* mesophyll protoplasts transfected with translational GUS fusions. Purified protoplasts **(A)** were mock-transfected **(D)** or challenged with the following plasmid DNAs: pRTL2-GUS which encodes the nonfused, authentic GUS protein **(B)**; pRTL2-GUS-IAA1 encoding a GUS::IAA1 nuclear fusion protein **(C)**; pRTL2-GUS-PS-IAA6 encoding a GUS::PS-IAA6 nuclear fusion protein **(E,F)**. Transfected protoplasts were cultured for 20 h, stained for GUS activity (B–E) and nuclei (F) as described *(11)*.

Plate 2 (Fig. 1 from Chapter 29). GUS and GFP reporter gene expression assays with *Arabidopsis* protoplasts. **(A)** GUS staining of PEG-transformed protoplasts derived from roots of *Arabidopsis* ecotype Columbia after incubation with X-gluc for 6 h at room temperature. **(B)** Leaf mesophyll protoplasts from *Arabidopsis* ecotype Columbia transformed with pCK–GFPs65c exhibit green fluorescence when illuminated with blue light. The chloroplasts emit red fluorescence, whereas the yellow fluorescence results from overlapping red and green areas.

Plate 3 (Fig. 1 from Chapter 36). (A) WISH on an *Arabidopsis thaliana* seedling hybridized with an antisense *cdc2a* probe detected by gold-labeled antibodies. The signal is visible as a black precipitate resulting from the silver amplification reaction. **(B)** In situ localization of *cdc2a* mRNA on a vibroslice of an *Arabidopsis* root infected with a root-knot nematode. The blue precipitate resulted from the histochemical reaction with the substrates X-phosphate and NBT. **(C)** Cotyledon hybridized with an antisense *rhal* probe. Hybrids were visualized by silver amplification of specifically bound gold-labeled antibodies seen as a dark precipitate. **(D)** Chicory root hybridized with an antisense nitrate reductase probe detected by AP-labeled antibodies. A strong expression is visible in the vascular cylinder and in the root meristem. The sample was cleared in CLP for better visualization of the signal.

Plate 4 (Fig. 3 from Chapter 38). X-Gluc reactions on *Arabidopsis* tissue. **(A)** Transverse section through a 2-wk-old root of a *pRPS18A-gus* transformed line *(15)*. **(B)** Detail of the vascular tissue; main GUS activity in vascular tissue. **(C)** Transverse section through a young (stage 9) flower of a line transformed with a *gus* gene fused to a stomium-specific tobacco promoter (provided by T. Beals and P. Sanders, Plant Molecular Biology Laboratory, University of California, Los Angeles, CA). **(D)** Detail of mature anther. Main GUS activity located at the stomium, the site of anther dehiscence. Abbreviations: al, anther locule; c, cortex; en, endodermis; ep, epidermis; g, gynoecium; po, pollen; rh, root hair; s, sepal; sc, stomium cells; se, septum; st, stamen; vt, vascular tissue. Visualization with Normansky interference microscopy. Bar = 50 μm (A), 20 μm (B,D), and 100 μm (C).

I

ARABIDOPSIS CULTURE

1

Growth of Plants and Preservation of Seeds

Randy Scholl, Luz Rivero-Lepinckas, and Deborah Crist

1. Introduction

Arabidopsis can be grown in a variety of environmental settings including growth rooms, window ledges, outdoors, growth chambers, and greenhouses. The growth media that can be employed include soil, commercial greenhouse mixes, vermiculite, and other relatively inert media watered with nutrient solutions, as well as agar. This chapter focuses on growth of plants on soil in various environmental settings and especially in growth chambers and greenhouses. Harvesting, seed quality, and seed preservation are also considered.

Whereas *Arabidopsis* is adaptable, it is not the equivalent of a house plant in that a number of environmental factors must be carefully controlled, or the culture may result in very unhealthy plants and even total failure. One of the main objectives of this chapter is to indicate which conditions are critical to healthy plant growth and development and to high quality and quantity of seeds. The plant and seed management methods are discussed in the chronological order in which they would normally be utilized.

2. Materials
2.1. Growth and Harvest

1. Plastic pots (e.g., 10-cm square) (Hummert, Earth City, MO).
2. Soil mixture (e.g., Metromix 350 or other peat moss-based potting mix).
3. Large autoclavable container.
4. Large spoon or trowel.
5. pH Meter.
6. Magnetic stirring device.
7. $2M$ KOH or $1M$ H_2SO_4.
8. Beakers (100 or 250 mL, 1 L)

9. Two magnetic stirring bars.
10. Distilled water.
11. Micronutrient stock (used in making nutrient solution): To 90 mL of stirring, distilled water add the following ingredients in order (concentrations shown are final concentration after volume has been adjusted). After the last salt has been added and is dissolved, adjust the volume to 100 mL. Ingredients: 70 mM H_3BO_3, 14 mM $MnCl_2$, 0.5 mM $CuSO_4$, 1 mM $ZnSO_4$, 0.2 mM Na_2MoO_4, 10 mM NaCl, 0.01 mM $CoCl_2$.
12. Nutrient: Add approx 700 mL of distilled water and a magnetic stirring bar to a 1-L beaker and stir. Add salts in the amounts (final concentration basis) and order shown: 5 mM KNO_3, 2.5 mM KH_2PO_4 (adjusted to pH 6.5), 2.0 mM $MgSO_4$, 2.0 mM $Ca(NO_3)_2$, 50 µM Fe-EDTA, 1 mL micronutrient stock **(step 11)**. After all salts have dissolved, adjust to pH 6.5 with 2M KOH (or 1M H_2SO_4) and bring the volume to 1 L. Solution may be stored for several weeks before use, but should be discarded if cloudiness or precipitates appear.
13. 8-cm round watch glass.
14. Pasteur pipet and latex bulb.
15. Labeling tape.
16. Permanent marker.

2.2. Postharvest Seed Management

1. Light-weight transparent plastic food storage bags (approx 4 L).
2. Hand sieves (mesh size = 0.425 mm).
3. Small glass jars (125 mL) or other storage containers (preferably not plastic).
4. Cryovials (1 mL) or other containers for permanent seed storage.
5. Forced draft oven.
6. Analytical balance.
7. Desiccator with silica gel.
8. Light-weight, heat-resistant dishes with cover (preferably aluminum).
9. Tongs or forceps.
10. Petri dishes (9-cm diameter) or other similar containers.
11. Absorbent paper (e.g., filter paper).
12. Distilled water.
13. Parafilm or tape.
14. Bunsen burner.

3. Methods
3.1. Plant Growth and Seed Harvest
3.1.1. Planting and Germination

Different mixtures and media can be utilized for growing *Arabidopsis*. Growth of plants on "soil" includes all media which can be successfully utilized for nonsterile growth of plants in pots or other similar containers. Kranz (*1*) recommended a 1:1:1 mixture of perlite, fine vermiculite, and sphagnum moss for growth of plants in pots. Mixtures of soil, which have substantial peat moss, can be used successfully. Inert substrates such as vermiculite can also be utilized if nutrients are supplied in the watering solution. These are useful for nutritional studies. Peat-based commercial mixes represent a convenient and reliable base for growing plants. Mixes such as Metromix 350 support healthy *Arabidopsis* growth and have fertilizer added, so that fertilization is not necessary in the early growth phases. Seeds can be planted in various ways. However, strict control of numbers of seeds planted is maintained, and separate rows of different lines can be planted in the same pot for critical comparisons with the technique described here. Preparation of pots and planting can be accomplished as follows:

1. Autoclave soil in covered steel pan for at least 30 min to kill any pests. Approximately fifteen 10-cm pots can be filled from 25 L of soil.
2. Thoroughly wet soil with tap water or nutrient solution. Mix well with trowel or large spoon.
3. Place soil in pots. The surface of the soil should be approx 1 cm below the top of the pot. Pots are ready for planting (*see* **Note 1**).
4. Draw out the end of a long Pasteur pipet to a point over a bunsen burner as follows: Grasp the wide end of the pipet with one hand and use tweezers to hold the narrow end. Concentrate the flame approx 2 cm from the narrow end of the pipet. When the glass begins to soften, pull the two ends apart until a constriction has been created. Allow the pipet to cool. Then, hold the constricted area of the modified pipet inside several layers of paper towel and break it so that the resulting opening size is slightly larger than an *Arabidopsis* seed. This requires trial and error as well as care to avoid injury. Place a latex bulb on the upper end of the modified Pasteur pipet.
5. Submerge the seeds to be planted in a watch glass, or other dish, with water (preferably distilled).
6. Exhaust air from the pipet, submerge its tip, and use slow release pressure on the bulb to suck a single seed into the end of the pipet. The seed can be transferred to the pot and dropped at the desired location by careful exhausting of the pipet. Be careful not to draw seeds beyond 1–2 cm into the pipet. Make certain that seeds are not planted below the soil surface. Repeated pipetings are used for the remainder of the seeds (*see* **Note 2**).
7. Cover pots with clear plastic wrap, and cut several small slits in the plastic with a knife (*see* **Note 3**). Place pots in the refrigerator for at least 2 d (*see* **Note 4**).
8. After cold treatment, place pots in growth area. Remove plastic wrap and maintain approx 2 cm of water around base of pots during the germination phase (4–6 d). Leave plastic wrap on for plants grown in growth chamber and do not add additional water.

3.1.2. Care of Plants During Growth

The rate of development, time to flowering, and time to harvest of *Arabidopsis* depend on several factors besides the genetic background:

1. Light (plants flower rapidly under continuous light or long days, whereas under short days, flowering is prevented or delayed).
2. Temperature (lower temperatures slow the growth, favoring the vegetative phase).
3. Nutrition (poor nutrition can lead to rapid flowering, shortened growth period, and low seed set).

Under continuous light, 25°C, adequate watering, and good nutrition, seeds of the commonly used ecotype Columbia germinate 3–5 d after planting. They form rosettes, bolt, and flower within 3–4 wk and can be harvested within 8 wk.

Management of light, temperature, water, and nutrition during the growth of plants will ensure that healthy plants develop that produce high quality and quantity of seeds. The effect of each factor is discussed separately.

3.1.2.1. Light

Optimum light is approx 150 $\mu E/m^2/s$. Cool white or very high output (VHO or SHO) fluorescent lamps supplemented by incandescent lighting are used for growth chambers. Higher light intensity, up to full sun, is tolerated by older plants, although the use of 60% shade cloth in summer greenhouses helps with light intensity control and temperature regulation. Supplemental evening and morning light is provided in the greenhouse during winter since the plants generally require a long photoperiod (at least 12 h) for flowering. Continuous light is well tolerated and can be used to accelerate the reproductive cycle. Photoperiods of 16 h work well for greenhouse growth.

3.1.2.2. Temperature

The optimum growth temperature for *Arabidopsis* is 25°C. Lower temperatures are permissible, but higher temperatures are not recommended for germination through early rosette development. Older plants tolerate higher temperatures, at least up to 30°C, but maintenance of 25°C is nevertheless desirable for the entire growth period. Growth chambers can be set at 25°C when used for seed production. It is advisable to set the greenhouse temperature at 23°C unless temperature control is exceptionally good, so that the temperature fluctuations do not cause problems. It is recommended that night temperatures be maintained 2–4°C lower than the day temperature.

3.1.2.3. Watering and Humidity

After germination, plants are watered as needed to avoid water stress. Water is best applied by subirrigation for the first few weeks and only when the soil begins to dry. Overwatering should be avoided due to the potential for algal or fungal growth on the soil surface (*see* **Note 5**). Overwatering of greenhouse plants also provides favorable soil conditions for fungus gnat larvae. More frequent watering may be necessary during the first few days, as it is necessary to avoid any drying before the first two true leaves begin expanding. After plants have developed true leaves, watering frequency may be reduced to as low as once or twice per week until the plants flower. The water requirement of plants increases dramatically during silique filling. Daily watering at this stage is necessary for good seed production.

Water requirement is strongly influenced by relative humidity. *Arabidopsis* plants, including seedlings, tolerate low humidity although increased humidity (e.g., 50–60%) greatly reduces the risk of accidental drying of the soil surface and subsequent desiccation of the fragile, germinating seedlings. Rosette-stage plants thrive at a wide range of humidities. For optimal seed quality, lower humidities (<50%) are desirable when siliques begin to mature.

3.1.2.4. Fertilization

Mild mineral nutrient solution can be applied to pots at 2-wk intervals. This solution can also be mixed with the soil when it is initially prepared. Fertilization is not absolutely required during the growth phase for seed production if the initial soil mix is well fertilized. However, the addition of nutrients in the later growth phases will increase seed set and result in healthier plants. Preparation and application of the fertilizer solution described in **Subheading 2.** has worked well for seed production (*see* **Note 6**) at the *Arabidopsis* Biological Resource Center (ABRC). This solution was adapted from Estelle and Somerville *(2)*.

3.1.3. Harvesting

The avoidance of seed mixing among adjacently growing lines, loss of seeds owing to shattering, and assurance of the quality of the harvested seeds are goals that pose some specific challenges. Various means and devices can be employed to achieve these. Watering of pots should be discontinued several days prior to harvest so that pots are dry when harvest is conducted. It should be noted that delays in harvesting following physiological maturation of the plant result in seed deterioration, especially under nonoptimal environmental conditions. Several seed collection strategies are compared in the following.

3.1.3.1. Bulk Production on the Open Bench

For bulk seed production, the best method is to simply grow the plants on the open bench, keep all accessions separated by adequate space, avoid disturbance of maturing inflorescences, and harvest when all siliques are dry. = yellow The entire inflorescence is cut off at its base and carefully placed into an approx 4-L transparent plastic bag. This is compatible with the goals of high seed quality, maximum seed yield, and good pest protection. Some seeds may be lost, but the remainder are almost always healthy, and result in vigorously germinating seedlings. After harvest, the entire contents of the bag are allowed to dry in preparation for threshing.

3.1.3.2. Early Harvest of Individual Siliques

Seeds from individual siliques can be harvested after the siliques have turned completely yellow, if rapid turnover is required. However, such seeds do have high levels of germination inhibitors. For normal seed production, seeds are harvested only after the siliques have completely browned, and if pressed with fingers, do not compress (if the silique has dried even further, the silique will shatter at this point). At this stage, seeds are completely formed.

Since formation and maturation of siliques occur over time, early siliques can be harvested before later ones mature to avoid seed loss. However, it is usually recommended to wait until the entire inflorescence has browned before harvest.

3.1.3.3. Commercial Seed Collectors

Aracons™ (Lehle Seed Co, Round Rock, TX) *(3)* are effective for single-plant harvesting, preventing cross pollination, and seed contamination, but are not necessary when simple bulk production is desired. Harvesting from Aracons (after pots have been allowed to dry) is accomplished easily by carefully cutting off the inflorescence under the device, placing the Aracon plus contents carefully into a large plastic bag (approx 4 L), removing the plant material from the plastic tube, and then shaking the seeds into the bag.

3.1.3.4. Bagging Inflorescences by Pot

Inflorescences of non-*erecta* lines can simply be trained into an approx 4 L transparent plastic bag before any siliques begin to brown. The bag, however, may potentially collect moisture from transpiration or careless watering and provides a haven for insects when greenhouses are sprayed. To reduce these possibilities, the bag should be kept widely open at all times. Wait until the inflorescence has browned before harvesting. This method is conducive to strict isolation of the lines, and the bag serves to collect shattered seeds. Harvesting is accomplished by carefully cutting the entire inflorescence off at its base after all seeds have matured and shaking the seeds into the plastic bag.

3.2. Postharvest Handling and Preservation of Seeds

The longevity of seeds can be affected by: genotype; pre-storage environment, such as conditions during seed maturation, harvesting, and seed handling; and seed storage conditions. The genotype and prestorage conditions are important because they determine the maximum potential for seed longevity. Our experience regarding effect of genotype is limited, although rapid deterioration of seeds has not been observed for the diverse collections currently maintained at ABRC. The abscisic acid mutants are one possible exception.

A slow process of deterioration begins as soon as seeds mature on a plant. Therefore, the sooner seeds are placed into storage, the better. Harvested seeds should be processed promptly (including threshing, cleaning, drying and packaging) and then placed into storage. The following procedures form a sequence that ensures that the seeds will be conserved in the best possible condition.

3.2.1. Threshing and Cleaning

The harvested plant material should be allowed to dry for a few days in the opened plastic bag before threshing, since threshing is easier when the inflorescences are dry. Seeds should be threshed when the moisture content is approx 10%, to minimize seed damage during threshing. This moisture content will be reached when all material in the bag appears to be dry. The plastic bags containing dried inflorescences can be gently hand-pressed from the outside, and the seeds will fall to the bottom of the bag (*see* **Note 7**). Most of the dry inflorescence can be removed from the bag by hand before seeds are sieved to separate them from chaff. Hand sieves with graded mesh sizes are recommended to remove large and small debris. Small, shrunken, and immature seeds are returned to the bulk, particularly for accessions which are genetically heterogeneous. After sieving, the seeds are still likely to be mixed with soil and residue. A combination of additional sieving, blowing, and visual inspection can be employed to clean the seeds completely. Small samples can be cleaned by hand with the aid of a pointed tool on an opaque glass plate illuminated from below. Cleaned seed samples are placed in open, carefully labeled glass jars (do not use plastic because of static effects) to allow seeds to dry.

3.2.2. Seed Drying

The moisture content of *Arabidopsis* seeds after threshing is usually approx 10%. The seeds should be dried to 5–6% moisture, prior to storage. Higher moisture content can cause seed deterioration. There are many methods available for drying seeds. The safest method is to air-dry the seeds at room temperature for 1–3 wk (*see* **Note 8**). Low relative humidity is necessary for seeds to reach the desired storage levels. The lower the humidity, the faster the seeds will dry, and the lower their final moisture content. If after testing, the moisture content is not low enough, continue to dry further and check again.

3.2.3. Seed Moisture Content Determination

Moisture testing is necessary to verify that seeds are dry enough for storage. Seed moisture content can be determined by several methods. The following method outlined is a destructive method, and the seeds employed for testing will no longer be viable. This is a "modified low constant temperature oven" method.

1. Preheat oven to 100–105°C (*see* **Note 9**).
2. Accurately weigh one clean, numbered dish and its cover to four decimal places using an analytical balance. Record the weight (W1; *see* **Note 10**).
3. Add approx 100 or 200 mg of seeds distributed evenly over the base of the dish, replace the cover, and accurately weigh the dish and cover. Record this weight (W2).
4. Place the dish in a safe place, and continue to prepare the second and/or third replicates in the same way.
5. When all samples have been weighed into numbered dishes, place each dish on top of its numbered lid in the oven. Heat the samples for 15–17 h.
6. Remove the dishes from the oven, replace their covers and immediately place in a desiccator at room temperature to cool for 30–45 min. After heating, the dishes must be placed directly into the desiccator so that the dry seeds do not absorb any moisture.
7. Remove the dishes one by one from the desiccator, immediately weigh each dish and cover, and record the weight (W3). Do not leave the desiccator open during the weightings.
8. Moisture content is calculated as the percentage loss in weight of the original weight of seeds. This is known as wet- or fresh-weight basis, and is expressed to one decimal place. Algebraically, if W1 is the weight of the dish, W2 the weight of dish and seed before drying, and W3 the weight of dish and seed after drying, then:

$$\% \text{ Moisture Content} = (W2 - W3)/(W2 - W1) \times 100$$

3.2.4. Seed Packaging for Storage

After seed moisture content is within the safe storage limits, dried seeds should be placed in tightly sealed, impermeable containers to prevent rehydration. There are many containers and equipment for storing plant seeds. Cryovials with volume markings are convenient. They hold large numbers of seeds, seal tightly, and can be resealed many times.

In packaging seeds, each container should be labeled with relevant information including date of storage using a waterproof permanent marker. In determining seed quantities, approximately 1250 seeds = 25 mg = 50 µL. Seal the container immediately after filling, and visually check. During storage, check the containers at regular intervals to assure that they remain in good condition.

3.2.5. Seed Storage and Preservation

The major factors influencing seed longevity are storage temperature and seed moisture content. The higher the value of either, the shorter the lifespan of the seeds. Seeds left at ambient temperature and relative humidity lose viability relatively quickly. The conditions which prolong viability during storage have been well defined for plant seeds *(4–6)*. Seed storage principles for *Arabidopsis* are similar to those for other plants. For active collections in which seeds are stored for short to medium terms and are accessed often, a convenient storage temperature is approx 4°C (regular refrigerator temperature). For base collections in which seeds are placed in long-term storage without disturbance, a temperature of –20°C is appropriate. Storage of vials in larger containers with desiccant or placement of gelatin capsules with desiccant in the storage vials are also possible. However, if vials are sealed tightly and not opened cold, these measures should not be necessary. The arrangements of vials in storage can vary, but it is important to record the exact location of each accession. Codes indicating boxes, racks, trays, freezers, and so on, can be used.

Removal of vials from storage represents a potentially very dangerous step. Vials must be warmed to room temperature before opening. Rapid rewarming (placing vial in a 37°C water bath for approx 10 min) serves to minimize freeze/frost damage that can occur during this process. Working in a low relative humidity environment, if possible, also aids in prevention of hydration. If it is suspected that condensation has occurred in a vial during storage or opening, the vial should be left open in a dry location until seeds have desiccated before returning them to cold storage.

3.2.6. Seed Viability and Testing

Seed viability is the condition in which seeds are alive, have the potential to germinate, and develop into normal, reproductively mature plants, given the appropriate conditions. Factors that affect viability include the initial viability of the seeds at the start of the storage, seed moisture content, and storage environment. Viability should be monitored at regular intervals *(7,8)*. It is anticipated that viability of *Arabidopsis* seeds should remain high for long storage periods, assuming proper conditions. The International Board for Plant Genetic Resources (IBPGR) recommends that seeds in long-term storage under the optimal preservation standards should be monitored at least every 10 yr. Seeds in long-term storage with either poor storage life or poor initial viability and seeds in short to medium-term storage should be monitored at least every 5 yr.

A viability test for *Arabidopsis* seeds can be conducted in 3–6 d. Tests should be carried out before seeds are packaged and stored, so that poor quality seeds can be recognized. A germination test is the best method of estimating seed viability. *Arabidopsis* seeds may fail to germinate because they are dormant or because they are dead. Dormant seeds typically remain firm and in good condition during the germination test, whereas dead seeds soften and are attacked by fungi. Dormancy can usually be broken by imbibing seeds at low temperatures (*see* **Note 4**). The following method is suitable for *Arabidopsis*:

1. Place two layers of filter paper (free from chemical residues that could interfere with the germination of the seeds) firmly in the bottom of a labeled 9-cm diameter Petri dish.
2. Moisten the paper with distilled water. The paper should be totally saturated, but no excess water should be left in the dish.
3. Distribute 100 seeds uniformly on the surface of the paper. Replace the lid and seal the dish with parafilm or clear tape, to preclude desiccation.
4. Cold treat seeds by placing Petri dishes in the refrigerator for 2–4 d.
5. Place the dishes on an illuminated shelf (or in a growth chamber) under standard light and temperature conditions (see **Note 11**).
6. After 3–6 d count germinated vs ungerminated seeds, and record germination percentage.

4. Notes

1. Prepared pots can be stored in deep, covered pans for several days before planting, although pot preparation and planting should be conducted on the same day if possible.
2. Seeds can be planted by various methods depending on the purpose of the plants and availability of seeds. Simple scattering of seeds from a folded piece of paper works well for relatively dense populations and when some variation of density can be tolerated. The density of planting depends on the genetic material and the purpose of the planting. For seed production, high yields are achieved utilizing densities of 10–20 plants/10-cm pot. Larger populations do not necessarily reduce yield, but production per plant is reduced inversely. Larger populations are necessary for maintenance of representative proportions in a segregating population, and this can be achieved with more dense plantings in one or two 10-cm pots. Planted seeds should not be covered with additional soil; *Arabidopsis* seeds need light for germination.
3. If several pots are planted, they can be placed in a tub or other similar container and covered with clear plastic wrap. In all cases, the plastic wrap should not be allowed to contact the soil surface. The wrap is perforated in order to provide some aeration. However in the greenhouse, this method can overheat the soil surface and kill the germinating seedlings. In this case, the pots are left uncovered and placed in pans filled with 1–3 cm of water, which is maintained continually until all plants germinate and have expanded cotyledons.
4. The use of a cold treatment (2–4°C for 2–4 d) to break dormancy is very important for plantings utilizing freshly harvested seeds, which have more pronounced dormancy. Most widely used strains have moderate dormancy, and cold treatment may not be required when planting older seeds of these accessions. For certain accessions, as many as 7 d of cold treatment is necessary. Cold treatment of dry seeds is usually not effective in breaking dormancy.
5. If algae cover the soil as a result of overwatering of young plants, the pots should be allowed to dry, which will kill algae before the plants become stressed. The algae can then be carefully scraped from the soil surface, if necessary.

6. Several mild commercial fertilizers sold at plant stores for use with house plants also can be substituted for the mineral solution. Apply these in amounts and intervals according to the labels. Timed-release fertilizers are a labor saving alternate fertilizer source. Careful observation of the plants as they grow is effective in determining when fertilizer is needed. Plants begin to get slightly lighter green as nutrients run out, and fertilizer should be applied immediately. Normal healthy plants are bright dark green in color.
7. Hand rather than machine threshing and cleaning of the small *Arabidopsis* seeds is recommended mainly because the threshing machines require rigorous cleaning between accessions to avoid sample cross-contamination, require very careful adjustment, and do not accommodate the variable size of *Arabidopsis* seeds well.
8. Since *Arabidopsis* seeds equilibrate to room conditions, it is possible to reliably predict the approximate moisture content of seeds stored in open containers by simply measuring the room humidity. The relationship for *Arabidopsis* is similar to others published for crop seeds having similar chemical composition *(4)*.
9. High temperatures should not be used because the oil in the seeds may also vaporize and give false results. Temperatures of just over 100°C evaporate water and minimize vaporization of oils.
10. Light-weight dishes should be used, so that the ratio of the weight of the seeds and the dish is not too disproportionate. It is suggested that a minimum of three replicates of 100 mg of seeds or two replicates of 200 mg of seeds per sample be used. Always work with care and finish one sample at a time. Do not leave the dishes open in the laboratory between weighings because the seeds will absorb moisture rapidly from the air, and small changes in weights can result in large differences in the calculations.
11. Environmental conditions for seed germination tests are the same as for growing plants. Two replicates of 100 seeds each provide reliable germination estimates. Cases in which observed germination is < 80% may warrant follow-up testing to verify the low rate before it is accepted.

References

1. Kranz, A. and Kirchheim, B. (1987) Genetic resources in *Arabidopsis. Arabidopsis Information Service*, Frankfurt.
2. Estelle, M. and Somerville, C. (1987) Auxin-resistant mutants of *Arabidopsis thaliana* with altered morphology. *Mol. Gen. Genet.* **206**, 200–206.
3. Lehle Seed Catalog. (1995) Lehle Seed Co., Round Rock, TX.
4. International Board for Plant Genetic Resources. (1976) Report of IBPGR Working Group on engineering, design and cost aspects of long-term seed storage facilities. IBPGR, Rome.
5. Justice, O. L. and Bass, L. N. (1978) Principles and practices of seed storage. *Agriculture Handbook*, **506**, 34–80.
6. Roberts, E. H. and Ellis, R. H. (1984) The implications of the deterioration of orthodox seeds during storage for genetic resources conservation in *Crop Genetic Resources: Conservation and Evaluation* (Holden, J. H. W. and Williams, J. T. eds.) George Allen & Unwin, London, pp. 19–37.

7. Ellis, R. H. and Roberts, E. H. (1984) Procedures for monitoring the viability of accessions during storage, in *Crop Genetic Resources: Conservation and Evaluation,*. (Holden, J. H. W. and Williams, J. T. eds.) George Allen & Unwin, London, pp. 63–76.
8. Ellis, R. H., Hong, T. D., and Roberts, E. H. (1985) *Handbook of Seed Technology for Genebanks. Volume I. Principles and Methodology.* International Board for Plant Genetic Resources (IBPGR), Rome.

2

Sterile Techniques in *Arabidopsis*

Peter McCourt and Kallie Keith

1. Introduction

One of the many reasons for the popularity of *Arabidopsis* as an experimental system is the ease with which mutant screens can be carried out on Petri plates under sterile conditions. Petri plate screens have been successful for a number of reasons. First, large numbers of mutagenized seed can be easily scored for variability, and secondly, plants in Petri plates can be grown constantly under completely defined conditions. These attributes open up the world of plant genetics to mutant screens similar to those described for micro-organisms. For example, an *Arabidopsis* mutant insensitive to some particular compound such as a herbicide or growth regulator can be easily identified in a background of uniformly inhibited wild-type plants. The alternatives, spraying plants or irrigating soil with an inhibitor, not only result in problems of uniform application, but also are complicated by evaporation and catabolism of the compound by other organisms. Aside from the ease of screening, Petri plate screens also have the advantage that retesting and genetic analysis of subsequent generations can be done under exactly the same conditions as the original screen.

Many different Petri plate screens have been done with *Arabidopsis*, but the methods used fall roughly into three categories. The first, alluded to in the aforementioned paragraphs, involves the identification of mutants by their altered response to some compound that has been added to the growth medium. An example of this approach has been the isolation of mutants insensitive to various growth inhibitors (e.g., chlorsulfuron) or phytohormones *(1–4)*. A second screening approach has been to identify mutants that show altered responses to specific environmental cues such as light or gravity *(5–7)*. The third type of screen includes those in which the growth and development of the plant has been followed on Petri plates to identify mutant phenotypes that are normally difficult to observe in soil, for example, mutants with altered root morphology *(8)*. In all of these cases, continual growth and development under sterile conditions was an important factor in the success of the screens.

2. Materials

2.1. Solutions

1. Sterilizing solution: 20% bleach and 0.05% Tween-20. This solution must be made fresh every 2 wk.
2. Growth media: 0.5X Murashige ald Skoog basal salt mix (MS) salts; 5 mM 2-[N-morpholinol ethanesulfonic acid (MES), pH 5.7; 0.8% agar (*see* **Notes 1** and **2**).
3. 0.8% Top agar.
4. 0.8% Top agar.

2.2. Other Equipment

1. Pots.
2. Promix B/X soil (Premiere Hortoculture, Red Hill, PA) (*see* Chapter 1).
3. #5 Forceps.
4. Petri plates.
5. 12- or 24-well, flat-bottomed dishes (Corning, Corning, NY).
6. Deep dish Petri plates (Falcon, Los Angeles, CA).
7. Plastic wrap.
8. Falcon tubes.
9. Vortex.
10. Aspirator.
11. Micropore surgical tape 1530-0 (3M, St. Paul, MN).

3. Methods

3.1. Sterilization and Plating of Seeds

This method is most appropriate for screening mutants. If smaller amounts of seed are used, volumes and container size can be reduced.

1. Place up to 10,000 seeds (200 mg; an *Arabidopsis* seed weighs approx 0.02 mg) in a 50-mL Falcon tube and add approx 25 mL of sterilizing solution. Vortex vigorously for 20–30 s or more as seeds are slightly hygroscopic and tend to float. With enough mixing, the majority of seeds will disaggregate and sink to the bottom of the tube.
2. Let stand for 5–10 min with occasional mixing. Draw off the liquid by aspiration and wash thoroughly with sterile distilled water five times (*see* **Note 3**). After each addition of water, mix the seeds vigorously and then draw off the water after the seeds settle. If the seeds do not settle well (some mutants may not), low speed centrifugation for 10 s can be used to keep the seeds in the bottom of the tube during aspiration. For the last aspiration, draw off the water to the level of the seeds.
3. Add 3 mL of molten top agar (42°C) for every 500–1000 seed you wish to plate (e.g., 10,000 seeds/30 mL) and invert the tube to mix the seed (*see* **Note 4**).

Sterile Techniques in Arabidopsis

4. Using a 5-mL disposable pipe to draw up approx 3 mL and distribute it onto a Petri plate, swirl contents to distribute the seed, after the agar has solidified (approx 5 min) wrap them with gas-permeable tape (*see* **Note 5**).

3.2. Transplantation of Plantlets from Petri Plates to Soil

If desired, plants can be grown to seed under sterile conditions, however, yield is usually poor. Plants can be moved to larger containers such as culture tubes containing agar or plant cons. Culture tubes stoppered with cotton plugs or loosely fitting caps reduce the humidity, which increases seed set. If possible, it is usually best to transplant seedlings to soil, in which the most healthy plants usually grow (*see* **Note 6**).

1. Drench soil in 0.5X MS so that the soil is reasonably damp but not waterlogged (when you squeeze a handful of soil, little or no water drains out). Transplant no more than four plants/10-cm pot, since mature plants will tangle and be difficult to harvest individually.
2. Using forceps, pinch the hypocotyl very gently under the cotyledons and slowly remove the plantlet with the complete root from the agar. Put the root into a small hole in the soil and cover the base of the plant with soil.
3. Cover the pots with plastic wrap to keep the atmosphere moist so that the plants can acclimate.
4. After 3–4 d, slit the wrap with a razor blade to decrease humidity, and after a week remove it completely.

3.3. Retesting Mutants

1. Prepare 12- or 24-well plates with agar for retests. These plates can hold up to 2 mL/well of agar (*see* **Note 7**).
2. Sterilize 10–50 seeds from each line as described in 1.5-mL Eppendorf tubes with 1 mL of sterilizing solution. Seeds can be centrifuged after each wash to reduce their volume for aspiration.
3. After the last wash, resuspend seeds in 50–100 µL of a 0.1% agar solution, and using a P1000 Pipetman plate (Nichiyro, Tokyo, Japan), an appropriate amount of seed into each well (*see* **Note 8**). Plates should be sealed with gas-permeable tape (*see* **Note 5**).

4. Notes

1. A simple way to make and store media is to weigh out separately enough 1X MS salts and 2X agar (1.6%) to make 1–2 L of each solution and distribute them into separate bottles of convenient sizes. For example, put 100 mL of 1X MS and 100 mL of 2X agar into two, 250 mL bottles. Autoclave these for 20 min and store the sterile solutions on the shelf. When plates are needed, melt the appropriate size bottle of 2X agar in a microwave and add an equal amount of sterile 1X MS, which cools the solution and results in 1X agar in 0.5X MS salts.

Add 500 mM MES pH 5.7 (100X) sterile stock solution to give a final concentration of 5 mM and pour the plates. Ready-mixed MS salts powder can be purchased from Sigma (St. Louis, MO).
2. 0.8% Agar allows the removal of plantlets without damaging the roots. Depending on the experiment being conducted, sucrose can be added as a carbon source (1–2%). Plants do grow more vigorously on sucrose containing media, but the risk of fungal contamination is much higher. If sucrose is required, then plating of seeds should be done in a sterile hood. However, for many screens, plants only need to germinate and grow for a short time and sucrose can be left out. Plants will grow autotrophically and set seeds if the light conditions are adequate (150–250 µE/m^2/s) at room temperature and the container is big enough.
3. The major cause of the contamination from seed lots is the result of debris that is carried along with the seed during harvesting. This material is difficult to sterilize and seed lots with large amounts of debris should be cleaned up as much as possible before sterilization. The addition of Tween-20 to the sterilizing solution produces a soap sud interface after vortexing which traps much of this debris at the surface. Then, debris can be removed with the aspirator after the seeds have settled and with each subsequent wash.
4. For most screens, good distribution of seed on the Petri plate is essential so that each seedling or plant can be easily scored after germination. The mixing of seeds with molten agar allows them to be distributed evenly over the plate in a similar fashion to plating lambda phage in top agar. After addition of molten agar to the seeds, the tube can be kept at 42°C for only about 15 min. After this, they lose viability. To increase the time in which to distribute the seeds, MS plates can be preheated for 20 min at 37°C which slows the solidification of the molten agar after plating. Plating 500–1000 seeds/100 × 15 mm Petri plate allows easy scoring of seedlings up to a week old. If larger plants are required, fewer seeds should be plated or the plants will become overcrowded. Also, deep dish Petri plates (100 × 20 mm) can be used if larger plants are needed. When the screen being conducted is designed to inhibit wild-type growth, the seed can be plated at much higher densities. The number of seeds/plate does affect the concentration of compound per seed, therefore when determining inhibitory concentrations for screens using wild-type seeds, the conditions should be worked out using the same number of wild-type seeds/plate as used in the actual screen. If the compound completely inhibits germination or is very toxic early in germination, up to 10,000 seeds per Petri plate can be used. It is also possible to use 150 × 15 mm Petri plates with 8 mL of top agar/plate. In many inhibitor screens, some sensitive plants appear normal based on seedling color and development, however, root development is usually retarded. By turning the plates over and scoring root growth into the agar under a dissecting scope (10–20X), it is possible to discriminate between true insensitive mutants and wild-type escapers.
5. Petri plates will dry out in many fan-driven incubators, especially plates on the edges of the incubator. Plates can be sealed with parafilm, but this inhibits gas

exchange for the plants. We have found gas-permeable surgical tape reduces drying, allows the plants to photosynthesize, and is impermeable to spores.
6. The time to remove plants after plating depends on the type of screen, but generally the sooner you can get the plants to soil, the better the chance of recovery and growth.
7. Retesting usually involves large numbers of lines and is best done on 12- or 24-well plates with one well per individual M3 family. Not only does this save on solutions and space, but it also makes keeping track of various lines less tedious.
8. 0.1% Agar is viscous enough to allow the distribution of seeds into small wells without clumping. When water is used, the seeds usually aggregate in a drop on top of the well.

References

1. Bleeker, A. B., Estelle, M. A., Somerville, C., and Kende, H. (1988) Insensitivity to ethylene conferred by a dominant mutation in *Arabidopsis thaliana. Science* **241**, 1086–1089.
2. Estelle, M. A. and Somerville, C. R. (1987) Auxin-resistant mutants of *Arabidopsis* with altered morphology. *Mol. Gen. Genet.* **206**, 200–206.
3. Haughn, G. and Somerville C. R. (1986) Sulfonylurea-resistant mutants of *Arabidopsis thaliana. Mol. Gen. Genet.* **204**, 430–434.
4. Koornneef, M., Reuling, G., and Karssen, C. M. (1984) The isolation and characterization of abscisic acid-insensitive mutants of *Arabidopsis thaliana. Physiol. Plant* **61**, 377–383.
5. Chory, J., Peto, C., Feinbaum, R., Pratt, L., and Ausubel, F. (1989) *Arabidopsis thaliana* mutant that develops as a light-grown plant in the absence of light. *Cell* **58**, 991–999.
6. Koornneef, M., Rolff, E., and Spruit, C. J. P. (1980) Genetic control of light-inhibited hypocotyl elongation in *Arabidopsis thaliana. Z. Pflanzenphys.* **100**, 147–160.
7. Okada, K. and Shimura, Y. (1990) Reversible root tip rotation in *Arabidopsis* seedlings induced by obstacle-touching stimulus. *Science* **250**, 274–276.
8. Benfey, P. N., Linstead, P. J., Roberts, K., Schiefelbein, J. W., Hauser, M. T., and Aeschbacher, R. A. (1993) Root development in *Arabidopsis*: four mutants with dramatically altered root morphogenesis. *Development* **119**, 57–70.

3

Control of Pests and Diseases of *Arabidopsis*

Mary Anderson

1. Introduction

When growing *Arabidopsis* as an experimental organism, the best standards of hygiene should be adopted to ensure uniform plant growth, representative populations are maintained for mapping analysis, and potentially useful new mutants are not lost in the early fragile stages of screening.

Arabidopsis is prone to many of the pests that commonly infect greenhouses and growthrooms *(1,2)*. These can cause severe damage to the plants and, in some cases, can kill them. In general, the best philosophy to adopt in pest management is that of prevention rather than cure.

2. Pest Management

It is advisable to adopt a strict regime for growing *Arabidopsis* so that infection can be minimized. Greenhouse/growthroom rules should be strictly enforced to ensure that even in communal growth areas these rules are observed. Measures should be in place to allow for the removal of material, to avoid experiments being put at risk by others' sloppy practices. This kind of strategy works best if someone is put in charge of the area, such as a greenhouse manager, who can monitor the growth area and act appropriately.

Before starting to grow *Arabidopsis*, purge the greenhouse area or growthroom to remove all potential pathogens. If chronic infections develop later, it is advisable to throw out infected material, clean the greenhouse, and replant. Regardless of infection, thoroughly clean out the greenhouse area at least once a year, replacing gravel or capillary matting. Greenhouses should be cleaned more frequently if infection levels rise above an acceptable level.

2.1. How to Maintain Clean Growth Conditions

1. Have greenhouse-specific laboratory coats that are laundered regularly. If plant material is kept in more than one greenhouse or growthroom, do not move between them in the same lab coat. Beware, infection can be spread between growth areas on clothing.
2. Maintain footwipes at the entrance to the greenhouse which contain a general biocide to prevent the movement of seed and potential contaminants between greenhouses from the soles of shoes.
3. Maintain screens over vents and windows to prevent the entry of flies, bees, and wasps.
4. Employ fly strips as insect traps to catch insects and give an indication of the type and level of infection.
5. Always use fresh, clean compost mixes. Avoid leaving bags of compost open, which can accumulate flies' eggs and fungal spores.
6. Never move infected material between greenhouses.
7. Do not keep growth conditions over-wet, or over-dry. Either of these regimes can encourage infections.
8. Stake the material to prevent lodging. It will also reduce the chance of fungal infection.
9. Do not let old plant material accumulate in the waste bins as this can act as a reservoir for infection.
10. Do not have stagnant pools of water on the benches or on the floor.
11. Regularly sweep the floors to remove plant debris.
12. If infection is suspected, identify the source of infection and act accordingly.

3. *Arabidopsis* Pests and Diseases

The best way to minimize infestation of any kind in the greenhouse or growthroom is to maintain good hygienic practice as just outlined. However, it is almost inevitable that at some time the greenhouse will become infected with some kind of *Arabidopsis* pest. The following sections list guidelines on how to identify and treat the types of infection that effect *Arabidopsis*.

3.1. Arabidopsis *Pests*

3.1.1. Aphids

Aphids, also known as greenflies are small, sap feeding insects, which characteristically cluster on the stems of plants. Primary damage to the plant results from feeding on young tissues, which has the effect of distorting and weakening new growth. In severe cases, plants can wilt and die. In addition, major infections can lead to the accumulation of honey dew on the stems and leaves which can encourage further infection from sooty molds.

Aphid infestations can become overwhelming in a short period of time owing to the fact that most females are parthenogenetic and can give birth to live young, which mature in about a week. Prompt action is therefore required. Regularly inspect the plants for aphids and at the first sign of infection apply treatment.

Infections can stem from air-borne aphids. In the case of infection, check that screens are not damaged. Aphids can also be easily caught on clothing and introduced to the greenhouse or growthroom from the outside. Wearing a lab coat will help reduce the risk.

To treat aphid infection, fumigate the greenhouse with 40% nicotine shreds (DowElanco, Hitchins, Herts, UK). Aphid infections in growthrooms are more problematical. It may be necessary to move the material from the growthroom to an area where it can be fumigated. Alternatively, spray with a proprietary systemic action insecticide.

3.1.2. Red Spider Mite

Red spider mite infections can be very destructive and most often occur under dry and hot growth conditions. Infection is characterized by a silvery speckled effect on the leaves (**Fig. 1A**). In heavy infections, plants suffer yellow discoloration followed by wilting and slow death. The underside of the leaves have the appearance of being dusted with a white powder which is the result of the empty cases of the young mites and eggs which are suspended from thin white strands of silk (**Fig. 1B**). The mites can build webs along the flowering stem and among the flowers. It is very easy to catch these silks on a lab coat and this can result in the transfer of eggs to other plants. Extreme caution should be observed.

To treat for infection, fumigate with 40% nicotine shreds (DowElanco) maintaining a temperature of 25°C for at least 8 h or spray plants with Pentac Aquaflow (Dow Agriculture) or Savona (Koppert, Wadhurst, East Sussex, UK).

3.1.3. Sciarid Fly (Mushroom Fly or Fungus Gnat)

Infection with sciarid fly is characterized by the plants lacking vigor and leaves turning yellow without the appearance of any visible damage above ground. Damage occurs through attack on the roots. Small white maggots can be observed among the root tissue where they feed, or moving through the soil. Small flies (gnats) gather around plant and are easily caught on fly strips.

The main cause of infection with sciarid fly comes from eggs in the compost. Compost can be treated with insecticide (Basudin 5G granules [Ciba-Geigy, Whittlesford, Cambridge, UK]) or autoclaved to kill the eggs and larvae. Care should be taken in autoclaving since this can cause the production of nitrite compounds which are toxic to the plant.

To treat sciarid fly infections, fumigate with Lindane smoke pellets (Octavius Hunt, Redfield, Bristol, UK) or nicotine 40% shreds (DowElanco) to eradicate flies and incorporate Basudin 5G granules (Ciba-Geigy) into the compost to kill the larvae. Compost and benches may be drenched with a solution of diazinon liquid (Murphy Chemicals, St. Albana, Herts, UK), or Savona (Koppert), to eradicate the larvae. Alternatively biological control with Nemasys (Koppert) can work, but is not suitable for severe infections.

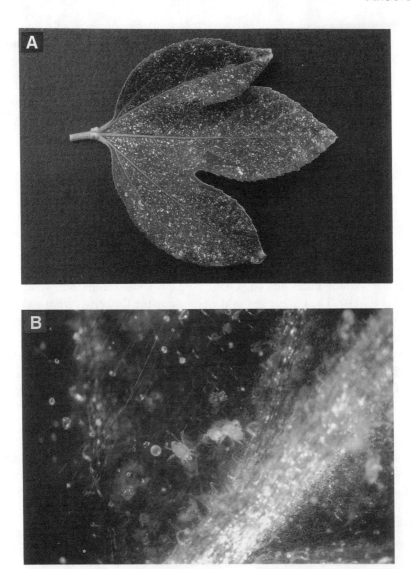

Fig. 1. **(A)** Characteristic speckling of a passion flower leaf caused by red spider mite. **(B)** Mites make silken webs in which they suspend their eggs. These are commonly observed on the underside of leaves. Magnification ×1.5.

3.1.4. Thrips

Thrips are small, elongate insects with two pairs of very narrow wings that have a distinct fringe of long fine hairs. The larvae resemble the adults, but lack the wings and are generally lighter in color. They tend to

Fig. 2. Thrips characteristically produce white or silvery flecks on the leaf surface. This illustration shows the effects of thrip infection on a tobacco leaf.

accumulate in the inflorescence and can be a major problem in the failure of crosses to take.

Thrips infections are characterized by the appearance of white or silvery flecks on the leaves (**Fig. 2**). The underside of leaves may also carry black spots as an indication of infection.

To treat thrips infections, fumigate with nicotine 40% shreds (DowElanco), or Lindane smoke pellets (Octavius Hunt) at 14-d intervals. If the infestation is severe, fumigation may be carried out three times at 7-d intervals, although this may result in some scorching of the plants. Alternatively, granules of Temik 10G (Union Carbide, Harrogate, West Yorkshire, UK), or Basudin 5G (Ciba-Geigy) may be sprinkled around and watered in.

3.1.5. Whitefly

Whitefly infections are characterized by small, snow-white, four-winged flies and very small pale green nymphs which cover the underside of leaves (**Fig. 3**). Damage is caused through the flies feeding on the sap of the plants. Heavy infections will result in plants lacking vigor, wilting, and dying. Whitefly build-up is not as rapid as in aphid infections, but regular checking of plant material will minimize whitefly damage. To combat whitefly infections, fumigate with Lindane smoke pellets (Octavius Hunt) or Propoxur smokes (Octavius Hunt) at weekly intervals.

Fig. 3. Whitefly adults and nymphs together on the underside of a petunia leaf. Magnification ×1.5.

3.2. Arabidopsis *Diseases*

In general, fungal infections are less common than insect infestation. The risk of infection is decreased by avoiding overwatering and allowing the surface of soil to partially dry between waterings. Allow good air circulation over soil, and avoid completely covering pots or trays, except when they are first established in soil after transfer from tissue culture. Also, avoid planting material too close together as this can cause problems, particularly with ecotypes which produce very large and dense rosettes.

3.2.1. Botrytis

Botrytis infection is characterized by plants developing gray, fluffy mold on the leaves which then rot (**Fig. 4**). Infected material should be cleared and surrounding plants treated. Spray or drench plants with a solution of Benlate systemic fungicide (ICI, Haslemere, Surreys, UK).

3.2.2. Powdery Mildew

Powdery mildew is characterized by a thin, powdery covering to leaves, stems, and siliques (**Fig. 5**). Severe infections can lead to plant wilt and death. Spray or drench plants with a solution of Benlate systemic fungicide (ICI).

4. Notes

The methods of control that are given have been found to be effective at Nottingham University but are not an endorsement of the products. They are

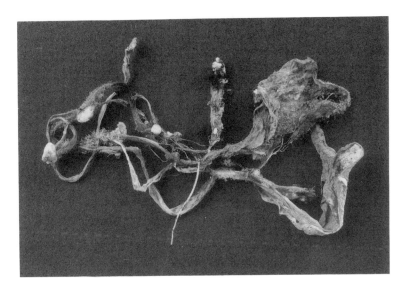

Fig. 4. A severe botrytis infection of lettuce.

Fig. 5. Powdery mildew infection of oil seed rape. Note the powdery covering to the leaf surface.

for information only and do not imply that other approaches or treatments may be inferior. The use of these chemicals should only be carried out by personnel who have received the necessary training to handle these materials. The manufacturers' instructions should be followed at all times in the application of the chemicals. All pesitcide and fungicide treatments should be applied during the evenings or on weekends. Do not apply during bright and sunny conditions because droplets can cause scorching. Warning signs indicating the time and type of treatment should be clearly posted. Areas that are undergoing treatment should be locked to prevent accidental exposure to personnel. Greenhouses and growthrooms should be thoroughly ventilated following treatment to flush with fresh air before personnel are allowed to re-enter the areas. Pesticides should be stored in an appropriately locked area.

Acknowledgments

I thank John Gaskin for his helpful advice and Brian Case for his excellent photography of the pests.

References

1. Metcalf, R. L. and Metcalf, R. H. (1993) *Destructive and Useful Insects Their Habits and Control*. 5th ed. McGraw Hill, New York.
2. Pegg, G. F. and Ayres, P. G. (eds.) (1987) *Fungal Infection of Plants*. Cambridge University Press, Cambridge, UK.

4

Establishment and Maintenance of Cell Suspension Cultures

Jaideep Mathur and Csaba Koncz

1. Introduction

Cell suspension cultures are rapidly dividing homogenous suspensions of cells grown in liquid nutrient media from which samples can be precisely removed *(1)*. Cell suspensions are used for generating large amounts of cells for quantitative or qualitative analysis of growth responses and metabolism of novel chemicals, as well as for studies of cell cycle under standard conditions *(2,3)*. In addition, cell suspensions serve as an ideal material for the isolation of protoplasts used in transient gene expression assays and *Agrobacterium*-mediated transformation *(4,5)*. The establishment of suspension cultures of *Arabidopsis thaliana* cells derived from leaf and hypocotyl calli has been reported *(6)*. In order to initiate *Arabidopsis* suspensions retaining a high regenerative potential, a procedure described by Ford *(7)* has been modified. Callus tissues derived from root or hypocotyl explants of *Arabidopsis* yield well proliferating cell suspensions capable of morphogenesis in liquid medium. The regeneration capability and genetic stability of suspension cells, however, decrease by the length of culture period. Therefore, it is recommended to use newly initiated cell suspensions when the applications require the regeneration of diploid fertile plants. The method described is used to establish morphogenic cell suspensions from the *Arabidopsis* ecotypes Columbia, C24, RLD, and Wassilewskija. Slight modification of the concentration of growth regulators (e.g., auxin) and/or the time of subcultures may be required for other ecotypes.

2. Materials

1. Dry seeds of *Arabidopsis thaliana*.
2. 10% Sodium hypochlorite solution containing 0.1% Triton X–100.

3. Eppendorf tubes.
4. Sterilized double-distilled water.
5. Sieves of 850 µm and 500 µm.
6. Sterile pipets.
7. Pair of forceps.
8. Scissors or scalpel blades.
9. Microfuge.
10. Pipet pump.
11. Sterile 250-mL Erlenmeyer flasks with plugs.
12. Rotary shaker set at 120 rpm.
13. Aluminium foil.
14. Suspension culture medium (pH 5.8, 1 L): 4.33 g MS medium (*8*, Sigma, St. Louis, MO), 2X B5 vitamins (*9*, consisting of 2 mg nicotinic acid, 2 mg pyridoxin HCl, 20 mg thiamine–HCl, and 200 mg myo–inositol), 30 g sucrose, 0.5 mg 2,4–dichloro–phenoxyacetic acid (2,4–D), 2.0 mg indole3–acetic acid (IAA), 0.5 mg 6–(γ,γ–methylallylamino)–purine riboside (IPAR, *see* **Note 1**).

3. Methods

3.1. Initiation of Cell Suspensions from root explants

3.1.1. Initiation of Root Cultures

1. Surface sterilize 0.1 g (approx 5000) seeds in Eppendorf tubes by adding 1 mL of 10% (v/v) sodium hypochlorite solution containing 0.1% Triton X–100 as surfactant and shaking for about 15 min (*see* **Note 2**).
2. Collect the seeds by slow centrifugation in a microfuge for a few seconds and remove the supernatant. Wash the seeds five times with 1 mL of sterile water.
3. Germinate the seeds in Petri dishes containing 0.5X MS medium with 0.8% agar using 16 h light to 8 h dark period at 25°C.
4. Place 15–20 1-wk-old seedlings into 250-mL Erlenmeyer flasks containing 35 mL of liquid MS medium with 3% sucrose.
5. Place the flasks on a shaker set for 120 rpm at 25°C under 16 h light to 8 h dark period.
6. Harvest roots after 15–20 d (*see* **Note 3**).

3.1.2. Initiation of Cell Suspensions

1. Place the roots collected from an Erlenmeyer flask into a 9-cm Petri dish and remove all green tissue.
2. Using a scalpel blade or scissors, cut the roots into fine pieces (approx 2 mm long).
3. Remove the rest of liquid medium and resuspend the root explants in approx 50 mL of suspension medium in an Erlenmeyer flask (*see* **Note 4**).
4. Place the flask on the shaker and cover with aluminium foil (*see* **Note 5**).
5. Harvest the cells released from root calli after 21 d; using a 25-mL pipet and pass them through a 850 µm sieve (*see* **Notes 6** and **7**).
6. Allow the cells to settle for about 30 min before gently removing the liquid medium from the top. Add fresh suspension medium (up to 50 mL) and place the

flask back onto the shaker. The roots left in the sieve can be resuspended in fresh medium to generate more cells in suspensions (*see* **Note 6**).

3.2. Initiation of Cell Suspensions from Seed–Derived Calli

1. Sterilize seeds as described in **Subheading 3.1.1.** and place approx 1000 seeds in 50 mL liquid suspension medium in an Erlenmeyer flask (*see* **Note 8**).
2. Place the flask on the shaker and cover with aluminium foil.
3. Examine the cultures at periodic intervals after 21 d by taking out 0.5-mL aliquots from the flasks and observing under a microscope.
4. Once the cultures contain large number of single cells and proliferating cell clumps, sieve the cells sequentially through 850-µm and 500-µm sieves to obtain a homogenous starting material for suspension culture.
5. Allow the cells to settle and then replace the medium as described in **Subheading 3.1.2.** (*see* **Note 6**).

3.3. Maintenance of Cell Suspensions

1. Sieve and subculture the cell suspensions every 7 d until a homogenous proliferating cell population is established (*see* **Notes 9** and **10**).
2. After the fifth subculture, the suspensions can be usually divided into three portions at the time of subculture. To achieve this, approx 100 mL of fresh suspension medium is added to 50-mL cell suspension in the flask. After swirling the flask, 50-mL aliquots are dispensed into new flasks.

4. Notes

1. A 100X stock of B5 vitamin is prepared in distilled water and added to the medium before autoclaving. Growth regulator stocks of 1 mg/mL are filter-sterilized and added to cooled medium. All stocks are stored at 4°C.
2. Do not sterilize too many seeds in one Eppendorf tube because their drying may cause a problem. In case the seeds would clump, use a sterile toothpick to break the clumps before plating the seeds.
3. After 15–20 d approx 3–5 g of roots (fresh weight) should be available in each flask. The growth rate of roots is important for obtaining fine suspensions. The roots should be white, actively growing and not yellow–brown or green.
4. For the *Arabidopsis* RLD ecotype the application of 4.0 mg/L of α–naphtheleneacetic acid instead of other hormones in the suspension medium should also result in fast cell proliferation.
5. A dark incubator may be useful, since we have found considerable differences between the growth rates of cell cultures grown in the dark in comparison to those grown in the light. The slower growth rate in the light may be the result of photodegradation of IAA.
6. Great care has to be taken to ensure aseptic conditions. Flame the neck of flasks well before opening them. Constant swirling of the flask and moving the pipet is necessary to avoid clogging.

7. Occassionally check the cultures for bacterial and/or fungal contamination by plating aliquot of the supernatant on bacterial growth media and/or subculturing a small aliquot of cells in a rich medium containing 1% glucose added to the suspension medium.
8. In the case of certain *Arabidopsis* mutants, the root development may be inhibited, and it is thus difficult to obtain suitable amount of starting root material. By initiating seed derived calli, usually this problem can be successfully overcome.
9. At this stage, usually 3 g (fresh weight) of cell suspension is subcultured in 50 mL of medium and a three to fourfold increase in fresh weight is obtained during a growth period of 7 d. Such cultures reach their exponential phase between d 2.5 and 5.5. A prolongation of culture period for more than 8 d results in rapid browning and cell death.
10. Cultures at this stage are composed of single cells (approx 10–25 µm in length) and clumps of 16–64 cells.

References

1. King, P. J. (1984) Induction and maintenance of cell suspension cultures, in *Cell Culture and Somatic Cell Genetics of Plants*. vol. 1 (Vasil, I.K. ed) Academic, New York, pp. 130–138.
2. May, M. J. and Leaver, C. J. (1993) Oxidative stimulation of glutathione synthesis of *Arabidopsis thaliana* suspension cultures. *Plant Physiol.* **103**, 621–627.
3. Magyar, Z., Bakó, L., Bögre, L., Dedeoglu, D., Kapros, T., and Dudits, D. (1993) Active *cdc2* genes and cell cycle phase specific cdc2–related kinase complexes in hormone–stimulated alfalfa cells. *Plant J.* **4**, 151–161.
4. Doelling, J. H. and Pikaard, C. S. (1993) Transient expression in *Arabidopsis thaliana* protoplasts derived from rapidly established cell suspension cultures. *Plant Cell Rep.* **12**, 241–244.
5. An, G. (1985) High efficiency transformation of cultured tobacco cells. *Plant Physiol.* **79**, 568–570.
6. Gleddie, S. (1989) Plant regeneration from cell suspension cultures of *Arabidopsis thaliana* Heynh. *Plant Cell Rep.* **8**, 1–5.
7. Ford, K. G. (1990) Plant regeneration from *Arabidopsis thaliana* protoplasts. *Plant Cell Rep.* **8**, 534–537.
8. Murashige, T. and Skoog, F. (1962) A revised medium for rapid growth and bioassays with tobacco tissue cultures. *Physiol. Plant.* **15**, 473–497.
9. Gamborg, O. L., Miller, R. A. and Ojima, K. (1968) Nutrient requirements of suspension cultures of soybean root cells. *Exp. Cell Res.* **50**, 151–158.

5

Callus Culture and Regeneration

Jaideep Mathur and Csaba Koncz

1. Introduction

Regeneration of plants by micropropagation of in vitro cultures can be achieved from organ primordia existing in shoot tips and axillary bud explants. Alternatively, plants can be regenerated from unorganized callus tissues derived from different explants by dedifferentiation induced by exogenous growth regulators. Plant regeneration from calli is possible by *de novo* organogenesis or somatic embryogenesis. Callus cultures also facilitate the amplification of limiting plant material. In addition, plant regeneration from calli permits the isolation of rare somaclonal variants which result either from an existing genetic variability in somatic cells or from the induction of mutations, chromosome aberrations, and epigenetic changes by the in vitro applied environmental stimuli, including growth factors added to the cultured cells *(1–3)*.

In *Arabidopsis thaliana,* one of the earliest studies on callus formation was conducted by Loewenberg *(4)*, who grew seedlings in a medium containing kinetin and parachlorophenoxyacetic acid. Subsequent studies (for review, *see* **ref. 5***)* of cell culture and regeneration demonstrated that it is relatively easy to regenerate *Arabidopsis* from callus cultures by many approaches. However, the efficiency of regeneration and the proportion of somaclonal variants are considerably influenced by the ecotype and the source of explants, as well as by the medium and growth regulators employed. Hypocotyl and root explants thus provide excellent materials for callus initiation and regeneration in contrast to stem and leaf explants. The tissue culture protocols include callus induction in auxin containing media. Shoot regeneration is induced by lowering the auxin content and increasing the cytokinin levels in the media. The elongated shoots are usually transferred into root induction media, but even in the absence of roots, flowering and seed setting can take place in test tubes in vitro.

The method described here utilizes root explants that can be cultured and efficiently regenerated in large quantities. The described technique can be applied equally well for the *Arabidopsis* ecotypes Columbia, C24, RLD, and Wassilewskija. Without changing the general protocol, some slight alterations may be required for other ecotypes or mutants of *Arabidopsis*.

2. Materials

1. Dry seeds of *Arabidopsis thaliana*.
2. 10% Sodium hypochlorite solution containing 0.1% Triton X–100.
3. Eppendorf tubes.
4. Sterilized double-distilled water.
5. Culture tubes.
6. 9 cm Petri dishes.
7. Pair of forceps.
8. Scissors or scalpel blades.
9. Sterile 250-mL Erlenmeyer flasks with plugs.
10. Rotary shaker set at 120 rpm.
11. Microfuge.
12. Basal medium (BM): Murashige and Skoog (MS) medium *(6)* or MS medium for *Arabidopsis* (MSAR) medium *(7)* with 3% sucrose (pH 5.8) can equivalently be used (*see* **Note 1**).
13. 0.5X BM: BM consisting of half concentration of MS macroelements and 0.5% sucrose.
14. MSAR I (Callus medium): Supplement BM with 0.5 mg/L 2,4–dichloro–phenoxyacetic acid (2,4–D), 2.0 mg/L indole–3–acetic acid (IAA), 0.5 mg/L 6–(γ,γ–dimethylallylamino)–purine riboside (IPAR), and gel using either 0.8% agar or 0.2% gelrite (*see* **Note 1**).
15. MSAR II (Shoot medium): Supplement BM with 2.0 mg/L 6–(γ,γ–dimethylallylamino)–purine riboside (IPAR), 0.05 mg/L α–naphtaleneacetic acid (NAA), and either 0.8% agar or 0.2% gelrite.
16. MSAR III (root inducing medium): Supplement BM with 1.0 mg/L IAA, 0.2 mg/L indole–3–butyric acid (IBA), 0.2 mg/L 6–furfuryl–aminopurine (kinetin), and either 0.8% agar or 0.2% gelrite.

3. Methods

1. Surface sterilize 0.1 g (approx 5000) seeds in Eppendorf tubes by adding a 1 mL of 10% (v/v) solution of sodium hypochlorite containing 0.1% Triton X–100 as surfactant and shaking for about 15 min (*see* **Note 2**).
2. Pellet the seeds by slow centrifugation in a microfuge for a few seconds and remove the supernatant. Wash the seeds five times with 1 mL of sterile water.
3. Germinate the seeds in Petri dishes containing 0.8% agar gelled 0.5X BM medium (BM medium containing half concentration of macroelements) using 16 h light and 8 h dark period at 25°C (*see* **Note 3**).
4. Place 15–20 1-wk-old seedlings into 250-mL Erlenmeyer flasks containing approx 35 mL of liquid BM medium with 3% sucrose.

5. Place the flasks on a rotary shaker at 120 rpm using 16 h light and 8 h dark period at 25°C.
6. Harvest the roots of these seedlings after 15–20 d of growth (*see* **Note 4**).
7. Place the roots into a Petri dish, remove all liquid and cut the roots into small pieces.
8. Transfer root explants onto MSAR I plates and either cover them with aluminium foil or place them in low light conditions.
9. After 3 wk, transfer the root-derived calli either to fresh MSAR I plates for maintenance or to MSAR II plates for shoot regeneration.
10. After 2 wk, pick the regenerating shoots from the calli and transfer them into glass jars with MSAR II medium for further elongational growth (*see* **Note 5**).
11. Shoots consisting of 4–6 leaves may be transferred further into MSAR III medium for 3–6 d to induce root development.
12. Transfer the elongated shoots (approx 2 cm) into culture tubes containing 0.5 BM agar medium with 0.5% sucrose for rooting, flowering, and setting seeds (*see* **Notes 6** and **7**).

4. Notes

1. Stocks of growth regulators are prepared at 1 mg/mL concentration, filter sterilized, and stored at 4°C. Growth regulators are added to autoclaved medium after it has cooled to about 60°C.
2. Do not sterilize too many seeds in one Eppendorf tube, as they may not be easily dried. In case seeds clump, use a sterile toothpick to break the clumps before plating the seeds.
3. Calli can also be obtained from germinating seeds that are placed on MSAR I plates after surface sterilization.
4. After 15–20 d of culture, 3–5 g (fresh weight) of roots should be produced in each flask. The roots should be white, actively growing, and not yellow–brown or green.
5. At this stage, it is possible to regenerate and amplify shoots from the regenerated leaves by cutting them into pieces and placing them on MSAR II Medium.
6. The culture tubes are capped with loose cotton to facilitate the aeration required for seed setting and maturation. Take care not to place the tubes too close to the light source because this will cause moisture condensation inside the tubes and result in a low rate of fertilization.
7. Rooted plants can be transferred to soil after washing the roots with water containing a fungicide, such as Benomyl (0.02%). The plants are gradually acclimated by reducing the humidity of transfer chambers stepwise.

References

1. Flick, C. E., Evans, D. A., and Sharp, W. R. (1983) Organogenesis, in *Handbook of Plant Cell Culture,* vol. 1 (Evans, D. A., Sharp, W. R., Ammitato, P. V., and Yamada, Y. eds.), MacMillan, New York, pp. 13–81.
2. Raghavan, V. (1986) *Embryogenesis in Angiosperms.* Cambridge University Press, Cambridge, UK.
3. Larkin, P. J. and Scowcraft, W. R. (1981) Somaclonal variation – a novel source of variability from cell culture for plant improvement. *Theor. Appl. Genet.* **60**, 197–214.

4. Loewenberg, J. R. (1965) Callus cultures of *Arabidopsis*. Arabidopsis *Inf. Serv.* **2**, 34.
5. Morris, P. C. and Altmann, T. (1994) Tissue culture and transformation, in *Arabidopsis*. (Meyerowitz, E. M. and Somerville, C. R., eds.), Cold Spring Harbor Laboratory, Cold Spring Harbor, New York, pp. 173–222.
6. Murashige, T. and Skoog, F. (1962) A revised medium for growth and bioassays with tobacco tissue cultures. *Physiol. Plant.* **15**, 473–497.
7. Koncz, C., Martini, N., Szabados, L., Hrouda, M., Bachmair, A., and Schell, J. (1994) Specialized vectors for gene tagging and expression studies, in *Plant Molecular Biology Manual*. (Gelvin, S. B. and Schilperoort, R. A., eds.), B2, Kluwer, Dordrecht, pp. 1–22.

6

Protoplast Isolation, Culture, and Regeneration

Jaideep Mathur and Csaba Koncz

1. Introduction

Protoplasts of *Arabidopsis thaliana* have been isolated from a variety of explant sources with varying degree of success *(1–3)*. Most workers have faced problems in achieving a high frequency of sustainable division of protoplasts in liquid culture. The problem can be alleviated to a certain extent by embedding mesophyll protoplasts in calcium alginate *(4)*. We have developed a technique allowing the culture of protoplasts derived from roots or cell suspensions in liquid medium and their regeneration to plantlets *(2)*. The yield of root-derived protoplasts capable of division and regeneration is considerably higher in comparison to protoplasts obtained from leaf mesophyll cells. The protoplast isolation requires a ready supply of root cultures maintained in auxin containing medium. Alternatively, a high yield of protoplasts, suitable for transient gene expression studies using direct DNA transformation, can also be obtained from cell suspension cultures, excluding any limitation of available starting material. Because the methods may have to be tailored to the needs of different experiments, here we describe protocols for the isolation and culture of protoplasts from leaf mesophyll tissue *(4)*, as well as from root and cell suspension cultures. These methods are equally effective for the *Arabidopsis* ecotypes Columbia, C24, RLD, and Wassiljewskija.

2. Materials

2.1. Materials and Equipment

1. Dry seeds of *Arabidopsis*.
2. Seed sterilizing solution: 10% sodium hypochlorite containing 0.1% Triton X–100.
3. Sterilized double-distilled water.

From: *Methods in Molecular Biology, Vol. 82: Arabidopsis Protocols*
Edited by: J. Martinez-Zapater and J. Salinas © Humana Press Inc., Totowa, NJ

4. 250 mL Sterile Erlenmeyer flasks.
5. 0.22 μm Filters for sterilization of solutions.
6. Clinical centrifuge (500g).
7. Scissors and razor blades.
8. Pair of forceps.
9. 25-, 50-, and 100-μm Sieves.
10. 12 mL Screw-capped glass centrifuge tubes.
11. 10 and 25 mL Sterile pipetes.
12. Pipet pump.
13. Pasteur pipets with cotton plugs and teats.
14. Hemocytometer.
15. Inverted microscope.
16. 55 and 90 mm Petri dishes.

All glassware should be sterilized prior to use and all operations are carried out using a sterile bench.

2.2. Media and Solutions

All media and solutions are adjusted to pH 5.8 with $1M$ KOH or $1M$ HCl. Protoplast medium (PM), and the enzyme solution are filter sterilized. Growth regulator stocks (1 mg/mL) are filter sterilized and added separately to sterilized media.

1. Basal medium (BM): MS medium *(5)* containing B5 vitamins *(6)* with and without gelling agents (0.8% agar or 0.2% gelrite), and 3% sucrose if not stated otherwise.
2. 0.5 BM medium: consisting of half concentration of MS macroelements *(5)*, B5 vitamin, and 3% sucrose with and without gelling agents (0.8% agar or 0.2% gelrite).
3. MSAR I medium *(7)*: BM medium containing 2.0 mg/L indole–3–acetic acid (IAA), 0.5 mg/L 2,4–dichloro–phenoxyacetic acid (2,4–D), 0.5 mg/L 6–(γ,γ–dimethylallylamino)purine riboside (IPAR).
4. MSAR II medium *(7)*: BM medium containing 2.0 mg/L IPAR, 0.05 mg/L a–naphtaleneacetic acid (NAA) with 0.2% gelrite.
5. MSAR III medium *(7)*: BM medium containing 1.0 mg/L (IAA), 0.2 mg/L indole–3–butyric acid (IBA), 0.2 mg/L 6–furfurylaminopurine (kinetin), and 0.2% gelrite.
6. Protoplast medium (PM): 0.5X BM medium containing MSAR I hormones with $0.45M$ sucrose or $0.45M$ mannitol.
7. Enzyme solution: 1.0% cellulase (Onozuka R–10; Serva, Heidelberg, Germany), 0.25% macerozyme (R10; Serva) dissolved in PM medium.
8. $0.45M$ Mannitol and $0.45M$ sucrose solutions.
9. Sodium alginate solution: 1% (w/v) solution in BM medium containing $0.45M$ sucrose.
10. Calcium agar plates: 20 mM calcium chloride, $0.45M$ sucrose, and 1% agar.

3. Methods
3.1. Isolation and Culture of Protoplasts
3.1.1. Leaf Mesophyll Protoplasts

1. Sterilize *Arabidopsis* seeds in an Eppendorf centrifuge tube for 15 min using 1 mL of seed sterilizing solution. Wash the seeds five times with sterile water (1 mL) and dry them by placing the open tubes in the sterile bench. Plate the seeds for germination on 0.5 BM agar plates.
2. To grow plants for mesophyll protoplast isolation, transfer 5–8 7-d-old seedlings into 250-mL jars containing BM medium and culture them at 25°C using 16 h light and 8 h dark period (*see* **Note 1**).
3. Use leaves from 3–5-wk-old plants (*see* **Note 2**).
4. Cut leaves in half through the mid vein and then into 4–6 pieces by cross–sectioning with a sharp blade (*see* **Note 3**).
5. Wash the cut leaves once with 0.45M sucrose solution. Transfer the leaf explants into a 9-cm Petri dish and add 15 mL of enzyme solution. Incubate the material for 12–16 h at room temperature in the dark (*see* **Note 4**).
6. Shake the Petri dishes gently to release the protoplasts and incubate them further for 30 min (*see* **Note 5**).
7. Collect the protoplast suspension using a broad-mouthed pipet using gentle suction, and filter the protoplast sequentially through 100 µm and 50 µm sieves (*see* **Note 6**).
8. Gently dispense the protoplast suspension into 12-mL screw-capped glass tubes and centrifuge at 50g for 5 min.
9. Remove the dark-green band of floating protoplasts concentrated at the top of the tubes and transfer them to a new glass tube (*see* **Note 7**).
10. Add 10 mL 0.45M mannitol solution and centrifuge the protoplasts at 60g for 5 min. Resuspend the protoplast pellet and wash the cells twice in a similar fashion with 0.45M mannitol solution.
11. Suspend the protoplast pellet obtained after the final wash in 1 mL of sodium alginate solution and count the protoplast yield using a hemocytometer. Adjust the protoplast density to $5-7 \times 10^5$ cells/mL by adding more alginate solution (*see* **Notes 8** and **9**).
12. Using a wide-bore pipet, take up the protoplasts dispersed in sodium alginate solution and create 250–500-µL drops on the surface of calcium–agar plates (*see* **Note 10**).
13. Allow the alginate to form a gel for 45 min. Transfer the individual droplets with a spatula into 55-mm Petri dishes containing approx 5 mL of PM medium and culture the protoplast in a growth chamber at 25°C under dim light.
14. Remove 2.5 mL of the PM medium and add 2.5 mL of fresh PM medium after 7 and 14 d (*see* **Note 11**).
15. Remove 1 mL of PM medium and add 1 mL of MSAR I medium to the protoplast cultures on d 21, 28, and 35 (*see* **Note 12**).

3.1.2. Root-Derived Protoplasts

1. Sterilize seeds for 15 min using 10% sodium hypochlorite solution and after washing and drying of the seeds, plate them on 0.5 × BM agar plates.
2. In order to obtain root cultures for protoplasts, transfer 15–20 7-d-old seedlings into 50 mL of BM in Erlenmeyer flasks and place the flasks on a rotary shaker at 120 rpm and 25°C using 16 h light and 8 h dark period (*see* **Note 13**).
3. After 10–14 d growth, remove plantlets from the flasks and separate the roots from the green tissue. Using a scalpel, cut the roots into small pieces (approx 2– 4 mm) and transfer the root segments into Petri dishes with MSAR I medium (*see* **Note 14**).
4. Place the Petri dishes on a shaker set at 100 rpm in the dark (or at low light conditions) for 7–12 d (*see* **Note 15**).
5. Remove the liquid medium from the Petri plates and wash the root explants once with 0.45M sucrose solution. Remove the washing solution and add approx 20 mL of enzyme solution (*see* **Note 4**).
6. Incubate the root explants for 12–16 h with occasional shaking (*see* **Note 16**).
7. Collect the protoplast with a broad-mouthed pipet using gentle suction and sieve them through 100-, 50-, and 25-µm sieves (*see* **Notes 6** and **17**).
8. Place the protoplast suspension in screw-capped 12 mL glass tubes and centrifuge for 5 min at 50g.
9. Collect the band of floating protoplasts concentrated at the top of the solution and transfer them into a new tube (*see* **Notes 7** and **18**).
10. Resuspend the protoplasts in 0.45M mannitol solution and pellet the cells at 60g for 5 min. Wash the protoplast pellet by repeating this step twice (*see* **Note 19**).
11. If the protoplasts are being cultured in alginate drops, suspend the cells in 1 mL of alginate solution. If the protoplasts are being cultured in liquid medium, resuspend the cells in 1 mL of PM medium.
12. Take 5 µL from the protoplast suspension and count the number of cells in a hemocytometer (*see* **Note 20**).
13. For embedding of the protoplasts, use a wide-bore pipet to take up the cells dispersed in sodium alginate solution (at a density of 3–5 × 10^5 cells/mL) and create 250–500-µL drops on calcium–agar plates (*see* **Notes 10** and **21**).
14. Transfer the drops of alginate gel-carrying protoplasts into 55-mm Petri dishes containing approx 5 mL of PM medium and culture the protoplasts in a growth chamber at 25°C under dim light.
15. For liquid culture, adjust the initial density of protoplasts to 1 × 10^6 cells/mL until cell divisions start and then dilute the protoplast suspension with PM medium to a density of 5 × 10^5 cells/mL (*see* **Note 22**).
16. Remove 2.5 mL of the spent medium and add 2.5 mL of fresh PM medium after 7 and 14 d (*see* **Note 11**).
17. Remove 1 mL of PM medium and add 1 mL of MSAR I medium on d 21, 28, and 35 (*see* **Note 12**).

3.1.3. Protoplasts from Cell Suspensions

1. Methods used for the initiation and maintenance of cell suspensions are described in Chapter 4.
2. Collect 50 mL of 2-d-old cell suspension in a sterile Falcon tube and allow the cells to settle for 5–7 min. Remove the supernatant.
3. Wash the cells once with $0.45 M$ sucrose solution and centrifuge them at $50g$ for 5 min.
4. Remove the supernatant and resuspend the pelleted cells in approx 50 mL of protoplasting enzyme solution (see **Note 4**).
5. Dispense the cells in three, 90-mm Petri dishes, seal the plates with parafilm, and place them on a shaker at 50 rpm.
6. After 4 h of incubation, collect the protoplast suspension and filter the cells through 50- and 25-µm sieves. Dispense the protoplasts to 12-mL, screw-capped glass tubes (see **Note 17**).
7. Centrifuge the protoplasts for 5 min at $50g$. Collect the band of floating protoplasts from the top of solution (see **Notes 7** and **18**).
8. Wash the protoplasts twice with $0.45 M$ mannitol solution by centrifuging at $60g$ for 5 min and resuspending the pellet each time (see **Note 19**).
9. Resuspend the protoplasts in 1 mL of PM medium or alginate solution (as described in **Subheading 3.1.2.**) and count the cell number in a hemocytometer (see **Note 20**).
10. For liquid culture, adjust the density of protoplasts in PM medium to 1×10^6 cells/mL. When the protoplasts start dividing, dilute the suspension with PM medium to 5×10^5 cells/mL (see **Note 22**).
11. For embedding the protoplast into alginate, adjust the cell density to $3-5 \times 10^5$ cells/mL and create alginate drops of 250– 500– µL on calcium–agar plates (see **Note 10**).
12. After 45 min, remove alginate drops carrying the embedded protoplasts and transfer them into 55-mm Petri dishes containing 5 mL of PM medium. Culture the protoplasts at 25°C under low light conditions (see **Note 21**).
13. Remove 2.5 mL of the spent medium and add 2.5 mL of fresh PM medium after 7 and 14 d (see **Note 11**).
14. Remove 1 mL of PM medium and add 1 mL of MSAR I medium on d 21, 28, and 35 (see **Note 12**).

3.2. Regeneration of Plants from Protoplast-Derived Calli

1. Remove the liquid medium using a pipet, and transfer the microcalli carried by the alginate beads, after sectioning the gel beads into four to five pieces, into MSAR II medium (see **Note 23**).
2. Place the Petri dishes in an illuminated culture room (3000 lx) using 16 light and 8 h dark period at 25°C.
3. Pick out green calli and regenerating shoots and transfer them into MSAR II medium until shoots attain a size of approx 2 cm (see **Note 24**).

4. Transfer the shoots into MSAR III medium to induce root formation for 3–6 d, then place the plantlets in culture tubes containing 0.5X BM agar medium with only 0.5% sucrose for flowering and seed setting (*see* **Note 25**).

4. Notes

1. Do not place the glass jars close to the light source because condensation inside the jars results in vitrification of the leaves.
2. The age and state of leaves is critical *(8)*. Do not use pale, discolored, curled, or vitrified leaves. Do not use leaves from plants that are already flowering.
3. In order to decrease the amount of debris in the protoplast preparation, the cut should be clean and the leaf should not be crushed. Allow the weight of the blade and handle alone to do the cutting and change the blade between different dishes.
4. The activity of enzyme preparations may differ from batch to batch and thus the optimal time of incubation may vary. It is advisable to determine the requirements for optimal protoplasting and store a particular enzyme batch in bulk.
5. At this stage of enzyme digestion, the cell suspension should be green in color and numerous spherical protoplasts should be seen floating under the microscope. A dirty preparation will contain many broken protoplasts and may result in low viability of surviving protoplasts.
6. We use steel sieves of different pore size available from Fastnacht (Bonn, Germany), but plastic mesh with the right pore size can be sewn into a cone and put in a glass funnel fitted onto a 100-mL Erlenmeyer flask. The whole assembly can be wrapped by aluminium foil and sterilized by autoclaving.
7. Although the band of floating protoplasts is removed carefully, some protoplasts are invariably lost. An alternative way is to use a thin, sterile, glass capillary tube connected to a pipet pump with a narrow tubing. The capillary is inserted to the bottom of the centrifuge tube, and the debris pellet and the enzyme solution are gently removed without disturbing the band of floating protoplasts.
8. An optimal preparation from leaves of five plants yields about $3-4 \times 10^6$ protoplasts.
9. Gelrite solution (0.2% dissolved in $0.45M$ sucrose solution) can be used instead of sodium alginate solution in the same fashion. This obviates the need of releasing the microcalli from the embedding matrix at a later stage, which may be necessary when alginate is used as embedding matrix.
10. Larger drops may be prepared and later cut into smaller pieces. However, we find it easier to use small drops since it does not create problems in spreading. Also, individual and uniform drops carrying roughly the same number of protoplasts can be used to test a wide range of culture media and conditions.
11. After 14 d of culture, the dividing cells should form colonies of 4–16 cells. For leaf mesophyll protoplasts, this time may be extended by another 7 d.
12. By day 35, microcalli are visible that may differentiate to roots and somatic embryo–like structures.
13. Roots should be actively growing, white in color, and not yellow–brown or green. The plants grown in the flask should not flower. The quality of root cultures affect the yield of protoplasts and the time required for subsequent cell proliferation.

14. Divide the roots from one 250-mL flask into three portions and, after cutting them into smaller pieces, transfer each portion of roots into approx 20 mL of MSAR I medium in a 90-mm Petri dish. More root material in this volume may delay the cell proliferation response.
15. Root explants exhibit numerous regions of cell proliferation, particularly at the sites of lateral roots, and assume a bulging appearance at this stage. They should be white in color and not brown. The liquid medium at this stage shall contain a few single cells.
16. Protoplasts start to be released after 4 h, and a periodic check is beneficial to reach an optimal yield (1.5–3×10^6 cells/mL). Alternatively, digestion of roots with 2.0% cellulase (Onozuka R–10, Serva) releases protoplasts within 3–5 h.
17. A glass Pasteur pipet with the tip broken off serves well since the protoplasts are not subjected to the suction force in the narrow diameter of the pipet tip.
18. Collect the protoplast bands from three tubes into one tube to obtain a sufficient pellet size in the subsequent steps.
19. When changing the solutions, try to remove as much of the previous solution as possible without disturbing the protoplast pellet. The soft transition from one osmoticum to another results in less protoplast bursting and loss between the steps.
20. If the number of protoplasts exceeds 100/square of the hemocytometer, dilute the protoplast suspension.
21. Protoplasts embedded in alginate may be slower in dividing by 1 or 2 d.
22. Depending on the quality of protoplast preparation, up to 75% of protoplasts survive and 20–40% of cells will undergo divisions during the first 3–5 d of culture. In liquid medium, an initial high density may result in aggregation and collapse of protoplasts. Therefore, care should be taken to properly dilute the protoplast suspension. Aggregation of protoplasts may also be observed if there was too much debris and/or dead protoplasts in the preparation, which may be a consequence of either inadequate removal of the enzyme during the washes or sudden osmotic changes.
23. Alternative approaches involve a depolymerization of the alginate gel in 20 mM sodium citrate solution, using an appropriate osmoticum *(4)*, or a removal of visible microcalli from the alginate gel and transferring them in the regeneration medium *(9)*. We found that the regeneration of calli comprising of more than 64 cells is not hindered by alginate embedding.
24. Care must be taken to remove dead cells from the regenerating cultures.
25. The culture tubes are capped with loose cotton to facilitate a proper aeration necessary for seed setting in vitro. Take care not to place the tubes too close to the light source because this will cause moisture condensation inside the tubes and result in low efficiency of fertilization. Alternatively, rooted plants can be transferred to soil after washing the roots gently with a dilute fungicidal solution (e.g., 0.02% Benomyl, DuPont, Boston, MA). After proper hardening, the plants will flower and set seed.

References

1. Morris, P. C. and Altmann, T. (1994) Tissue culture and transformation, in *Arabidopsis*. (Meyerowitz, E. M. and Somerville, C. R. eds.), Cold Spring Harbor Laboratory, Cold Spring Harbor, NY, pp. 173–222.

2. Mathur, J., Szabados, L. and Koncz, C. (1995) A simple method for isolation, liquid culture, transformation and regeneration of *Arabidopsis thaliana* protoplasts. *Plant Cell Rep.* **14**, 221–226.
3. Wenck, A. R. and Marton, L. (1995) Large–scale protoplast isolation and regeneration of *Arabidopsis thaliana*. *Bio/Techniques* **18**, 640–643.
4. Damm, B. and Willmitzer, L. (1988) Regeneration of fertile plants from protoplasts of different *Arabidopsis thaliana* genotypes. *Mol. Gen. & Genet.* **213**, 15–20.
5. Murashige, T. and Skoog, F. (1962) A revised medium for rapid growth and bioassay with tobacco tissue cultures. *Physiol. Plant.* **15**, 473–497.
6. Gamborg, O. L., Miller, R. A. and Ojima, K. (1968) Nutrient requirements of suspensions cultures of soybean root cells. *Exp. Cell Res.* **50**, 151–158.
7. Koncz, C., Schell, J. and Redei, G.P. (1992) T–DNA transformation and insertional mutagenesis, in *Methods in* Arabidopsis *Research*. (Koncz, C., Chua, N–H., and Schell, J., eds.), World Scientific, Singapore, pp. 224–273.
8. Mason, J. and Paszkowski, J. (1992) The culture response of *Arabidopsis thaliana* protoplasts is determined by the growth conditions of donor plants. *Plant J.* **2**, 829–833.
9. Altmann, T., Damm, B., Halfter, U., Willmitzer L., and Morris, P. C. (1992) Protoplast transformation and methods to create specific mutants in *Arabidopsis thaliana*, in *Methods in* Arabidopsis *Research*. (Koncz, C., Chua, N–H., and Schell, J., eds.), World Scientific, Singapore, pp. 310–330.

II

PURIFICATION OF SUBCELLULAR ORGANELLES AND MACROMOLECULES

7

Preparation of Physiologically Active Chloroplasts from *Arabidopsis*

Ljerka Kunst

1. Introduction

Only structurally intact chloroplasts with a functional envelope exhibit metabolic activities comparable to those of the original tissues. Therefore, the most important objectives that should guide the isolation of chloroplasts from any plant species are their morphological and physiological integrity. The use of intact chloroplasts is absolutely essential for studying processes like light-driven CO_2 fixation, CO_2-dependent O_2 evolution, *in organello* RNA and protein synthesis, import of cytoplasmically made polypeptides into chloroplasts, or incorporation of labeled precursors to dissect the enzymatic machinery involved in fatty acid biosynthesis and fatty acid incorporation into chloroplast lipids. In some plant species, like spinach or pea, highly active chloroplasts can be obtained by following several simple rules:

1. Use of young, fresh plant material free from starch.
2. Disruption of cells in a medium containing buffered sorbitol.
3. Rapid separation of chloroplasts from the remaining cell constituents.

Unfortunately, the standard protocols for chloroplast isolation *(1,2)* typically result in poor chloroplast yields, or chloroplasts of low activity, when applied to *Arabidopsis*. The reasons for this apparent sensitivity of *Arabidopsis* chloroplasts are not known. However, concerted efforts of several research groups to optimize established chloroplast isolation protocols for *Arabidopsis* and the discovery that the inclusion of high levels of EDTA or EGTA in the isolation medium is critical for the recovery of active chloroplasts *(3)*, led to the development of two reliable procedures: chloroplast isolation by homogenization of leaf tissue using polytron and chloroplast isolation from protoplasts.

Although both procedures yield intact chloroplasts capable of high rates of CO_2-dependent O_2 evolution, they differ in several important ways. Polytron homogenization is simpler and more rapid, but the recovery of intact chloroplasts is rarely >30% of that obtained by the protoplast isolation method. On the other hand, in our experience, chloroplasts prepared from protoplasts exhibit very low rates of lipid biosynthesis from radioactivelly labeled precursors, owing perhaps to a relatively long time required for the enzymatic digestion of the leaves. Thus, the method of choice will ultimately depend on the intended use of the isolated chloroplasts.

2. Materials

2.1. Plant Material Preparation

1. Two-to-three-week-old *Arabidopsis thaliana* (L.) Heynh. plants (*see* **Note 1**).
2. Razor blade.
3. Balance.

2.2. Polytron Homogenization

1. 20 g Freshly harvested *Arabidopsis* rosette leaves.
2. Polytron.
3. Cut-off 500-mL plastic cylinder.
4. Centrifuge with SS-34 and HB-4 rotors (Sorvall, DuPont, Wilmington, DE) or equivalent ($43,400g$).
5. Ice bucket.
6. 1-L Beaker.
7. 500-mL Beaker.
8. Miracloth (Calbiochem, La Jolla, CA).
9. 8–10 50-mL Polycarbonate tubes.
10. Paper tissue.
11. Natural bristle paint brush.
12. Aspirator.
13. Glass pipet.
14. Spectrophotometer.
15. Two cuvets.
16. Phase-contrast microscope *(optional)*.
17. Clark-type electrode (Hansatech, Hardwick Industrial Estate, Kings Lynn, Norfolk, UK; *optional*).
18. Homogenization buffer (HB): $0.45M$ sorbitol, 20 mM Tricine-KOH, pH 8.4, 10 mM EDTA, 10 mM NaHCO$_3$, 0.1% bovine serum alumin (BSA; fatty acid-free, Sigma, St. Louis, MO).
19. Resuspension buffer (RB): $0.3M$ sorbitol, 20 mM Tricine-KOH, pH 7.6, 5 mM MgCl$_2$, 2.5 mM EDTA.
20. Percoll gradient mix (PGM): 50% Percoll ([v/v,] Pharmacia, Uppsala, Sweden) in RB. Mix equal volumes (15 mL each) of 100% Percoll and cold 2X RB.

2.3. Chloroplast Isolation from Protoplasts

1. 10 g Freshly harvested *Arabidopsis* rosette leaves.
2. Aspirator.
3. Vacuum flask with a rubber stopper.
4. Forceps *(optional)*.
5. Cheesecloth.
6. Ice bucket.
7. 250-mL Beaker.
8. Four 30-mL corex tubes.
9. Clinical centrifuge ($100g$).
10. Four 50-mL polycarbonate tubes.
11. Glass pipet.
12. 5-mL Syringe.
13. 15-μm Mesh nylon net (Tetko, Elmsford, NY).
14. Centrifuge with SS-34 rotor (Sorvall) or equivalent ($280g$).
15. Spectrophotometer.
16. Two cuvets.
17. Phase-contrast microscope *(optional)*.
18. Clark-type electrode (Hansatech; *optional*).
19. Protoplast isolation buffer (PIB): $0.5M$ sorbitol, 10 mM MES, pH 6.0, 1 mM $CaCl_2$, 1.6% (w/v) pectinase (Macerozyme Onozuka R-10, Calbiochem, La Jolla, CA), and 1.6% (w/v) cellulase (Onozuka RS).
20. Protoplast storage buffer (PSB): $0.5M$ sorbitol, 10 mM MES, pH 6.0, 1 mM $CaCl_2$.
21. Protoplast percoll buffer (PPB): 50% percoll ([v/v], Pharmacia) in PSB. Mix equal volumes (15 mL each) of 100% percoll and cold 2X PSB.
22. Protoplast lysis buffer (PLB): $0.3M$ sorbitol, 20 mM Tricine-KOH. pH 8.4, 10 mM EDTA, 10 mM $NaHCO_3$, 0.1% BSA (Fatty acid free, Sigma).
23. Chloroplast resuspension buffer (CRB): $0.3M$ sorbitol, 20 mM Tricine-KOH, pH 7.6, 5 mM $MgCl_2$, 2.5 mM EDTA.

3. Methods
3.1. Preparation of Chloroplasts Using Polytron Homogenizer

1. Prepare percoll gradients by combining equal volumes (typically 15 mL) of cold 100% percoll and cold 2X RB and centrifuging at $43,400g$ for 30 min in a Sorvall SS-34 rotor, or equivalent, with the brake off. Store gradients on ice until use *(4)* (*see* **Note 2**).
2. Harvest 20 g of leaves at the rosette stage (50–90% fully expanded) and float them on ice-cold water in an 1-L beaker for 20–30 min.
3. Blot the leaves and homogenize them by three bursts of 4–5 s each at full polytron speed in 200 mL of cold HB in a cut-off plastic cylinder.
4. Proceeding rapidly, filter the homogenate through Miracloth (Calbiochem) into a 500-mL beaker submerged in ice. Applying gentle hand pressure on the brie significantly increases the final chloroplast yield.

5. Distribute the filtrate into 50-mL precooled tubes and centrifuge for 90 s (including deceleration) at 280g in the SS-34 rotor (*see* **Note 3**).
6. Carefully decant the supernatant, and holding the tubes upside-down, wipe their insides with paper tissue. Resuspend the crude chloroplast pellet in 1 mL of ice-cold RB using a large, natural bristle paint brush. Pool the resuspended chloroplasts and rinse all the tubes and brushes with 1 mL of RB. Add this wash to the chloroplast suspension (*see* **Note 4**).
7. Carefully overlay the resuspended chloroplasts on top of the cold, preformed percoll gradients and centrifuge at 13,300g for 6 min in a swinging bucket rotor (for example HB-4, Sorvall) with the brake off. The crude chloroplast suspension gets resolved into two bands, an upper band of broken chloroplasts, and a lower band of intact chloroplasts (*see* **Note 5**).
8. To recover intact chloroplasts, aspirate the gradient down to the lower band and gently collect the chloroplasts using a glass pipet.
9. Pool the intact chloroplasts, dilute them with three volumes of RB, and collect by centrifugation at 3300g for 90 s in a swinging bucket rotor (HB-4, Sorvall) (*see* **Note 6**).
10. Resuspend the washed chloroplasts in 2 mL of cold RB and determine the chlorophyll concentration. Adjust chlorophyll concentration to 500 µg/mL and store chloroplasts on ice until use (*see* **Note 7**).
11. Perform the functional test of interest, i.e. CO_2-dependent O_2 evolution, protein or RNA synthesis, metabolic enzyme assays, protein import, and so on.

3.2. Chloroplast Isolation from Intact Protoplasts

1. Vacuum infiltrate intact leaves with 40 mL of PIB and incubate at room temperature for 1–2 hrs (*see* **Note 8**).
2. Filter protoplasts through several layers of cheesecloth into a 250-mL ice-cold beaker to remove debris, transfer the suspension into cold 30-mL tubes, and harvest protoplasts by centrifugation for 5 min at 100g in a clinical centrifuge.
3. Discard the supernatant and resuspend protoplasts in 5 mL of cold PSB by gently swirling the tubes.
4. Pool the protoplasts from different tubes, rinse the tubes with an additional 1 mL of PSB, and add to the protoplast suspension.
5. Carefully transfer the suspension to a clean, 50-mL tube already containing 30 mL of cold PPB and band intact protoplasts by centrifugation at 100g for 10 min.
6. To recover intact protoplasts, aspirate the buffer down to the protoplast band and collect the protoplasts with a pipet.
7. Pool the intact protoplasts, resuspend them in 10 mL of PSB, and store on ice until use.
8. Before isolating chloroplasts, collect the protoplasts by centrifugation for 5 min at 100g and resuspend at 100 µg chlorophyll/mL in cold PLB.
9. Take an aliquot of the protoplast suspension (1 mL) and gently lyse the protoplasts by passing them through a 15-µm mesh nylon net (*see* **Note 9**).
10. Collect the chloroplasts by centrifugation at 280g for 90 s (including deceleration) in a Sorvall SS-34 rotor (*see* **Note 3**).

Preparation of Arabidopsis Chloroplasts

11. Resuspend the chloroplasts in 2 mL of cold CRB and determine the chlorophyll concentration. Adjust chlorophyll concentration to 500 µg/mL and store chloroplasts on ice until use (*see* **Note 10**).
12. Perform the functional test of interest.

4. Notes

1. *Arabidopsis* plants are germinated and grown under continuous fluorescent illumination (100–150 µE/m^2/s) at 22°C on a perlite:vermiculite:sphagnum (1:1:1) mixture at a density of about 50 plants/100 cm^2. Approximately 12 h before harvesting transfer the plants in the dark to reduce the leaf starch content.
2. For routine experiments 1–2 30-mL gradients are sufficient.
3. It is essential to take as little time as possible to carry out this step—decelerate the rotor with the brake on maximum. Prolonged contact of chloroplasts with potentially toxic secondary metabolites released from the vacuole may result in physiologically inactive preparations.
4. At this point, we usually examine the chloroplast preparation in a phase-contrast microscope *(1)*. The preparation should contain at least 20% intact chloroplasts and no clumps. Intact chloroplasts appear as bright objects, whereas ruptured ones appear dark. It should also be demonstrated that the bright chloroplasts disappear after exposure to hypotonic solutions. Finally, this measure of intactness is reliable only with starch-free chloroplasts, because starch grains also appear bright like chroroplasts.
5. We overlay each 30-mL gradient with up to 10 mL of crude chloroplast suspension.
6. This step is important for removal of the percoll silica particles from the chloroplasts.
7. We routinely determine chloroplast intactness by measuring ferricyanide-dependent O_2 evolution before and after an osmotic shock *(5)*. Briefly, two aliquots of isolated chloroplasts (10–20 µg chlorophyll, each) are resuspended in fresh RB or distilled water, respectively, with 1.5 m*M* $K_3Fe(CN)_6$ and 10 m*M* D,L-glyceraldehyde (to inhibit CO_2-dependent O_2 evolution by intact chloroplasts) in a total volume of 2 mL. Oxygen evolution is measured in a twin Clark-type electrode (Hansatech) as described by Delieu and Walker *(6)*. To ensure that electron transport is fully uncoupled, 2.5 m*M* NH_4Cl is added to the reaction mixture after approx 45 s and oxygen evolution monitored for subsequent 45 s. Percentage of intact chloroplasts is determined by comparing the recorded O_2 evolution rates of the untreated and the osmotically shocked chloroplasts. For example, if the recorded rate of RB resuspended chloroplasts is 20% of that of the distilled water treated chloroplasts, an apparent intactness estimate is 80% *(5)*.
8. Alternatively, the lower epidermis of the leaves can be stripped using fine forceps, and protoplasts released by floating leaves on the PIM for 1–2 h *(3)*. The yield of intact protoplasts obtained by the leaf-stripping method is typically greater and does not require purification of protoplasts by flotation on percoll. On the other hand, removal of leaf epidermis may take some practice. Both, the leaf epidermis removal method and the vacuum infiltration method result in chloroplasts of comparably high activity.

9. We transfer the protoplasts into a 5-mL syringe from which the end has been removed and replaced by a 15-µm mesh net, and apply gentle pressure to rupture them. Protoplast lysis in the presence of 10 mM EDTA at high pH (8.4) seems essential for isolation of intact, highly active chloroplasts *(3)*.
10. The preparation can be examined in a phase-contrast microscope and should contain at least 80% intact chloroplasts. Intact chloroplasts will appear bright. On exposure to hypotonic solutions, chloroplasts should lyse and appear dark. In addition, we routinely determine chloroplast intactness by measuring ferricyanide-dependent O_2 evolution before and after a hypotonic shock (*see* **Note 7**).

References

1. Ellis, R. J. and Hartley, M. R. (1982) Preparation of higher plant chloroplasts active in protein and RNA synthesis, in: *Methods in Chloroplast Molecular Biology* (Edelman, M., Hallick, R. B., and Chua N.-H., eds.), Elsevier, Amsterdam, pp. 170–188.
2. Rathnam, C. K. M. and Edwards, G. E. (1976) Protoplasts as a tool for isolating photosynthetically active chloroplasts from grass leaves. *Plant Cell Physiol.* **17**, 177–186.
3. Somerville, C. R., Somerville, S. C., and Ogren, W. L. (1981) Isolation of photosynthetically active protoplasts and chloroplasts from *Arabidopsis thaliana*. *Plant Sci. Lett.* **21**, 89–96.
4. Cline, K., Andrews, J., Mersey, B., Newcomb, E. H., and Keegstra, K. (1981) Separation and characterization of inner and outer envelope membrane of pea chloroplasts. *Proc. Natl. Acad. Sci. USA* **78**, 3595–3599.
5. Lilley, R. Mc C., Fitzgerald, M. P., Rienits, K. G. (1975) Criteria of intactness and the photosynthetic activity of spinach chloroplast preparations. *New Phytol.* **75**, 1–10.
6. Delieu, T., and Walker, D. A. (1972) An improved cathode for the measurement of the photosynthetic oxygen evolution by isolated chloroplasts. *New Phytol.* **71**, 201–225.

8

Purification of Mitochondria from *Arabidopsis*

Mathieu Klein, Stefan Binder, and Axel Brennicke

1. Introduction

Plant cells contain different subcellular compartments, which serve distinct physiological functions. One of these organelles, the mitochondrion, provides most of the nonphotosynthetic energy required in the cell. Mitochondria contain their own genome(s), but encode only a small number of vital gene products. Similar plastids, the majority of the organellar proteins are nuclear encoded, and imported into these organelles.

Highly purified mitochondria are important as a source of many biochemical, physiological and molecular applications. The preparation of mitochondria as a prerequisite to analyze mitochondrial DNA is described in Chapter 12. Purified mitochondria are also required for protein and RNA import studies *(1,2)*, enzymatic activity measurements *(3,4)*, *in organello* protein analysis *(5,6)*, to prepare protein extracts and protein complexes from mitochondrial membranes *(7)* and for the isolation of mitochondrial RNA *(8)*. Most biochemical investigations of mitochondrial functions also depend on the purity of the organelles. A major obstacle in the purification procedure of mitochondria from plants is the contamination of mitochondrial fractions by plastids and stored metabolites such as starch or phenolic compounds. An easy way to overcome at least part of these problems is the purification of mitochondria from dark grown cell cultures in which the amount of plastids is very low, and mitochondria are the unique energy source of the cell. From these tissues, the use of a percoll step gradient-based procedure allows purification of clean and active mitochondria. The following method is designed for the purification of mitochondria from plant suspension cell culture.

2. Materials

2.1. Plant Material

Dark grown *Arabidopsis thaliana* cell cultures should be used. However, mitochondria from a wide range of nonphotosynthetic plant tissues have also been successfully purified by this method.

2.2. Equipment

1. Centrifugation bottles (400 and 250 mL) and ultracentrifuge tubes (Beckman, Fullerton, CA, 1 × 3.5 in. or equivalent).
2. Microfuge, centrifuge, ultracentrifuge, and corresponding rotors (Beckman JA-10, JA-20, SW28, or equivalents).
3. Waring blender.
4. Cheesecloth.
5. Nylon membranes, 100- and 52-µm pore size.
6. Dounce homogenizer.
7. Fine painting brush.
8. Tips, gloves, glass pipets, and standard laboratory material.
9. A coat for the cold-room operator.

2.3. Stock Solutions for Purification of Mitochondria

The volumes of buffers indicated below are sufficient for 200–250 g fresh weight material.

1. Percoll (Pharmacia, Uppsala, Sweden). Autoclave and store at 4°C. Precaution should be taken to avoid contamination of this solution.
2. Stock I: 1M MOPS solution (without adjusting pH).
3. Stock II: 5M KOH.
4. Stock III: 100 mM EGTA-KOH pH 7.2.
5. Stock IV: 200 mM tricine-KOH pH 7.2.
6. Extraction buffer (EB) (500 mL): 450 mM sucrose, 15 mM MOPS, 1.5 mM EGTA-KOH pH 7.2. Adjust the pH to 7.4 (with 5M KOH), autoclave, and store at 4°C. Before use, add dithiotreitol (DTT) to a final concentration of 10 mM, bovine serum albumin (BSA) to 2 g/L and polyvinylpyrrolidon (PVP) to 6 g/L.
7. Washing buffer (WB) (500 mL): 300 mM sucrose, 10 mM MOPS, 1 mM EGTA-KOH pH 7.2. Adjust the pH to 7.2 (with 5M KOH), autoclave, and store at 4°C. Before use, add BSA to 1 g/L.
8. Resuspension buffer (1X R) (1 L): 400 mM mannitol, 10 mM tricine-KOH pH 7.2, EGTA-KOH pH 7.2. Adjust pH to 7.2 (with 5M KOH), autoclave, and store at 4°C.
9. Gradient buffer (5X G) (100 mL): 1.5M sucrose, 50 mM MOPS. Adjust pH to 7.2 with 5M KOH solution, autoclave, and store at 4°C.
10. Solutions for the percoll gradient steps are summarized in **Table 1**. We recommend preparation of these solutions freshly before starting with the procedure.

Table 1
Solutions for Percoll Step Gradient

	Percoll steps[a]		
Solution	18%, mL	23%, mL	40%, mL
5X Gradient buffer	6	6	6
Percoll	5.5	7	12
Double-distilled water	18.5	17	12
Final volume	30	30	30

[a]The final volumes given here are sufficient for up to four-step gradient tubes.

3. Methods

The isolation procedure should be performed at 4°C or on ice with sterile and cold solutions, tips, and tubes (*see* **Notes 1** and **2**).

3.1. Cell Disruption and Differential Centrifugation

1. Harvest cells on a 100-μm mesh nylon membrane (*see* **Note 3**).
2. Wring out the membrane until all culture medium has been removed quantitatively. Add extraction buffer (EB) to a volume of 2 mL/g fresh weight material.
3. Transfer to a blender and disrupt cells three times by 15-s strokes (first at high speed and then twice at low speed) with 30 s waiting delay between each burst. Longer disruption can dramatically decrease the amount of purified mitochondria.
4. Remove cell debris by filtering the homogenate through a 52-μm pore size nylon membrane (*see* **Note 3**) and filter through one layer of Miracloth.
5. Transfer the supernatant to 400-mL centrifugation bottles.
6. Centrifuge at 2000g for 5 min at 4°C.
7. Repeat **step 5** and centrifuge at 6000g for 5 min at 4°C (*see* **Note 4**).
8. Repeat **step 5** and pellet mitochondria at 16,000g for 10 min at 4°C.
9. Resuspend mitochondria with a fine hair brush in 3–5 mL of WB and adjust the volume to 400 mL.
10. Repeat **steps 7** and **8**.
11. Resuspend mitochondria in 4–5 mL WB (*see* **Note 5**) and transfer to a dounce homogenizer. Carefully disperse mitochondria by applying 3–5 strokes.

3.2. Percoll Density Gradient Centrifugation

1. Prepare percoll step gradients with the solutions as described in **Table 1**. One gradient tube is needed per 100 g starting material. For pouring the gradient, use a 10-mL glass pipet, and carefully layer the solutions under each other, starting with the 18% percoll layer.
2. Carefully pipet the mitochondrial suspension onto the top of the gradient (no more than 3 mL per gradient). Centrifuge at 12,000g for 45 min at 4°C.

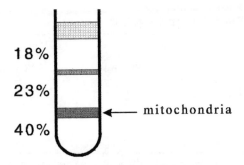

Fig. 1. Schematic representation of a percoll step gradient after centrifugation. The mitochondria (yellow-brown) collect at the 23–40% interface.

3. Carefully remove the top half of the gradient with a 10-mL pipet, and collect the mitochondria from the 23–40% interface in a 250-mL centrifugation bottle (**Fig. 1**). Add resuspension buffer (1X R) to a final volume of 200 mL and centrifuge at 14,000g for 10 min at 4°C.
4. Carefully remove most of the supernatant, resuspend the pellet by gentle shaking or loosening with a paint brush, and wash with 200 mL 1X R buffer to completely remove the percoll (*see* **Note 6**).
5. Discard the supernatant and use a pipet to resuspend the pellet in a maximum volume of 1.4 mL of 1X R buffer. Transfer the mitochondrial suspension into a preweighed microfuge tube and centrifuge for 15 min at 4°C at maximum speed.
6. Pour off the supernatant and weigh the tube to estimate the yield of fresh weight material (*see* **Note 7**).
7. At this step, mitochondria can be directly used for further applications or can be stored at –80°C (*see* **Note 8**).

4. Notes

1. All procedures are performed at 4°C to minimize nucleic acid and protein degradation. Note also that the equipment should not be washed with detergents, but should only be rinsed with double-distilled water.
2. DTT or β-mercaptoethanol should be present in the extraction buffer to prevent oxidative damage of the biological compounds by secondary metabolites. It can be added to all buffers at concentrations up to 100 mM depending on the nature of the plant material used and/or the further applications.
3. A nylon membrane is preferred to harvest suspension callus cultures since it is easier to wring and to retrieve the cells. Two layers of cheesecloth and one layer Miracloth, however, also give good results.
4. If the pellet of cell debris is too voluminous, resuspend it in 100–200 mL WB and reiterate step 6 to increase the recovery rate. The amount of this pellet varies depending on the tissue and/or culture used. In some cases, when the pellet

represents 1/3 of the total volume for example, it could be necessary to increase this volume considerably or to wash the pellet several times.
5. Since the volume of the mitochondrial suspension you will load onto the percoll gradient must not exceed 3 mL/tube, the total volume of the suspension should be in the range 5–7 mL. If the suspension is too diluted, you will need more gradients and will lose considerable amounts of mitochondria during the percoll gradient purification step.
6. At this step, the mitochondrial pellet is generally very loose. Most of the supernatant should be removed by vacuum suction.
7. One gram of wet mitochondria corresponds to approx to 100 mg total mitochondrial protein.
8. Purified mitochondria can also be stored in a cold room on ice for several hours without serious loss of activity. For a longer storage at -80°C, some cryopreservatives should be added to keep the organelles active for later physiological or biochemical studies *(9)*.

References

1. Chaumont, F., O'Riordan, V., and Boutry, M. (1990) Protein transport into mitochondria is conserved between plant and yeast species. *J. Biol. Chem.* **265**, 16,856–16,862.
2. Maréchal-Drouard, L., Weil, J. H., and Guillemaut, P. (1988) Import of several tRNAs from the cytoplasm into the mitochondria in bean *Phaseolus vulgaris*. *Nucleic Acids Res.* **16**, 4777–4788.
3. Struglics, A., Fredlund, M., Rasmusson, A. G., and Møller, I. M. (1993) The presence of a short redox chain in the membrane of intact potato tuber peroxysomes and the association of malate dehydrogenase with the peroxysomal membrane. *Physiologia Plantarum* **88**, 19–28.
4. Lin, T., Sled, D. D., Ohnishi, T., Brennicke, A., and Grohmann, L. (1995) Analysis of the iron-sulfur clusters within the complex I (NADH: ubiquinone oxidoreductase) isolated from potato tuber mitochondria. *Eur. J. Biochem.* **230**, 1032–1036.
5. Colas des Frances-Small, C., Ambard-Bretteville, F., Darpas, A., Sallantin, M., Huet, J. C., Pernollet, J. C., and Remy, R. (1992) Variation of the polypeptide composition of mitochondria isolated from different potato tissues. *Plant Physiol.* **98**, 273–278.
6. Hahne, G., Maier, J. E., and Lörz, H. (1988) Embryogenic and callus-specific proteins in somatic embryogenesis of the grass *Dactylis glomerata L*. *Plant Science* **55**, 267–279.
7. Herz, U., Schröder, W., Liddell, A., Leaver, C. J., Brennicke, A., and Grohmann, L. (1994) Purification of the NADH:ubiquinone oxidoreductase (complex I) of the respiratory chain from the inner mitochondrial membrane of *Solanum tuberosum*. *J. Biol. Chem.* **269**, 2263–2269.
8. Binder, S. and Brennicke, A. (1993) Transcription initiation sites in mitochondria of *Oenothera berteriana*. *J. Biol. Chem.* **268**, 7849–7855.
9. Schieber, O., Dietrich, A., and Maréchal-Drouard, L. (1994). Cryopreservation of plant mitochondria as a tool for protein import or *in organello* protein synthesis studies. *Plant Physiol.* **106**, 159–164.

9

Preparation of DNA from *Arabidopsis*

Jianming Li and Joanne Chory

1. Introduction

Successful extraction of DNA is an essential and time consuming step in many plant molecular biology procedures. The classical approaches of plant DNA isolation are often designed to produce large amounts of DNA of high molecular weight with sufficient purity. Since the polymerase chain reaction (PCR) is the method of choice in modern plant genetic analyses such as map-based breeding and positional gene cloning, many recently developed protocols are aimed at rapid extraction of a minute quantity of plant DNA that is suitable for PCR. This allows large numbers of samples to be analyzed in a short period of time.

In order to extract DNA from plant cells, both cell walls and cell membranes must be broken. Whereas the cell walls can be broken either by mechanical forces (e.g., grinding with pestle and mortar) or chemical reactions (e.g., treatment with potassium ethyl xanthogenate, *(1)*, the cell membranes can be dissolved by using detergents such as sodium dodecyl sulfate (SDS) and cetyltrimethylammonium bromide (CTAB). Once DNA is released from plant cells, it must be protected from the endogenous nucleases. The most common strategy to reduce nuclease activities is to keep the plant tissue frozen until the addition of extraction buffer containing detergents (e.g., SDS) and high concentration of EDTA.

The isolation of plant DNA is affected by many factors. These include the amount and the type of tissues, species chosen, the number of steps involved, the time required to complete the procedure, the number of samples to be processed, the requisite purity of the extracted DNA for a specific application, and the use of expensive and/or toxic chemicals. Of these factors the type of plant tissues is the most important one. The structure of the cell walls as well as the molecular composition of the cytoplasm may influence the efficiency of a particular protocol. The final DNA preparation must be digestible with many restriction enzymes and/or amplifiable by thermo-stable DNA polymerase.

The method we describe here is based on that of Dellaporta et al. *(2),* and can be applied to many different plant species. Plant DNA extracted from *Arabidopsis* seedlings using this protocol should be larger than 25 kb in length and can be used not only for PCR-based analyses but also for enzyme digestion and Southern-blot analysis.

2. Materials

1. Latex gloves.
2. Mortar and pestle.
3. 1.5-mL Eppendorf tubes.
4. 40-mL Oak Ridge centrifuge tubes.
5. VWR Vortex mixer.
6. Sorval RC-5B refrigerated superspeed centrifuge.
7. Eppendorf microcentrifuge.
8. Speed-Vac.
9. 65°C Water bath.
10. Liquid N_2.
11. Autoclaved deionized water.
12. Extraction buffer: 100 mM Tris-HCl, pH 7.5, 500 mM NaCl, 50 mM EDTA, 10 mM β-mercaptoethanol, 1% SDS (v/v).
13. Tris-EDTA (TE) buffer: 50 mM Tris-HCl, pH 8.0, 10 mM EDTA, pH 8.0.
14. 3M sodium acetate, pH 4.8.
15. 5M potassium acetate.
16. 20% (w/v) SDS.
17. 75% (v/v) ethanol.
18. Isopropanol.

3. Methods

The following protocol is for a single *Arabidopsis* seedling (10–12-d-old). A 10X scaled-up version works quite well for up to 5 g of plant tissues. A modified micropreparation protocol is also described (*see* **Note 1**).

1. Freeze a single plant (100–300 mg) in liquid N_2 (*see* **Notes 2–4**). Grind to a very fine powder in a precooled (-70°C) mortar and pestle (*see* **Notes 5 and 6**).
2. Transfer the powder with liquid N_2 to a 40-mL Oak Ridge tube. After N_2 evaporates, add 5 mL extraction buffer (preheated to 65°C) to the frozen powder, mix the solution thoroughly and quickly by vigorous shaking, and incubate the tube at 65°C for 10 min with occasional shaking.
3. Add 1.67 mL 5M potassium acetate (no need to adjust pH) and shake vigorously. Incubate the tube on ice for 20 min.
4. Spin tube(s) at 25,000g (in SS34 rotor) for 20 min (*see* **Note 7**).

Arabidopsis DNA Preparation

5. With a 5-mL pipet, carefully transfer the supernatant into a new, 40-mL Oak Ridge tube (*see* **Note 8**), and add 3.33 mL isopropanol. Mix the solution and incubate the tube at -20°C for at least 30 min.
6. Pellet nucleic acids by centrifugation at 20,000g (in a Sorvall SS34 rotor) for 15 min.
7. Carefully remove supernatant and lightly dry DNA pellet by inverting the tube on paper towels for 30 min (*see* **Note 9**).
8. Resuspend the DNA pellet in 0.5-mL TE buffer.
9. Transfer the DNA solution to a 1.5-ml Eppendorf tube. Spin in a microcentrifuge for 10 min to pellet any insoluble materials.
10. *(Optional)* Transfer the supernatant to a fresh Eppendorf tube. Add 10 µL of RNase stock solution (10 mg/mL) and incubate at 37°C for 1 h. Add equal volume of 1:1 phenol/chloroform, mix thoroughly, and spin in microcentrifuge for 10 min. Carefully transfer the upper layer to a new tube and extract the DNA solution again with an equal volume of chloroform (*see* **Notes 10–12**).
11. Transfer the aqueous phase to another fresh Eppendorf tube, add 50 µL of 3M sodium acetate (NaOAc, pH 4.8) and 0.35 mL of isopropanol. Mix and incubate at -20°C for at least 1 h (*see* **Note 13**).
12. Pellet DNA by spinning tube(s) in a microcentrifuge for 10 min. Wash the pellet with 0.5 mL of 75% ethanol kept at -20°C (*see* **Notes 14–16**).
13. Dry DNA pellet in a Speed-Vac (*see* **Note 17**) and resuspend it in 100 µL TE buffer at 4°C overnight.
14. Measure the DNA concentration and check the quality of the preparation (*see* **Notes 18–20**). Store the DNA at 4°C.

4. Notes

1. A modified scaled-down version of this protocol can be performed entirely in an Eppendorf tube very quickly and the resulting DNA preparation is good for PCR analysis (*see* **Note 21**).
   a. Snip one young leaf from a plant and place into the bottom of a 1.5-mL Eppendorf tube.
   b. Macerate the tissue using a disposable pestle (VWR Scientific, San Diego, CA) at room temperature for 15 s without buffer.
   c. Add 700 µL of extraction buffer (200 mM Tris-HCl, pH 8.0, 250 mM NaCl, 25 mM EDTA, and 0.5% SDS) and vortex the sample for 5 s (at this stage, the DNA solution can be left at room temperature).
   d. Centrifuge the extracts for 1 min at 18,000g in a microcentrifuge.
   e. Remove 600 µL of the supernatant and transfer it to a fresh Eppendorf tube.
   f. Add 600 µL of isopropanol, vortex briefly, and immediately centrifuge at 18,000g for 5 min in the microcentrifuge.
   g. Carefully remove the supernatant using a Pasteur pipet, leaving the pellet in the bottom of the tube.
   h. Dry the DNA pellet in a Speed-Vac to remove any of the remaining liquid and resuspend in 100 µL of sterile deionized water.

2. Since most plant DNA preparations are often contaminated by polysaccharides, dark adapting the plants for 2–3 d before collecting tissues is usually enough to reduce their content in plants. Younger leaves or other young tissues are the preferred sources for plant DNA extraction since they normally contain low amount of polysaccharides.
3. Wear gloves to handle plant tissues to avoid possible contamination.
4. After collecting tissues in liquid N_2, one can allow liquid N_2 to evaporate and then store the samples at -70°C for later use. Alternatively, plant tissues can be dried overnight in a Speed-Vac, stored dry at room temperature, and then ground to powder.
5. It is extremely important that plant materials should be kept frozen at this stage by repeatedly adding excess liquid N_2. This is because plant cells are rich in nucleases. Also plant cells contain many polyphenolic compounds which can be easily oxidized and irreversibly react with nucleic acids, thus making the extracted DNA indigestible. Keeping plant tissues frozen prior to mixing with the extraction buffer can inhibit nuclease activities and significantly reduce the formation of oxidized polyphenolic compounds.
6. For maximum DNA yields, the plant cells should be further broken by grinding the mixture with a polytron (Brinkmann Instruments, Westburg, NJ) at a low setting (about 3) to maximize the release of DNA from nuclei.
7. At this step, most polysaccharides and proteins precipitate from the DNA solution by forming coarse complexes with SDS.
8. Sometimes it is necessary to filter the supernatant through two layers of Miracloth (Calbiochem, La Jolla, CA) to get rid of some floating contaminants.
9. Do not let the DNA pellet dry completely at this stage or it will be very difficult to redissolve it into solution.
10. The RNase stock solution contains 10 mg/mL RNase A and 100 U/mL RNase T1 and should be heated in a boiling water bath for at least 15 min to destroy any contaminating DNases. It can then be kept frozen at -20°C until needed. The RNase treatment during purification, followed by a phenol/chloroform (1:1 [v/v]) and a chloroform extraction, produces pure DNA that is both RNA- and protein-free.
11. The phenol/chloroform extraction is also often used to remove contaminated polysaccharides from the DNA preparation. These carbohydrates, which coprecipitate with DNA during alcohol precipitation, inhibit many enzymes commonly used in molecular cloning procedures and affect electrophoresis during further uses of DNA. Many other methods have been developed to remove these contaminated compounds from DNA preparations including CTAB extraction *(3,4)*, high-salt treatment *(5)*, DEAE-cellulose method *(6)*, Sephacryl S-1000 chromatography *(7)*, and Caylase M3 digestion *(8)*.
12. Addition of $3M$ sodium acetate before the phenol/chloroform extraction, rather than at **step 11**, can result in a better recovery of DNA in the aqueous phase during the organic extraction *(9)*.
13. It has been reported that some of the contaminated polysaccharides can be separated from the DNA by precipitation of DNA from $0.3M$ sodium acetate with small volume (0.6 vol) of isopropanol *(10)*.

14. DNA can be further purified by two successive isopropanol precipitations and a 70% ethanol wash. To 100 μL DNA solution, add 10 μL of $3M$ NaOAc (pH 4.8) and 70 μL of isopropanol, mix the solution, incubate at -20°C for 15 min, and pellet DNA by centrifugation at the highest setting in a microcentrifuge. The DNA pellet is washed with 70% (v/v) ethanol, dried in a Speed-Vac, and resuspended in 100 μL TE buffer.
15. One can stop the procedure at this stage and store the DNA pellet in 70% (v/v) ethanol at -20°C indefinitely.
16. Great care should be taken at this step not to disturb the DNA pellet when removing the supernatant because the pellet may not stick well to the wall of Eppendorf tubes. Therefore, DNA could be easily lost during these final stages of the procedure.
17. It is necessary to dry the DNA pellet thoroughly before resuspension in TE buffer since the alcohol residue in the final DNA preparation may inhibit restriction digestion.
18. Measure the absorbance of the final DNA solution at wavelengths of 230, 260, and 280 nm. Generally, an A_{260}/A_{280} ratio of 1.8–2.0 combined with an A_{260}/A_{230} ratio of about 2, indicates good quality of the final DNA preparation. Although the A_{260}/A_{280} ratio is the most widely used criterion for judging the purity of a DNA preparation, it is insensitive to many potential contaminants which may have similar A_{260}/A_{280} ratios to DNA. Many of these compounds, including phenol, also strongly absorb light at wavelengths < 260 nm (e.g., 230 nm). It is therefore essential to use both A_{260}/A_{280} and A_{260}/A_{230} ratios to evaluate the quality of a DNA preparation. Multiply the A_{260} reading by 50 μg/mL (1 mL solution with an A_{260} of 1 contains approx 50 μg of DNA) to determine the final concentration of the DNA preparation.
19. For restriction digestion, use about 20 μL (~0.5–1 μg DNA) of the DNA preparation in a 100 μL digestion reaction containing 1 mM spermidine and 10 μg/mL RNase. Whereas spermidine is added to overcome the inhibitory effect of contaminating polysaccharides in the reaction, the addition of RNase is to digest away any residual RNAs which may also interfere with many enzyme reactions.
20. For PCR analysis, 1 μL of the final DNA solution is sufficient to amplify single copy genomic sequences. Occasionally, the final DNA solution may still contain some contaminating compounds that are inhibitory to PCR. Simple dilution of the final DNA solution can significantly reduce such inhibitions. Alternatively, one can add 1–5% BLOTTO (10% skim milk powder and 0.2% NaN_3) to the PCR reactions to eliminate the effect of inhibitory compounds that coextract with DNA *(11)*.
21. This modified micropreparation protocol is designed only for PCR analysis. Recently, several other simple DNA isolation methods have also been described. They include the NaOH extraction method *(12)*, the ROSE method *(13)*, and a heat protocol *(14)*.

References

1. Williams, C. E. and Ronald, P. C. (1994) PCR template-DNA isolated quickly from monocot and dicot leaves without tissue homogenization. *Nucleic Acids Res.* **22**, 1917,1918.

2. Dellaporta, S. L., Wood, J., and Hicks, J. B. (1983) A plant DNA minipreparation: version II. *Plant Mol. Biol. Rep.* **1**, 19–21.
3. Murray, M. G. and Thompson, W. F. (1980) Rapid isolation of high molecular weight plant DNA. *Nucleic Acids Res.* **8**, 4321–4325.
4. Paterson, A. H., Brubaker, C. L., and Wendel, J. F. (1993) A rapid method for extraction of cotton (*Gossypium* spp.) genomic DNA suitable for RFLP or PCR analysis. *Plant Mol. Biol. Rep.* **11**, 122–127.
5. Fang, G., Hammar, S., and Grumet, R. (1992) A quick and inexpensive method for removing polysaccharides from plant genomic DNA. *BioTechniques* **13**, 52–55.
6. Maréchal-Drouard, L. and Guillemaut P. (1995) A powerful but simple technique to prepare polysaccharide-free DNA quickly and without phenol extraction. *Plant Mol. Biol. Rep.* **13**, 26–30.
7. Li, Q.-B., Cai, Q., and Guy, C. L. (1994) A DNA extraction method for RAPD analysis from plants rich in soluble polysaccharides. *Plant Mol. Biol. Rep.* **12**, 215–220.
8. Rether B., Delmas, G., and Laouedj, A. (1993) Isolation of polysaccharide-free DNA from plants. *Plant Mol. Biol. Rep.* **11**, 333–337.
9. de Kocho, A. and Hamon, S. (1990) A rapid and efficient method for the isolation of restrictable total DNA from plants of the genus *Abelomoschus*. *Plant Mol. Biol. Rep.* **8**, 3–7
10. Marmur, J. (1961) A procedure for the isolation of deoxyribonucleic acid from microorganisms. *J. Mol. Biol.* **3**, 208–218.
11. De Boer, S. H., Ward, L. J., and Chittaranjan, S. (1995) Attenuation of PCR inhibition in the presence of plant compounds by addition of BLOTTO. *Nucleic Acids Res.* **23**, 2567,2568.
12. Wang, H., Qi, M., and Cutler, A J. (1993) A simple method of preparing plant samples for PCR. *Nucleic Acid Res.* **21**, 4153,4154.
13. Steiner, J. J., Poklemba, C. J., Fjellstrom, R. G., and Elliott, L. F. (1995) A rapid one-tube genomic DNA extraction process for PCR and RAPD analyses. *Nucleic Acid Res.* **23**, 2569,2570.
14. Thomson, D. and Henry, R. (1995) Single-step protocol for preparation of plant tissue for analysis by PCR. *BioTechniques* **19**, 394–400.

10

High Molecular Weight DNA Extraction from *Arabidopsis*

David Bouchez and Christine Camilleri

1. Introduction

Large DNA molecules (>100 kb) are extremely sensitive to mechanical shearing in aqueous solution. Therefore, classical DNA extraction procedures from living tissues generally do not allow recovery of DNA fragments larger than 50–100 kb in size. In recent years, technical improvements have made possible to purify and analyze large DNA molecules in vitro. Techniques such as pulsed-field agarose gel electrophoresis (PFGE) *(1)* or cloning in yeast artificial chromosomes (YACs) *(2)* have been extremely useful in map-based gene cloning strategies. A large number of genes have been isolated by positional cloning in several model species, including plants like tomato and *Arabidopsis (3)*. These megabase DNA techniques also allow initiation of large-scale physical mapping of complex genomes like *Arabidopsis (4,5)*. Physical mapping of eukaryotic genomes generally involves genome reconstruction using ordered clones from large-insert genomic libraries in YACs. Several *Arabidopsis* YAC libraries are available *(6–9)*.

However, the availability of high amounts of high molecular weight DNA (HMW DNA) is an absolute prerequisite for the analysis of large DNA fragments, and the construction of large insert clone libraries. Methods have been devised to avoid mechanical shearing of DNA, by inclusion of living cells in a solid agarose matrix (agarose plugs or microbeads), followed by *in situ* lysis and proteinase K treatment *(1,10)*. The immobilized DNA can subsequently be subjected to enzymatic treatment (for example by restriction endonucleases), since most enzymes can diffuse through the matrix.

In plants, the presence of the cell wall reduces the yield and the quality of the DNA extracted in agarose plugs. Therefore, many methods for preparation of megabase DNA from plants require the preparation of protoplasts *(11–15)*.

The first *Arabidopsis* YAC libraries, which were constructed from DNA in solution, have a relatively small insert size—160 kb, *(7)* to 250 kb *(8)*—and contain many chimeric clones that can seriously hamper physical mapping experiments *(16)*. At least part of these problems can be attributed to the DNA isolation methods used for the library construction.

In this chapter, we present a method for preparing agarose plugs of megabase DNA from *Arabidopsis* protoplasts. This method has been successful for the construction of a large-insert YAC library of the Columbia ecotype with a mean insert size of nearly 500 kb *(9)*. In our hands, the size of the DNA obtained is very high, since it appears to be well beyond the resolution limit of the CHEF PFGE system (>5 Mb). However, given the small DNA content of *Arabidopsis* and the average volume of protoplasts, there is clearly a physical limit to the DNA concentration one can achieve with DNA plugs made from protoplasts. Using this method, it is difficult to achieve DNA concentrations higher than 20–30 ng/µL.

For some applications in which a higher concentration is needed, we describe an alternative approach based on the protocol of Liu and Whittier *(17)*. This procedure involves isolation of nuclei from whole plants, followed by embedding in agarose and detergent/proteinase K treatment. It has been used to construct a high-quality P1 library of *Arabidopsis (18)*. This method is simpler and gives higher yields of HMW DNA (100–200 ng/µL), but it appears that the DNA is significantly smaller and more difficult to digest than the one obtained from the protoplast method. For most applications, the nuclei method will give satisfactory results. If one wants to analyze or clone very large DNA fragments, we recommend starting from protoplasts.

2. Materials
2.1. Plant Culture
2.1.1. Materials

1. Growth chamber.
2. Cold-room or refrigerator.
3. Laminar flow hood.
4. Gas-permeable chirurgical tape (e.g., Urgopore™, Laboratories Urgo, Chenove, France).
5. Eppendorf tubes.
6. Petri dishes (14 cm diameter).

2.1.2. Media and Solutions

1. Seed sterilization solution: Dissolve one tablet of Bayrochlor (Bayrol GMBH, D-800 München, Germany) in 40 mL H_2O, add a few drops of Teepol. Mix 5 mL of this solution with 45 mL 95% ethanol.

2. Microelements 1000X:

	Amount/L (1000X)	Final concentration (1X)
H_3BO_3	4328 mg	70 µM
$MnCl_2, 4H_2O$	2770 mg	14 µM
$CuSO_4, 5H_2O$	125 mg	0.5 µM
$Na_2MoO_4, 2H_2O$	50 mg	0.2 µM
NaCl	584 mg	10 µM
$ZnSO_4, 7H_2O$	288 mg	1 µM
$CoCl_2, 6H_2O$	2.5 mg	0.01 µM

Autoclave at 120°C for 20 min.

3. Vitamins 500X:

	Amount/L(500X)
Myo-inositol	50 g/L
Ca Panthotenate	0.5 g/L
Niacin	0.5 g/L
Pyridoxine	0.5 g/L
Thiamine HCl	0.5 g/L
Biotin	5 mg/L

Keep at -20°C.

4. *Arabidopsis* culture medium (AtM, *19*):

	Stock	Amount/L	Final concentration
KNO_3	1M	5 mL	5 mM
KH_2PO_4	1M	2.5 mL	2.5 mM
$MgSO_4$	1M	2 mL	2 mM
$Ca(NO_3)_2$	1M	2 mL	2 mM
Microelements	1000X	1 mL	1X
Vitamins	500X	2 mL	1X
Bromcresol purple	0.16%	5 mL	0.0008%
MES pH 6.0	14%	5 mL	0.07%
Sucrose		10 g	1%
Agar		7 g	0.7%

Autoclave at 120°C for 20 min, then add:

| Ferric ammonium citrate | 1% | 5 mL | 0.005% |

(autoclaved separately).

2.2. Protoplast Isolation

2.2.1. Materials

1. Agitator.
2. Water bath.

3. Lab-top centrifuge, swinging bucket rotor with adaptors for 30-mL Corex tubes.
4. Hemocytometer and microscope.
5. 30-mL Corex tubes.
6. 20-mL sterile glass pipets.
7. Scalpel.
8. Petri dishes (9 cm diameter).

2.2.2. Solutions

1. Gamborg B5 salts 10X *(20)* : 25 g/L KNO_3, 1.34 g/L $(NH_4)_2SO_4$, 1.5 g/L NaH_2PO_4, $2H_2O$, 1.5 g/L $CaCl_2$, $2H_2O$, 2.5 g/L $MgSO_4$, $7H_2O$.
2. Protoplast digestion medium (GSG):

	Stock	Amount/L	Final concentration
Gamborg B5 salts	10X	100 mL	1X
Sorbitol		70 g	7%
Glucose		20 g	2%
Glycine		15 g	1.5%
MES pH 6.0		5 mL	0.07%
Bromcresol blue	0.16%	5 mL	0.0008%
Tween-80	1%	1 mL	0.001%
Vitamins	500X	2 mL	1X

   Autoclave at 120°C for 20 min.
3. Protoplasting enzymes:
   Solution A: 0.2% Macerozyme R-10 (Yakult Pharmaceutical, Tokyo, Japan), 1% cellulase Onozuka RS (Yakult). Solution B: 10% Driselase (Fluka, Buchs, Switzerland). Dissolve in H_2O, then centrifuge 10 min (300g), and filter supernatant under vacuum with three superposed filters (1.2-μm, 0.8-μm, and 0.45-μm); filter-sterilize with a 0.2-μm disposable filter.
4. Protoplast washing buffer (PWB):

	Stock	Amount/L	Final concentration
$CaCl_2$, $2H_2O$		44.1 g	$0.3M$
MES pH 6	14%	5 mL	0.07%
Bromcresol purple	0.16%	5 mL	0.0008%

2.3. Nuclei Preparation

2.3.1. Materials

1. Chemical hood.
2. Centrifuge with swinging bucket rotor and adaptor for 50-mL Falcon tubes.
3. Liquid nitrogen.
4. Mortar and pestle.
5. 50-mL Falcon tubes or equivalent.
6. 500-mL Polypropylene disposable beakers.
7. Nylon sieves: 250-μm, 90-μm, 50-μm, and 20-μm mesh size.

2.3.2. Media and Solutions

1. Nuclei isolation buffer (NIB):

	Stock	Amount/L	Final concentration
Tris-HCl pH 9.5	1 M	10 mL	10 mM
EDTA pH 8.0	0.5 M	20 mL	10 mM
KCl	1 M	100 mL	100 mM
Sucrose		171.15 g	0.5 M
Spermidine	1 M	4 mL	4 mM
Spermine	1 M	1 mL	1 mM
ß-mercaptoethanol		1 mL	0.1%

2. NIB-T: NIB plus 10% Triton X-100.

2.4. DNA Plug Preparation

2.4.1. Materials

1. Water bath.
2. 100 µL Plastic molds for agarose plugs (Pharmacia, Uppsala, Sweden).

2.4.2. Solutions

1. Incert agarose (FMC) 1.2% in PWB (protoplasts) or 10 mM Tris-HCl, pH 9.5, 10 mM EDTA (nuclei).
2. NDS: 10 mM Tris-HCl, pH 8.0, 0.5M EDTA, pH 8.0, 1% laurylsarcosine.
3. Proteinase K (Boehringer, Mannheim, Germany): dissolve directly in NDS.
4. 50 mM EDTA, pH 8.0.

2.5. Restriction and PFGE Analysis

2.5.1. Materials

1. Water bath
2. PFGE apparatus (e.g., CHEF system, BioRad, Richmond, CA).
3. UV transilluminator.

2.5.2. Solutions

1. TE: 10 mM Tris-HCl, pH 8.0, 1 mM EDTA, pH 8.0.
2. Pefablock (Pentapharm AG, CH-4002 Basel, Switzerland): make a 100X stock solution (100 mM in TE) and store at -20°C.
3. Restriction enzyme buffers (BRL, Gaithersburg, MD), spermidine (25 mM stock solution), ß-mercaptoethanol, BSA (molecular biology grade).
4. Agarose (e.g. BRL Ultrapure).
5. TBE 1X: 89 mM tris-borate, 89 mM boric acid, 2 mM EDTA.
6. Ethidium bromide stock solution 10 mg/mL in H_2O.

3. Methods
3.1. Plant Culture
3.1.1. Seed Sterilization

1. Prepare 20 mg of *Arabidopsis* seeds (approx 1000 seeds) in an Eppendorf tube and add 1 mL of sterilizing solution. One tube is enough for four large (14-cm) Petri dishes. For large-scale preparation of protoplasts, about 40 plates are necessary.
2. Incubate at room temperature for 10 min with constant agitation (*see* **Note 1**).
3. Let the seeds sediment and decant supernatant; rinse the seeds twice with 1 mL 95% ethanol.
4. Leave the tubes open under a sterile laminar flow until completely dry (usually overnight).

3.1.2. Plant Culture

1. Sprinkle approx 200–250 surface-sterilized seeds on a 14-cm agar plate containing *Arabidopsis* culture medium (AtM). Seal the plates with a gas-permeable chirurgical tape.
2. Synchronize germination by a cold treatment at 4°C for 48 h.
3. Place in the growth chamber under the following conditions: photoperiod 16 h day (100–150 $\mu E/m^2/s$)/8 h night; temperature 20°C day/15°C night; humidity 70%. Inspect the plates daily to remove contaminations.
4. After 3 wk culture, plantlets (four-leaf rosettes) are ready for protoplast or nuclei isolation (*see* **Note 2**).

3.2. Protoplast Isolation

1. Cut plantlets from one plate, leaving the root system in the agar and transfer to two, 9-cm Petri dishes containing 9 mL digestion medium (GSG).
2. Using a scalpel, chop the plantlets into small pieces (about 1 mm across; *see* **Note 3**).
3. Add protoplasting enzymes (1 mL solution A + 50 µL solution B per plate) and incubate overnight at 20°C with very gentle agitation.
4. The next day, filter the suspension through two superposed stainless steel sieves (140-µm and 80-µm mesh diameter) into 30-mL Corex tubes (three plates into one tube) (*see* **Note 4**).
5. Pellet the protoplasts by centrifugation (300g, 10 min) and gently resuspend in 1/2 vol of PWB with a glass pipet. Pool two tubes into one.
6. Pellet the protoplasts by centrifugation (300g, 5 min), resuspend in a small volume and pool all the tubes into one. Adjust the volume to 30 mL with PWB and count the protoplasts with a hemocytometer. Expect 200–300 × 10^6 protoplasts when starting from 40 large plates. The protoplasts must be round, of variable size, and few cellular debris must be visible.
7. Pellet the protoplasts by centrifugation (300g, 5 min), and remove the supernatant as completely as possible without disturbing the pellet. Estimate the volume of the final pellet. At this stage, protoplasts are ready for DNA plug preparation.

3.3. Preparation of Arabidopsis *Nuclei*

1. After 3 wk culture, it is preferable to keep the plants for 3–5 d in the dark to decrease the level of starch contamination in the final DNA plugs. Wrap the plates in aluminium foil, and replace in the growth chamber.
2. Remove the *Arabidopsis* plantlets from the agar with their root system, in order to obtain about 20 g fresh weight (6–8 plates). Take care to remove any agar from the plantlets.
3. Grind in liquid nitrogen to a fine powder, and transfer to a 500 mL polypropylene disposable beaker kept on ice. From this stage on, all the manipulations are performed on ice and under a chemical hood (however, it is not necessary to work in the cold room).
4. Wait until all the liquid nitrogen has evaporated, then add 200 mL NIB and resuspend by gentle mixing.
5. On ice, filter through 250-µm nylon mesh into a new disposable beaker. Repeat this step successively with 90-, 50-, and 20-µm nylon mesh (*see* **Note 5**).
6. Measure the volume of the suspension, add 1/20th vol of NIB-T, and mix gently in order to lyse the chloroplasts. Aliquot into 50-mL Falcon polypropylene tubes, and pellet the nuclei ($2000g$, 4°C, 10 min) in a swinging bucket rotor (*see* **Note 6**).
7. Resuspend the pellet into NIB to give a final volume of 1–1.2 mL.

3.4. DNA Plug Preparation

3.4.1. Plug Preparation

1. Prepare a solution of low melting temperature agarose Incert (FMC), 1.2% in PWB (protoplasts) or 10 mM Tris-HCl, pH 9.5, 10 mM EDTA (nuclei), melt, and keep at 40°C in a water bath.
2. Equilibrate the tube containing protoplasts or nuclei at 40°C in the water bath.
3. Mix an equal volume of Incert agarose with the cell/nuclei suspension and gently resuspend. Care must be taken to insure that protoplasts/nuclei are well resuspended. Use a 1-mL Pipetman with a cut tip to mix the suspension.
4. Once the cells/nuclei are resuspended, replace the tube at 40°C, and aliquot into 100-µL plastic molds using a 1-mL Pipetman with a cut tip. Place at 4°C for 10 min to solidify.

3.4.2 DNA Extraction

1. Transfer the plugs from the molds to 10 vol of NDS solution containing 2 mg/mL proteinase K, and place in a water bath at 50°C for 6 h (*see* **Note 7**).
2. Incubate the plugs twice in 10 vol of NDS containing 1 mg/mL proteinase K at 50°C for 20 h (*see* **Note 8**).
3. Rinse the plugs in several changes of 50 mM EDTA . They can be stored in 50 mM EDTA at 4°C for several months.

3.5. Restriction and PFGE Analysis

1. Inactivate proteinase K by treatment with Pefablock-SC: twice in 10 vol of 1mM Pefablock in TE, 1 h at 50°C. Rinse plugs in several changes of 10 vol of TE at room temperature (*see* **Note 9**).

2. Equilibrate 1/4 of a plug (25 µL) in 250 µL restriction enzyme buffer (BRL) supplemented with spermidine (5 mM), ß-mercapthoethanol (7 mM), and BSA (500 µg/mL) for 1 h before digestion.
3. Remove buffer and add 100 µL restriction buffer with 5–20 U of the appropriate restriction enzyme, incubate 2–16 h at 37°C (*see* **Note 10**). As a control, always include native DNA incubated in restriction enzyme buffer.
4. Native and restricted DNA are analyzed by PFGE, along with size standards (lambda concatemers and yeast chromosomes). Several systems are commercially available. We use a BioRad CHEF-DRII apparatus in the following conditions: 1% agarose (BRL), 0.5X TBE, 14°C, 190 V, pulse time 60 – 90 s, 18 h run time (*see* **Note 11**).
5. Stain the gel in ethidium bromide (0.2 µg/mL in H_2O) for 20 min and photograph on a UV transilluminator (*see* **Note 12**).
6. For Southern hybridization studies, the PFGE gel should be processed using standard protocols, except that the depurination step in 0.25N HCl should be extended to 20 min.

4. Notes

1. Sterilization time should be adjusted depending on the freshness and viability of the seed batch used.
2. At this stage, the plants should be very healthy and not induced to flowering, since in this case it will be difficult to isolate viable protoplasts and get good DNA.
3. Tissues can be chopped as finely as possible to improve the final yield of protoplasts.
4. Protoplasts are very sensitive to mechanical and osmotical shocks. Therefore, they must be manipulated with extreme care, using only sterile glass vessel and pipets.
5. Cut pieces of nylon sieve about 25 × 25 cm^2. Fold into four and pour the suspension directly into the filter. Press gently to improve recovery (wear gloves!).
6. At this stage, the pellet should be white-yellow, and devoid of chloroplast contamination.
7. The plugs can simply be blown out of the mold using a small bulb.
8. Dark-green protoplast plugs must clear gradually to become almost translucid and white or slightly yellow. Brown plugs must be discarded. Nuclei plugs must also become translucid.
9. This inactivation step should not be omitted given the high concentration of proteinase K used.
10. Nuclei plugs necessitate more enzyme to give a complete digestion.
11. Instead of inserting the pieces of plugs in the wells, it is more convenient to place them directly on the comb before pouring the gel. Then pour the gel, taking care not to disturb the plug pieces. When the gel has solidified, remove the comb and fill the wells with the agarose solution.
12. For native DNA, a large part of the fluorescence should stay in the original plug. This presumably corresponds to very large DNA molecules (>5 Mb) and/or DNA molecules complexed with polysaccharides. However, upon digestion, the fluorescence generally almost disappears in the plug. The HMW DNA migrates as a

unresolved "compression zone." One can easily distinguish the mono- and multimeric forms of the chloroplast genome (monomere size 185 kb) in protoplast plugs.

References

1. Schwartz, D. C. and Cantor, C. R. (1984) Separation of yeast chromosome-sized DNAs by pulsed field gradient gel electrophoresis. *Cell* **37**, 67–75.
2. Burke, D. T., Carle, G. F. and Olson, M. V. (1987) Cloning of large segments of exogenous DNA into yeast by means of artificial chromosome vectors. *Science* **236**, 806–812.
3. Giraudat, J., Hauge, B. M., Valon, C., Smalle, J., Parcy, F., and Goodman, H. M. (1992) Isolation of the *Arabidopsis* ABI3 gene by positional cloning. *Plant Cell* **4**, 1251–1261.
4. Hwang, I., Kohchi, T., Hauge, B. M., Goodman, H. M., Schmidt, R., Cnops, G., Dean, C., Gibson, S., Iba, K., and Lemieux, B. (1991) Identification and map position of YAC clones comprising one-third of the *Arabidopsis* genome. *Plant J.* **1**, 367–374.
5. Schmidt, R., West, J., Love, K., Lenehan, Z., Lister, C., Thompson, H., Bouchez, D., and Dean, C. (1995) Physical map and organization of *Arabidopsis thaliana* chromosome 4. *Science* **270**, 480–483.
6. Grill, E. and Somerville, C. (1991) Construction and characterization of a yeast artificial chromosome library of *Arabidopsis* which is suitable for chromosome walking. *Mol. Gen. Genet.* **226**, 484–490.
7. Ward, E. R. and Jen, G. C. (1990) Isolation of single-copy-sequence clones from a yeast artificial chromosome library of randomly-sheared *Arabidopsis thaliana* DNA. *Plant Mol. Biol.* **14**, 561–568.
8. Ecker, J. R. (1990) PFGE and YAC Analysis of the *Arabidopsis* genome. *Methods* **1**, 186–194.
9. Creusot, F., Fouilloux, E., Dron, M., Lafleuriel, J., Picard, G., Billault, A., Paslier, D. L., Cohen, D., Chabouté, M.-E., Durr, A., Fleck, J., Gigot, C., Camilleri, C., Bellini, C., Caboche, M., and Bouchez, D. (1995) The CIC library : a large insert YAC library for genome mapping in *Arabidopsis thaliana*. *Plant J.* **8**, 763–770.
10. Cook, P. R. (1984) A general method for preparing intact nuclear DNA. *EMBO J.* **3**, 1837–1842.
11. Ganal, M. W. and Tanksley, S. D. (1989) Analysis of tomato DNA by pulsed field gel electrophoresis. *Plant Mol. Biol. Rep.* **7**, 17–27.
12. Daelen, R. A. J. v., Jonkers, J. J., and Zabel, P. (1989) Preparation of megabase-sized tomato DNA and separation of large restriction fragments by field-inversion gel electrophoresis (FIGE). *Plant Mol. Biol.* **12**, 341–352.
13. Cheung, W. Y. and Gale, M. D. (1990) The isolation of high molecular weight DNA from wheat, barley and rye for analysis by pulse-field gel electrophoresis. *Plant Mol. Biol.* **14**, 881–888.
14. Honeycutt, R. J., Sobral, B. W. S., McClelland, M., and Atherly, A. G. (1992) Analysis of large DNA from soybean (*Glycine max* L. Merr.) by pulsed-field gel electrophoresis. *Plant J.* **2**, 133–135.

15. Wing, R. A., Rastogi, V. K., Zhang, H.-B., Paterson, A. H., and Tanksley, S. D. (1993) An improved method of plant megabase DNA isolation in agarose microbeads suitable for physical mapping and YAC cloning. *Plant J.* **4**, 893–898.
16. Schmidt, R., Putterill, J., West, J., Cnops, G., Robson, F., Coupland, G., and Dean, C. (1994) Analysis of clones carrying repeated DNA sequences in two YAC libraries of *Arabidopsis thaliana* DNA. *Plant J.* **5**, 735–744.
17. Liu, Y.-G. and Whittier, R. F. (1994) Rapid preparation of megabase plant DNA from nuclei in agarose plugs and microbeads. *Nucl. Acids Res.* **22**, 2168,2169.
18. Liu, Y. G., Mitsukawa, N., Vazquez-Tello, A., and Whittier, R. F. (1995) Generation of a high-quality P1 library of *Arabidopsis* suitable for chromosome walking. *Plant J.* **7**, 351–358.
19. Estelle, M. A. and Somerville, C. R. (1987) Auxin-resistant mutants of *Arabidopsis thaliana* with an altered morphology. *Mol. Gen. Genet.* **206**, 200–206.
20. Gamborg, O. L., Murashige, T., Thorpe, T. A., and Vasil, I. K. (1976) Plant tissue culture media. *In Vitro* **12**, 473–478.

11

Chloroplast DNA Isolation

George S. Mourad

1. Introduction

The size of the chloroplast DNA molecule of *Arabidopsis thaliana* has been determined to be about 153 kb *(1)*. In this chapter, the goal is to provide a step–by–step laboratory procedure for isolating chloroplast DNA from *Arabidopsis thaliana* with minimal nuclear or mitochondrial DNA contamination. In general, a first step in isolating chloroplast DNA is the homogenization of the plant material followed by a filtration step to remove large-sized cell debris and cell fragments. The filtrate is then centrifuged at low speed to precipitate nuclei and chloroplasts. The intact mitochondria, being smaller in size than chloroplasts, remain in the supernatant. To get rid of nuclear DNA, one of two methods is usually used. In the first, the pellet containing nuclei and chloroplasts is treated with DNase. The latter has access to the nuclear DNA, via nuclear pores of the nuclear membrane, leading to its digestion, but has no access to chloroplast DNA because intact chloroplasts have a nonporous envelope. In the second method, the intact chloroplasts are banded in a sucrose– or a percoll–density–gradient, while the nuclei pellet. Banded chloroplasts are carefully removed. Intact chloroplasts, obtained by either a DNase method or a gradient method, are lysed and their proteins digested with a protease. Digested proteins are removed by organic (phenol/chloroform/isoamyl alcohol) extractions, and the nucleic acids (DNA and RNA) of the chloroplasts precipitated by ethanol. The chloroplast RNA can be removed by digestion with a DNase–free RNase. An alternative method for getting rid of the proteins and RNA is to band the chloroplast DNA in a CsCl density gradient by ultracentrifugation. After removal from the CsCl density gradient, the chloroplast DNA can be purified by dialysis against TE buffer to get rid of CsCl or can be ethanol-precipitated then washed with 70% ethanol to get rid of the salt.

The use of DNase in extracting pure chloroplast DNA from intact chloroplasts has been described by several laboratories *(2–5)*. In these published procedures, a relatively large amount of green material is needed and the yield of chloroplast DNA is quite small. The reason for this is that a good fraction of the isolated chloroplasts have cracked envelopes, and thus, DNase finds its way inside the leaky chloroplasts and digests a large amount of their DNA. On the other hand, protocols that employ density gradients to isolate the chloroplasts as a first step in isolating chloroplast DNA *(6)* yielded higher amounts of chloroplast DNA when compared to the DNase methods.

In this chapter, I describe a method for isolating chloroplast DNA from *Arabidopsis thaliana*. This simple method is a modification of published procedures *(6–8)*. In this method, the plants are kept in the dark for 2–3 d before they are harvested for extraction. The plants are then homogenized in extraction buffer. The homogenate is then filtered to get rid of large cellular debris, centrifuged at low speed, and the pellet is suspended in a suspension buffer. The suspended pellet, containing the nuclei and chloroplasts, is loaded onto a sucrose gradient. After centrifugation, the banded chloroplasts are carefully removed, washed and lysed. The lysate is loaded onto a CsCl density gradient and ultracentrifuged at high speed. The chloroplast DNA is pulled out of the gradient, and ethanol precipitated.

2. Materials
2.1. Equipment and Supplies

1. Waring blender.
2. Cheesecloth and Miracloth.
3. Small, soft, camel–hair paintbrush.
4. 5–mL Syringe with an 18–gauge needle.
5. Long–wavelength UV hand–held lamp.
6. Gradient mixer.
7. Dialysis tubing and clamps.
8. High-speed, refrigerated centrifuge such as Sorvall (Newtown, CT), Beckman (Fullerton, CA), table–top refrigerated centrifuges (e.g., Eppendorf [Westbury, NY], Heraeu [South Plainfield, NJ]), or any equivalent centrifuge.
9. Fixed–angle rotor such as the Sorvall GSA, Sorvall SS34, or any equivalent rotor.
10. Beckman ultracentrifuge or any equivalent ultracentrifuge.
11. Swinging bucket rotor such as the Beckman rotors SW28, SW28.1, SW40Ti, or SW41Ti. Equivalent rotors can be used when using a different ultracentrifuge.
12. Small size swinging bucket rotor such as the Beckman rotors SW50.1, SW55Ti, SW65Ti, or SW60Ti. Equivalent rotors can be used when using a different ultracentrifuge.

Chloroplast DNA Isolation

2.2. Buffers and Reagents

1. Extraction buffer: $0.35 M$ sorbitol, 50 mM Tris–HCl, pH 8.0, 5 mM disodium EDTA, pH 8.0, 0.1% bovine serum albumin, 15 mM β–mercaptoethanol. The buffer is mixed fresh from sterile stock solutions and kept on ice for an hour before use. β–mercaptoethanol is added to the ice–cold buffer just before grinding the plants.
2. Suspension buffer: $0.35 M$ sorbitol, 50 mM Tris–HCl, pH 8.0, 5 mM disodium EDTA, pH 8.0. The buffer is mixed fresh then kept on ice for an hour before use.
3. Sucrose gradients: Step gradients and continuous gradients are prepared by dissolving the correct amount of sucrose in 50 mM Tris–HCl, pH 8.0 and 5 mM disodium EDTA, pH 8.0. Sucrose gradients were prepared and kept on ice for an hour before loading them with plant homogenate.
4. Lysis buffer: 5% (w/v) sodium sarcosinate, 50 mM Tris–HCl, pH 8.0, 25 mM disodium EDTA, pH 8.0.
5. High TE buffer (HTE): 50 mM Tris–HCl, pH 8.0, 25 mM disodium EDTA, pH 8.0.
6. Low TE buffer (LTE): 10 mM Tris–HCl, pH 8.0, 0.1 mM disodium EDTA, pH 8.0.

3. Methods

1. Grow *Arabidopsis thaliana* wild-type plants in the greenhouse at 24°C under short days (400–500 foot–candle illumination for 10 h light/d). Two days before harvesting, place 3–4-wk-old plants in the dark for 2–3 d (*see* **Note 1**). Harvest 100–200 g of the etiolated whole plants, wash with distilled water, and gently blot with paper towels.
2. Homogenize the plants in a precooled Waring blender in 1000–1500 mL of ice–cold extraction buffer for 5 s at full speed three times (*see* **Note 2**).
3. Filter the homogenate through four layers of cheesecloth then four layers of Miracloth (*see* **Subheading 2.1., item 2**).
4. Centrifuge the homogenate at 1000g in a Sorvall GSA rotor for 10 min at 4°C. Equivalent centrifuges and rotors can also be used (*see* **Subheading 2.1.**).
5. Using a soft, camel–hair paintbrush resuspend the pellet on ice in about 5 mL of ice–cold suspension buffer, then load using a plastic Pasteur pipet onto an ice–cold 55:40:20% sucrose step–gradient (*see* **Note 3**). Centrifuge, with brakes off, in an SW28 rotor of a Beckman utracentrifuge at 83,000g at 4°C for 1 h. Equivalent centrifuges and swinging bucket rotors can also be used (*see* **Subheading 2.1.**).
6. Using a plastic Pasteur pipet, remove the dark green chloroplast band at the interphase between 55 and 40% sucrose (**Fig. 1**).
7. Dilute the chloroplasts with 20–30 mL of suspension buffer and centrifuge in a Sorvall HB4 rotor (or an equivalent rotor, *see* **Subheading 2.1.**) at 2000g at 4°C for 10 min. Carefully resuspend the pellet with a camel–hair brush in 30 mL suspension buffer and repeat the centrifugation once more. Wash the pelleted chloroplasts one more time.
8. Using a camel–hair brush, very carefully suspend the pellet in about 2 mL of HTE buffer then add 0.5 mL (or 1/10 vol) of 10 mg/mL proteinase K (freshly prepared) and mix the tube very gently for 15 min at room temperature.

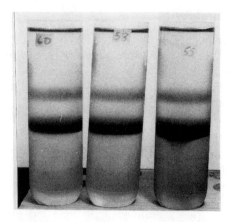

Fig. 1. *Arabidopsis* chloroplasts isolated by sucrose step gradients. Left is 60:40:20%; middle is 58:40:20%, and right is 55:40:20%. Note that in the gradient on the right the chloroplast band is not as tightly packed as in the other two gradients.

9. Gently add 1 mL (or 1/5 vol) of lysis buffer and mix the tube gently by inverting several times then incubate at room temperature for 30–60 min.
10. To a 5-mL ultraclear centrifuge tube of the SW50.1 rotor of a Beckman ultracentrifuge, add 3.75 g of solid CsCl then add the chloroplast lysate. Cover the tube with a piece of parafilm and dissolve the CsCl by gently inverting the tube several times. Add ethidium bromide to a final concentration of 200 µg/mL. If needed, fill up the centrifuge tubes using HTE buffer. Centrifuge at 189,000g for 16–24 h at 19°C. Equivalent rotors can also be used (*see* **Subheading 2.1.**).
11. After centrifugation, visualize the chloroplast DNA band using a long wavelength UV hand–held lamp and pull out the band using an 18–g syringe needle. Remove the ethidium bromide by 4–6 extractions with isopropanol or n–butanol saturated with NaCl and water. Dialyse the DNA at 4°C against 2 L of LTE buffer for a period of 24 h. Change the dialysis buffer once during this period. Instead of dialysis, the chloroplast DNA can be desalted by using purification columns such as Promega's Wizard™ DNA Clean–Up System (Madison, WI). Equivalent DNA desalting columns can also be used.
12. The chloroplast DNA is ready for endonuclease digestion after dialysis or column purification. Alternatively, after dialysis or column purification the chloroplast DNA can be concentrated by adding 1/10 vol of 3M sodium acetate, pH 5.0, and 2 vol of absolute ethanol or 0.6 vol of isopropanol, store overnight in –20°C. and finally precipitate the DNA by centrifugation at 11,000g in a Sorvall SS34 rotor at 4°C. Equivalent rotors can also be used (*see* **Subheading 2.1.**). Wash the DNA with 70% ethanol to remove salts, then allow to air dry and finally suspend in 200 µL LTE buffer and store in –20°C.

Figure 2 shows the banding pattern of *Arabidopsis thaliana* chloroplast DNA isolated by this method, digested with endonuclease restriction enzymes,

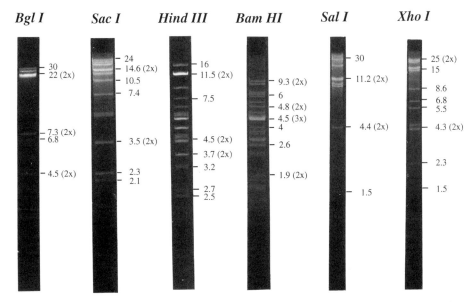

Fig. 2. Chloroplast DNA of *Arabidopsis thaliana* Columbia wild–type digested with six endonuclease restriction enzymes and electrophoresed in 0.8% TAE agarose gels. The molecular size in kilobases of some of the chloroplast DNA fragments for each of the endonuclease restriction digests are shown. Note that different endonuclease restriction digests were electrophoresed in different agarose gels.

and electrophoresed in agarose gels. For each of the endonuclease restriction digests, the molecular size of some of the chloroplast DNA fragments are shown. Within experimental error, the total size of the chloroplast DNA molecule of *Arabidopsis thaliana*, calculated by summing up the size of the individual fragments produced by 11 different endonuclease restriction enzymes (6-bp recognition–site restriction enzymes: *Bam*HI, *Bgl*I, *Eco*RI, *Hin*dIII, *Sac*I, *Sal*I, *Xba*I, and *Xho*I, and 4-bp recognition–site restriction enzymes: *Alu*I, *Cfo*I, and *Hae*III) averaged about 152 kb, with the largest estimate being about 154 kb for *Xho*I and the smallest being 151 kb for *Sal*I. These slight differences could be owing to errors in assigning stoichiometry (doublet fragments) to some DNA fragments or due to the fact that some endonuclease restriction enzymes such as *Eco*RI, *Hin*dIII, *Xba*I, *Bam*HI, *Cfo*I, *Hae*III, and *Alu*I produce a large number of fragments, some of which are of a very small size that either could not be resolved on the gel or run off the gel into the electrophoresis buffer. The size of the chloroplast DNA molecule of *Arabidopsis thaliana* calculated by this method and reported here (152 kb) is very close to the 153 kb published by Palmer et al. (1994), and calculated by hybridization of blots carrying digested chloroplast DNA of *Arabidopsis* to either tobacco or *Brassica* chloroplast DNA probes.

4. Notes

1. During the dark treatment, the chloroplasts consume their stored starch grains. This treatment allowed the purification of a larger fraction of intact chloroplasts with no cracks in their envelope which was reflected in a higher yield of chloroplast DNA.
2. The homogenate was allowed to settle in the blender between homogenizations. This allows the leaf fragments to settle and come closer to the blades of the blender for a second maceration. This improved the yield of chloroplast DNA.
3. Several sucrose step–gradients were tested:
    a. 60:30%
    b. 58:30%
    c. 55:30%
    d. 60:40:20%
    e. 58:40:20%
    f. 55:40:20%

    The dark green band at the interface of 60:30%, 58:30%, 55:30% as well as of the 60:40%, 58:40%, and 55:40% was carefully removed with a wide–mouthed plastic transfer pipet and examined with the light microscope. The best quality chloroplasts with the least nuclear fragments or broken organelles were obtained in the interface of the 55:30% two–step gradient and the interface of 55:40%; and of the three–step gradient 55:40:20% **(Fig. 1)**. It was observed that the higher the concentration of sucrose in the bottom step of the gradient, the more packed the green band becomes in the interface and the more nuclei and nuclear fragments are trapped with the chloroplasts. Agarose gel electrophoresis of digested chloroplast DNA isolated from such packed green bands had a high background of nuclear DNA smear. On the other hand, decreasing the sucrose concentration in the bottom step of the gradient to < 55% will greatly lower the yield of chloroplast DNA, since quite a bit of the chloroplasts will pellet with the nuclei in the bottom of the gradient. Good quality chloroplast DNA was also obtained by using a continuous sucrose gradient, in which the plastids do not go through sudden changes in the osmoticum, but rather travel smoothly in the continuous gradient and band at their proper density. The continuous gradient used was a 20–57% with 2 mL of 60% sucrose cushion. The plastids were carefully removed from only the upper half of the lower dark green band. The nuclei and nuclear fragments pelleted in the bottom of the 60% cushion.

Acknowledgments

I thank Deborah Ross for her critical reading of this manuscript.

References

1. Palmer, J. D., Downie, S. R., Nugent, J. M., Brandt, P., Unseld, M., Klein M., Brennicke, A., Schuster, W., and Börner, T. (1994) Chloroplast and mitochondrial

DNAs of *Arabidopsis thaliana*: conventional genomes in an unconventional plant, in *Arabidopsis* Meyerowitz, E. M. and Somerville, C. R., eds.) pp. 37–62.
2. Bogorad, L., Gubbins, E. J., Krebbers, E., Larrinau, I. M., Mulligan, B. J., Muskuvitch, K. M. T., Orr, E. A., Rodermel, S. R., Schantz, R., Steinmetz, A. A., DeVos, G., and Ye, V. K. (1983) Cloning and physical mapping of maize plastid genes. *Methods Enzymol.* **97**, 524–555.
3. Hermann, R. G., Bohnert, H. J., Kowallik, K. V., and Schmitt J. M. (1975) Size, confirmation and purity of chloroplast DNA from some higher plants. *Biochim. Biophys. Acta* **378**, 305–317.
4. Kolodner, R. and Tewari, K. K. (1975) The molecular size and confirmation of the chloroplast DNA from higher plants. *Biochim. Biophys. Acta* **402**, 372–390.
5. Mourad, G. and Polacco, M. (1989) Mini–preparation of highly purified chloroplast DNA from maize. *Plant Mol. Biol. Rep.* **7**, 78–84.
6. Palmer, J. D. (1982) Physical and gene mapping of chloroplast DNA from *Atriplex triangularis* and *Cucumis sativa*. *Nucleic Acids Res.* **10**, 1593–1605.
7. Fluhr, R. and Edelman, M. (1981) Physical mapping of *Nicotiana tabacum* chloroplast DNA. *Mol. Gen. Genet.* **181**, 484–490.
8. Tewari, K. K. and Wildman, S. G. (1966) Chloroplast DNA from tobacco leaves. *Science* **153**, 1269–1271.

12

Purification of Mitochondrial DNA from Green Tissues of *Arabidopsis*

Mathieu Klein, Rudolf Hiesel, Charles André, and Axel Brennicke

1. Introduction

Plant mitochondrial genomes vary in size and complexity. Both can be rapidly estimated by analysis of the restriction pattern of purified mitochondrial DNA (mtDNA). A prerequisite to purify mtDNA with little contamination of nuclear and also of plastid DNA is the isolation of mitochondria from cells. Mitochondria from green plants are generally difficult to purify because of the low amount of these organelles and the high number of chloroplasts in photosynthetic tissues. For a large number of plants, it is possible to isolate comparatively pure mitochondria from nonphotosynthetic tissues (1) or cell cultures (2). The absence of mature chloroplasts reduces the plastid contamination and increases the purification rate. In many investigations, however, it is necessary to use green tissues such as leaves as sources, when as for *Arabidopsis,* only few nonphotosynthetic tissues or certain cell cultures are available. Furthermore, tissue cultures are not always the ideal source for investigations of mitochondrial genome structure, since frequent reorganizations have been reported for such tissue cultures, which distort the *in planta* structure of these genomes (3,4).

Here we present an easy protocol to purify mitochondrial DNA of *Arabidopsis thaliana* suitable for restriction enzyme or pulsed-field gel electrophoresis (PFGE) analysis (5). This protocol has also successfully been applied to other plants including tobacco and potato and even notoriously difficult plants such as ferns (O. Malek, personal communication). In short, mitochondria are enriched by differential centrifugation steps and digested with DNase I to reduce contaminating nuclear DNA adhering to the outer mitochondrial membrane. In addition, we describe a convenient method to prepare mitochondrial DNA for PFGE analysis of the entire mitochondrial genome complexity.

2. Materials

2.1. Plant Material

Arabidopsis thaliana plantlets grown from seeds for 30–35 d under standard conditions. Plants should be harvested before chlorosis of the leaves.

2.2. Equipment

1. Centrifuge and corresponding rotors (Beckman, Fullerton, CA, JA-10, JA-20, or equivalents).
2. 400-mL centrifugation bottles.
3. Waring blender.
4. Miracloth and cheesecloth.
5. Dounce homogenizer.
6. Fine painting brush.
7. Sterile 15- and 50-mL polycarbonate tubes.
8. Tips, gloves, glass pipets, and standard laboratory material.
9. CHEF gel apparatus system (BioRad, Richmond, CA, DR2 or equivalent) for PFGE analysis.

2.3. Stock Solutions

1. Extraction buffer (EB): $0.44M$ sucrose, 50 mM Tris-HCl pH 7.5, 3 mM EDTA, pH 8.0. Just before use, add 100 mM β-mercaptoethanol (β-ME)., 0.2 % bovine serum albumin (BSA), and 0.1% polyvinylpyrrolidone (PVP). Given are the final concentrations.
2. Washing buffer (WB): $0.44M$ sucrose, 50 mM Tris-HCl pH 7.5, 20 mM EDTA, pH 8.0. Just before use, add 100 mM β-ME, 0.1% BSA, and 0.1% PVP (final concentrations).
3. CTAB lysis buffer: 2 % N-Cetyl-N,N,N-trimethyl-ammoniumbromide (CTAB), $1.4M$ NaCl, 100 mM Tris-HCl pH 8.0, 20 mM EDTA pH 8.0. Store at room temperature. Add β-ME to 100 mM just before use.
4. Agarose embedding solution (AE): 1% Incert Agarose (F.M.C., Rockland, MD) made up freshly in WB buffer without BSA and without PVP, but with 100 mM β-ME.
5. Plug lysis buffer (PL): 100 mM EDTA, pH 8.0, 1% N-laurylsarcosine, 0.5 mg/mL proteinase K, 100 mM β-ME.
6. TE buffer: 10 mM Tris-HCl pH 7.5, 0.1 mM EDTA pH 8.0.
7. $0.5M$ EDTA pH 8.0.

3. Methods

3.1. Isolation of Mitochondria

Perform all procedures at 4°C or on ice with sterile and cold solutions, precooled tips, tubes, and so on (*see also* **Note 1**).

Purification of Mitochondrial DNA

1. Collect green leaves and wash them with distilled water (*see* **Note 2**).
2. Cut the plants into small pieces with scissors and add EB using a ratio of 1 L buffer for 100 g fresh plant material.
3. Homogenize in a Waring blender once for 20 s at high speed, and after waiting for 2 min, twice for 15 s at low speed.
4. Filter through four layers of cheesecloth and one layer of miracloth (squeeze the cheesecloth, but not the Miracloth).
5. Centrifuge the filtrate for 15 min at 2200g to pellet cell debris.
6. Transfer the supernatant to new tubes and centrifuge for 15 min at 3600g.
7. Repeat **step 6** to remove remaining cell debris and chloroplasts.
8. Transfer the supernatant to new bottles and centrifuge for 25 min at 17,700g.
9. Pour off the supernatant and resuspend the pellet in 3 mL volume of EB solution for each 100 g starting material using a fine painting brush. If more than two centrifugation bottles have been used, collect resuspended mitochondria into two bottles.
10. Add $MgCl_2$ to 25 mM and DNase I to a final concentration of 200 µg/mL, and keep on ice for 30 min (*see* **Note 3**).
11. Add $MgCl_2$ to 50 mM and DNase I to 300 µg/mL as final concentration, and keep on ice for another 30 min.
12. Stop the reaction by adding EDTA to a final concentration of 70 mM.
13. Dilute the mitochondria in 500 mL WB and centrifuge for 25 min 17,700g.
14. Pour off supernatant and resuspend the pellet in 1 mL WB solution with a fine painting brush.
15. Repeat **step 13** in order to remove DNase I quantitatively (*see* **Note 4**).
16. Resuspend the pellet in 1–2 mL WB solution, transfer to a sterile 15-mL tube, and complete the volume to 10–12 mL with WB solution.
17. Centrifuge for 25 min at 14,600g.
18. At this step the mitochondrial pellet may be frozen and stored at –20°C for further processing or can alternatively be embedded in agarose for PFGE analysis (*see* **Note 5**).

3.2. Extraction of Mitochondrial DNA

1. Resuspend mitochondria in 5 mL preheated (60°C) CTAB lysis buffer and incubate for 30 min at 60°C.
2. Add 1 vol chloroform/isoamylalcohol (24:1) and mix gently for 5 min.
3. Centrifuge for 15 min at 12,000g.
4. Transfer the aqueous phase to new tubes and repeat **steps 2** and **3** once.
5. Precipitate nucleic acids by adding 2/3 vol of isopropanol, mix gently, and let the mixture stand at room temperature for 5 min.
6. Centrifuge for 15 min at 12,000g.
7. Discard supernatant, wash the pellet with 70% ethanol and dry the pellet under vacuum.
8. Redissolve the pellet in an appropriate volume of TE (50–100 µL), incubate on ice for a few hours, and store at –20°C.
9. Estimate the concentration by loading an aliquot of this undigested DNA on a 1% agarose gel; 100 g of fresh weight material should yield about 10–50 µg of mtDNA.

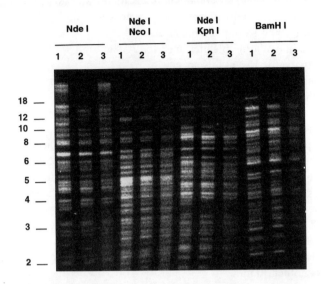

Fig. 1. Restriction digests of mitochondrial DNA from *Arabidopsis*. MtDNA was purified as outlined in the protocol and digested with the enzyme(s) indicated above. Lanes labeled 1 show mtDNA from *Arabidopsis thaliana* C24, lanes 2 contain mitochondrial DNA from *Arabidopsis thaliana* Columbia green plants, and lanes 3 show mtDNA from a nonphotosynthetic tissue culture of *Arabidopsis* Columbia. Migration of length standards is indicated in the left margin in kilobasepairs. Mitochondrial DNAs were separated on a 0.8 agarose gel.

10. About 3-4 µg DNA will give a nice restriction pattern on a 1% agarose gel (**Fig. 1**). Contaminating RNA may be eliminated by simply adding some RNase A to the restriction digest (*see* **Note 6**).

3.3. Preparation of Agarose-Embedded Mitochondria for PFGE Analysis

1. Preheat the AE solution at 50°C and cool the mold on ice.
2. Resuspend the mitochondrial pellet in a minimal volume of WB solution with a fine painting brush and homogenize completely in a Dounce homogenizer to prevent organellar aggregations.
3. Aliquot resuspended mitochondria on ice in 1.5-mL microfuge tubes (100, 250, and 500 µL, for example) and bring each volume up to 500 µL with WB solution.
4. Incubate one tube at 37°C for a few seconds and add 500 µL AE solution. Using a sterile glass Pasteur pipet, mix quickly but gently, and transfer to the mold, avoiding air bubbles.
5. Repeat **step 4** for the other aliquots, but do not treat more than one tube at a time.

6. Cut the plugs into 5-mm pieces and transfer them to 20 mL plug lysis buffer (PL). Use new Falcon tubes for each dilution.
7. Incubate at 50°C under gentle shaking for 6–12 h.
8. Change lysis buffer and incubate for another 16 h. Repeat this step twice.
9. Discard lysis buffer and wash the plugs for 30 min to 1 h at room temperature four to five times in PW buffer under gentle shaking.
10. Store the plugs at 4°C in 5–10 mL EDTA (500 mM pH 8.0), and change this buffer once several hours after the preparation. Under these conditions, embedded DNA should be stable for several months.
11. Run one plug of each dilution on a PFGE gel. A DNA smear should be visible in the 50–200 kb range and a large amount of the mtDNA does not migrate into the gel at all. This a typical PFGE gel analysis pattern of plant mitochondrial DNA *(6–8)*.
12. If restriction digests of the embedded mtDNA are desired, plugs should be treated prior to digestion with phenylmethyl sulfonyl (PMSF) to remove the protease (*see* **Note 7**).
    a. Rinse the plugs three times with sterile TE.
    b. Incubate twice in TE containing 0.04 mg/mL PMSF for 30 min each at 55°C. The plugs can be used immediately for restriction enzyme digestion or can be returned to storage in 0.5M EDTA pH 8.0. Conditions for restriction enzyme digestion of the agarose plugs have been described elsewhere (*see* **Note 8** and refs. *9,10*).

4. Notes

1. The isolation procedure is carried out at 4°C or on ice in order to minimize degradation of mitochondrial nucleic acids by nucleases. A large amount of β-mercaptoethanol is required in all buffers, because some secondary metabolites of the plant can modify the DNA and make it undigestable by restriction enzymes.
2. As starting material, at least 50–100 g green leaves (100–200 *Arabidopsis* plants) are required to obtain sufficient mtDNA (10–50 µg). However, increasing this amount up to 400–500 g can present some inconveniences in the first purification steps (harvesting time, volume of extraction buffer needed, centrifugation volume, and so on) up to the first mitochondrial pellet. We recommend the use of 200–250 g starting material, which presents the advantage to be treated easily and to yield up to 100 µg mtDNA.
3. The concentrations of $MgCl_2$ and DNase I indicated here are optimized for *Arabidopsis* tissues. However, because of the exceptionally small size of this nuclear genome, concentrations and incubation times should be increased several fold when using other plants. As an example, purification of mtDNA from pea requires a doubled amount of nuclease and a 50% longer incubation time to remove the nuclear DNA background quantitatively.
4. It is important to remove the nuclease completely for the subsequent steps. Increasing the number of washings can become necessary, particularly when the amount of nuclease is increased (refer to **Note 3**).
5. It is recommended to embed mitochondria directly after the mitochondrial purification step in the agarose plugs. In this case, mtDNA of at least one aliquot

of the purified organelles should be extracted and analyzed on a conventional agarose gel as a control (**Fig. 1**).
6. Purity of the extracted mtDNA can be checked easily by comparing the restriction enzyme patterns of purified mtDNA and total plant DNA. In total plant DNA, weak bands appear corresponding to the restriction pattern of the nuclear ribosomal gene cluster. These bands should be absent from the mtDNA restriction pattern. Also, if a nice mtDNA restriction pattern is observed with a background smear in the range of 50–200 bp, the DNase I treatment step performed during the extraction was not complete. In this case, the amount of nuclease should be increased in the mtDNA extractions.
7. PMSF stock is prepared by dissolving the compound at 40 mg/mL in isopropanol at 37°C and storing aliquots at -20°C. Caution should be taken when handling PMSF.
8. Experimental conditions for restriction enzyme digestion of agarose-embedded DNA have been investigated. Buffers and enzymes suitable for argose-embedded DNA restriction are frequently given in the manufacturer's instructions.

References

1. Binder, S. and Grohmann, L. (1995) Isolation of mitochondria, in *Methods in Molecular Biology, vol. 49. Plant Gene Transfer and Expression Protocols* (Jones, H. ed.), Humana, Totowa, NJ, pp. 377–381.
2. Schuster, W., Hiesel, R., Wissinger, B., Schobel, W., and Brennicke, A. (1988) Isolation and analysis of plant mitochondria and their genomes, in *Plant Molecular Biology. A Practical Approach.* (Straw, C.H., ed.), IRL, Oxford, pp. 79–102.
3. Gu, J., Dempsey, S., and Newton, K. J. (1994) Rescue of a maize mitochondrial cytochrome oxidase mutant by tissue culture. *Plant J.* **6**, 787–794.
4. Hartmann, C., Recipon, H., Jubier, M. F., Valon, C., Delcher-Besin, E., Henry, Y., De Buyser, J., Lejeune, B., and Rode, A. (1994) Mitochondrial DNA variability detected in a single wheat regenerant involves a rare recombination event across a short repeat. *Curr. Genet.* **25**, 456–464.
5. Klein, M., Eckert-Ossenkopp, U., Schmiedeberg, I., Brandt, P., Unseld, M., Brennicke, A., and Schuster, W. (1994) Physical mapping of the mitochondrial genome of *Arabidopsis thaliana* by cosmid and YAC clones. *Plant J.* **6**, 447–455.
6. Andre, C. P. and Walbot, V. (1995) Pulsed-field gel mapping of maize mitochondrial chromosomes. *Mol. Gen. Genet.* **247**, 255–263.
7. Narayanan, K. K., Andre, C. P., Yang, J., and Walbot, V. (1993) Organization of a 117-kb circular mitochondrial chromosome in IR36 rice. *Curr. Genet.* **23**, 248–254.
8. Bendich, A. J. (1993) Reaching for the ring: the study of mitochondrial genome structure. *Curr. Genet.* **24**, 279–290.
9. Smith, C. L., Lawrence, S. K., Gillespie, G. A., Cantor, C. R., Weissman, S. M., and Collins, F. S. (1987) Strategies for mapping and cloning macroregions of mammalian genomes, in *Molecular Genetics of Mammalian Cells.* (Gottesman., M. M., ed.), Academic, London, pp. 461–489.
10. Birren, B. and Lai, E. (1993) *Pulsed Field Gel Electrophoresis: A Practical Guide.* Academic, London.

13

Preparation of RNA

Clifford D. Carpenter and Anne E. Simon

1. Introduction

The preparation of high quality (i.e., intact) total RNA from biological samples is the primary step in the study of gene expression. It is the procedure that bridges the interface between the experimental manipulation of the living system, and the subsequent analysis of effects through molecular techniques. *Arabidopsis* is an ideal system for such analyses. Large quantities of plant material are easily obtained, from which milligram quantities of total RNA may be isolated. Conversely, small scale preparations from one or a few plants can be used to rapidly obtain multiple experimental samples. The procedure, a variation of a method originally proposed by Palmiter *(1)*, uses detergent sodium dodecyl sulfate (SDS) and phenol for lysis and denaturation of cellular constituents. The predominant alternative method of Chirgwin et al. *(2)* employs guanidinium isothiocyanate as a denaturant. This latter method has also spawned variations, most notably a single-step method that eliminates an ultracentrifugation step *(3)*. Although this method may compare favorably with the procedures presented herein with regard to technical simplicity, our protocol consistently results in high yields of intact RNA of a quality suitable for such manipulations as RNA blot hybridizations, RACE polymerase chain reaction (PCR), in vitro translation, RNA sequencing and cDNA cloning.

2. Materials

1. Plastic food storage bags.
2. Gloves.
3. Mortars and pestles or polypropylene beakers (*see* **Note 1**).
4. Scoopulas.
5. Liquid nitrogen.

6. 45-mL Screw-capped polypropylene centrifuge tubes (e.g., Oakridge tubes, Nalge Nunc, Rochester, NY).
7. RNA extraction buffer: $0.2M$ Tris-HCl, pH 9.0, $0.4M$ LiCl, 25 mM EDTA, 1% SDS (*see* **Note 2**).
8. H_2O-saturated phenol (*see* **Note 3**).
9. Chloroform.
10. $3M$ Sodium acetate, pH 5.3.
11. 95% Ethanol.
12. 75% Ethanol.
13. $2M$ LiCl.
14. Distilled, deionized, autoclaved H_2O (ddH_2O).
15. 1.5 mL Microcentrifuge tubes.
16. High-speed centrifuge (e.g., Sorval RC5B, Sorval, Newtown, CT) and rotor (e.g., Sorval SS34) for large-scale RNA preparation.
17. Analytical tabletop centrifuge (e.g., IEC model HN-SII with swinging bucket rotor) for large-scale RNA preparation.
18. Microcentrifuge ($\geq 10,000g$).
19. Vortex.
20. Low temperature freezer (–70°C).

3. Methods
3.1. Large-Scale Preparation

1. Harvest at least 2 g of plant materials and immediately place at –70°C in a plastic food storage bag.
2. Weigh 2–5 g of plants into a mortar that has been prechilled with a liquid nitrogen rinse (*see* **Note 4**).
3. Add a small amount of liquid nitrogen and grind with pestle to a fine powder, using additional nitrogen to keep the tissue frozen, if necessary.
4. Scoop powder into a 45–mL screw-capped tube (Oakridge). Do not fill to more than 1/3 the volume of the tube. Add 10 mL of RNA extraction buffer. Vortex vigorously until mixed.
5. Add 10 mL of H_2O-saturated phenol. Vortex vigorously until mixed.
6. Spin for 5 min at room temperature in a tabletop centrifuge at $750g$.
7. Remove the aqueous phase and re-extract the phenol phase with 5–7 mL of RNA extraction buffer. Vortex until mixed and centrifuge as in **step 6**.
8. Pool aqueous phases and re-extract with an equal volume of phenol by vortexing until well mixed, and centrifuging as in **step 6**.
9. Add an equal volume of chloroform to the aqueous phase. Vortex until mixed and centrifuge as in **step 6**.
10. To the aqueous phase, add 1/10 vol of $3M$ sodium acetate pH 5.3 and 2 vol 95% ethanol. Chill at –70°C for 30 min.
11. Centrifuge at $12,000g$ for 10 min at 4°C. Decant supernatant.
12. Add 10 mL of $2M$ LiCl to the pellet. Vortex well, chill on ice for 10–30 min and then centrifuge at $12,000g$ at 4°C for 10 min (*see* **Notes 5** and **6**).

13. Dissolve the pellet into 0.4 mL of ddH$_2$O and transfer to a 1.5-mL microcentrifuge tube.
14. Ethanol precipitate by adding 1/10 vol of 3M sodium acetate pH 5.3 and 2 vol 95% ethanol. Chill for 5 min at −70°C and centrifuge at ≥10,000g in a microcentrifuge at 4°C for 5 min.
15. Rinse the pellet with 0.5 mL of 75% ethanol. Remove all traces of ethanol (*see* **Note 7**).
16. Resuspend in 500 µL ddH$_2$O. Ascertain yield spectrophotometrically and check quality by electrophoresis (*see* **Note 8**).
17. Store at −70°C. From 2–5 g of plant material, the yield should be 0.5–2 mg of total RNA.

3.2. Small-Scale Preparation

1. Harvest whole plants or individual tissue types (rosette leaves, flowers, bolts, or siliques) in sufficient quantity to produce a 0.5-mL vol of frozen ground material. Freeze in plastic food storage bags at −70°C.
2. Remove one sample from the −70°C freezer and place the contents into a 50-mL plastic beaker. Immediately add a small amount of liquid nitrogen and grind to a fine powder with a metal scoopula, adding additional nitrogen if necessary to keep the tissue frozen (*see* **Note 4**).
3. Scoop the powder into a 1.5 mL microcentrifuge tube. Do not exceed a volume of 0.5 mL. Add 0.55 mL of RNA extraction buffer. Vortex vigorously until mixed by inverting the tube and spinning cap down.
4. Add 0.55 mL of H$_2$O-saturated phenol. Vortex until mixed.
5. Place tube on ice and proceed to the next sample, repeating **steps 2–4**. If more than four samples are being prepared concurrently, vortex all occasionally to maintain the phenol-aqueous emulsion.
6. Centrifuge all samples for 2 min at ≥10,000g in a room temperature microcentrifuge.
7. Remove the aqueous phase to a new tube. Repeat phenol extraction (**steps 4** and **6**) (*see* **Note 10**).
8. Remove aqueous phase to a new tube and extract once with 0.55 mL chloroform. Vortex until mixed and spin as in **step 6**.
9. Remove aqueous layer to a new tube, add 55 µL of 3M sodium acetate, pH 5.3, and fill the remaining volume in the tube with 95% ethanol. Chill 5 min at −70°C.
10. Centrifuge at ≥10,000g for 5 min at 4°C. Discard supernatant.
11. Add 0.3 mL of 2M LiCl to pellet. Partially resuspend by pipeting up and down and vortexing vigorously (*see* **Note 5**). Chill on ice for 10–30 min. Spin as in **step 10** (*see* **Note 11**).
12. Resuspend pellet in 0.3 mL of ddH$_2$O. Pipet up and down and vortex vigorously to aid in resuspension.
13. Add 30 µL of 3M sodium acetate, pH 5.3, 0.7 mL of 95% ethanol, and chill 5 min at −70°C. Centrifuge as in **step 10**.
14. Rinse pellet with 0.5 mL of 75% ethanol. Remove all traces of ethanol (*see* **Note 7**).
15. Resuspend in 50 µL of ddH$_2$O. Ascertain yield spectrophotometrically and check quality by electrophoresis (*see* **Note 8**).

16. Store at –70°C. Depending on the tissues used, this procedure should yield 50–200 µg of total RNA (*see* **Note 9**). At least 12 preparations may be performed concurrently.

4. Notes

1. The materials for large- and small-scale preparations are essentially the same, except that it is usually more convenient to grind tissue for multiple small preparations using scoopulas and 50-mL plastic beakers. Beakers are also appropriate for the large preps, although yields may be 10–20% better using a mortar and pestle.
2. A major factor influencing the success of any RNA isolation or manipulation is RNase contamination. Human skin is a primary source of external contamination. Therefore, it is important to start with clean materials and solutions and avoid contamination throughout all procedures. Since autoclaving is not adequate for RNase inactivation, steps involving treatment of solutions with diethylpyrocarbonate (DEPC) or other commercially available reagents to inactivate RNases, and baking glass and plasticware at high temperatures are generally appropriate. Although the authors do not dispute the efficacy of such precautions, our experience has shown that for extraction of plant RNA, solutions made with good quality ddH$_2$O do not require DEPC-treatment. Furthermore, since the initial step of the RNA isolation procedure involves grinding frozen plant material which itself contains RNases, the beakers, mortars, pestles and scoopulas used in this step need not be autoclaved. Following chloroform extraction, all materials should be autoclaved and baked prior to use. Commercially available RNase inhibitors are not effective against plant-derived RNases and are therefore not used in the procedure. Individual technique and local conditions may prompt reassessment of the appropriate level of caution regarding potential RNase problems.
3. High quality redistilled phenol is now commercially available, obviating the laborious task of distillation. An equal volume of sterile ddH$_2$O is added to the solidified phenol, which is then warmed to 37°C. On melting, the bottle is shaken to mix and the phases allowed to separate at room temperature. More ddH$_2$O can be added if phase separation does not occur. Water-saturated phenol can be stored at 4°C in a dark bottle for several months, but should be discarded if yellowing occurs.
4. It is crucial to maintain the frozen state of the plant material until the addition of the RNA extraction buffer in order to avoid endogenous plant RNase activity. It is also good practice to keep the RNA on ice throughout subsequent steps with the exception of organic extractions that are generally performed in a room temperature centrifuge.
5. Double-stranded nucleic acids are soluble in 2*M* LiCl, whereas single-stranded RNA... seldom possible to completely dissolve the pellet in this step, but... well dispersed by vortexing vigorously and pipetting up and... material that can adhere to t...

6. The duration of the ice incubation in $2M$ LiCl is somewhat arbitrary. Following the subsequent centrifugation, it is possible to save the supernatant and reincubate it overnight on ice or at 4°C. The supernatant is then recentrifuged and treated as an additional preparation that can yield up to 30% of the amount of RNA in the main preparation. This RNA should be kept separate from the main preparation until its quality is assayed, since it can exhibit some degradation.
7. It is not necessary to dry the final pellet completely, as this will greatly inhibit resuspension. Removal of any liquid from the sides and bottom of the tube using brief centrifugation and removal with an appropriate pipeting device is sufficient.
8. Although it is beyond the scope of this chapter to give detailed procedures for either spectrophotometric measurements or electrophoresis, a good rule of thumb is to check the absorbance (OD_{260}) of a 250-fold dilution of the RNA preparation. A value of 1.0 corresponds to an RNA concentration of 40 µg/mL. For electrophoresis, 2.5 µg can be loaded into 0.5-cm wells of either a 1% agarose gel or a 5% denaturing polyacrylamide gel, and easily visualized by ethidium bromide staining.
9. The small-scale preparation is designed primarily as a scaled down version of the large preparation for multiple batches of whole plants. It can, however, be adapted for single plant or single tissue types with few modifications. Generally, the only requirement is for starting material equivalent to approx 0.5 mL after grinding in liquid nitrogen. Additional modifications are noted in **Notes 10** and **11**. Roots are not amenable to this technique.
10. Certain tissues will benefit from additional extractions. Two additional phenol extractions are recommended for preparation of stem RNA and one additional extraction is recommended for preparation of silique RNA.
11. Two additional resuspensions, incubations, and centrifugations in LiCl are recommended for silique and stem RNA preparations.

References

1. Palmiter, R. D. (1974) Magnesium precipitation of ribonucleoprotein complexes. Expedient techniques for the isolation of undegraded polysomes and messenger ribonucleic acid. *Biochemistry* **13**, 3606–3615.
2. Chirgwin, J. M., Przbyla, A. E., MacDonald, R. J., and Rutter, W. J. (1979) Isolation of biologically active ribonucleic acid from sources enriched in ribonuclease. *Biochemistry* **18**, 5294–5299.
3. Chomczynski, P. and Sacchi, N. (1987) Single step method of RNA isolation by acid guanidinium thiocyanate-phenol-chloroform extraction. *Anal. Biochem.* **162**, 156–159.

III

MUTAGENESIS AND GENETIC ANALYSIS

14

Seed Mutagenesis of *Arabidopsis*

Jonathan Lightner and Timothy Caspar

1. Introduction

The generation of mutations is the most basic element of genetic analysis. Mutagenesis is the process by which heritable alterations in the genome of an organism, mutations, are produced. In order to conduct genetic analysis, at least two alleles for a given locus must exist and the facile production of these alleles through mutagenesis has been the focus of considerable research since the early part of this century. Today, the common mutagens available to geneticists fall into three general classes: chemical, physical, and biological, and the use of these in *Arabidopsis* has been reviewed recently *(1,2)*. The use of introduced T-DNA and heterologous transposons such as Ac/Ds and suppressor mutator (SPM) as biological mutagens are treated in detail in subsequent chapters. The focus of this chapter is on the production of mutations in *Arabidopsis* seed with chemical mutagens. In seed mutagenesis the individuals actually treated with the mutagen are referred to as the M_1 generation. The progeny derived from the self-fertilization of these individuals are referred to as the M_2 generation. The M_2 generation is the first generation following mutagenesis in which homozygous recessive mutations can be detected, and for this reason, it is the generation most frequently used in screening for mutants.

Due to primarily biological considerations, *Arabidopsis thaliana* is an ideal organism for conducting mutation experiments *(3,4)*. Of particular utility in terms of conducting mutation experiments are the small size and short generation time of the organism and its natural tendency to self-fertilize, even when grown quite densely. Because of these characteristics, tens to hundreds of thousands of plants can be readily grown to maturity and selfed in a single walk-in growth chamber or on a greenhouse bench. Thus, mutation experiments in *Arabidopsis* can easily involve tens or hundreds of thousands of M_1

individuals, a number that would be quite daunting in other well-developed plant genetic systems. Seeds are the most commonly mutagenized *Arabidopsis* tissue, because of their ready availability, robustness in mutagenesis, and ease of propagation. *Arabidopsis* seeds are thought to contain only two genetically effective cells *(5)*, that is cells that will contribute to the germ-line of the next generation, thus there is little incentive to harvest seeds as individual seed pods or in sectors as is advantageous in some species. There are a multitude of ecotypes and mutant lines of *Arabidopsis* available which can be used as backgrounds for mutagenesis experiments (*see* **Note 1**). Mutant seeds can often be obtained from other investigators and are available commercially. However, many investigators may wish to conduct their own mutagenesis either to optimize the mutagen or dosage, or in order to generate back mutations in a specific mutant background (*see* **Note 2**). Conducting seed mutagenesis in *Arabidopsis* requires few specialized materials, and with proper attention to safety and a modest amount of effort, can be accomplished successfully in any laboratory.

1.1. Choice of Mutagen

The choice of mutagen depends on several factors including the desired type of mutation, the desirability of facile cloning of the identified mutant genes, and the frequency of mutation. These requirements can frequently be at odds with one another. Biological insertional mutagens such as T-DNA or transposons provide one of the most straightforward ways to clone the identified mutant gene. However, these insertional mutations generally occur at quite low frequencies, thus greatly increasing the number of plants that must be screened in order to identify the mutation of interest. Physical mutagens, particularly ionizing radiation, frequently produce deletion or chromosomal rearrangement mutations, which may be of use in cloning the identified gene *(6)* and are also useful in that they are more likely to provide true null alleles. Chemical mutagens have high mutation rates, but often favor single basepair changes rather than gross genetic changes. These rather modest genetic changes provide little or no handle, other than the mutant phenotype, for cloning identified genes and the only currently available technology for cloning these genes are chromosome walking strategies (*7*, and Chapter 28). Single base changes, however, are useful in that they can yield both null alleles and mutations with varying degrees of functionality. Because single base changes can result in single amino acid changes in the encoded protein, chemical mutagens are also a good choice for production of temperature-sensitive alleles. From a practical standpoint, when initiating work on identifying a new group of mutations, it is often best to take advantage of the high mutation rates of chemical mutagens to identify the first mutants. If tagged alleles are desired for gene cloning, characterization of the initial, chemically induced, mutants often provides insight for simpler or more efficient screens which can then be employed to screen the higher numbers of individuals required for biologically mutagenized populations. The most popular chemical mutagen for use with *Arabidopsis* is ethyl-methane sulfonate (EMS) and a detailed protocol will only be given for its use (*see* **Note 3**). EMS is an alkylating agent thought to produce mutations by donating an ethyl group to a nucleic acid *(8)*. This modification can result in DNA strand breakage or mispairing. Based on the results of reversion studies *(9)* and molecular characterization of a small number of EMS-induced alleles (summarized in **ref. 2**), EMS is thought to produce primarily GC→AT transitions. Alkylating agents such as EMS are extremely powerful carcinogens and they must be handled with extreme care and foresight.

1.2. Population Size and Organization

Another consideration in a mutagenesis experiment is the size of the population to be mutagenized and subsequently screened. A related consideration is the relative size of the M_1 and M_2 generations. Because most mutations will be recessive and because M_1 plants are chimeric it is usually necessary to self these M_1 plants to obtain the M_2 seed that will be the generation that is actually screened. The relationship of the M_1 to M_2 generation has been well described both theoretically and experimentally *(5)*. In essence, the probability of a given mutation occurring is a function of the size of the M_1 generation; however, the probability of actually observing this given mutation in a screen is a function of the number of M_2 individuals screened from each M_1 plant. The relative amount of effort required for the production of the M_2 seed (i.e., the mutagenesis of M_1 seeds and the growth and harvest of these plants) vs the screening of the M_2 seed for the desired mutants will bear on the most useful strategy for sampling. One of the great advantages of *Arabidopsis* biology is that the cost of producing M_2 seed is usually small compared to the cost of conducting a mutant screen. When this is true it will be advantageous to reduce repetitive screening effort by sampling in such a way that few siblings are assayed. This can be accomplished by screening only 1 or 2 M_2 plants from each M_1 pedigreed plant or by screening a relatively small number of plants from a large pooled M_2 population. In the less typical case, where the cost of producing the M_2 seed is high relative to the cost of screening the M_2 plants, (e.g., in the case where a mutant with an obvious morphological phenotype that also has extremely low fecundity is to be remutagenized to screen for suppressors) it may be more advantageous to screen many progeny from each M_1 plant in order to raise the probability of finding the desired mutant from a limited number of M_1 plants.

Closely related to consideration of the M_1 and M_2 population sizes is the means of harvesting of the M_2 generation. The M_2 can either be pedigreed, that is harvested from single M_1 individuals and maintained as discreet groups of siblings, or pooled to some extent. Pooling greatly simplifies the harvesting and subsequent handling and screening of the M_2 seed. Pedigreeing, in contrast, because it requires harvesting the M_1 as discreet individuals and maintaining and screening the progeny in discreet units involves substantially more effort. Pedigreeing has the advantage that mutations that are infertile or lethal can be easily recovered as heterozygotes. It also provides a ready means of knowing that two allelic mutations isolated in the M_2 screen are in fact unique alleles at the same locus rather than multiple isolates of the same mutation event. Finally, pedigreeing is advantageous if the desired mutations might impact the fitness of the plant in the heterozygous state because pooling strategies usually involve growing plants in circumstances that are fairly competitive. Thus, if a mutation of interest reduces the ability of a heterozygote to compete effectively, the seed of that heterozygote will be underrepresented in the pooled M_2 progeny further increasing the difficulty in identifying such a plant. Unfortunately, harvesting M_2 seed from thousands or tens of thousands of individual *Arabidopsis* plants can be daunting, as can handling these seeds during the subsequent screening process. However, it is worth bearing in mind that a mutant population can be used for many screens and by many investigators. If this is true, then the extra time spent either pedigreeing or in harvesting smaller pools during the propagation of the M_2 seed will be a better investment than if the population is being produced for a very narrow purpose (i.e., to isolate revertants or suppressors in a specific mutant background).

The two extremes of harvesting strategy are shown diagramatically in **Fig. 1**. From a practical standpoint, pedigreeing is most useful in situations in which the desired mutations are known or suspected to be lethal or to affect fertility. If pedigreeing is to be used, we recommend harvesting a minimum of 1870 M_1 plants in order to give a reasonable likelihood of identifying mutations that occur at normal frequencies (*see* **Note 4**). Pooling will likely be necessary when the frequency of occurrence of the desired mutation is so low that progeny of tens of thousands of M_1 plants need to be screened to identify a mutant. Pooling is the most efficient strategy for making large numbers of unique mutations in processes that do not substantially affect plant viability or fertility. It is also an excellent strategy to employ when selective screens (for example herbicide resistance screens) are to be used. Finally, pooling can be employed to a more modest degree than is shown in **Figure 1**. Pools of 10–100 M_1 individuals can significantly reduce the amount of labor used in harvesting the M_2, while retaining much of the utility of pedigreeing. As a useful pooled population for most routine work, we recommend an M_1 population containing

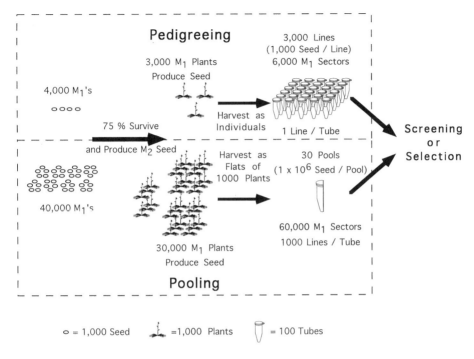

Fig. 1. Pedigreeing vs pooling in seed mutagenesis. The extremes of M_2 harvest strategy are shown. By employing the pooling strategy, the experimenter has created 10-fold more mutations for less investigator effort. Alternatively, by pedigreeing, the investigator has retained the ability to easily identify heterozygotes of mutations that affect either plant viability or reproduction, while sacrificing the overall number of mutations created for a given amount of effort. The dynamics of the process being investigated and the ease of the screen will determine which strategy, or what mixture of the two, is best for a particular experiment.

25,000–100,000 individuals harvested as pools of 1000 M_1 plants (*see* **Notes 4** and **5**).

1.3. Mutagen Dose

The final consideration is the dose of the mutagen to be used. For most purposes, it is advantageous to have as high a mutation frequency as possible in order to allow small mutagenized populations to contain large numbers of mutations. However, as the mutagen dose is increased to produce high mutation frequencies, seedling lethality produced by toxic (as opposed to mutagenic) effects of the mutagen can become problematic. High mutation rates can also lead to the production of very little or no M_2 seed by apparently perfectly healthy M_1 plants. This is likely to be caused primarily by the generation of large numbers of gametic lethal mutations. In this case, essentially every gamete will contain at least one lethal mutation leading to sterile plants. Excessively high mutation rates can also produce large numbers of embryo lethals that will also limit M_2 seed yield, although usually not so drastically as caused by the gametic lethals. The mutagenic efficiency of EMS treatments is variable, owing to factors such as the purity of the EMS itself, the temperature of the EMS solution during mutagenesis, the physiological condition of the seeds, and so on. Because of this, it is difficult to specify an exact EMS treatment that will always be suitable. Rather, it is worthwhile to empirically determine the appropriate mutagenic treatment using the conditions given in the **Subheading 3.** as a starting point. Pilot experiments in which small numbers of seeds are treated with a range of mutagen doses and then grown to maturity can be used to determine the highest dose of mutagen that can be used and still obtain sufficient M_2 seed for the desired uses.

2. Materials
2.1. Plant Materials
1. Seeds for *Arabidopsis* lines (*see* **Note 1**) are available from the *Arabidopsis* Biological Resource Center at Ohio State (e-mail: arabidopsis+@osu.edu) and the Nottingham Stock Centre, (e-mail: arabidopsis@notingham.uk). Seed from several wild-type lines can also be purchased in bulk quantities from Lehle Seeds (Round Rock, TX). A typical mutagenesis experiment will require about 5000–200,000 seeds.

2.2. Plant Growth
1. Methods for standard growth of *Arabidopsis* have been treated in Chapter 1 and in general no modifications are necessary for this procedure. Large tubs (approx 40 × 50 × 13 cm) are most useful for growing large numbers of M_1 plants.
2. For production of a pedigreed population, plants will be harvested as individuals so we suggest providing at least 10 cm²/plant. In order to harvest seeds from individual plants, siliques may be stripped by hand from single inflorescences. However, during growth, plants usually become intertwined, making it difficult to unambiguously separate one plant from the next. In most cases, a small amount of cross-contamination or double-sampling of the M_1 plants is not a serious problem. However, in those cases in which it might be, Lehle seeds markets the Arasystem which facilitates harvesting seeds from individual plants. Alternatively, Neff and colleagues *(10)* described an inexpensive seed harvesting system that can be improvised from used beverage containers. In experiments in which the M_2 seed will be pooled, the plants can be sown at a higher density of about 1–2 cm²/M_1 plant ($\approx$1000/tub) to provide adequate space for growth and to not too severely disadvantage the more physiologically compromised plants in the M_1 generation.

2.3. Glassware and Experimental Apparatus

Disposable labware should be used whenever possible for mutagenesis experiments.

1. 30 × 60-cm tray to contain spills.
2. Standard laboratory magnetic stir plate and stir bar.
3. Several 250-mL disposable beakers.
4. Several absorbent diapers to contain spills.
5. Disposable 10 and 25 mL plastic pipettes.
6. Micropipetor and filter tips capable of accurately pipeting 100–500 µL.
7. A closeable plastic chemical waste disposable jug, preferably about 5 L.
8. A fully functional and inspected fume-hood equipped with a magnahelic gauge is required for this procedure.

2.4. Chemicals

1. Laboratory-grade deionized water.
2. $3M$ Sodium Hydroxide (2 L).
3. EMS (methanesulfonic acid ethyl ester); EMS can be purchased inexpensively from most laboratory chemical suppliers. We recommend purchasing sufficient quantities (1 g) to conduct several pilot experiments and a large mutagenesis of *Arabidopsis* seed. When ordering EMS be sure to specifically request a current material safety data sheet (MSDS). EMS is usually shipped bottled inside a metal secondary container. On receiving EMS, this entire container should be transferred to a fume-hood for storage.
4. Seed suspending solution: 0.15% agar in water. Prepare a 1.5% solution by heating agar and water in a microwave or autoclave and then dilute 10-fold with water. Once gelled, the solution can be pipeted but the seeds will remain suspended in this solution allowing uniform sowing. Prepare the 0.15% agar at least a day before the mutagenesis, and check its effectiveness by using an aliquot to suspend a small number of untreated seed. Properly prepared agar solution will suspend the seed with no settling over extended times (several hours). At least 50 mL of 0.15% agar is needed for each 1000 M_1 seed to be sown.

3. Methods
3.1. Safety

1. EMS is highly carcinogenic and volatile. Experimenter contact with EMS must be completely avoided. Obtain and read a current MSDS sheet from your chemical supplier before carrying out this procedure. EMS should only be handled in a fully functional fume-hood equipped with a magnahelic gage so that proper operation of the hood can be assured throughout the experiment. Hood sashes should be closed as much as possible to provide only the minimum necessary room for manipulation of the chemical and should be secured in a fully closed position when no manipulations are being performed. The experimenter

should wear a laboratory coat, safety glasses, and a double layer of chemical resistant gloves when conducting the experiment.
2. All solutions containing EMS and objects contaminated with EMS solutions should be collected and stored in the fume-hood until decontaminated and disposed of.
3. Before performing the experiment, inform all laboratory occupants of the hazards of EMS. Identify the hood in which the mutagenesis is to be performed by posting warning signs and remove any unrelated materials from the hood. Develop a plan to respond in the event of a chemical spill or exposure and inform all laboratory workers of this plan. Finally, arrange to have the building roof or other venting area for the hood system secured against entry for the period during which the mutagenesis is to be performed.
4. EMS–treated seeds can retain mutagen even after extensive rinsing, so the treated seed should be handled with gloves and never pipeted by mouth.

3.2. EMS Mutagenesis

1. We recommend performing a small-scale pilot mutagenesis before conducting the full scale experiment to assess lethality and mutagen effectiveness (*see* **Note 6**). The pilot mutagenesis should follow the same procedure that follows with limited changes as indicated in **Subheading 4**.
2. Place the large tray (30 × 60 cm) near the back of the hood and line with absorbent diapers. Place a magnetic stir plate and the chemical waste container on the diaper paper in the tray.
3. Place the desired number of *Arabidopsis* seeds (*see* **Notes 7** and **8**) into a disposable 250-mL beaker, and add 100 mL of deionized water and a stir bar (a large paper clip can also be used). Place a 1-cm thick piece of flat Styrofoam on top of the stir plate to insulate the beaker thermally, and place the beaker on top of this piece of Styrofoam. Adjust the stirring action to as slow as possible to just keep the seeds well mixed (*see* **Note 9**).
4. The standard mutagen dosage for *Arabidopsis* seeds is 0.3% EMS (v/v). Using a filter-plugged pipet tip, add 300 µL of EMS into the beaker. Rinse the tip in the water to fully transfer the EMS. Place the rinsed tip into the waste container. Cover the beaker with aluminum foil.
5. After the appropriate time (typically 15 h), begin the wash procedure. Turn off the stir plate and allow seeds to settle. Carefully decant the EMS solution into the waste container. Add about 120 mL of distilled deionized water to the beaker, and stir for about 1 min. Allow the seeds to settle and decant the water into the waste container. Repeat the wash procedure seven additional times.
6. Wash the seeds one more time with 120 mL of water and transfer the seeds and water to a fresh 1-L flask. Transfer the stir bar to the waste container. Add 500 mL of water and a fresh stir bar to the 1-L flask and allow the seeds to stir in the hood for 30 min. Turn off the stir plate, allow the seeds to settle, and decant the last rinse into the waste container.
7. Rinse the disposable plastic 250-mL beaker three times with $3M$ NaOH and add these washes to the waste container.

8. Decontaminate the EMS solutions and materials using NaOH. To the waste container, which should contain about 1.5 L of 0.02% EMS, add 750 mL of $3M$ NaOH. Loosely cap and place on the stir plate for 30 min in the hood. Ensure that the stirring action is sufficient to fully mix the waste without splashing. EMS is destroyed within several minutes in $1M$ NaOH.
9. Cap the waste container (leaving the stir bar in the container) and dispose of the liquid ($1M$ NaOH) using normal chemical disposal procedures. Wipe off stir plate and tray with paper towels and place along with the disposable beaker, diaper paper, and any other solids into a plastic bag and dispose of using normal chemical disposal procedures.
10. Spills and Emergencies: If any EMS solution is spilled into the tray; decontaminate it by soaking in $1M$ NaOH for at least 30 min. Dispose of this NaOH with the liquid waste. Rinse the tray twice with water and add this to the liquid waste for disposal.

    If EMS is spilled outside the tray but remains in the hood; wipe the area with paper towels and then wash with $1M$ NaOH and then with water. Dispose of all towels into the liquid waste so that they can be decontaminated.

    In the event of hood failure; immediately close the hood doors and evacuate the lab. Close the lab doors and mark with barricade tape and caution signs. Notify your installation's emergency response team and wait for their instructions.
11. Add 10 mL of 0.15% agar/p 1000 seeds initially used. Mix well and then withdraw 2 mL, spread in rows onto a filter paper or paper towel, and count the number seeds in order to estimate the number remaining after washing. Also examine the seeds with a dissecting microscope to determine whether they have been damaged by the EMS treatment. If the seed coats are torn or badly discolored, the seeds probably will not germinate. Add sufficient 0.15% agar to dilute the seeds to 1000 viable seeds/50 mL of 0.15% agar.
12. Distribute the agar solution containing the seeds evenly across a tub of soil. Use a 25 mL pipet attached to a pipet bulb and slowly empty the agar solution containing the seeds from the pipette onto the soil surface. While this is being done, the pipet should be very rapidly moved around the surface of the tub to evenly distribute the seeds (*see* **Note 10**). Also, sow a single tub of the unmutagenized parent seeds as a control at the same density as the mutagenized seeds. For pooled M_2 populations, estimate the number of surviving plants after about 10 d of growth as an approximation of the size of the M_1 generation (*see* **Note 7**). Grow the plants to maturity and harvest using standard practices (*see* **Note 11** and **12**).
13. Following harvesting, evaluate the degree of mutagenesis (*see* **Note 13**).

4. Notes

1. Mutagenesis experiments are best conducted in one of the three common backgrounds, Columbia, Landsberg *erecta*, or Wassilewskija, unless there is a compelling reason to use a nonstandard ecotype. Comparisons between isolated mutants are more definitive if the mutations are in common genetic backgrounds and the availability of physical genetic markers (*see* Chapters 18 and 19) in these backgrounds is advantageous.

2. One common reason for performing your own mutagenesis is to produce revertants or suppressors of a characterized mutation of interest. It is critical in back-mutation experiments to include markers in the M_0 plant (the parent mutant line) that can be used to rule out contamination by wild-type seeds or pollen. Ideally, the parent mutant line should be homozygous for two marker mutations that flank the mutation that you intend to revert or suppress. Back-mutation experiments often require substantially larger numbers of M_1 individuals than is necessary to isolate the original mutation particularly if true revertants, rather than suppressors, are desired.

3. Chemical mutagens are by far the most popular mutagens currently employed with *Arabidopsis*. Of these, EMS is the most popular because of its high mutagenicity and low lethality. Other popular chemical mutagens that have been successfully used are methyl methanesulfonic acid (MMS), nitrosomethylurea (NMU), and diepoxybutane (DEB) (reviewed in **ref. *1***). Physical mutagens are frequently used when deletion or chromosomal rearrangement mutations are desired. Redei and Koncz *(1)* discussed dosage and effectiveness of x-ray, gamma-ray, and fast neutron mutagenesis with *Arabidopsis*. Although the specific protocols for mutagenesis will differ for each of these mutagens, the underlying considerations governing the generation of the mutagenized population will be the same as those described for EMS.

4. Mutation rates for different genes in *Arabidopsis* vary widely due to differential mutagen damage or subsequent repair of the genes. The phenomenon of mutagenic hot-spots is well known in simple genetic systems *(11)*. Similarly, overall mutation rates differ from one batch of mutagenized seed to the next (*see* **Note 13** for ways to estimate overall mutation rates). However, in typical EMS-mutagenized M_2 populations derived from large M_1 pools, loss of function, recessive mutations in many genes occur at a frequency of roughly 1 in 2000 to 1 in 5000 M_2 plants. In order to convert this frequency into a number informative for pedigreed populations, one must take into account the fact that the genetically effective cell number of *Arabidopsis* is two and that only one-quarter of the progeny from a heterozygous sector will be homozygous for a given mutation. Thus only 1/8 of the M_2 progeny from an M_1 plant with a mutation in one of the two genetically effective cells will be homozygous for the mutation. Assuming that the progeny from each M_1 pedigreed line will be sampled sufficiently to identify this class, (screening 17 M_2 individuals/ M_1 line gives a probability of 90% of identifying at least one) then screening the progeny from one M_1 pedigreed line is equivalent to screening eight M_2 plants from a large pooled population. (However, this degree of sampling is inherently repetitive, so that if the screen is relatively difficult, it may be more advantageous to increase the number of M_1 pedigreed lines and screen fewer M_2 plants from each.) Thus, the conservative estimate of mutation frequency of 1/5000 translates to a frequency of about 1/625 pedigreed lines containing the mutation of interest. In order to have a 95% probability of producing at least one mutation given this frequency, pedigreed populations should contain a minimum of 1870 M_1 lines, although more

would be even better. Finally, it is worth noting that many mutant phenotypes can be produced by mutations in more than one gene. The elaboration of a complex morphological character or the production of a particular biomolecule are both likely to be influenced by several genetically controlled steps. In such cases, the likelihood of identifying mutants in a given population size increases with the complexity of the process being examined in the screen.

5. Haughn and Somerville *(12)* estimated that a typically mutagenized EMS M_1 population of 125,000 individuals will contain all possible EMS-induced mutations throughout the genome. This estimate, calculated by extrapolating from microbial mutation results, should be considered rough, but offers at least an approximation from which to plan large, pooled M_2 populations.

6. Pilot experiment: For any mutagen, including EMS at the standard dosage, we recommend conducting a small pilot mutagenesis on 300–500 seeds to assess lethality and sterility in the M_1 generation. This pilot-scale experiment can be conducted by varying the mutagen concentration or the time of exposure in order to estimate the optimum mutagen dosage. The pilot mutagenesis should use exactly the same procedure as the full-scale mutagenesis, only with a limited number of seeds. (In our experience, some batches of EMS can give excessively high lethality [>75 %] even at the standard concentration of 0.3% [v/v] and we recommend ordering a fresh sample of EMS if this is observed.)

7. The number of seeds to be mutagenized will depend on the intended use of the M_2 seed, whether it will be harvested in pools or in pedigrees, the availability of the seed, and how much growth space is available. It is valuable to know the size of the M_1 generation (i.e., the number of M_1 plants which produce M_2 progeny). For a pooled population, this can be difficult to determine and usually the number of M_1 plants that grow to an early rosette stage is used as an estimate of this number. (After the rosette stage, the M_1 plants become so crowded that counting is impossible.) General considerations and recommendations for the size of the M_1 population are discussed in the introduction. However, allowances should also be made for the loss of 10–20% of the seeds during rinsing in **step 5** of the protocol and for up to 25% lethality caused by the EMS treatment. Since mutagenized M_1 seeds can be dried and stored after rinsing, it may be advantageous to treat more than the minimal number of seeds and store the excess for future use. At least 200,000 seeds can be mutagenized in 100 mL of EMS solution.

8. Some investigators use a cloth bag to suspend seed during the mutagenesis and rinsing which prevents the problems of seed damage by stirring and loss during rinsing. We do not use this technique, because of potential difficulties with mutagen both diffusing into the seed mass, and later being rinsed out of the seed mass effectively. In our view, a gentle stirring action will give a more uniform treatment of the M_1 seeds and provide more effective rinsing.

9. Rapid stirring may initially be required to fully wet the seeds. However, the stirring rate should be turned down as soon as the seeds are wetted. If stirring action is too vigorous, seeds will be damaged and will not germinate.

10. Even distribution is critical to good growth of M_1 plants. Because the desired density will depend on the harvest strategy to be employed, it is a good idea to optimize the sowing conditions (in terms of number of seeds and amount of agar solution per pot) with untreated seed beforehand.
11. During the early growth of the M_1 plants, it is helpful to cover the flats with clear plastic to prevent the seedlings from desiccating. However, flats covered with plastic will overheat if placed in very bright light. For this reason, it is best to start the plants in dim light (50–100 μmol/m^2/s) and only increase the light after the plastic has been removed after about 7–10 d. For growth in greenhouses, shading may be necessary during this first period.
12. During the growth of the plants, watch for chlorotic somatic sectoring indicative of successful mutagenesis. During seed maturation, check the degree of elongation of the siliques. If the siliques on the majority of the M_1 plants (but not the unmutagenized control plants) fail to elongate, the plants are likely to be sterile owing to excessive mutagenesis (see **Subheading 1.**).
13. The rigorous estimation of mutation frequency is described by Li and Redei *(13)* and is beyond the scope of our discussion. However, the effectiveness of the mutagenic treatment can be monitored in several ways. Chlorotic sectors caused by somatic segregation of organellar mutations should be present in about 0.1–1% of the M_1 plants. However, these sectors are difficult to quantify and so are useful only for a general confirmation that the M_1 seeds were successfully mutagenized. More quantitative measures of the degree of mutagenesis can be performed on the M_2 generation by measuring the frequency of one or more known classes of mutants. By comparing these rates with other mutagenized populations, the relative degrees of mutagenesis can be determined. Absolute levels of mutagenesis, however, can only be crudely estimated *(12,13)*. Mutations in the adenine phosphoribosyl transferase gene can be positively selected by growth on 0.1 m*M* 2,6-diaminopurine *(14)*. In well-mutagenized populations, these resistant mutants occur at a frequency of roughly 1 in 5000. Similarly, positive screens for mutations in alcohol dehydrogenase *(15)* and in the pathway of nitrate utilization *(16)* can also be used for calibration. Phenotypic screens that measure mutation rates in many genes at once have advantages in that they give an estimate of the average mutation rate, which is less susceptible to the variation caused by differential mutability of specific genes. The easiest of these screens involves scoring for pigment phenotypes (i.e., white, yellow, and pale green) in young seedlings germinated in the light on minimal agar. Mutation frequencies scored by this assay for well-mutagenized M_2 populations should be in the range of 2–10%. Because of the subjective nature of the mutant phenotypes, this rate may vary from lab to lab. However it can still be useful in comparing different mutagenized populations.

References

1. Redei, G. P. and Koncz, C. (1992) Classical mutagenesis, in *Methods in* Arabidopsis *Research,* (Koncz, C., Chua, N.-H., and Schell, J., eds.) World Scientific Publishing, Singapore, pp. 16–82.

2. Feldmann, K. A., Malmberg, R. L., and Dean, C. (1994) Mutagenesis in *Arabidopsis*, in *Arabidopsis* (Meyerowitz, E. M. and Somerville, C., eds.), Cold Spring Harbor Laboratory, Cold Spring Harbor, NY, pp. 137–172.
3. Estelle, M. A., and Somerville, C. R. (1986) The mutants of *Arabidopsis*. *TIG* **2**, 89–93.
4. Meyerowitz, E. M. (1989) *Arabidopsis*, a useful weed. *Cell* **56**, 263–269.
5. Redei, G. P., Acedo, G. N., and Sandhu, S. S. (1984) Mutation induction and detection in *Arabidopsis*, in *Mutation, Cancer and Malformation* (Chu, E. H. Y. and Generoso, W. M., eds.), Plenum, New York, pp. 689–708.
6. Sun, T.-P., Goodman, H. M., and Ausubel, F. M. (1992) Cloning the *Arabidopsis GA1* locus by genomic subtraction. *Plant Cell* **4**, 119–128.
7. Gibson, S. I. and Somerville, C. Chromosome walking in *Arabidopsis thaliana* using yeast artificial chromosomes, in *Methods in* Arabidopsis *Research*, (Koncz, C., Chua, N.-H., and Schell, J., eds.), World Scientific Publishing, Singapore, pp. 119–143.
8. Lawley, P. D. (1974) Some chemical aspects of dose-response relationships in alkylation mutagenesis. *Mutat. Res.* **23**, 283–295.
9. Prakash, L. and Sherman, F. (1973) Mutagenic specificity: Reversion of iso-1-cytochrome c mutants of yeast. *J. Mol. Biol.* **79**, 65–82.
10. Neff, M., Ewing, D. and Comai, L. *Arabidopsis* seed harvester, in *The Compleat Guide,* available via the World Wide Web at: http://www-genome.stanford.edu/Arabidopsis.
11. Benzer, S. (1961) On the topography of the genetic fine structure. *Proc. Natl. Acad. Sci. USA* **47**, 403–415.
12. Haughn, G. and Somerville, C. R. (1987) Selection for herbicide resistance at the whole-plant level, in *Biotechnology in Agricultural Chemistry* (LeBaron, H. M., Mumma, R. O., Honeycutt, R. C., and Duesing, J. H., eds.), American Chemical Society, Washington, DC, pp. 98–107.
13. Li, S. L. and Redei, G. P. (1969) Estimation of mutation rate in autogamous diploids. *Radiation Biology* **9**, 125–131.
14. Moffat, B. and Somerville, C. R. (1988) Positive selection for male-sterile mutants of *Arabidopsis* lacking adenine phosphoribosyl transferase activity. *Plant Physiol.* **86**, 1150–1154.
15. Jacods, M., Dolferus, R., and VandenBossche, D. (1988) Isolation and biochemical analysis of ethyl-methanesulfonate induced alcohol dehydrogenase null mutants of *Arabidopsis thaliana* (L.) Heynh. *Biochem. Genet.* **26**, 105–122.
16. Braaksma, F. J. and Feenstra, W. J. (1982) Isolation and characterization of nitrate reductase-deficient mutants of *Arabidopsis thaliana*. *Theor. Appl. Genet.* **64**, 83–90.

15

Genetic Analysis

Maarten Koornneef, Carlos Alonso-Blanco, and Piet Stam

1. Introduction

The genetic behavior of *Arabidopsis* is not different from that of other diploid organisms, and therefore one can find the topics of this chapter in almost any handbook of genetics. It is only owing to the specific (reproductive) biology of a species that sometimes a certain analytical procedure is favored over another one. For *Arabidopsis* the fact that it is a self-fertilizing species and that it has relatively small flowers are important in this respect.

A classical genetic analysis of a certain trait implies that observed genetic variation for such a trait is described in terms of certain genetic parameters. Genetic variation, the starting material for such an analysis, can be induced by mutagenic treatments or can be found as "natural" variation present within and among ecotypes.

The analysis of genetic variation should answer the following questions:

1. How many genes control the observed variation?
2. What are the dominance relationships of the various alleles?
3. Is the gene (as a genetic entity) allelic to previously described genes?
4. Which are the epistatic relationships with other genes in the same pathway?
5. Where is the gene located on the linkage map?

At the moment genetic variation has been identified (e.g., a mutant with a specific phenotype), the variant should be crossed with its wild-type. The resulting F_1 hybrid is used to determine if either the wild-type or the variant is dominant. In addition, it can be used to obtain the F_2 generation by harvesting (selfed) seeds from the F_1 plant. The detailed scoring of individual F_2 plants for the trait of interest and the analysis of the obtained numbers of each phenotypic class, allow the determination of the mono-, di-, or polygenic nature of the variant phenotype that one identified. In case the mutant has a complicated phenotype (i.e., several characters seemed to be changed simultaneously), the segregation analysis furthermore indicates which of these changed properties are the result of mutations in the same gene (pleiotropic) and which are the result of mutations in other genes. The F_1 plants can also be crossed (back) to the original variant to give the backcross (BC) generation, but in *Arabidopsis* often only an F_2 analysis is performed. From the F_2 generation, a number of plants with the variant phenotype (and without other visible mutations) are selected and selfed and the individual F_3 progenies are grown to be retested for being "true type" and for the absence of segregation of other unlinked mutations. Preferentially, this backcross and reselection procedure is repeated several times so that a "clean" mutant line showing monogenic inheritance is obtained. In case the mutant resembles a mutant phenotype isolated or described before, it should be crossed with such mutants to test for allelism with these "old" mutants. In many situations, several loci are identified that are involved in the same process or pathway. The phenotype of the double mutants is often indicative of the interaction between the two respective genes. To identify these double mutants in segregating generations several precautions should be taken since the double mutant phenotype is not known *a priori.*

A next step in a genetic analysis is the location of the new gene (as identified by its mutant phenotype) on the genetic map of *Arabidopsis*. For this, the mutant (backcrossed and reselected) has to be crossed with a genotype that differs from the original mutant for several previously mapped loci. These can be mutants with a defined map position, which should have a phenotype that does not interfere with the scoring of the mutant to be tested. A number of genotypes that are recessive for a number of mapped genes have been constructed *(1)*. These tester lines are predominantly in L*er* background and facilitate the detection and subsequently the measurement of linkage. A procedure that can be followed is the analysis of populations segregating for the gene with an unknown map position and a number of markers at different chromosomes. Genotypes such W100, with one or two markers on each chromosome can be used for this. When the position of a new locus is approximately known, a more detailed mapping can be performed with other "morphological" or molecular markers in that region. To determine the location of the new locus in relation to molecular markers (RFLPs, RAPDs, SSCPs, ARMs, CAPs, AFLPs; *see* Chapters 18, 19, and 30), which is the most commonly procedure used nowadays, a cross has to be made with another ecotype. The use of widely available lab ecotypes (L*er*, Col, Ws, and so on) is preferred because, in that case, information on the various polymorphisms is known already. Genes are located on the same chromosome when they are genetically linked to other genes located on that chromosome. Linkage of two genes will be detected by a deviation from independent segregation in any type of segregating population such as an F_2, a backcross, a set of recombinant inbred lines, and so on. The detailed map position of a locus to be mapped relies on the identification of a locus (or loci) with which it shows a close (preferentially almost absolute) linkage.

The quantification of genetic linkage between two loci is based on the estimation of the fraction *(p)* of recombinant gametes (with respect to the total number of gametes) originating from a meiosis of a double heterozygote for the loci under study. Recombination between these two loci is a direct consequence of crossing over, the recombinant frequency *(r)* being the basis of the genetic distance. Genetic distances between loci are given as map distances *(D)*, usually expressed in centiMorgans (cM). Genetic distance is defined as the average number of crossover events between two loci during a single meiosis and per chromatid. Recombinant frequency *(r)* increases with map distance but a doubling of map distance does not result in a doubling of the recombinant frequency. This is due to the fact that a second cross over between two loci may or may not involve the same chromatid of a given chromosome. A second crossover involving the same chromatid "cancels" the effect of the first one and therefore gives rise to a parental like gamete.

The mathematical relationship between map distance and recombinant frequency is given by the genetic mapping function. Its form depends on the degree of interference in crossing over. In the absence of interference, the recombination events in adjacent regions are independent, i.e., a crossover at a given position does not influence the probability of another crossover in its neighborhood.

When estimates of the map distances between at least three genes are available, these loci can be ordered in a linear linkage group. In most cases, the order of genes that segregate in the same population can be determined unambiguously both from the estimates of the recombination percentages and from the presence or absence of joint recombination of the different markers. The order of genes that are closely linked often cannot be determined with certainty, when recombination data originate from different populations, for several reasons: the criterium of joint recombination can not be applied, standard deviations of the estimates of r are often relatively high, and variations in the recombinant frequency and the distribution of recombination events along chromosomes can occur among different mapping populations.

Software packages, such as MAPMAKER *(2)*, JOINMAP *(3)* and GMENDEL *(4)*, have been developed for the construction of genetic maps either based on crude data or on pairwise recombination data.

2. Materials

1. A pair of sharp-pointed tweezers.
2. A stereo microscope.

3. Labels for plants and crosses (e.g., colored threads).
4. The growing conditions are standard, but should allow the handling of individual plants (for making crosses). Specific mutants may need selective growing conditions to be able to score the various phenotypes.
5. Computers and specific software to estimate recombinant frequencies (e.g., LINKAGE, RECF2) and to construct linkage maps (MAPMAKER, JOINMAP, GMENDEL).

3. Methods
3.1. Crossing *Arabidopsis*
1. Select the parents for the cross. These can be plants that just started to flower (*see* **Notes 1** and **2**).
2. Remove all young flower buds and flowers with petals from the female parent from the main inflorescence until three to four large buds (white petals not yet visible) are left (*see* **Note 3**).
3. Open the buds carefully, and remove the six anthers without damaging the pistil. This emasculation should be performed under a low magnification of a stereo microscope (*see* **Note 3**).
4. When all buds are emasculated, carefully check that no anthers are left.
5. Take a fresh, fully open flower from the male parent and "brush" the stigma with this flower so that the pollen is visible on the stigma as yellow powder (*see* **Note 3**).
6. Mark the group of flower buds with a label attached to the stem just under the buds, which indicates the cross and the parent that was used as female parent, and take note of the crossing date (*see* **Note 3**).
7. Three to four days later, the pistil should have developed as a young silique indicating that the cross was successful.
8. Seeds should be harvested when the siliques are brown and have not dehisced. This is 2–3 wk after pollination, depending on the growing conditions.

3.2. Segregation Analysis

The analysis of the genetic segregation of the trait of interest in the studied population (usually an F_2) will allow the determination of the number of genes that segregates, and in addition will give information on the dominance relationship between alleles.

1. To obtain F_2 generations, the F_1 seeds should be planted, checked whether they are the result of unintentional selfing, and harvested individually (*see* **Note 4**).
2. The phenotype of the F_1 already gives an indication about dominance, which will be confirmed by the type of segregation in the subsequent generations (*see* **Note 5**).
3. Make the segregating population large enough to be able to distinguish between the different segregation ratios that one theoretically may expect (*see* **Note 5**).
4. Score each individual plant of the population, taking specific precautions to allow the identification of all possible genotypes.

Genetic Analysis

5. Determine the number of plants with each phenotype and test if the data are not in conflict with the expected ratio for the simplest genetic explanation (e.g., 3:1 in case of a single gene controlling the trait) using a Chi-square test (*see* **Note 5**).
6. In case the Chi-square test indicates that the observed data are not in agreement with this simple genetic explanation, test if the obtained data agree with a di-genic genetic model, and so on (see **Note 5**).
7. In order to identify the F_2 individuals that are either homozygous or heterozygous, progeny testing of individual F_2 plants with a wild-type phenotype can be performed (*see* **Note 6**).

3.3. Allelism Tests

Allelism tests will tell whether the new mutant allele corresponds to (is allelic with) an already known gene or whether it represents a new locus.

1. Cross the new mutant with all available independent mutants with a similar phenotype.
2. Compare the phenotype of the F_1 hybrids with that of the wild-type and the parental mutants. If two recessive allelic mutants are crossed, their F_1 hybrid has a mutant phenotype: *a–1/a–1* × *a–2/a–2* → *a–1/a–2* (mutant phenotype).
   When two nonallelic mutants are crossed the F_1 hybrid has a nonmutant phenotype, that is the two mutants complement each other: *aaBB* × *AAbb* → *AaBb* (wild-type phenotype; *see* **Note 7**).
3. In the case of dominant alleles, the F_1 hybrid of two homozygous parents is not informative, and one has to self this hybrid or backcross it with a recessive genotype to find, in case of nonallelism, recessive (wild-type) phenotypes in the progeny (*see* **Note 8**).

3.4. The Isolation of Double Mutants to Analyze Epistatic Relationships

The isolation of double mutants is usually started from an F_2 population segregating for two genes controlling similar or related traits. The difficulty of identifying the double mutant depends on its phenotype, which is *a priori* not known. The double-mutant phenotype can be similar to one of the parents, can be a novel phenotype, or sometimes can resemble the wild-type. Additional work will be required to isolate it (larger number of plants to analyze) when both mutations are genetically linked.

1. When in an F_2 generation, in case two recessive mutants, a novel (often more extreme) phenotype appears with a frequency of approx 1/16, this most likely is the double recessive. When the mutants are not recessive, other frequencies such as 9/16 (in case of two dominant mutants) and 3/16 (in case of one recessive and one dominant mutant) can be expected for the novel phenotype.
2. Cross the putative double mutants with both parental mutants and check for the absence of complementation in the hybrids. In case of dominant mutants, see the previous section.

3. In case of complete epistasis, the phenotype of one mutant is not visible in the double mutant, and one has to take specific precautions to detect this genotype.
    a. When the single recessive mutants have a distinguishable phenotype, plants with either one or the other mutant phenotype can be identified in the F_2 and the corresponding F_3 lines can be checked for segregation of the second phenotype (*see* **Note 9**). Allelism tests with the parent mutants should confirm the genotype of the double mutants identified in such F_3 lines.
    b. In case both recessive mutants cannot be distinguished phenotypically and no novel phenotypes are observed in an F_2, this usually results in a 9:7 segregation in an F_2. Testcrossing of F_2 plants with mutant phenotype to both parental mutants will be necessary to identify the double mutant (*see* **Note 10**).

3.5. Linkage Analysis and the Construction of Linkage Maps

Linkage analysis is performed to detect which loci are linked. Subsequently, the frequency of recombinant gametes is estimated from the observed segregation ratios and these estimates serve as basis for the construction of genetic linkage maps.

3.5.1. Linkage Analysis

1. Make an F_1, which is a multiple heterozygote for the markers to be analyzed, and derive from this F_1 a segregating F_2 or BC generation.
2. Score all individual plants for each marker.
3. Use this scoring to determine, for each pair of segregating markers, the number of plants within each phenotypic class (*see* **Note 11**).
4. Test the independent segregation of both markers using the Chi-square test for a contingency table (*see* **Notes 12** and **13**).

3.5.2. Estimation of Recombination Frequencies and Map Distances

Once the Chi-square test for independence of segregation has revealed linkage, the recombinant frequency and map distance between the loci are estimated.

1. Estimation is usually done by maximum likelihood (ML). Apart from the simple backcross population, where recombinants can be counted directly, computer programs are used for this purpose (LINKAGE, RECF2). In addition to the estimate itself, the standard error (SE) of the estimate and/or the LOD score is produced. Both SE and LOD score can be seen as measures of "linkage information" (*see* **Notes 14–17**).
2. Recombination estimates and their standard errors can be converted into genetic map distances. Usually either Haldane's (assuming absence of interference) or Kosambi's (assuming interference) mapping functions are used (*see* **Note 18**).

3.5.3. The Construction of Linkage Maps

The pairwise map distances for three or more linked genes allow the ordering of these genes into a linear linkage map. When more than five loci have to be ordered this is better done by using the software packages developed for this purpose. These packages implicitly perform the estimation and calculation of LOD scores.

1. Prepare a crude data file to be used as input for the mapping program. The format of these data files depends on the software package that is being used; however packages require a similar layout. Crude data files can be prepared with any text editor or spreadsheet program that enables export of plain text files (*see* **Notes 19** and **20**).
2. Depending on the software that is being used, intermediate results are produced before a final map is calculated. JOINMAP, for example, produces a suggested assignment of genes to linkage groups, as well as a list of pairwise estimates and LOD scores. It also allows testing for segregation distortion for each gene (*see* **Notes 19** and **20**).
3. Graphical representations of the maps can be produced with specific graphic software such as Drawmap *(5)* or with any other computer- or artist-based method.

4. Notes

1. Whenever it is possible, use the genotype with a recessive trait as the female parent, since this will allow you to check for unwanted selfing of the parent in the F_1 generation. This recessive parent is often the mutant, but it can also be the parent that carries another recessive allele such as the *erecta* mutation (in L*er* and its mutants) when the other parent carries the dominant allele of this gene (in the other ecotypes).
2. When a mutation is lethal or leads to complete sterility, crosses have to be made with heterozygous individuals which can be identified by progeny testing. It may happen that at the moment the cross is made, this information is not available (e.g., embryo lethals can often be identified in older siliques of the same plant). In this situation, several putative heterozygous plants (wild-type sister plants in a line that segregates for the mutant) have to be crossed and the seed of these parents has to be harvested for progeny testing.
3. Although all flower buds can be used for crosses, the flowers at the lower part of the main stem are often preferred because the flower buds seem "stronger." The use of these early flowers allows you to remake the cross when it has failed. Removing (accidentally) petals is not a problem, however, the pistil should have no damage. To check for pollen availability, touching a nail with the open flower should show some yellow pollen. *Arabidopsis* buds that are emasculated just before the petals become visible (1 d in practice) can be pollinated on the same day although the stigma is also receptive 1 d later. The transfer of an overdose of pollen on a stigma prevents cross-pollination by open flowers from the same or neighboring plants and, therefore, "bagging" of the cross-pollinated buds is not necessary.

   When several crosses have to be made, use different plants or inflorescences. Making different pollinations in a group of buds easily leads to mistakes because of mislabeling or problems with harvest.

4. Although the different F_1 plants from a cross of two homozygous parents should all be genetically identical, it is advised not to bulk seeds from different F_1 plants but to test the individual progenies. When one F_1 is wrong due to selfing, unwanted cross-pollination, seed contamination, and so on, this can be recognized by its unexpected segregation and the progeny of such a plant can be eliminated from the analysis. In case of recently isolated mutants, they can be heterozygous for other mutations, and when such a plant has been crossed, half of the F_1 will be heterozygous for this second mutation and half will not. The data of progenies that are identical can always be added, but data from a contaminated bulk of F_1 plants are useless.
5. An 1:1 segregation in the BC with the recessive parent (variant) and a 3:1 (wild-type:variant) in the F_2 indicates monogenic inheritance and recessiveness of the variant. Dominance of the variant yields 1:3 (wild-type:variant) ratios for the F_2 and a 1:1 segregation only in the BC with wild-type. Other monogenic ratios are caused by intermediate expression of the heterozygotes (1:2:1) in the F_2 and/or reduced viability or even lethality of certain genotypes (often the homozygous mutant). Another factor that may lead to a deficit of a class of genotypes is a reduced transmission of certain (mostly mutant) alleles through the gametophyte. This phenomenon, which is called certation, is assumed to occur only in pollen and is the main cause for recessive deficits in induced mutants *(6)*. Digenic inheritance of a certain character gives ratios in the F_2 that are variations of the classical 9:3:3:1 *(A_B_:aaB_:A_bb:aabb)* ratio for two genetically independent genes. These ratios are, depending on the epistatic relations between the two genes, 9:7; 13:3; 15:1.

When traits (e.g., seed traits) are controlled by the genotype of mother plant (maternal inheritance) one has to be aware that such a trait has to be analyzed in the progeny of the mother plant.
6. Use at least 16 F_3 plants per selfed progeny (= line or family) from an F_2 plant to distinguish between a progeny that originates from either a heterozygous genotype or from a progeny of a homozygous plant with 99% certainty ($0.75^n < 0.01$ at $n = 16$). Using <16 plants may result in some progenies that contain, because of chance, only wild-type plants, although they should segregate 25% recessive ones.
7. The outcome of the complementation test can be complicated when allelic complementation occurs. A well-documented case of this in *Arabidopsis* is some alleles at the *py* locus *(7)*. However, in contrast to nonallelic complementation of true recessive mutations, allelic complementation is often only partial, and does not occur in all allele combinations and therefore incorrect conclusions are rare.
8. Very close linkage of two loci with a dominant allele for gene 1 in one parent and a dominant allele of gene 2 in the second parent (resulting in similar phenotype) cannot be distinguished from allelism until a recombinant is found.
9. The phenotype of the epistatic mutant will be deduced from the segregation in F_3 lines from F_2 plants having either one mutant phenotype or the other. Most likely, one type of mutant phenotype F_2 plants will not segregate in F_3, indicating that

this phenotype class includes the double mutant phenotype. Two out of each three F_2 plants of the other type of mutant phenotype F_2 plants will segregate in F_3 lines for plants with either one or the other mutant phenotype (in the progeny of *Aabb*, the *aabb* segregants will resemble the *aaBB* parent and not the *AAbb* parent when the *a* mutant is epistatic over the *b* mutant). In case that the double mutant phenotype resembles wild-type, both types of single-mutant phenotype F_2 plants will segregate for wild-type and the corresponding mutant phenotypes.

10. The double mutant will comprise 1/7 of the mutants when no novel phenotypes are observed and the two parents cannot be distinguished. In this case, at least 20 plants $[(6/7)^n < 0.05$ at $n = 20]$ have to be tested to be sure that the double mutant will be identified.

11. In case of dominance, four phenotypic classes are expected in an F_2, whereas one expects six classes when one of the markers is codominant and nine when both are codominant.

12. In the case of dominance of linked markers, the deviation from the "classical" independent segregation ratio in an F_2 (i.e., 9:3:3:1 for $A\_B\_:aaB\_:A\_bb:aabb$) depends on the so called "phase" of the recessive and dominant alleles in the double heterozygous F_1 parent. We distinguish "coupling phase" (F_1 derived from a cross such as *aabb* x *AABB*) and "repulsion phase" (F_1 derived from the cross *aaBB* × *AAbb* or reciprocal). With absolute linkage, the F_2 ratio for coupling phase is 12:0:0:4, whereas for repulsion phase this is 8:4:4:0. Since it is easier to tell the difference between 9:3:3:1 and 12:0:0:4 than between 9:3:3:1 and 8:4:4:0, it is intuitively clear that it is harder to detect linkage from a repulsion phase F_2 than from a coupling phase F_2.

It is obvious that linkage will be detected easier, when recombination is rare. The minimal population size required to detect linkage for different types of segregating populations has been published elsewhere *(1)*.

Population sizes around 100 are in general, sufficient to detect linkage, but with $r > 35\%$ this population size often results in observed segregation ratios that are not significantly different from independent segregation.

13. Independence of segregation of two genes can be tested with the Chi-square test. It is better to use a contingency table analysis instead of a test of deviation of the 9:3:3:1 for F_2 or 1:1:1:1 for BC generations (or any other "classical" ratio), since the expected ratios can be distorted when the monogenic ratios are not Mendelian (e.g., due to certation or reduced viability of certain genotypes).

14. Information on linkage can be expressed in a LOD value, the logarithm of the likelihood odd ratio.

Mathematically the LOD reads:

$$\text{LOD} = \log[(\text{likelihood of observed data with } p = \hat{p})/(\text{likelihood of observed data with } p = 0.5)]$$

A LOD score of, e.g., 5.2 means that given the data, the true recombination frequency is $10^{5.2} = 158489$ times more likely to be equal to the estimated $\hat{p}$, than to be equal to 0.5.

The use of LOD scores is especially useful when the recombination estimate equals zero. If this estimate is obtained from a backcross, i.e., from a binomial sample, its variance formally equals zero, irrespective of the sample size. The LOD score, however, also indicates that finding no recombinants in a sample of 100 is more informative than finding no recombinants in a sample of 10 (LOD = $n$log2, in this case; n is sample size). Usually a LOD score >3 is taken as evidence for linkage. A separate test for independent segregation is often omitted when a large number of markers are tested and LOD scores, recombinant frequencies, and linkage maps are estimated directly with the appropriate software.

15. The fraction of recombinant gametes (which is used as the estimate of recombination frequency) can be derived from the frequencies of the various phenotypes in segregating generations. The analysis of the segregation in a test cross, that is the progeny of a cross of a double recessive with the diheterozygote ($aabb \times AaBb$) is an effective way to determine the recombination percentage between the two loci. The number of nonparental (recombinant) progeny gives a direct estimate of the recombination fraction p of which standard deviation can be derived from the binomial distribution. A drawback of test crosses is that they require double recessive genotypes and accurate crossings, which is not easy in some genotypes of *Arabidopsis*. For these practical reasons and the fact that selfed seeds can be easily obtained in large quantities in *Arabidopsis*, mapping has been performed mainly with F_2 and F_3 populations, which can also be very efficient and which averages the often significant differences of male and female meiosis *(8)*.

16. From the observed F_2 segregation frequencies, the recombination fraction, p, can be estimated by the product ratio (PR) method. The PR is the product of the two nonparental phenotypic classes divided by the product of the two parental classes. This PR relates to recombination fractions, which are listed in Tables *(9–10)*. These tables also give values from which the standard deviations can be estimated taking into account the total number of F_2 plants. The PR method can only be used for the "classical" F_2 (dominance, all phenotypes equally viable, resulting in a 9:3:3:1 ratio in case of independent inheritance). In more complex situations, such as F_2's with dominance or recombinant inbred lines, the recombination rate is usually estimated by maximum likelihood. Allard *(11)* and Ritter et al. *(12)* compiled a large number of ML estimation equations for many different situations.

Computer programs such as LINKAGE *(13)*, RECF2 *(1)* and also the mapping programs can be used to estimate recombination frequencies with standard deviations.

17. The analysis of the selfed (F_2) progeny of a diheterozygote, in which both markers are fully recessive, is much more efficient when the genes are linked in coupling phase as compared to linkage in repulsion phase.

The problem with F_2 repulsion data, especially in the case of close linkage, is the statistical inaccuracy of the estimate of recombination frequencies. The analysis of F_3 lines derived by selfing from specific F_2 plants allows an accurate estimate of the recombinant fraction (p) in this case. Progeny testing is more efficient than increasing the size of the F_2 population at $p < 0.11$. Since the aim of this progeny test is to determine whether an F_2 plant was either homozygous or heterozygous for the dominant allele, 16 plants per line are sufficient to be sure about the classification with 99% certainty.

Progeny tests are more efficient when the F_3 line can be screened for an early and easy-to-score marker. For example, when one is interested in linkage between a seed coat gene and a trichome gene, the seed coat recessive plants are harvested in the F_2 and their progeny is screened for the trichome mutation at the seedling stage. Recombinant fractions (p) can be estimated by combining the F_3 data with F_2 data using the maximum likelihood procedure or by the following formulas:

If: $x = (aaBb)/n$ = number of segregating lines/number of aa lines tested

$$p = x/(2-x) \text{ with } S_p = 2[x(1-x)]^{1/2}/[(2-x)^2 (n^{1/2})] \tag{1}$$

With low recombination values the number of segregating F_3 lines is almost the same as the number of recombinants found among the number of gametes that are tested, which is twice the number of progenies tested (each F_2 plant comes from two gametes).

18. Recombination data can be converted into map distances using the Haldane *(14)* or Kosambi *(15)* mapping functions.

The Haldane mapping function reads:

$$\begin{aligned} p &= 1/2[(1-e^{-2x})] \\ x &= -1/2[\ln(1-2p)] \end{aligned} \tag{2}$$

with x in Morgans; or expressed with centiMorgans *(D)* and recombination percentages *(r)*:

$$\begin{aligned} r &= 50(1 - e^{(-D/50)}) \\ D &= -50 \ln[(50-r/50)] \end{aligned} \tag{3}$$

A commonly used mapping function that accounts for a certain degree of interference is the Kosambi function *(14)*.

$$\begin{aligned} p &= 1/2[\tanh(2x)] \\ x &= 1/4 \ln[(1+2p)/(1-2p)] \end{aligned} \tag{4}$$

or

$$\begin{aligned} r &= 50 \tanh(D/50) \\ D &= 25 \ln[(100+2r)/(100-2r)] \end{aligned} \tag{5}$$

Transforming estimates of recombination percentages into map distances requires transforming of standard errors as well. Writing S_r for the standard error of an estimate of recombinant percentage, the corresponding standard error of the map distance approximately equals:

$$S_D = [50/(50-r)] \, S_r \text{ (Haldane)} \qquad (6)$$

$$S_D = [50/(2500-r^2)] \, S_r \text{ (Kosambi)} \qquad (7)$$

The best mapping function is that one giving the best additivity of the calculated map distances.

19. The mapping programs differ in the type of data they can handle. Both MAPMAKER and GMENDEL can only use crude data, i.e., phenotypic scores for the markers to be mapped. For these programs, the phenotypes must have been scored in a single mapping population, e.g., a backcross, an F_2 or a set of recombinant inbred lines. JOINMAP uses crude data to produce a list of pairwise recombination estimates. This list is used to construct the linkage map. Various pairwise recombination lists, obtained from different mapping populations (not necessary of the same type, e.g., from literature) can be merged into a single input list. So, with JOINMAP linkage maps obtained from various data sets can be combined to calculate an integrated map *(3)*.

    The various software packages for map construction differ furthermore in the way that they are being used and the methods of finding the "best fitting" gene order. MAPMAKER can be used in an interactive way, i.e., the user can specify any gene order for which a map and the corresponding joint likelihood are calculated. This allows "manual" insertion by trial-and-error of additional genes to an established map. The map construction part of JOINMAP is completely noninteractive. However, JOINMAP consists of a number of distinct modules, each one performing a specific task (calculating pairwise estimates and LOD scores, assignment of genes to linkage groups, testing for segregation distortion, testing for heterogeneity among estimates, ordering the loci, etc.), allowing flexibility and data checking. Furthermore, JOINMAP in contrast to MAPMAKER, can deal with all types of mapping populations that are being used in plants.

20. The MAPMAKER program can be obtained from MAPMAKER (c/o Dr. Eric Lander, Whitehead Institute for Biomedical Research, 9 Cambridge Center, Cambridge, MA 02142; internet: mapmaker@genome.wi.mit.edu or through WWW by anonymous FTP from host genome.edu.wi.mit.edu).

    The JOINMAP and Drawmap programs can be requested from Dr. J. W. van Ooijen (CPRO-DLO, P.O. Box 16, 6700 AA Wageningen. The Netherlands; E-mail: mapping@cpro.dlo.nl).

    The updated classical and RFLP maps, as well as information concerning all kinds of markers, can be found through the *Arabidopsis thaliana* Database (*At*DB) by WWW browsing to the URL: http://genome-www.stanford.edu or via Gopher at URL: gopher://genome-gopher.stanford.edu.

Acknowledgment

Carlos Alonso-Blanco was supported by a fellowship from the FICYT, Principado de Asturias, Spain.

References

1. Koornneef, M. and Stam, P. (1992) Genetic analysis, in *Methods in* Arabidopsis *Research* (Koncz, C., Chua, N. H., and Schell, J., eds.) World Scientific, Singapore, pp. 81–99.
2. Lander, E. S., Green, P., Abrahamson, J., Barlow, A., Daly, M. J., Lincoln, S. E., and Newberg, L. (1987) MAPMAKER: an interactive computer package for constructing primary genetic linkage maps of experimental and natural populations. *Genomics* **1**, 174–181.
3. Stam, P. (1993) Construction of integrated linkage maps by means of a new computer package: JOINMAP. *Plant J.* **3**, 739–744.
4. Liu, B. H. and Knapp, S. J. (1990) GMENDEL: a program for Mendelian segregation and linkage analysis of individual or multiple progeny populations using log-likelihood ratios. *J. Heredity* **81**, 407.
5. van Ooijen J. W. (1994) Drawmap: a computer program for drawing genetic linkage maps. *J. Heredity* **85**, 66.
6. Dellaert, L. M. W. (1980) Segregation frequencies of radiation-induced viable mutants in *Arabidopsis thaliana* (L.) Heynh. *Theor. Appl. Genet.* **57**, 137–143.
7. Li, L. and Rédei, G. P. (1969) Direct evidence for models of heterosis provided by mutants of *Arabidopsis* blocked in the thiamine pathway. *Theor. Appl. Genet.* **39**, 68–72.
8. Vizir, I. Y. and Korol, A. B. (1990) Sex differences in recombination frequency in *Arabidopsis*. *Heredity* **65**, 379–383.
9. Stevens, W. L. (1939) Tables of the recombination fraction estimated from the product ratio. *J. Genet.* **39**, 171–180.
10. Rédei, G. P. (1982) *Genetics*. Macmillan, New York.
11. Allard, R. W. (1956) Formulas and tables to facilitate the calculation of recombination values in heredity. *Hilgardia* **24**, 235–278.
12. Ritter, E., Gebhardt, C., and Salamini, F. (1990) Estimation of recombination frequencies and construction of RFLP linkage maps in plants from crosses between heterozygous parents. *Genetics* **125**, 645–654.
13. Suiter, K. A., Wendel, J. F., and Case, J. S. (1983) LINKAGE-1: a PASCAL computer program for detection and analysis of genetic linkage. *J. Heredity* **74**, 203,204.
14. Haldane, J. B. S. (1919) The combination of linkage values, and the calculation of distance between linked factors. *J. Genet.* **8**, 299–309.
15. Kosambi, D. D. (1944) The estimation of map distances from recombination values. *Ann. Eugen.* **12**, 172–175.

16

Cytogenetic Analysis of *Arabidopsis*

John S. Heslop-Harrison

1. Introduction

The cytogenetical analysis of a species enables the correlation of the structure, number and morphology of its chromosomes with heredity and variation. Most accessions of *Arabidopsis thaliana* L. have a diploid (2n) chromosome number of 10 (**Fig. 1**). The small genome size, convenient for many aspects of molecular genetics, makes the chromosomes average only 1.5 µm in length *(1,2)* and intractable to analysis using many of the cytogenetic methods applied to other genera. However, despite the small size of chromosomes, the rapid advances in chromosome preparation methods, fluorescent microscopy, and staining since 1990 mean that analysis is now straightforward (*see* **refs.** *3–5*).

Among the most important cytogenetic analysis is the determination of plant ploidy and examination for aneuploidy. Polyploid, particularly tetraploid (2n = 4x = 20), *A. thaliana* accessions are common and may be as fertile as other material under some conditions. Polyploids may occur naturally (e.g., the ecotype Stockholm), or be induced or arise spontaneously during tissue culture and regeneration *(6)*. Variation in ploidy may be significant for the success of molecular procedures. For example, regeneration from diploid cells may be more reliable than from tetraploid, where as scoring of somatic recombination events or DNA sequence excision may depend heavily on the ploidies found in particular tissues. Many methods enable measurement of ploidy; flow cytometry following fluorescent staining of nuclei, is extremely rapid and accurate *(7-9)*, but beyond the scope of this chapter. Quantitative measurement of staining intensity enables the ploidies of individual cells to be determined within a tissue. Here, a quick method for determining the basic ploidy of a whole plant is presented, using examination of the number of apertures of pollen (colpi). Methods for chromosome preparation and counting, valuable for determination of ploidy are also given in detail.

From: *Methods in Molecular Biology, Vol. 82: Arabidopsis Protocols*
Edited by: J. Martinez-Zapater and J. Salinas © Humana Press Inc., Totowa, NJ

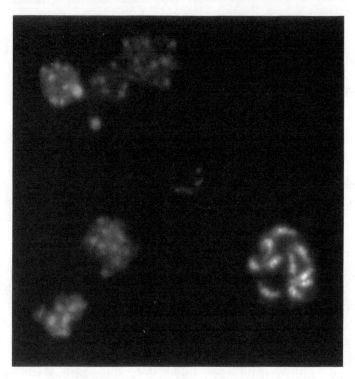

Fig. 1. A prometaphase and several interphase nuclei from a bud preparation of *Arabidopsis thaliana* stained with DAPI. The 10 prophase chromosomes are visible, with brightly staining centromeric regions, and they are extended enough that their morphology can be distinguished. Interphase nuclei show brightly staining chromocentres, corresponding to the bright centromeres at metaphase and prometaphase.

Accurate chromosome numbers are often required for genetical analysis of *A. thaliana*. As with many other species, trisomic lines, with an additional (third), copy of one chromosome, are important for genetic analysis, and both plant morphology and chromosome counting is valuable to check individual plants. Many aneuploid lines arise in the progeny of hybrids between diploid and tetraploid lines, and knowledge of chromosome number is valuable to understand the behavior in these progeny. Aneuploidy may arise with high or low frequency, following almost any other treatment of plants, and it is often essential to prove that differences seen are not simply due to different dosages of individual chromosomes. Finally, with interest increasing in plant evolution and biodiversity, there is a need to determine chromosome numbers in relatives of *A. thaliana* and hybrids. Because of the high level of plasticity of phenotype

Cytogenetic Analysis

in such ephemeral species, and the variability of chromosome number with the Crucifereae, chromosome counting is important in such studies. The methods for chromosome preparation presented here will enable simple preparation of large numbers of metaphase chromosome spreads to enable accurate counting of chromosome numbers.

Although methods for molecular cytogenetic analysis of *A. thaliana* are outside the scope of this chapter, fluorescently stained chromosomes can be used subsequently for DNA:DNA *in situ* hybridization, giving detailed information about the distribution of defined DNA sequences along chromosomes. Further details of in situ hybridization methods and results will be found in other references *(10–13)*. Meiotic analysis using similar staining methods to those presented here is straightforward once anthers at meiosis have been found (techniques to find meiotic material are presented in ref. *14*.) Analysis of synaptonemal complexes using light and electron microscopy has given valuable results of relative chromosome sizes *(15,16)*, and would easily detect interchanges and similar chromosomal rearrangements, as well as assisting with characterization of meiotic mutants.

The present chapter presents optimized protocols for making chromosome preparations from *Arabidopsis*. Many alternative methods for chromosome preparation are possible, in particular acid hydrolysis followed by staining with chromatic chemicals such as Feulgen's reagent. However, our experience is that aspects of these methods require greater optimization in a particular laboratory, and the final preparation shows less contrast and is much more difficult to interpret than a fluorescently stained preparation. Most biology departments now have fluorescent microscopes available, and the total cost of the reagents required is no more than a single tube of many restriction enzymes. Furthermore, the preparations are immediately available for *in situ* hybridization or other immunocytochemical techniques, if required.

2. Materials
2.1. Analysis of Ploidy Using Pollen Morphology

1. Plant material: mature anthers from unopened or recently opened flowers.
2. Compound microscope with through light illumination.
3. 15% Sucrose solution in water.
4. Dissecting instruments.
5. Microscope slides and cover slips.

2.2. Sources of Metaphases and Materials for Fixation

1. Plant material: The optimal material for analysis of *A. thaliana* chromosomes is young buds (*see* **Note 1**; *13*). Roots from seedlings may also be used (*see* **Note 2**). For nonflowering plants, very young leaves or roots from plant grown in moist

soil or hydroponics can be used, although preparations are normally poor and have much cytoplasm.
2. Metaphase arresting agent: 2 mM 8-hydroxyquinoline is a good solution to use for species with small chromosomes (*see* **Note 3**). Colchicine (0.05%) can be used, but the chromosomes become very condensed (looking like dots on the slide), so little morphological information is available.
3. Fixative: freshly prepared, 3 parts 100% (or 96%) ethanol or methanol to 1 part glacial acetic acid. Cheap chemical grades can be used; 50 mL is kept at –20°C for chromosome preparations.

2.3. Chromosome Preparation

1. Enzyme buffer (pH 4.6): Make a 10X stock solution with 4 parts 0.1M citric acid and 6 parts 0.1M tri-sodium-citrate. Dilute to 1X enzyme buffer with distilled water for use.
2. Enzyme solution (1X): 1% cellulose (from *Aspergillus niger*, Calbiochem, La Jolla, CA, 21947, 0.1–0.25% can be substituted by Onozuka RS) and 10% pectinase (from *Aspergillus niger*, solution in KCl and Sorbitol, Sigma, St. Louis, MO, P9179) in 1X enzyme buffer (*see* **Note 4**).
3. Cover slips (preferable No. 1 thickness, 18 mm square) and clean microscope slides: Place slides into chromium trioxide solution in 80% (w/v) sulphuric acid (Merck, Darmstadt, Germany) for at least 3 h at room temperature. Wash slides in running water for 5 min, rinse them thoroughly in distilled water and air-dry. Place slides in 100% ethanol, remove, and dry with a tissue immediately prior to use.
4. 45% and 60% Aqueous acetic acid: Solutions can be stored on the bench indefinitely.
5. Fine forceps (No. 5) and dissecting needles.
6. Stereo microscope.
7. Dry ice (block, or metal plate placed on pieces), liquid nitrogen, or metal plate placed in a –70°C freezer.

2.4. Chromosome Staining and Examination

1. Stain buffers: Many buffers are suitable including McIlvaine's solution (18mL 0.1M citric acid, 82 mL 0.2M Na$_2$HPO$_4$, pH 7.0) or phosphate buffered saline, (PBS, 0.13M NaCl, 0.007M Na$_2$HPO$_4$, 0.003M NaH$_2$PO$_4$, adjusted to pH 7.4). SSC (salt-sodium citrate; 20X solution: 3M NaCl, 0.3M NaCitrate) is described here, as it is available in many laboratories.
2. Stains for chromosomes:
   a. DAPI (4',6-diamidino-2-phenylindole): Prepare a stock solution of 100 µg/mL in water (*see* **Note 5**).
   b. PI (propidium iodide): Prepare a stock solution of 100 µg/mL in water (*see* **Note 6**).
3. Antifade mountant: after staining, slides can be mounted in glycerol (a pure grade that shows no autofluorescence; *see* **Note 7**).
4. Glass cover slips: No. 0 (i.e., thinner than the standard), 24 × 24 mm or longer.
5. Fluorescent photomicroscope with filter blocks for UV (DAPI staining) and green (PI) excitation, fluorescence objective lenses; ideally 40X and 100X oil immersion.

3. Methods
3.1. Pollen Examination of Ploidy Determination

1. Dissect anther from mature flower (under a stereo dissecting microscope) and place in a drop of 15% sucrose (which stops pollen grains bursting) on a microscope slide. Squash cover slip gently to extrude pollen grains from the anthers.
2. Examine under a compound microscope. Sufficient contrast will normally be obtained by lowering the condenser. Phase-contrast illumination will give improved resolution, and fixative stains such as carmine in 45% acetic acid (acetocarmine) can also be used.
3. Count the number of germination apertures (colpi) on each pollen grain. Haploid pollen (n) derived from a diploid (2n) plant has three colpi (most usual), where as diploid pollen (from a tetraploid plant) has four colpi and tetraploid pollen (from an octoploid plant) has six colpi.

3.2. Accumulation and Fixation of Nuclei and Metaphases for Chromosome Preparations

1. Young buds: Excise for pretreatment. The whole tip of the floral stem may be used if ample material is available. Otherwise, very young, unopened buds, in which the petals have not yet grown longer than the pistil are dissected from the plants. Seedling root tips: Germinate seeds on 1% agar plates, optionally made using Murashige and Skoog mixture. Use when roots are 3–6 mm long.
2. Transfer to 8-hydroxyquinoline solution warmed to the same temperature as the plants. (Alternatively, place in 0.05% colchicine at the plant temperature for 3 h before fixation.)
3. Leave for 30–90 min at the temperature of the plants, then transfer to 4°C for a further 30–90 min.
4. Drain material and plunge into fixative for 2 h at room temperature. If fixed material is to be kept (up to 3 months for in situ hybridization, indefinitely for chromosome preparation), leave for 2 h at room temperature and then transfer to –20°C. Labeling is best done by placing small pieces of paper with pencil writing in the fixative.

3.3. Chromosome Preparation

Preparations should aim to have many well spread metaphases with clearly separated chromosomes and the minimum amount of cytoplasm. Glass embryo dishes or 5-cm Petri dishes are convenient. Material is transferred by forceps or, if small roots or buds are used, with a pipet.

1. Wash fixed material 2X 10 min in enzyme buffer to remove fixative.
2. Transfer to 2X enzyme solution (0.3–1 mL), incubate at 37°C for 60–90 min (buds) or 30–60 min (root tips).

3. Transfer to enzyme buffer, leave for 15 min to 3 h (*see* **Note 9**).
4. Place enough material for 3–5 preparations (typically one floral stem), in 45% acetic acid for a few minutes.
5. Under a stereo dissecting microscope, dissect out the pistil from a young bud, in which the pistil is visible emerging from the undeveloped petals and sepals (approx 3 mm long). Place pistil in a small drop of 45% acetic acid on a clean slide (*see* **Note 10**).
6. Squashing: Tease the material to fragments with a fine needle. Remove any clumps of cells that are visible (most metaphase cells will fall out from clumps). Apply a cover slip and tap slip with the needle, and then gently squash the material. Place slide onto dry ice, into –70°C freezer or dip into liquid nitrogen. When frozen, flick off cover slip using a single-edged razor blade. Drop-spreading: Tease material apart in the drop of acetic acid on the slide. Drop 2–3 drops of 3:1 fixative at –20°C onto the preparation.
7. Allow slide to air-dry.
8. Slides can be screened dry under phase contrast microscopy. However, screening is not efficient as the small metaphase chromosomes are very hard to see, although nuclei and excessive cytoplasm are visible (*see* **Note 11**).
9. Slides can be stored desiccated in the fridge or freezer for several weeks.

3.4. Chromosome Staining and Photomicroscopy

1. Add 50 µL of DAPI or PI solution per slide and place a cover slip over the preparation (*see* **Note 12**).
2. The preparation can now be examined in the epifluorescent microscope with the relevant filter block (*see* **Notes 11** and **13**).
3. Longer lasting preparations: Make by washing the slide briefly in 50 mL of 2X SSC and allowing the cover slip to fall off.
4. Drain slide and add a drop of glycerol over the preparation. Apply a glass cover slip and squeeze firmly between sheets of filter paper to remove excess mountant. The slide may then be stored at 4°C for 1–2 y (*see* **Note 14**).
5. Photography: In a microscope with a 50 W fluorescence bulb, exposure times will be typically 0.5–1 s with DAPI and 1–2 s with PI on 400 ASA film (*see* **Note 15**).

4. Notes

1. Young buds have many divisions, collection does not harm the plant, and material from one plant is available at all times for 2 wk or more.
2. Seedling roots generally have very few divisions (10 or less each) and are less convenient; the roots are too small to excise conveniently, so the seedling is usually killed.
3. The powder may take 20 h to dissolve and gives a pale yellow solution. The solution may be stored in a dark refrigerator for many months.
4. Enzyme solution can be stored in 0.3–1 mL aliquots in microcentrifuge tubes at –20°C. The enzyme solution can be reused several times.
5. DAPI is a potential carcinogen, so order small quantities and use the whole vial to make the stock solution. Aliquot and store indefinitely at –20°C. Prepare a working solution of 2 µg/mL by dilution in 2X SSC, aliquot, and store at –20°C.

6. PI is a potential carcinogen, so order small quantities and use the whole vial to make the stock solution. Aliquot and store indefinitely at $-20°C$. Dilute with 2X SSC to 2–5 µg/mL prior to use. PI is not stable in diluted form.
7. More complex antifade solutions are unnecessary for DNA stains, but are available commercially.
8. We prefer color print films. These are easy to use, very cheap to buy and print commercially, and they have high exposure latitude. Adjustment of contrast and color is easy during printing, and reproduction as plates or slides is excellent. We use Fujicolor 400 film. Other suitable films include Kodak TMAX 400 for black-and-white photography, and Agfa or Fuji slide films. Avoid films labeled 'professional,' which translates to mean low contrast! It can be very difficult to obtain uniform dense blacks in small dark rooms. Automated color equipment, found in high street photography shops, can easily obtain these, although it is important to talk to the operator. Many machines place codes of the back of the prints; a well exposed, landscape-type scene will normally include the letters "NNNN," in which the first three refer to color balance and the last print exposure. For a black background fluorescence microscopy, the last character will normally be a number between four and nine, and require manual intervention by the operator.
9. With root-tip material, the enzyme digestion time is critical. Ten minutes under-digestion will give poor spreads with too much cytoplasm, where as over-digestion gives a "soup" of chromosomes. Floral material seems much less affected by digestion time.
10. When squashing material, the size of the liquid droplet is quite important; a little but not large volume of liquid should be squashed out from the edge of the cover slip. If there is too much, material will be lost; if too little, spreading will not be efficient.
11. If there is too much cytoplasm surrounding nuclei and chromosomes (seen either by phase-contrast microscopy, or by weakness of DAPI staining), either increase the time the material stands in 45% acetic acid, either before or after dissection, or use 60% acetic acid in the drop on the slide, and leave the teased material for several minutes.
12. The fluorochrome stains are very specific for nucleic acids. However, under some circumstances, weak staining of DAPI may be observed. *Reducing* the concentration may improve contrast since DAPI has two modes of staining DNA, base pair intercalation and minor groove binding. The latter occurs at higher concentrations and gives less fluorescence.
13. Although an epifluorescent microscope has suitable filters for DAPI, many lenses and immersion oils are unsuitable since they do not transmit UV light. If no image is seen with DAPI, but PI is satisfactory, check the lenses and oil used.
14. Fluorescent stains are not completely stable, and will fade considerably after 20 s exposure to UV light without glycerol, or 1 min with glycerol. Photograph good chromosome spreads quickly, and turn the illumination off when leaving the microscope. Most objective lenses are also damaged by extended (1 yr) UV exposure.

15. Exposure indications are very inaccurate on the microscope, since exposure meters cannot cope with monochromatic fluorescence and black backgrounds. If necessary, take several exposures and choose the best negative. Typical exposure times are 1 s to 2 min.

Acknowledgments

The author is grateful to Gill Harrison, Jola Maluszynska, Trude Schwarzacher, Minoru Murata and F. Motoyoshi for enormous help with development of molecular cytogenetic methods with *Arabidopsis* over the last 10 years.

References

1. Ambros, P. and Schweizer, D. (1976) The Giemsa C-banded karyotype of *Arabidopsis thaliana.* Arabidopsis *Info. Serv.* **13,** 167–171.
2. Steinitz-Sears, L. M. (1963) Chromosome studies in *Arabidopsis thaliana. Genetics* **48,** 483–490.
3. Maluszynska, J. and Heslop-Harrison, J. S. (1991) Localization of tandemly repeated DNA sequences in *Arabidopsis thaliana. Plant J.* **1,** 159–166.
4. Maluszynska, J. and Heslop-Harrison, J. S. (1993) Molecular cytogenetics of the genus *Arabidopsis: in situ* localization of rDNA sites, chromosome numbers and diversity in centromeric heterochromatin. *Ann. Bot.* **71,** 479–484.
5. Heslop-Harrison, J. S. and Maluszynska, J. (1994) The molecular cytogenetics of *Arabidopsis*, in *Arabidopsis* (Meyerowitz, E. M. and Sommerville, C. R., eds.) Cold Spring Harbor Laboratory Press, Cold Spring Harbor, New York, pp. 63–87.
6. Maluszynska, J., Maluszynski, M., Rebes, G., and Wietrzyk, E. (1990) Induced polyploids of *Arabidopsis thaliana.* Abstracts of the Fourth International Conference on *Arabidopsis* Research, Vienna, University of Vienna, pp. 4.
7. Brown, S. C., Bergounioux, C., Tallet, S., and Marie, D. (1991) Flow cytometry of nuclei for ploidy and cell cycle analysis, in: *A Laboratory Guide for Cellular and Molecular Plant Biology* (Negrutiu, I., and Gharti-Chhetri, G., eds.), Birkhauser, Basel, pp. 1–11.
8. Galbraith, D. W., Harkins, K. R., and Knapp, S. (1991) Systemic endopolyploidy in *Arabidopsis thaliana. Plant Physiol.* **96,** 985–989.
9. Heslop-Harrison, J. S. and Schwarzacher, T. 1996. Flow cytometry and chromosome sorting, in: Plant Chromosomes: Laboratory Methods, (Fukui, K. and Nakayama, S., eds.), CRC, Boca Raton, FL, pp. 87-108.
10. Schwarzacher, T. and Heslop-Harrison, J. S. (1994) Direct fluorochrome labelled DNA probes for direct fluorescent *in situ* hybridization to chromosomes, in *Protocols for Nucleic Acid Analysis by Nonradioactive Probes. Methods in Molecular Biology.* vol. 28 (Isaac, P.G., ed.), Humana, Totowa, NJ, pp. 167–176.
11. Leitch, I. J. and Heslop-Harrison, J. S. (1994) Detection of digoxigenin-labeled DNA probes hybridized to plant chromosomes *in situ*, in *Protocols for Nucleic Acid Analysis by Nonradioactive Probes. Methods in Molecular Biology.* vol. 28 (Isaac, P.G., ed.), Humana, Totowa, NJ, pp. 177–186.

12. Schwarzacher, T. and Leitch, A. R. (1994) Enzymatic treatment of plant material to spread chromosomes for *in situ* hybridization, in *Protocols for Nucleic Acid Analysis by Nonradioactive Probes. Methods in Molecular Biology.* vol 28 (Isaac, P.G., ed.), Humana, Totowa, NJ, pp. 153–160.
13. Murata, M. and Motoyoshi, F. (1995) Floral chromosomes of *Arabidopsis thaliana* for detecting low-copy DNA sequences by fluorescence *in situ* hybridization. *Chromosoma* **104,** 39–43.
14. Vieira, M. L. C., Briarty, L. G., and Mulligan, B. J. (1990) A method for analysis of meiosis in anthers of *Arabidopsis thaliana. Ann. Bot.* **66,** 717–719.
15. Albini, S. M., Jones, G. H., and Wallace, B. M. N. (1984) A method for preparing two-dimensional surface-spreads of synaptonemal complexes from plant meiocytes for light and electron microscopy. *Exp. Cell Res.* **152,** 280–285.
16. Albini, S. M. (1994) A karyotype of the *Arabidopsis thaliana* genome derived from synaptonemal complex analysis at prophase I of meiosis. *Plant J.* **5,** 665–672.

17

PCR-Based Identification of T-DNA Insertion Mutants

Rodney G. Winkler and Kenneth A. Feldmann

1. Introduction

With large-scale sequencing projects *(1,2)*, thousands of unique sequences (ESTs; expressed sequence tags) have been generated. In addition, hundreds of sequences have been identified based on their homology with a gene of known interest or from subtractive hybridization procedures. For the vast majority of these genes, the sequence indicates little or nothing about its function in the plant. In this protocol, we describe a reverse genetics procedure which will be useful for identifying an insertion mutant in *Arabidopsis* for any sequence of interest.

Tens of thousands of T-DNA generated transformants have now been produced in *Arabidopsis (3-6)*. These transformants contain insertions which are randomly distributed in the gene and genome *(7)*.

We have used T-DNA border-specific primers in combination with sequence specific primers to identify insertion mutants based on the generation of a PCR product among 6000 transformed lines (*8*; unpublished results). The identification and isolation of these insertion mutants is being made even more rapid with a three-dimensional pooling strategy (row × column × integer) that we are now employing. As this pooling strategy is important for expedient mutant identification, it will be explained in detail.

For a set of 1000 transformed lines, seeds ($n = 200$/line) from the first 10 lines (T1-T10), the next ten lines (T11-T20), and so on, are mixed together into individual subpools (SP10, SP20, and so on), so that 100 subpools are created for the 1000 lines (**Fig. 1**). Next, seeds ($n = 200$/subpool) from the first 10 subpools (SP 10-SP100), the next 10 (SPl 10-SP200), and so on are pooled together creating 10 pools of seeds in <u>rows</u> for the 1000 lines (**Fig. 2**). These

T1	T2	T3	T4	T5	T6	T7	T8	T9	T10	**SP10**
T101	T102	T103	T104	T105	T106	T107	T108	T109	T110	**SP110**
T201	T202	T203	T204	T205	T206	T207	T208	T209	T210	**SP210**
...	...	...	...	...	...	...	...	...	...	...
T901	T902	T903	T904	T905	T906	T907	T908	T909	T910	**SP910**

Fig. 1. Strategy for pooling seeds of individual lines into subpools containing 10 segregating lines. T represents a single transformed line numbered chronologically as it was generated. SP refers to a subpool of 10 specific transformed lines.

SP10	SP20	SP30	SP40	SP50	SP60	SP70	SP80	SP90	SP100	PR100
SP110	SP120	SP130	SP140	SP150	SP160	SP170	SP180	SP190	SP200	PR200
SP210	SP220	SP230	SP240	SP250	SP260	SP270	SP280	SP290	SP300	PR300
SP310	SP320	SP330	SP340	SP350	SP360	SP370	SP380	SP390	SP400	PR400
SP410	SP420	SP430	SP440	SP450	SP460	SP470	SP480	SP490	SP500	PR500
SP510	SP520	SP530	SP540	SP550	SP560	SP570	SP580	SP590	SP600	PR600
SP610	SP620	SP630	SP640	SP650	SP660	SP670	SP680	SP690	SP700	PR700
SP710	SP720	SP730	SP740	SP750	SP760	SP770	SP780	SP790	SP800	PR800
SP810	SP820	SP830	SP840	SP850	SP860	SP870	SP880	SP890	SP900	PR900
SP910	SP920	SP930	SP940	SP950	SP960	SP970	SP980	SP990	SP1000	PR1000
PC910	PC920	PC930	PC940	PC950	PC960	PC970	PC980	PC990	PC1000	

Fig. 2. Pooling strategy of rows and columns for PCR-based identification of insertion mutants. SP represents seeds from 10 lines, where as PR and PC represent DNA from 100 lines. The stipuled box (SP600) represents the subpool where the *act2-1* positive control will be observed.

pools are labeled PR100 (pooled row), PR200, and so on. In the next thousand lines, the pools are labeled PR1100, PR1200, and so on. For the second dimension (columns), seeds (n = 200/subpool) from SP10, SP110, SP210 through SP910 are pooled creating PC910 (pooled column) and so on. In the next 1000 lines, the pooled columns are labeled as PC1910, PC1920, and so on. Finally, for the third dimension (integer) seeds are pooled from the 100 single lines, in each population of 1000 lines, ending in 1, 2, 3, and so on to make 10 additional pools labeled PI991 (pooled integer), PI992, and so on (**Fig. 3**). In the next 1000 lines, they are labeled as PI1991, PI1992, and so on.

The pools of seeds were planted in soil in individual trays, allowed to grow for 3 wk (16:8 h light:dark cycle) and plants we e harvested and stored at −80°C until DNA was extracted. A modified cetyltrimethyl-ammonium bromide (CTAB) procedure was used to extract approx 3 mg of genomic DNA per 60 g of fresh weight tissue. To check the quality of the DNA, it was restricted with *Hin*dII and resolved on a 0.8% agarose gel. The DNA was diluted to approx 0.5 mg/mL and transferred to the *Arabidopsis* Biological Resource

PCR Identification of T-DNA Mutants

T1	T11	T21	T31	T41	T51	T61	T71	T81	T91
T101	T111	T121	T131	T141	T151	T161	T171	T181	T191
T201	T211	T221	T231	T241	T251	T261	T271	T281	T291
...	...	...	...	...	...	...	...	...	...
T901	T911	T921	T931	T941	T951	T961	T971	T981	T991

PB991

Fig. 3. Pooling strategy of blocks for PCR-based identification of insertion mutants. Ts represent single transformed lines. PI represents pooled integer of 100 lines, where each transformed line ends in a 1, 2, 3, and so on.

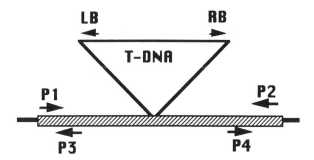

Fig. 4. Strategy for PCR-based identification of insertion mutants in T-DNA generated transformants. LB, primer homologous to LB of T-DNA (**Table 1**); RB, primer homologous to right border of T-D N A (**Table 1**); P1, forward primer homologous to 5' end of gene; P2, reverse primer homologous to 3' end of gene; P3, reverse primer homologous to 5' end of gene, and P4, forward primer homologous to 3' end of gene.

Center (ABRC) at Ohio State. The ABRC further dilutes the samples to 50 mg/ μL for distribution (*see* **Note 1**).

The pooling has been conducted to minimize the number of polymerase chain reactions, (PCRs), that need to be conducted, but yet to maximize sensitivity. When there is an insertion mutant in one of a thousand lines, 30 PCR reactions (10 rows, 10 columns, and 10 integers) will produce a positive (PCR product) in each of three dimensions and will pinpoint the insertion mutant for the sequence being tested (*see* **Notes 2** and **3**).

Figure 4 diagrams a number of possible primer combinations that can be used for this procedure (*see* **Notes 4** and **5**).

Performing PCR can be somewhat problematic for untested primers and substrates. To test various pooling strategies to optimize our own screening strategy and to allow comparative testing of laboratory-specific PCR conditions, we will describe a positive control that should be used before unique primers are employed. An insertion allele has been isolated for *act2* (*8*) and this mutant was included in the pooling strategy (*see* **Note 6**).

Table 1
Primers Used for Actin Screening

Primers	Sequence
ACT2Rev2007	CCATCGGGTAATTCATAGTTCTTCTCG
ACT2Rev2656	CGCAGACGTAAGTAAAAACCCAGAGA
RB 16843	GCTCATGATCAGATTGTCGTTTCCCGCCTT
LB 102	GATGCACTCGAAATCAGCCAATTTTAGAC

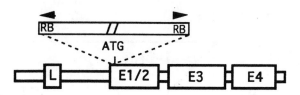

Fig. 5. Map of *act2-1*. E1/2, E3, and E4 indicate exons. ATG indicates translation start site. This insertion line contained a concatameric T-DNA insert with sequences homologous to the RB at each plant junction *(8)*.

2. Materials

1. DNA from pools of DNA (35–50 ng/µL dilutions, *see* NOTE 7).
2. Primers for left and right T-DNA borders and gene specific primers **(Table 1)**.
3. Thermocycler *(see* **Note 8**).
4. Mineral oil is necessary if a thermocycler without a heated lid is being used.
5. Taq enzyme.
6. 10X buffer: 200 mM Tris-HCl, pH 8.4, 500 mM KCl, 15 mM MgCl$_2$ (*see* **Note 9**).
7. 0.1M dNTPs.
8. Thin-walled PCR tubes.
9. Equipment and reagents for electrophoresis.

3. Methods

3.1. PCR Reactions

3.1.1. Use of act2-1 Mutant DNA in Control PCR Reactions

Sequence-specific primers have Seen designed to identify the *act2-1* mutant allele in the population **(Table 1)**. As *act2-1* possesses right borders on each side of the insert *(8)*, primers homologous to the right border are necessary **(Fig. 5** and **Table 1)**. A product should be identified in PR600 and PC1000 **(Fig. 2)**.

1. Program PCR machine *(see* **Note 8**).
2. Make 1X PCR reaction mixture with all components except DNA, make 30 µL per reaction, keep on ice *(see* **Notes 9** and **10)**.

3. Add 1 μL pooled DNA to PCR tubes on ice (*see* **Note 11**).
4. Transfer tubes to PCR block set at 4°C (*see* **Note 9**).
5. Add PCR reaction mixture to PCR tubes containing DNA.
6. Add mineral oil if necessary.
7. Start cycles.

3.2. Analysis of PCR Products

As these lines contain an average of 1.5 inserts each, and as the vast majority of the inserts are concatamers, there is a reasonably high probability of generating border–border products. As such, we recommend transferring the PCR products to a hybridization filter and probing it with DNA homologous to the plant-specific primers being tested. Adding this second step eliminates the possibility of false-positives. In addition, even if a product cannot be visualized on an EtBr-stained gel, it will still be identified with autoradiography.

1. Load 15 μL of the PCR reaction on an 2% agarose gel and electophorese at 100 V (*see* **Note 12**).
2. Transfer the products to a nylon membrane and fix to membrane.
3. Hybridize the filter with a gene-specific probe and evaluate (*see* **Note 13**).

3.3. Characterization of Positive Products

If DNA from all three dimensions is available, a single insertion mutant will be identified by the PCR reactions (*see* **Notes 14 and 15**). The positive family should be ordered and confirmed by PCR and Southerns. If only the rows and columns of DNA are available, seeds from the 10 lines contained in the positive subpool should be ordered, grown singly for DNA preparations and genotyped by PCR and Southerns as above to identify the exact family with the insertion of interest.

When a new insertion is being characterized, we select the positive reactions by hybridization, gel-purify the positive bands, reamplify these, and directly sequence the PCR products (*see* **Notes 16 and 17**).

Lastly, once the family has been identified and the site of insertion characterized by sequence, the mutant phenotype can be analyzed in detail.

4. Notes

1. The ABRC should be contacted to ascertain the number of pools that are available. The DNA from rows and columns for the first 6000 lines has been available since June 1996. The DNA from the pooled integers for the first 6000 will be available later. If necessary, the ABRC will make available seeds from the 10 lines in a positive subpool for screening. These transformants were generated in the ecotype Wassilewskija with *Agrobacterium tumefaciens* containing the 3850:1003 Ti plasmid *(8)*. As thousands of additional transformed lines are being made available to the ABRC, the number of pools available for screening will increase rapidly.

2. The pooled DNA can be mixed into much larger pools if numerous gene-specific primers are being tested. Gridded arrays can be developed with these pooled DNAs, which will allow the investigator to optimize the screening and sib-selection process (i.e., what is the most efficient strategy to go from a large pool of DNA to a single mutant?).
3. An alternative to ordering DNA from the stock centers is to order seeds in pools of 100 (CS3115 and CS6500) that are available and isolate DNA from these pools. When a positive PCR product is observed among a pool of 100, subpools of 20 representing the positive pool can be ordered. Upon identification of the subpool containing the insertion mutant, individual plants will need to be screened from that subpool to identify a plant, either homozygous or heterozygous, for the insert.
4. The primers that will be used will depend on what is known about the gene(s) of interest. For example, when screening for mutants from EST sequences, it will only be possible to use 3' or 5' primers depending on which end of the cDNA was sequenced. Alternatively, the EST clone of interest can be ordered from the stock center and sequenced. Forward and reverse primers can then be generated to screen the complete gene sequence.
5. From our experience, left border-plant DNA junctions are more prevalent in these T-DNA lines than right border-plant DNA junctions. However, left borders an be truncated. Still, there is a higher probability of finding a mutant using a left border primer (**Table 1**).
6. Additional positive controls may be helpful. Forward and reverse primers should be made from a gene of interest and these should be tested on the pools of DNA. Further, as many genes have now been cloned from T-DNA-tagged mutants from this population, we would recommend obtaining seeds from one or more of these mutants, isolating DNA, and mixing (spiking) this DNA into one of the pools. We describe the use of the *act2-1* mutant allele in this paper.
7. The address of the *Arabidopsis* Biological Resource Center at Ohio State is: ABRC, The Ohio State University, 1735 Neil Avenue, 309 Botany and Zoology Bldg., Columbus, OH 43210; email for DNA orders: dna@genesys.cps.msu.edu.
8. We used an Ericomp thermocycler with a heated lid. We keep the PCR block at 4°C while the reagents are being added. For actin controls we performed an initial denaturation at 95°C for 1 min and then 30 cycles of 94°C, 30 s; 60°C, 30 s; 72°C, 2 min, and then a 10 min 72°C extension cycle.
9. 1X PCR buffer contains 20 mM Tris-HCl, pH 8.4, 50 mM KCl, 1.5 mM MgCl$_2$, 0.1 mM dNTPs. We typically use the 10X buffer supplied with the *Taq* polymerase. We used 30 µL reactions with 15 pmol of each primer per tube. It is sometimes necessary to vary the Mg^{2+} concentration to achieve optimal amplification.
10. Two sets of negative controls were included in the experiment. As control for false priming, all primers were used singly with PR600 DNA in PCR reactions. Additionally, 50 ng of PR100 or PR200 DNA were included in a separate set of control PCR reactions; these are not expected to give a positive PCR product for the *act2-1* insertion.
11. PR600 DNA, which contains the *act2-1* mutant, was used in a dilution series containing either 50, 2.5, 0.4, or 0.1 ng PR600 DNA per PCR reaction.

12. We use a custom gel tray from BRL that is 30 cm long and has slots for four combs. We use 21 or 42 teeth combs; these can be loaded directly with a multichannel pipeter.
13. PCR amplification of PR600 DNA with both the act2Rev2007/RB 16843 and the act2Rev2656/RB16843 primer pairs gave a positive product by Southern hybridization with an *ACT 2* probe. Although less product was obtained when the target DNA was more dilute (0.1 ng PR600), the detection limit of the method was not yet reached as the positive bands were all visible after a 5 min exposure. (We routinely use chemiluminescent detection of digoxigenin-labeled probes following essentially the protocols of Boehringer-Mannheim, Mannheim, Germany). No products were seen with the control reactions.
14. To date, a number T-DNA insertion mutants have been identified by this reverse genetics approach *(9)*. The methods we describe should allow other laboratories q to quickly determine if their PCR conditions are sensitive enough for their screening strategy.
15. We expect that the optimal pooling strategy will vary depending on number of targets that the primers being used will likely amplify. The system is currently set up so in 180 PCR reactions the 6000 lines can be screened by a three dimensional strategy. In the isolation of the *act2-1* mutant *(8)*, degenerate primers were used and pools of 100 families were screened in each PCR reaction. Our reconstruction experiment with the *act2-1* mutant suggest that for many cases pools of DNA from 1000 families (10 pooled rows, columns, and integers) or more could be screened effectively with gene-specific primers.
16. To eliminate secondary bands, we run a minigel of the positive product, cut out the positive band (or "poke" the positive band with a pipet tip to get some of the product), melt in 1 mL TE, and dilute 1 to 1000 and use this "purified" band for reamplification.
17. We purify three, 90 µL reactions using Promega's Wizard (Madison, WI) system prior to directly sequencing the PCR product. Alternatively, the product can be cloned and sequenced.

References

1. Newman, T., de Bruijn, F.J., Green, P., Keegstra, K., Kende, H., McIntosh, L., Ohlrogge, J., Raikhel, M., Somerville, S., Thomashow, M., Retzel, E., and Somerville, C. (1994) Genes galore: a summary of methods for accessing results from large-scale partial sequencing of anonymous *Arabidopsis* cDNA clones. *Plant Physiol.* **106**, 1241–1255.
2. Cooke, R., Raynal, M., Laudie, M., Grellet, F., Delseny, M., Morris, P-C., Guerrier, D., Giraudat, J., Quigley, F., Clabault, G., Li, Y-F., Mache, R., Krivitzky, M., Gy, I., Kreis, M., Lecharny, A., Parmentier, Y., Marbach, J., Fleck, J., Clemnet, B., Philipps, G., Herve, C., Bardet, C., Tremousaygue, D., Lescure, B., Lacomme, C., Roby, D., Jourjon, M-F., Chabrier, P., Charpenteau, J-L., Desprez, T., Amselem, J., Chiapello, H., and Hofte, H. (1996) Further progress towards a catalogue of all *Arabidopsis* genes: analysis of a set of 5000 non-redundant ESTs. *Plant J.* **9**, 101–124.

3. Feldmann, K. A. (1991) T-DNA insertion mutagenesis in Arabidopsis: mutational spectrum. *Plant J.* **1,** 71–82.
4. Koncz, C., Nemeth, K., Redei, G. P., ancl Schell, J. (1992) T-DNA insertional mutagenesis in Arabidopsis. *Plant MoL Biol.* **20,** 963–976.5.
5. Fortsthoefel, N. R., Wu, Y., Schulz, B., Bennett, M. J., and Feldmann, K. A. (1992) T-DNA insertion mutagenesis in *Arabidopsis:* prospects and perspectives. *Aust. J. Plant Physiol.* **19,** 353–366.
6. Bechtold, N., Ellis, J., and Pelletier, G. (1993) *In planta* Agrobacterium mediated gene transfer by infiltration of adult *Arabidopsis thaliana* plants. *C.R. Acad. Sci, Paris* **316,** 1188–1193.
7. Feldmann, K. A., Malmberg, R., and Dean., C. (1994) Mutagenesis *inArabidopsis* in *Arabidopsis*. (Meyerowitz, E. and Somerville, C., eds.) Cold Spring Harbor Laboratory, Cold Spring Harbor, NY, pp. 137–172.
8. McKinney, E. C., Ali, N., Traut, A., Feldmann, K. A., Belostotsky, D. A., McDowell, J. A., and Meagher, R. B. Sequence based identification of T-DNA insertion mutations in *Arabidopsis:* actin mutants act2- 1 and act4- 1. *Plant J.* **7,** 613–622.
9. Azpiroz-Leehan, R. and Feldmann, K. A. (1997) T-DNA insertion mutagenesis in Arabidposis: going back and forth. *Trends Genet.* **13,** 152–156.

IV

GENE MAPPING IN ARABIDOPSIS

18

The Use of Recombinant Inbred Lines (RILs) for Genetic Mapping

Carlos Alonso-Blanco, Maarten Koornneef, and Piet Stam

1. Introduction

To locate (map) genes a population segregating for such genes is required. In many cases, a specific population is constructed to map a particular trait or locus, e.g., identified as a mutant. However, in other situations the gene(s) to be mapped do not show a specific phenotype and only polymorphisms at the DNA level can be studied. This is the case when cloned sequences have to be mapped. The mapping of a specific DNA polymorphism is more efficient when one can use populations that have already been scored for many markers. Classical mapping populations are backcross and F_2 populations in which individual plants must be analyzed. When the amount of DNA required per marker is high (for example in traditional restriction fragment length polymorphism [RFLP] analysis) the small size of *Arabidopsis* plants limits the number of markers that can be tested in these types of populations. This problem has been partially solved by analyzing the pooled progeny of individually selfed plants and, on the other hand, by developing new strategies in marker technology (such as *Arabidopsis* RFLP mapping set [ARMS] or polymerse chain reaction [PCR]-based markers amplified fragment length polymorphism [-AFLPs] and codominant cleaved amplified polymorphic sequence [CAPSs-]; *see* Chapters 19, 21, and 22) with a much lower requirement of DNA per marker. Nevertheless, due to heterozygosity, the mapping population is lost when one runs out of DNA and seeds. Permanent (immortal) mapping populations, which can be reproduced indefinitely, do not suffer from this and, furthermore, the data obtained by different researchers come from the same gamete sample and contribute to the same database. This advantage is important to establish the order of closely linked genes, since for statistical and biological reasons this is difficult when segregation data are collected in different populations.

In plant species, vegetative propagation can be used to immortalize a mapping population including not only homozygous but also heterozygous plants. However, in species that can be easily selfed and do not suffer from inbreeding depression (like *Arabidopsis* or rice), recombinant inbred lines (RILs) can be obtained relatively easy. The RILs are produced by successively selfing the progeny of individual F_2 plants (single seed descent method), from which the F_8 generation and onwards results in a number of practically homozygous lines that will produce further progeny that is essentially identical to the previous generation. Obtaining RILs, takes longer than the other types of mapping populations but the short generation time and the high seed production of *Arabidopsis* (maintenance and distribution of seeds are much easier than that of vegetatively propagated plant material) reduces this main disadvantage of RILs to a minor problem compared to most other plant species. Homozygous recombinant lines can also be obtained in one single generation when haploids can be induced in the F_1 generation as has been used in barley *(1)*. Since the induction of haploids is not possible or very difficult in *Arabidopsis*, doubled haploid lines (DHLs) are not a realistic alternative for permanent mapping populations in this species. Furthermore, in addition to being permanent and stable, an extra advantage of RILs is that during the process of inbreeding, the genetic material has undergone more than one round of meiosis, which implies that closely linked markers have more chance of being separated by a recombination event than, for instance, in a backcross population of similar size. In contrast, DHLs reach their "fixed genotypes" after a single meiotic cycle, thus RILs are more efficient for mapping closely linked markers *(2,3)*.

Besides qualitative (monogenic) markers, RILs can be used very efficiently for the location of quantitative trait loci (QTL). The determination of genetic map positions of loci involved in quantitative traits, usually identified as "natural" genetic variation, is more difficult because these traits are often controlled by several loci (polygenic) and relative large environmental effects on the phenotype can occur. The combination of pure lines that can be tested in many replications and environments, together with a dense and accurate molecular map that in general is available, contributes to the usefulness of RILs for this purpose *(2,4,5)*. In addition to QTL mapping and environment-genotype interaction analysis, RILs provide a good starting genetic material to obtain, after establishing the number of loci involved and their location, near isogenic lines (NILs) with monogenic inheritance for a single QTL affecting the trait of interest. Near isogenic lines are obtained by repeated backcrossing to one of the original parents; by monitoring the segregation of flanking markers a specific genome segment is introgressed into the genetic background of the recurrent parent. In an ideal set of NILs these introgressed regions, taken together, cover the whole genome *(6)*. Such a set of NILs is a useful tool for detailed genetic and molecular analysis, also for traits that are currently not of interest, but may become so in the future. The accurate location of QTLs is the first step for the cloning of a QTL by map-based techniques (*see* Chapters 30 and 31).

2. Materials

1. Seeds from existing RIL sets and/or homozygous parent lines (*see* **Note 1**).
2. Equipment and material for the analysis of the molecular marker to be mapped. (*See* protocols of ARMS, CAPSs, and AFLPs in Chapters 22, 21, and 19, respectively.)
3. Equipment and material to measure the quantitative trait of interest to be mapped. Specific traits may require specific growing conditions of the plants.
4. Computer and specific software to construct linkage maps (e.g., MAPMAKER—**ref. 7**, JOINMAP—**ref. 8**), to map QTLs (e.g., MAPMAKER/QTL—**ref. 9** or MapQTL—**ref. 10**, and a statistic software package), and data files of the available RI lines that are being used.

3. Methods

3.1. The Construction of RILs

The generation of a new set of RILs is often needed to map specific traits or QTLs and this will require one to obtain a new molecular marker map based on these RILs.

1. Choose the appropriate parents (*see* **Notes 1-3**).
2. Produce the F_1 generation and an F_2 population derived from one single F_1 plant (*see* **Notes 4 and 5**).
3. Harvest (selfed) seeds from each individual F_2 plant and grow one randomly chosen F_3 seed to produce from each F_3 plant an F_4 progeny (*see* **Note 6**).
4. Repeat this single-seed descent procedure up to the F_8/F_9 generation plants, whereafter the seeds of many plants per line might be bulk harvested (*see* **Note 7**).

3.2. The Use of RILs to Map DNA Markers

To genetically locate a cloned DNA sequence, one can use existing RIL sets, taking benefit of the available data.

1. Prepare DNA from parent lines of the different available sets of RILs (*see* **Note 1**).
2. Test the DNA sequence to be mapped for polymorphism between the parents of the RILs and choose the RIL set to be used (*see* **Notes 8 and 9**).
3. Prepare DNA from the individual RILs (*see* **Notes 10 and 11**).
4. Analyze the DNA of each line for the presence of the allele from one parent or from the other (*see* **Note 12**).
5. Add the scoring of the alleles for each RIL to the data file (*see* **Note 13**).
6. Use the new data file to generate a new linkage map using the appropriate software (e.g., MAPMAKER, JOINMAP; *see* **Notes 14 and 15**).

7. Compare for each RIL the scoring of the marker to be mapped with those of the two most closely linked flanking markers (on both sides) and recheck the original data when the new marker differs from both outside markers since this may indicate misscoring.

3.3. The Location of Quantitative Trait Loci (QTL) with RILs

1. Set up a statistically accountable test or assay to measure the quantitative trait of interest.
2. Grow the parent lines (10–15 plants of each one) from existing RIL sets or from a new set, in the appropriate test conditions, measure them for the trait(s) of interest and choose the RIL population to be used (*see* **Note 16**).
3. Grow 4–10 individuals of each RIL (and parent lines) in the appropriate test conditions (*see* **Notes 17** and **18**).
4. Analyze 4–10 plants per line for the trait of interest and average the different plant phenotype values in a single line phenotype estimate.
5. Add the data to a spread sheet file and analyze the phenotype frequency distribution of the trait (*see* **Note 19**).
6. It is also a good practice to perform a "classical" quantitative analysis on the data (*see* **Note 20**).
7. Combine the quantitative data file with the DNA marker file and process the data in the appropriate way according to the software instructions (e.g., MapQTL, MAPMAKER/QTL; *see* **Notes 21–24**).
8. The analysis can be completed by studying the epistatic interactions among the different loci identified, and further by repeating the analysis under different environments (*see* **Notes 25–27**).

4. Notes

1. In *Arabidopsis*, two sets of RILs have been described at present and are available through both *Arabidopsis* stock centers: one set of 153 RILs was derived from the cross W100 (mainly L*er* background) x Ws *(3)* and another set of 300 RILs was derived from the cross L*er* x Col *(4)*. The 153 Ws/W100 RILs and parent lines can be ordered quoting CS 2225(ABRC) or N 2225 (NASC). The L*er*/Col and parent lines can be ordered quoting CS 1899 or N 1899 (first set of 99 RI lines), CS 4858 or N 4858 (a second set of 201 RILs) or CS 4859 or N 4859 (to order the complete set of 300 RILs). Moreover, a large collection of different ecotypes are also available from the stock centers. Information concerning RIL sets (mapping data, the most common polymorphic restriction enzymes, subset of lines with highest frequency of recombination over the five chromosomes, references, and so on) and ecotypes (morphological characteristics, growing habits, habitat distribution, and so on) can be found in the A*t*DB (URL=http://genome-www.stanford.edu or via Gopher at URL=gopher://genome-gopher.stanford.edu) with the query of "Germplasm Resources" and down.
2. Distributing seed to Europe, Australia, Asia, and Africa is done by the Nottingham Arabidopsis Stock Centre (NASC) (Dept. of Life Science, Nottingham University,

Nottingham, NG7 2RD, UK); distributing to North America, by the *Arabidopsis* Biological Resource Center (ABRC) (Ohio State University, 1735 Neil Avenue, 309 Botany and Zoology Bldg., Columbus, OH 43210). RIL and ecotype seeds can be ordered by mail, e-mail (arabidopsis@nottingham.ac.uk or seeds@genesys.cps.msu.edu), fax (44 115 9513251 -NASC- and 614 292 0603 -ABRC-) or through the WWW servers (URL=http://nasc.nott.ac.uk/ or URL=http://genesys.cps.msu.edu:3333/).

3. The parents should be selected for the purpose of each experiment and should show enough polymorphism at the DNA marker level (which is correlated with the phylogenetic distance of the parents). In order to obtain RILs in a faster way, the use of late flowering ecotypes or parents with a strong seed dormancy should be avoided unless these are the traits of interest.

4. Although *Arabidopsis* is a strict self-fertilizing species, one cannot be sure that ecotypes are completely homozygous. Therefore, the advantage of starting with only one F_1 plant is that, at most, two alleles will segregate per locus. When one is interested in including cytoplasmic inheritance, two different F_1 plants can be used, both coming from the same parent plants but from reciprocal crosses.

5. The number of F_2 plants (and ultimately the number of RILs) is a matter of dispute. Increasing the number will increase the statistical accuracy of the mapping, but also the amount of work to obtain and analyze the RILs. One hundred lines will allow the detection of linkage of $r < 20\%$ (at a LOD of 3.0; for concept of LOD, *see* **Note 23** and Chapter 15). Finding no recombinants among 100 RILs gives an upper limit of the recombination percentage estimate of 1.5%. These values indicate that a commonly used number of 100 RILs is sufficient for most mapping purposes.

6. Most likely, "natural" selection will operate even when working under very controlled environmental conditions, but to reduce this effect, one should take all possible precautions when picking a single plant from each line to produce the next generation (*see* **Note 3**). Skewed allele frequencies in the population will reduce the effective size of the sample as well as the statistical power for mapping and QTL detection.

7. Both bulk harvesting propagation and single-plant propagation of the individual RILs will lead after several further generations to "fixation" of the very low level of heterozygosity that still might be present in the F_8 generation (0.8% per locus). However, bulk propagation will produce, at the end, more heterogeneous RILs.

8. RILs can be used for all type of molecular markers (RFLPs, RAPDs, CAPSs, SSCPs, AFLPs, and so on). Cloned DNA sequences usually are converted into genetic markers by looking for restriction polymorphisms after a Southern blot hybridization (RFLP), or polymorphisms after PCR amplification (e.g., CAPS) (when part of the sequence is known to allow design of primers; *see* Chapters 21 and 22; *see* **Note 1**). The existence of several sets of RILs involving different parents facilitate the chance to find such a polymorphism.

9. In contrast to F_2 populations in which dominant markers are less informative, RILs are equally informative with either dominant or codominant markers.

Nevertheless, when possible, it is advisable to select a codominant marker and preferably one that allows distinction between incomplete restriction and heterozygous phenotype. This will reduce possible misscoring, and will allow identification of markers that still segregate (*see* **Note 7**).

10. The number of RILs to analyze depends on the accuracy desired (*see* **Note 5**). The RILs can be classified with respect to the number of (recognizable) recombination events in each chromosome. One can take advantage of this variation between RILs when new markers are to be added to an established saturated marker map. For a rough mapping, a subset of 20–30 RILs can be selected, each having a number of recombination events in a particular chromosome (or subset of chromosomes) that is well above the average. When one wants to analyze only part of a set of RILs, the subsets established by the authors of the RIL population (*see* **Note 1**) should be used. When subsequently analyzing a larger set of RILs, it is advised to select them in a random way in order to retain the same order (on the map) of the markers that are common to the subset and the total RIL population.

11. Depending on the type of marker to be mapped, the quantity and quality of DNA to isolate might be different, but a larger amount is needed from the parent lines than from the RILs, in order to perform a satisfactory polymorphism screening producing a reliable marker. For most markers, a minipreparation of genomic DNA per RIL from a few harvested leaves is sufficient to map several markers (*see* Chapter 9).

12. Doubtful scores of a marker should be included as missing data. Heterozygous RILs should be scored just as they are, thus all the information will be used.

13. Files with the segregation data from the W100/Ws RILs and segregation data from the *Ler*/Col RILs are available through *At*DB (*see* **Note 1**). When adding a new marker to the existing file, carefully check the order of the RILs in the file and the nomenclature for both alleles according to software program instructions (see information given by the RIL authors in the AtDB). When the W100 x Ws lines are used, data can be sent to Pablo Scolnik (E.I. Du Pont de Nemours, P.O. Box 80402, Wilmington, DE 19880-0402).

14. RIL data can be analyzed with the software packages MAPMAKER *(7)* or JOINMAP *(8)*. The analysis will result in the assigning of the new marker to a linkage group and its genetic location within that linkage group. Pairwise recombination data will be generated with JOINMAP program together with the LOD score values (*see* Chapter 15). JOINMAP can be obtained from J. W. van Ooijen, CPRO-DLO, P.O. Box 16, 6700 AA Wageningen, The Netherlands; e-mail: mapping@cpro.dlo.nl.
MAPMAKER can be obtained from MAPMAKER, c/o Eric Lander, Whitehead Institute for Biomedical Research, 9 Cambridge Center, Cambridge, MA 02142 (Internet: mapmaker@genome.wi.mit.edu) or through WWW by anonymous FTP from host "genome.edu.wi.mit.edu."

15. Estimates of recombinant fraction for each pair of markers can be obtained manually using the equation of Haldane and Waddington *(11)*:

$$r = \widehat{R}/(2-2\widehat{R})$$

The standard error of r is:

$$Sr = 1/2\sqrt{\{\widehat{R}/[N(1-\widehat{R})^3]\}}$$

where r is the recombination rate (in a single meiosis), $\widehat{R}$ is the fraction of RILs with a recombinant (nonparental) phenotype for a given pair of markers, and N is the total number of RILs analyzed. For closely linked markers ($r < 5\%$), the observed fraction of recombinant RILs ($\widehat{R}$) is twice the r value, which makes the information per individual in a RIL mapping population as efficient as in F_2 population with co-dominant markers (*3*).

16. Even in cases in which the phenotypic differences between the parental lines are small, a QTL analysis may be worthwhile. The parental lines may carry alleles of opposite effect at a number of QTLs (dispersion of QTL alleles), their joint effect being unnoticed when considering the parental lines, but in a segregating offspring generation, such QTLs may cause considerable genetic variation (transgression).

17. In principle, any trait that can be measured is amenable to QTL analysis. The main factors determining the power of QTL detection and the accuracy of QTL location are: the size of the population, i.e., the number of RILs; and the accuracy with which the trait mean value of each RIL is estimated. Working with a given set of RILs, the first factor cannot be manipulated (*see* **Note 18**), but the second can. By using multiple observations per RIL, one reduces the error of the RIL means. When the measurement of the trait is laborious, it should be kept in mind that more than three replications will improve the accuracy of the measurement only marginally. In order to reduce the nongenetic variation, it is advised to use a randomized block design. In case of a very large number of RILs, one might use incomplete blocks. In all cases, parental lines should be included in the experiment.

18. The power of QTL detection increases with the number of RILs analyzed. The chance to detect a QTL depends also on the magnitude of its effect, relative to the environmental and error variation. Very roughly, QTLs that cause some 5% of the phenotypic variance among RILs are detected with a chance of 80% in a set of 100 RILs. This should be taken as a crude indication since detection power also depends on several other factors such as linkage of QTLs and the statistical technique applied (*12;* see **Note 21**).

19. Most statistical QTL detection methods assume a normal distribution of the trait measurements, and equal variances within marker genotype classes. Diagnostic tests should be performed in order to asses these requirements. For many quantitative traits, these assumptions do not hold. For example, traits relative to developmental differences often show an increase of the variance as the mean increases. Also, traits that are expressed as a proportion or percentage do not satisfy these assumptions. In such cases one should transform the original data, e.g., arcsin or log, in order to comply with the requirements of the statistical test.

The general shape of the phenotype frequency distribution can provide some information about the genetic control of the trait. For example a bimodal (1:1) distribution might suggest the presence of a major QTL. A skewed distribution may arise from interaction among different QTLs, although this also can be the result of skewed allele frequencies.

20. Such an analysis may comprise the estimation of the heritability of the trait, the estimation of the general dominance relationship of the trait (when an F^1 hybrid between parent lines is included in the assay) and estimation of correlation between traits *(13)*.

21. The analysis will result in the identification of one or several genomic regions "loci" affecting the quantitative trait of interest, their location within the corresponding linkage groups, and in an estimate of the proportion of the phenotypic variance that each locus accounts for.

    The power of QTL detection and mapping accuracy depends also on the statistical method employed (*12; see* **Notes 17** and **18**). Single-marker analysis and interval mapping (IM) *(9)* methods, map QTLs one-by-one, ignoring the effect of other (mapped or unmapped) QTLs. The IM uses the information of flanking markers to locate and asses the effect of a QTL more accurately within a genome interval, though when working with dense molecular maps both methods are similarly powerful. Recently, more complex and accurate methods have been developed by combining different statistic strategies, such as in Multiple-QTL model (MQM) mapping *(14)* which enables simultaneous mapping of multiple QTLs.

    The program MapQTL *(10)* handles these three different methods. MAPMAKER/ QTL *(9)* uses IM (*see* **Note 14** for requesting these programs).

22. Single-marker mapping methods can be performed with standard statistical software packages using a parametric method (one-way analysis of variance) to test at each marker whether the two genotypic classes have significantly different phenotypic values for the trait. A significant value of the test statistic indicates the presence of one or more QTLs linked to the marker. When the trait is not normally distributed, one can apply a transformation (*see* **Note 19**) or apply a nonparametric test (e.g., Kruskal-Wallis). Nonparametric tests are usually applied to ordinal or nominal data such as disease rates on an (arbitrary) scale. Applying a nonparametric test to normally distributed data will result in a decreased detection power.

    The phenotypic "effect" of an individual marker is calculated as the mean phenotypic difference between the two groups of RILs classified according to the genotype at the corresponding marker.

    The significance threshold (P value) for the identification of a QTL in "single marker mapping" is advised to be stringent to reduce false QTL identifications; P values of, at least, < 0.005 should be used.

23. In interval mapping information on the strength of the data supporting presence of a QTL in a defined interval of the genome is given by a likelihood ratio statistic, the so-called LOD score:

$$\text{LOD} = \log \frac{\text{likelihood for the presence of a segregating QTL}}{\text{likelihood for no segregating QTL}}$$

In order to control the overall type I error (chance of a false positive), a threshold value for LOD score must be set. This value depends on the number of linkage groups, the type of mapping population, and the total genetic map length. For RILs of *Arabidopsis*, a LOD threshold of 2.4 approximately corresponds to an overall level of 5% *(15)*.

A so-called support interval for the map position of a QTL can be constructed by taking the positions to the left and right of the maximum of the LOD profile corresponding with a LOD value of two less than the maximum *(16)*.

24. The percentage of the phenotypic variance of the trait explained by an individual marker, or a QTL detected by IM, is usually given as the R^2 value (= ratio of the sum of squares "explained" by the marker locus to the total sum of squares). When the experimental setup also allows estimation of the total genotypic variance and the nongenetic variance, the R^2 value can be used to estimate the contribution of the QTL (or marker) to the total genotypic variance.

25. Epistatic interactions between pairs of significant markers can be tested by two-way analyzes of variance (Anova). Alternatively, a multiple regression model with (multiplicative) interaction terms can be used for this purpose. With the latter approach, the number of interaction terms should not be large, in order to retain degrees of freedom for testing. A multiple regression model can also be used to estimate the total genotypic variance explained by the markers (R^2, *see* **Note 24**).

26. When the RILs are grown in different environments (years, locations, treatments) this allows studying genotype by environment interaction. The simplest way to do this is by performing a separate QTL analysis for each environment. A more sophisticated and powerful approach is described by Jansen et al. *(15)*. Both approaches may reveal that some QTLs are expressed in a specific environment only, whereas others display their effect in all environments.

27. The genomic regions (QTL locations) that are identified are relatively large and, therefore, further analyses are often needed to characterize them. The follow-up analyses can be:
    a. Analysis of the dominance relationships at the different QTLs (RILs do not provide this information directly) and for that, one can analyze F_1 hybrids between RILs differing only in one of the QTLs identified;
    b. Obtention of genotypes with single QTL (monogenic) differences affecting the trait ("Mendelising" QTLs);
    c. Fine mapping of genomic regions containing a major QTL, which will allow one to determine whether one is dealing with a single locus or several closely linked QTLs, to precise its/their location, and finally the (map-based) cloning of an interesting QTL.

Acknowledgments

C. Alonso-Blanco was supported by a fellowship from the FICYT, Principado de Asturias, Spain.

References

1. Keinhofs, A., Kilian, A., Saghai Maroof,M. A., Biyashev, R. M., Hayes, P., Chen, F. Q., Lapitan, N., Fenwick, A. Blake,, T. K., Kanazin, V., Ananiev, E., Dahleen, L., Kudrna, D., Bollinger, J., Knapp, S. J., Liu, B., Sorrels, M., Heun, M., Franckowiak, J. D., Hoffman, D., Skadsen, R., and Steffenson, B. J. (1993) A molecular, isozyme and morphological map of barley (*Hordeum vulgare*) genome. *Theor. Appl. Genet.* **86**, 705–712.
2. Burr, B. and Burr, F. A. (1991) Recombinant inbreds for molecular mapping in maize: theoretical and practical considerations. *Trends Genet.* **7**, 55–60.
3. Reiter, R. S., Williams, J. G. K., Feldmann, J. A., Rafalski, J. A., Tingey, S. V., and Scolnik, P. A. (1992) Global and local genome mapping in *Arabidopsis thaliana* by using recombinant inbred lines and random amplyfied polymorphic DNAs. *Proc. Natl. Acad. Sci. USA* **89**, 1477–1481.
4. Lister, C. and Dean, C. (1993) Recombinant inbred lines for mapping RFLP and phenotypic markers in *Arabidopsis thaliana*. *Plant J.* **4**, 745–750.
5. McCouch, S. R. and Doerge, R. W. (1995) QTL mapping in rice. *Trends Genet.* **11**, 482–487.
6. Eshed, Y. and Zamir, D. (1995) An introgression line population of Lycopersicon pennellii in the cultivated tomato enables the identification and fine mapping of yield associated QTL. *Genetics* **141**, 1147–1162.
7. Lander, E. S., Green, P., Abrahamson, J., Barlow, A., Daly, M. J., Lincoln, S. E. and Newberg, L. (1987) MAPMAKER: An interactive computer package for constructing primary genetic linkage maps of experimental and natural populations. *Genomics* **1**, 174–181.
8. Stam, P. (1993) Construction of integrated linkage maps by means of a new computer package: JOINMAP. *Plant J.* **3**, 739–744.
9. Lander, E. S. and Botstein, D. (1989) Mapping mendelian factors underlying quantitative traits using RFLP linkage maps. *Genetics* **121**, 185–199.
10. Van Ooijen, J. W. and Maliepaard, C. (1995) MapQTL (tm) 3.0: software for the calculation of QTL positions on genetic maps. CPRO-DLO, Wageningen, The Netherlands.
11. Haldane, J. B. S. and Waddington, C. H. (1931) Inbreeding and linkage. *Genetics* **16**, 357–374.
12. Jansen, R. C. (1995). Mapping of quantitative trait loci by using genetic markers: an overview of biometrical models used. *9th Conference of the Eucarpia. Biometrics in Plant Breeding*, pp. 116–124.
13. Mather, K. and Jinks, J. L. (1982) *Biometrical Genetics*, 3rd ed. Chapman and Hall, London.
14. Jansen, R. C. (1994) Controlling the type I and type II errors in mapping quantitative trait loci. *Genetics* **138**, 871–881.
15. Jansen R. C., Van Ooijen, J. W., Stam, P., Lister, C., and Dean, C. (1995) Genotype-by-environment interaction in genetic mapping of multiple quantitative trait loci. *Theor. Appl. Genet.* **91**, 33–37.
16. Van Ooijen, J. W. (1992) Accuracy of mapping quantitaive trait loci in autogamous species. *Theor. Appl. Genet.* **84**, 803–811.

19

AFLP™ Fingerprinting of *Arabidopsis*

Pieter Vos

1. Introduction

AFLP™ (KeyGene, Wageningen, The Netherlands), is a DNA fingerprinting technique that visualizes DNA restriction fragments by polymerase chain reaction (PCR) amplification *(1,2)*. Routinely, 50–100 restriction fragments are amplified simultaneously and detected on denaturing polyacrylamide gels. The AFLP technique consists of three major steps:

1. Restriction of the DNA and ligation of double–stranded adapters,
2. Selective amplification of a subset of the restriction fragments,
3. Gel analysis of the amplified restriction fragments.

The AFLP technique is a random fingerprinting technique that may be applied to DNAs of any origin or complexity. The technique differs importantly from other random fingerprint techniques by its robustness and reproducibility *(1–6)*. The AFLP protocol for genomic DNA of *Arabidopsis* is described in detail below.

The first step of the AFLP procedure is restriction of the DNA with two different restriction enzymes, a rare cutter and a frequent cutter, and the ligation of double–stranded adapters to the ends of the restriction fragments (**Fig. 1**). *Mse*I is used as frequent cutter: it recognizes a four-base cleavage sequence (TTAA) which cuts very frequently in *Arabidopsis* DNA, the average fragment size being about 150 bp. *Eco*RI is used as rare cutter restriction enzyme, predominantly because it is a reliable, low–cost enzyme. The adapter and restriction site sequences serve as primer binding sites in the subsequent amplification steps. The size range of the restriction fragments allow efficient amplification and subsequent separation on denaturing polyacrylamide gels. The use of two different restriction enzymes permits optimal flexibility in

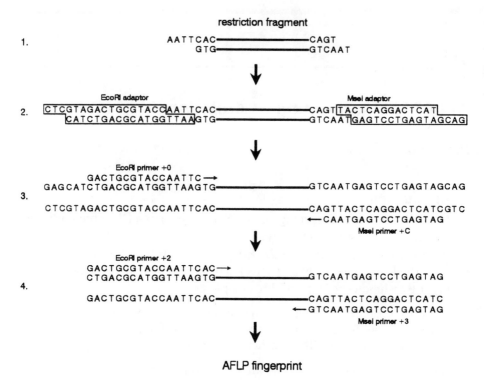

Fig. 1. Schematic representation of the AFLP protocol. 1. EcoRI–MseI restriction fragment. 2. Same fragment after ligation of the EcoRI and MseI adapters (boxed), creating the template DNA for AFLP amplification. 3. Target sites of the EcoRI and MseI preamplification primers on the template DNA; the arrows indicate the direction of DNA synthesis. 4. Target sites of AFLP primers on the preamplified DNA.

primer selection, which determines the number of fragments that will be coamplified.

The second step of the AFLP procedure is the amplification of subsets of EcoRI–MseI fragments using selective AFLP primers. The amplification is carried out in two consecutive steps, a first step called preamplification and a second step called selective AFLP amplification (**Fig. 1**). The amplification in two steps warrants optimal primer selectivity and allows low amounts of initial template DNA. In both amplification steps, primers are used that correspond to the adapter and restriction site sequences and that have additional nucleotides at the 3' ends extending into the restriction fragments (**Fig. 1**). These 3' extensions (called the selective nucleotides) assure that only a subset of the restriction fragments is amplified, i.e., those fragments of which the sequences adjacent to the restriction sites match the 3' primer extensions. The AFLP primers and amplification conditions are designed to result in preferential amplification of *Eco*RI–*Mse*I fragments with respect to *Eco*RI–*Eco*RI and *Mse*I–*Mse*I fragments *(1)*. Detection of the AFLP products occurs by labeling one of the two AFLP primers, generally the *Eco*RI primer.

The final step of the AFLP technique is the analysis of the fingerprints. For this purpose, the labeled reaction products are separated on denaturing polyacrylamide gels. These gels are very similar to those used for sequencing. After electrophoresis, gels are dried and exposed to X–ray films to visualize the AFLP fingerprints. Alternatively, AFLP images may be generated using phosphoimaging technology. Typically, fingerprints will contain 50–100 amplified restriction fragments. In the following paragraphs, a detailed technical description is provided of the AFLP fingerprinting of *Arabidopsis* DNAs.

2. Materials

2.1. Laboratory Equipment and Materials

1. 0.2-mL PCR tubes (Perkin Elmer, Norwalk, CT).
2. Base for 0.2-mL PCR tubes (Perkin Elmer).
3. Perkin Elmer 9600 thermal cycler.
4. Gel unit for "sequencing" gels (Sequi–Gen 38 × 50 cm, BioRad, Richmond, CA).
5. High voltage power supply (e.g., BioRad PowerPac 3000).
6. X–ray film (Sakura) or phosphoimaging device (Fujix BAS 2000).
7. *Arabidopsis* DNA (*see* **Note 1**).
8. *Eco*RI–adapter, 5 pMol/µL:
   5–CTCGTAGACTGCGTACC
       CATCTGACGCATGGTTAA–5
9. *Eco*RI–primer +0, 50 ng/µL: 5–GACTGCGTACCAATTC–3
10. *Eco*RI–primers +2, 50 ng/µL: 5–GACTGCGTACCAATTCNN–3
11. *Mse*I–adapter, 50 pMol/µL:
    5–GACGATGAGTCCTGAG
        TACTCAGGACTCAT–5
12. *Mse*I–primer +C, 50 ng/µL: 5–GATGAGTCCTGAGTAAC–3
13. *Mse*I–primers +3, 50 ng/µL: 5–GATGAGTCCTGAGTAACNN–3

2.2. Solutions for AFLP Reactions

1. $1M$ Tris-HAc, pH 7.5.
2. $1M$ Tris-HCl, pH 8.0.
3. $1M$ Tris-HCl, pH 8.3.
4. 0.1 mM MgCl$_2$.
5. $1M$ KCl.
6. $0.5M$ EDTA, pH 8.0.

7. 1M KAc.
8. 10X TE: 100 mM Tris-HCl, 10 mM EDTA, pH 8.0.
9. TE$_{0.1}$: 10 mM Tris.HCl, 0.1 mM EDTA, pH 8.0.
10. Milli–Q or double–distilled water (referred to as H$_2$O).
11. 100 mM DTT.
12. 10 mM ATP.
13. 5X RL (restriction/ligation) buffer: 50 mM Tris-HAc, 50 mM MgCl$_2$, 250 mM KAc, 25 mM DTT, pH 7.5.
14. 10X T4–buffer: 250 mM Tris-HCl, pH 7.5, 100 mM MgCl$_2$, 50 mM DTT, 5 mM spermidine (3 HCl–form).
15. 10X PCR buffer: 100 mM Tris-HCl, pH 8.3, 15 mM MgCl$_2$, 500 mM KCl.
16. 5 mM of a mix of all four dNTPs (Pharmacia, Uppsala, Sweden).
17. ^{32}P–γ-ATP (approx 3000 Ci/mmol) or ^{33}P–γ-ATP (approx 2000 Ci/mmol), (Amersham, Arlington Heights, IL).
18. Restriction enzymes: *Eco*RI, *Mse*I (New England Biolabs, Beverly, MA).
19. T4 DNA ligase (Pharmacia).
20. T4 polynucleotide kinase (Pharmacia).
21. Taq DNA polymerase (Amplitaq, Perkin Elmer).

2.3. Solutions for Gel Electrophoresis

1. Formamide (deionized and filtered, Merck, Darmstadt, Germany).
2. Urea (Gibco-BRL, Gaithersburg, MD).
3. 10X TBE (concentrated electrophoresis buffer): 1M Tris base, 1M Boric Acid, 20 mM EDTA.
4. 20% acrylamide, 1% methylene bisacryl in H$_2$O (Sequagel™, ready for use gelmix, National Diagnostics).
5. Solution of 4.5% acrylamide/methylene bisacryl (20:1) and 7.5M Urea in 0.5X TBE.
6. Ammonium persulphate (APS, Merck).
7. N, N, N', N', tetra methyl ethylene diamine (TEMED, Pharmacia).
8. Bind silane (Pharmacia).
9. Repel silane (Pharmacia).
10. Bromophenol blue.
11. Xylene cyanol.
12. Loading dye: 98% formamide, 10 mM EDTA, pH 8.0, and trace amounts of bromo phenol blue and xylene cyanol.
13. 10% Acetic acid solution.

3. Methods

3.1. Modification of DNA and Template Preparation

1. Digest *Arabidopsis* genomic DNA with the corresponding restriction enzymes in a 40 µL reaction containing 8 µL 5X RL buffer, 5 U *Eco*RI, 1 U *Mse*I, H$_2$O, and *Arabidopsis* DNA (50 ng) to 40 µL (*see* **Note 1**). Mix well and incubate for 2 h at 37°C.

2. To prepare the *Mse*I adapter, take 8 µg of the top strand and 7 µg of the bottom strand oligonucleotide (1500 pmol of both oligonucleotides) in 30 µL of H_2O, this gives you 50 pmol/µL of *Mse*I–adapter (*see* **Note 2**).
3. To prepare the *Eco*RI–adapter, take 8.5 µg of the top strand and 9.0 µg of the bottom strand oligonucleotide (1500 pmol of both oligonucleotides) in 300 µL of H_2O, this gives you 5 pmol/µL of *Eco*RI–adapter.
4. To ligate the adaptors to the digested genomic DNA, add the following components to the reaction mix and incubate another 2 h at 37°C: 1 µL *Eco*RI adapter (5 pmol), 1 µL *Mse*I adapter (50 pmol), 1 µL 10 mM ATP, 2 µL 5X RL buffer, 1 U T4 DNA ligase, 5 µL H_2O (*see* **Note 3**).
5. After ligation the reaction mixture is diluted 10X with $TE_{0.1}$. Generally 10 µL of the reaction mixture is diluted to 100 µL, and the remaining 40 µL is stored as backup. The diluted reaction mixture is used as template DNA for the AFLP preamplification reactions. Both the diluted DNA and undiluted DNA are stored at –20°C.

3.2. AFLP Amplification

1. The preamplification reaction is performed in 50 µL. A typical preamplification reaction contains the following components: 5.0 µL template–DNA, 1.5 µL *Eco*RI–primer +0 (75 ng), 1.5 µL *Mse*I–primer +C (75 ng), 2.0 µL 5 mM dNTPs, 0.2 µL Taq–polymerase (1 U), 5.0 µL 10X PCR–buffer, 34.8 µL H_2O (*see* **Note 4**). For preamplification 20 PCR cycles are performed using the following cycle profile: 30 s at 94°C; 60 s at 56°C; 60 s at 72°C (*see* **Note 5**).
2. After preamplification, 10 µL of the reaction is diluted with 90 µL of $TE_{0.1}$ to 100 µL which is sufficient for 20 AFLP–reactions +2/+3. The diluted reaction mix and the rest of the preamplification reaction is stored at –20°C. If necessary new dilutions of the preamplification reactions may be made to give additional template for the AFLP reactions.
3. Primers for selective AFLP amplification are labeled by phosphorylating the 5' end of the primers with ^{32}P- or ^{33}P–γ-ATP and polynucleotide kinase (*see* **Note 6**). Prepare the following primer labeling mixes using either ^{32}P–γ-ATP or ^{33}P–γ-ATP to generate sufficient labeled primer for 100 AFLP reactions. Labeling mix (40 µL) for ^{32}P: 20.0 µL ^{32}P–γ-ATP (approx 3000 Ci/mmol), 5.0 µL 10X T4–buffer, 2.0 µL T4–kinase (10 U/µL), 13.0 µL H_2O. Labeling mix (40 µL) for ^{33}P: 10.0 µL ^{33}P–γ-ATP (approx 2000 Ci/mmol); 5.0 µL 10X T4–buffer, 2.0 µL T4–kinase, 23.0 µL H_2O.
4. For primer labeling add the labeling mix to 10 µL *Eco*RI–primer +2 (50 ng/µL) to give 50 µL (*see* **Note 7**) and incubate 60 min at 37°C, followed by incubation at 70°C for 10 min (inactivation of the kinase). This gives a labeled primer with a concentration of 10 ng/µL.
5. Prepare the following AFLP reaction mixes for 10 reactions (*see* **Note 8**). Primers/dNTPs mix (50 µL): 5.0 µL labeled *Eco*RI–primer +2 (10 ng/µL), 6.0 µL unlabeled *Mse*I–primers +3 (50 ng/µL), 8.0 µL 5 mM dNTPs, 31.0 µL H_2O. Taq–polymerase mix (100 µL): 20.0 µL 10X PCR–buffer, 0.8 µL Taq–polymerase (4 U), 79.2 µL H_2O.

6. The PCR reaction is assembled by adding 5.0 µL primers/dNTPs mix and 10.0 µL Taq–polymerase mix to 5.0 µL preamplified template DNA in the PCR tubes (*see* **Notes 9–11**).

The AFLP amplification is carried out in a PE–9600 thermocycler with the following cycle profile. Cycle 1: 30 s 94°C, 30 s 65°C, 60 s 72°C; Cycles 2–13: Same PCR profile as cycle 1, except for a stepwise decrease of the annealing temperature in each subsequent cycle by 0.7°C during 12 cycles, Cycle 14–36: 30 s 94°C, 30 s 56°C, 60 s 72°C (*see* **Notes 12** and **13**).

3.3. Gel Analysis of AFLP Reaction Products

1. The AFLP reaction products are analyzed on 4.5% denaturing polyacrylamide gels (*see* **Note 14**). The gels should be cast at least 2 h before use and should be prerun for $1/2$ hour just before loading the samples, 1X TBE is used as running buffer (*see* **Note 15**). Prerunning and running of the gel is performed at 110 W with the BioRad system (no limits for voltage and current). This warrants an even heat distribution during electrophoresis, which is crucial for good quality fingerprints.
2. Mix AFLP reaction products with an equal volume (20 µL) of loading dye after the PCR is finished.
3. Heat the resulting mixtures for 3 min at 90°C, and then quickly cool on ice. (A similar denaturing procedure may be performed in the thermal cycler).
4. Rinse the surface of the gel well with 1X TBE and push carefully two, 24–well sharktooth combs ± 0.5 mm into the gel surface to create the gel slots.
5. Rinse the gel slots formed in this way with 1X TBE and load 2 µL of each sample per well. Each gel allows for simultaneous electrophoresis of 49 samples.
6. After electrophoresis disassemble the gel cassette. The gel will stick to the front glass plate because of the silane treatment.
7. Fix the gel by soaking in 10% acetic acid for 30 min and dry it subsequently at room temperature in a chemical hood for 10–20 h (gel drying is faster at elevated temperatures).
8. Autoradiograph the gel. Generally, exposure of ^{32}P-gels to standard X–ray film overnight, without intensifying screens, gives good results. ^{33}P-Gels take 2–3 d exposure to generate similar band intensities. Exposure times are reduced at least $2^{1}/_{2}$-fold using phosphoimaging technology.

4. Notes

1. DNA preparations of *Arabidopsis* are often of poor quality and have low yields, however, only 50 ng of DNA is used for AFLP template preparation. Complete restriction is crucial for good quality AFLP fingerprints. We have found that purification with glass beads (BIO 101) generally provides a good way to remove impurities which cause incomplete restriction.
2. The *Eco*RI and *Mse*I adapters have double–stranded parts of 14 and 12 base pairs, respectively. It appears unnecessary to perform a specific denaturation–renaturation procedure to anneal the two strands of the adapters.

3. DNA is incubated for a total of 4 h with restriction enzymes, the last 2 h in the presence of T4 DNA ligase and oligonucleotide adapters. The adapters do not restore the restriction sites and, hence, the presence of the restriction enzymes in the ligation step results in almost complete adapter–to–fragment ligation because of the restriction of fragment concatamers formed by ligation. However, prolonged incubation with restriction enzyme is not recommended because of possible star activity of EcoRI giving reduced cleavage specificity and, ultimately, aberrant AFLP fingerprints.
4. Working with mixes of reagents is preferred. The following procedure is suggested: 5.0 µL template–DNA + 25 µL mix 1 + 20 µL mix 2.

   Mix-1 (250 µL)

   15 µL *Eco*RI–primer + 0 (75 ng)
   15 µL *Mse*I–primer + C (75 ng)
   20 µL 5 mM dNTPs
   200 µL H$_2$O

   Mix-2 (200 µL)

   2.0 µL *Taq*–polymerase (1 U)
   50 µL 10X PCR–buffer
   148 µL H$_2$O

   500 µL

5. Nonphosphorylated adapters have been used for adapter ligation. Therefore, only one strand of the adapter binds to the DNA, which means that the adapters are ligated to opposite strands. This would prevent amplification, if the template–DNA would be denatured prior to the start of PCR. This should never be done. The recessed 3' ends of the template DNA are filled–in by the Taq polymerase during heating to 94°C of the first PCR–cycle.
6. Only one of the two primers of the AFLP reaction should be labeled (there is no preference which one), generally the *Eco*RI–primer is selected. Mobility of the two strands of a DNA fragment on sequencing gels is generally slightly different. Therefore,care should be taken in comparing AFLP fingerprints obtained with different primers labelled. ^{33}P–labeled primers are preferred, because they give a better resolution of the PCR products on polyacrylamide gels. Also, the reaction products are less prone to degradation due to autoradiolysis. However, the use of ^{33}P–labeled primers is more expensive compared to ^{32}P–labeled primers.
7. It is also possible to divide the reaction mix over several primers if preferred, e.g., 5 × 8 µL of the reaction mix may be added to 5 × 2 µL (100 ng primer) of five different primers.
8. Working with AFLP reaction mixes is important for the reliability and reproducibility of AFLP reactions. It also facilitates the assembly of the AFLP reactions. Reaction mixes should be prepared for a minimum of 10 reactions to warrant accurate pipeting. Generally, more AFLP reaction will be performed simultaneously and the quantitities of the reaction mixes should be adjusted accordingly.

9. The AFLP fingerprints on *Arabidopsis* DNA utilize an *Eco*RI–primer with two selective nucleotides and an *Mse*I–primer with three selective bases. Two mixes are added to the preamplified template DNA: a mix with AFLP–primers and dNTPs and a mix with Taq polymerase (including the 2X concentrated PCR buffer).
10. Working with mixes is not only convenient but also essential to allow a rapid start of the AFLP reactions after all reagents have been pipeted together. The reaction is assembled at room temperature and, although the activity of the Taq polymerase is low at ambient temperatures, the AFLP fingerprint quality suffers from a long lag time between assembly of the AFLP reaction mixtures and start of the PCR.
11. The PCR tubes are held by a "base plate" in a 8 × 12 format (96 tubes). The template DNA is pipeted first followed by the two mixes. The use of multichannel pipetors is advised to allow rapid pipeting. The reagents are mixed by tapping the base plate with the tubes on the bench. After mixing of the reagents, the AFLP reaction is started as soon as possible.
12. The AFLP PCR is started at a very high annealing temperature, to obtain optimal primer selectivity. In the following steps, the annealing temperature is lowered gradually to a temperature at which efficient primer binding occurs. This temperature is then maintained for the rest of the PCR cycles. The Perkin Elmer PE–9600 thermocycler allows easy programming of the PCR profile described below.
13. The PE9600 thermocycler is favored for performing AFLP reactions. Other thermocyclers may be used, but may not have the possibility to program the AFLP PCR cycle profile. Alternatively, the annealing temperature may be decreased 1°C in each subsequent cycle during the first 10 cycles, gradually decreasing the annealing temperature from 65°C in the first cycle to 56°C from cycle 10 onwards.
14. These gels are essentially normal sequencing gels *(7)*, with the exception that a lower percentage of polyacrylamide is used. Gels are cast using the instructions of the manufacturer of the gel system. We prefer a BioRad SequiGen sequencing gel system (38 × 50 × 0.04 cm), but there is no reason why other sequencing gel systems should not work equally well. The back plate of the gels, the so-called IPC (integrated plate chamber) is treated with 2 mLs of repel silane. The front plate is treated with 10 mLs of bind silane solution (30 µL acetic acid and 30 µL bind silane in 10 mLs ethanol, freshly made immediately before use). The silane treatments cause the gels to stick to the front plate upon disassembly of the gel cassette after electrophoresis. The Biorad SequiGen sequence gels require ± 100 mL of gel solution to which 500 µL of 10% APS (freshly made before use) and 100 µL of TEMED is added immediately before casting the gels.
15. The purpose of the prerun is to "warm up" the gel to about 55°C, which temperature is maintained throughout the electrophoresis.

References

1. Vos, P., Hogers, R., Bleeker, M., Reijans, M., van de Lee, T., Hornes, M., Frijters, A., Pot, J., Peleman, J., Kuiper, M., and Zabeau, M. (1995) AFLP: a new technique for DNA fingerprinting. *Nucleic Acids Res.* **23,** 4407–4414.

2. Zabeau, M., and Vos, P. (1993) selective restriction fragment amplification: a general method for DNA fingerprinting. European Patent Application, EP 0534858.
3. Thomas, C. M., Vos, P., Zabeau, M., Jones, D.A., Norcott, K. A., Chadwick, B. and Jones, J. D. G. (1995) Identification of amplified restriction fragment polymorphism (AFLP) markers tightly linked to the tomato Cf–9 gene for resistance to *Cladosporium fulvum. Plant J.* **8,** 785–794.
4. Becker, J., Vos, P., Kuiper, M, Salamini, F. and Heun, M. (1995) Combined mapping of RFLP and AFLP markers in barley. *Mol. Gen. Genet.* **249,** 65–73.
5. Meksem, K., Leister, D., Peleman, J., Zabeau, M., Salamini, F. and Gebhart, C. (1995) A high–resolution map of the vicinity of the R1 locus on chromosome V of potato based on RFLP and AFLP markers. *Mol. Gen. Genet.* **249,** 74–81.
6. Van Eck, H. J., Rouppe van der Voort, J., Draaistra, J., van Zandvoort, P., van Enckevort, F., Segers, B., Peleman, J., Jacobsen, E., Helder, J. and Bakker, J. (1995) The inheritance and chromosomal localization of AFLP markers in a non–inbred potato offspring. *Molecular Breeding*, **1,** 397–410.
7. Maxam, A. M. and Gilbert, W. (1980) Sequencing end-labeled DNA with base-specific chemical cleavages. *Methods Enzymol.* **65,** 499–560.

20

Building a High-Density Genetic Map Using the AFLP™ Technology

Martin T. R. Kuiper

1. Introduction

The AFLP™ (Keygene, Wageningen, The Netherlands) technology allows the simultaneous amplification of many restriction fragments in a single polymerase chain reaction (PCR). Thus, the technology can be used to very efficiently detect restriction fragment polymorphisms. As an example of the vast numbers of markers that can be detected, the author describes here a protocol to construct a high-density genetic map in *Arabidopsis*. The map may be used to identify primer combinations that will target fragments mapped to a specific area of the genome.

As explained in the previous chapter, the AFLP technology hinges on the use of two restriction enzymes and, hence, the use of two different primers to obtain PCR-amplified restriction fragments. The requirements for obtaining good quality fingerprint patterns are simple. Many different combinations of restriction enzymes will give good AFLP templates *(1)*. Attractive as this may seem, there is a potential pitfall: the choice of restriction enzyme combinations is so large, that it may ultimately impede correlating experimental results obtained in different crosses of a species. Therefore we promote the use of a single enzyme combination, *Eco*RI/*Mse*I, in most plant species. A set of preferred primer combinations for general use in genetic research will further facilitate the comparison of marker and mapping data.

To standardize the use of the AFLP technology in plant genetic research we at KeyGene have identified most favored primer/enzyme combinations for two major plant groups differing in genome size. For general use in crops with genome sizes ranging from 100 megabase pairs to 6000 megabase pairs, we have chosen

Table 1
List of Inbred Lines from the Landsberg X Columbia Cross Used to Construct the High-Density AFLP Map Described

CL4	CL62	CL175	CL253	CL345
CL5	CL67	CL177	CL257	CL349
CL13	CL68	CL179	CL264	CL350
CL17	CL71	CL180	CL266	CL351
CL19	CL79	CL181	CL267	CL356
CL25	CL84	CL190	CL279	CL358
CL29	CL90	CL191	CL283	CL359
CL30	CL107	CL199	CL284	CL363
CL32	CL113	CL209	CL288	CL367
CL33	CL115	CL214	CL295	CL370
CL34	CL123	CL217	CL296	CL377
CL35	CL125	CL231	CL297	CL378
CL36	CL131	CL232	CL302	CL386
CL37	CL160	CL235	CL303	CL390
CL46	CL161	CL238	CL311	CL395
CL52	CL166	CL240	CL321	CL398
CL54	CL167	CL242	CL332	
CL59	CL173	CL245	CL342	

the enzyme combination *Eco*RI/*Mse*I. Using the enzyme combination *Eco*RI/*Mse*I, a huge number of different primer combinations (1024–4096) can be used to generate fingerprints (*see* **Note 1**). For species with large genomes (500–6000 megabase pairs) we have identified a matrix of suitable primer combinations that have a total of six selective nucleotides. For small genomes like the *Arabidopsis* genome, a specific set of primer combinations should be used, to accommodate for the smaller genome size. In general, smaller genomes (100–500 megabase pairs) can be analyzed efficiently using five selective nucleotides. Kits based on this design, for specific use in plants with a small genome size (e.g., *Arabidopsis*, Rice, Cucumber, and so on) are commercially available from Life Technologies (Bethesda, MD, kit for radiolabeling), and from Perkin Elmer (Norwalk, CT, fluorescently labeled kit). The design of the primer combination matrix is given below.

2. Materials

1. This protocol is an application of the protocol described in Chapter 19. Therefore, all the materials described there are also required here.
2. *Arabidopsis thaliana* recombinant inbred lines: The lines used to prepare the high–density AFLP map shown in this protocol are listed in **Table 1** (*see* **Note 2**).

Building a High-Density Genetic Map

Table 2
Primer Combinations and Approximate Numbers of Fragments in *Arabidopsis* Fingerprints

		CAA M47	CAC M48	CAG M49	CAT M50	CTA M59	CTC M60	CTG M61	CTT M62
AA	E11	100	81	64	92	73	56	59	80
AC	E12	80	40	44	60	50	30	34	61
AG	E13	68	50	39	46	47	44	41	59
AT	E14	ᵃ65	55	43	73	64	74	49	57
TA	E23	27	ᵃ24	43	78	56	53	45	56
TC	E24	83	39	43	64	44	50	51	ᵃ39
TG	E25	65	44	49	71	49	38	44	47
TT	E26	75	48	63	74	70	54	51	66

The individual primers are indicated by codes (E11, E12, E13, E14, E23, E24, E25, E26, and M47, M48, M49, M50, M59, M60, M61, M62) and their corresponding 3' extensions in front (E–numbers) or above the primer codes (M–numbers).

ᵃThese combinations give rise to the detection of a highly–abundant DNA fragment (combinations E14/M47, E23/M48, and E24/M62, *see* also **Note 3**).

Table 3
Number of Polymorphic Bands Observed in the Columbia X Landsberg RIls Using 32 Primer Combinations

		CAA M47	CAC M48	CAG M49	CAT M50	CTA M59	CTC M60	CTG M61	CTT M62
AA	E11	20	30	20	20	27	23	18	25
AC	E12	23	17	11	19	24	16	19	18
AG	E13	30	18	19	24	10	18	13	11
AT	E14	30	20	8	32	29	31	17	22

Primer coding is as in **Table 2**.

3. The primer combinations we recommend for use in *Arabidopsis* are assembled from eight *Eco*RI primers and eight *Mse*I primers. The matrix with 64 possible primer combinations is given in **Table 2**. The complete panel of 64 primer combinations was used on the parental lines of the Columbia X Landsberg recombinant inbred population. The number of fragments that each primer combination will typically produce in an F_1 of these parents is also given in **Table 2** (*see* **Note 3**). With the complete set of 64 primer combinations from the 8 × 8 matrix, close to 3600 *Eco*RI/*Mse*I fragments (**Table 2**) can be detected (*see* **Note 4**).
4. Here, we used 32 primer combinations. These 32 primer combinations detect a total of 1880 fragments, of which 662 (**Table 3**) are polymorphic (*see* **Note 5**).

Thus, the percentage of AFLP bands that is polymorphic in fingerprints of the Columbia and Landsberg recombinant inbreds is approx 35%.
5. 10–base ladder (Sequamark™, Huntsville, AL), commercially available from Research Genetics.
6. Computer programmes JoinMap, DrawMap.

3. Methods

3.1. Preparation of AFLP Reactions

The sample preparation is described in Chapter 19. An example of a mapping fingerprint is given in **Fig. 1** (See **Notes 6–8**).

3.2. AFLP Marker Identification

The nomenclature for an AFLP marker is derived from the enzyme combination, the primer combinations, and the relative molecular weight, a typical AFLP marker name then will run as follows: E13M46.486. This name indicates an AFLP marker amplified with the primer combination E13 (E for *Eco*RI) and M46 (M for *Mse*I), running at 486 nucleotides. When analyzing markers in a segregating population, parental assignments can be included, as we have done in **Table 4** (*see* **Notes 9** and **10**). This table contains in detail the marker information for 32 primer combinations described in **Table 3**.

3.3. Analysis of the Data

The computer program JoinMap™ version 2.0 *(3)*, an updated version of the original package described in 1993 *(4)* can be used for the statistical analysis of the data. Using this package, we calculated the five *Arabidopsis* linkage groups with the AFLP markers, and an additional 56 RFLP markers retrieved from the *Arabidopsis* database (HTTP://GENOME–www.Stanford.edu). Of the 411 AFLP markers that were analyzed in our gels, 400 could be placed in the map, with sufficient confidence. All markers in the map have a good fit with respect to neighboring markers, there is little indication for fault–scores resulting in double "recombinants" (*see* **Note 11**). The resulting map is presented using the computer program DrawMap, developed by J. van Ooijen, CPRO-DLO, Wageningen, Netherlands. The map is shown in **Fig. 2**. From the map and the

---

Fig. 1. *(opposite page)* AFLP fingerprinting pattern: A phosphorimage of an AFLP fingerprint using the primer combination E14/M60 is shown. The image shows 48 lanes, from left to right containing 10 recombinant inbreds, parents Landsberg and Columbia, the SequaMark™ 10–base ladder, and an additional 35 recombinant inbreds. Apart from the conspicuous monomorphic bands, over 25 polymorphic bands can be observed.

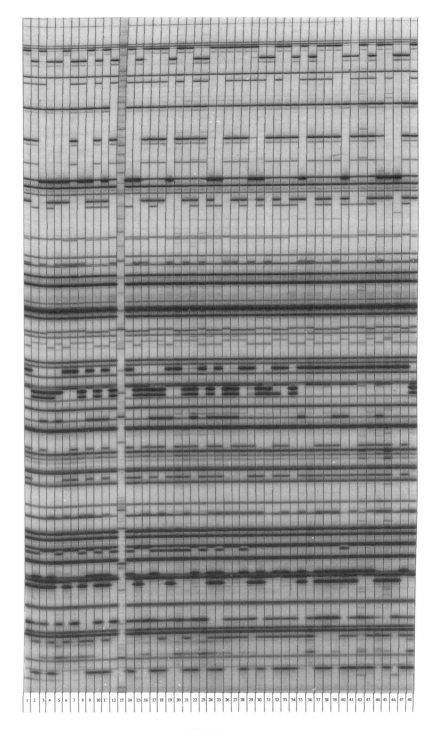

Fig. 1

Table 4
Listing of AFLP Markers Generated with 32 Primer Combinations

Name	P	C#	cM	Name	P	C#	cM	Name	P	C#	cM	Name	P	C#	cM
E11M47.M6	1	1	21.5	E11M48.M1	2	3	27.4	E11M49.144	2	2	37.9	E11M50.M2	2	4	36.8
E11M47.M5	1	1	69.0	E11M48.513	1	5	49.6	E11M49.120	1	2	37.9	E11M50.M1	1	4	56.2
E11M47.M4	1	1	67.7	E11M48.426	1	2	21.7	E11M49.117	2	3	30.3	E11M50.400	2	5	58.0
E11M47.M2	1	3	33.9	E11M48.397	2	2	21.7	E11M49.105	2	4	51.3	E11M50.292	1	5	5.3
E11M47.M1	1	1	24.7	E11M48.385	2	1	11.5	E11M49.081	1	1	49.3	E11M50.202	1	3	33.9
E11M47.292	2	3	12.7	E11M48.376	2	5	36.6	E11M49.066	2	2	59.5	E11M50.201	2	1	54.8
E11M47.259	1	3	32.4	E11M48.276	2	4	10.2	E11M49.048	2	1	64.4	E11M50.199	1	1	54.8
E11M47.258	2	2	10.6	E11M48.202	1	4	59.8					E11M50.189	2	4	19.0
E11M47.252	2	3	23.1	E11M48.162	1	5	19.0					E11M50.108	1	3	34.7
E11M47.206	1	5	56.1	E11M48.155	1	3	34.7					E11M50.093	1	3	26.1
E11M47.158	2	4	10.8	E11M48.149	2	5	17.3					E11M50.092	1	3	24.5
E11M47.150	1	5	36.6	E11M48.146	1	4	10.8								
E11M47.147	1	1	56.3	E11M48.137	2	4	7.2								
E11M47.130	2	5	58.5	E11M48.134	1	4	7.2								
E11M47.080	1	5	64.2	E11M48.129	1	3	11.1								
				E11M48.114	1	3	39.0								
				E11M48.096	1	3	27.4								
				E11M48.094	1	2	37.2								
				E11M48.092	1	4	30.6								
				E11M48.073	2	1	65.3								
				E11M48.065	2	2	7.0								
E11M59.468	1	1	69.0	E11M60.457	2	4	10.8	E11M61.369	1	4	22.6	E11M62.453	1	2	12.3
E11M59.459	2	5	36.6	E11M60.434	2	3	54.5	E11M61.247	2	2	15.6	E11M62.383	1	4	30.6
E11M59.427	2	3	33.9	E11M60.425	1	3	34.7	E11M61.210	2	4	10.2	E11M62.378	1	3	34.7

E11M59.377	1	4	10.8	E11M60.297	1	4	10.8				E11M62.326	1	1	12.9
E11M59.249	2	4	10.2	E11M60.249	2	2	18.2				E11M62.300	2	3	33.4
E11M59.246	2	4	66.8	E11M60.196	1	3	23.7				E11M62.297	2	2	19.3
E11M59.231	2	5	70.6	E11M60.190	2	2	12.3				E11M62.232	1	3	7.8
E11M59.211	1	4	15.2	E11M60.155	1	3	23.7				E11M62.188	2	3	34.7
E11M59.208	2	5	27.5	E11M60.132	1	4	40.0				E11M62.177	2	3	34.7
E11M59.206	1	5	27.5	E11M60.131	1	5	35.3				E11M62.162	1	5	4.1
E11M59.196	2	3	23.7	E11M60.115	1	1	62.4				E11M62.133	2	1	37.4
E11M59.174	2	4	10.8					E11M61.120	2	1	43.4	E11M62.131	2	3	18.9
E11M59.161	2	5	47.6					E11M61.090	2	3	34.7	E11M62.130	1	3	18.9
E11M59.126	1	1	20.1					E11M61.069	1	3	34.7	E11M62.126	1	3	33.9
E11M59.117	2	5	36.6								E11M62.116	1	4	10.8
E11M59.111	1	1	41.9								E11M62.111	1	5	4.1
E11M59.097	2	3	34.7								E11M62.076	2	5	38.9
E11M59.096	1	3	34.7								E11M62.063	1	1	41.9
E11M59.095	2	3	37.4								E11M62.055	2	2	15.6
E11M59.070	1	5	42.6											
E12M47.M5	1	3	33.9	E12M48.449	2	5	25.4	E12M49.485	2	3	33.4	E12M50.280	2	3	32.4
E12M47.M4	1	3	8.4	E12M48.416	1	5	25.4	E12M49.317	2	2	16.3	E12M50.245	2	4	19.0
E12M47.M3	2	1	56.8	E12M48.257	2	3	39.0	E12M49.284	2	4	32.8	E12M50.216	1	1	20.9
E12M47.M2	2	1	15.7	E12M48.231	1	5	67.9	E12M49.278	2	2	17.0	E12M50.210	1	2	60.5
E12M47.M1	2	1	15.1	E12M48.225	1	4	16.2	E12M49.126	2	1	42.5	E12M50.192	2	1	54.0
E12M47.322	2	1	65.3	E12M48.191	1	3	34.7	E12M49.055	1	5	78.7	E12M50.175	1	1	54.8
E12M47.208	2	3	33.9	E12M48.162	2	3	35.6	E12M49.054	2	2	17.6	E12M50.128	1	1	74.1
E12M47.186	1	1	95.9	E12M48.159	2	5	36.6				E12M50.122	2	1	74.1
E12M47.177	1	4	30.6	E12M48.155	1	3	35.6				E12M50.099	1	5	4.1

(continued)

Table 4 (continued)

Name	P	C#	cM	Name	P	C#	cM	Name	P	C#	cM	Name	P	C#	cM
E12M47.170	2	1	96.6	E12M48.122	2	4	6.6					E12M50.091	2	1	43.4
E12M47.159	1	1	91.4	E12M48.105	2	5	20.0								
E12M47.146	2	2	27.7	E12M48.092	1	2	48.8								
E12M47.139	2	4	0.9	E12M48.065	1	2	59.5								
E12M47.136	2	2	28.9	E12M48.057	2	3	34.7								
E12M47.127	2	1	20.1	E12M48.055	1	5	48.3								
E12M59.534	1	4	11.9	E12M60.307	1	3	38.4	E12M61.529	1	3	33.9	E12M62.M1	2	1	41.9
E12M59.495	1	5	36.6	E12M60.164	2	5	71.4	E12M61.405	1	5	26.5	E12M62.323	2	3	31.3
E12M59.369	1	2	26.3	E12M60.151	1	4	10.8	E12M61.255	1	2	42.9	E12M62.240	2	3	31.3
E12M59.302	1	2	11.5	E12M60.134	2	3	33.4	E12M61.253	2	2	42.9	E12M62.217	1	3	39.0
E12M59.288	1	2	16.3	E12M60.123	2	3	38.4	E12M61.215	1	2	19.3	E12M62.165	2	3	37.4
E12M59.269	2	3	48.2					E12M61.163	2	4	42.4	E12M62.162	2	1	43.4
E12M59.168	1	2	53.2					E12M61.146	1	2	47.8	E12M62.156	2	4	10.8
E12M59.162	2	2	53.2					E12M61.091	2	3	35.6	E12M62.147	1	1	6.5
E12M59.160	2	1	29.2					E12M61.084	1	4	68.8	E12M62.144	1	5	7.5
E12M59.157	1	1	24.7					E12M61.077	2	2	18.2	E12M62.125	1	5	22.1
E12M59.116	1	4	34.8					E12M61.074	1	3	8.4	E12M62.124	1	4	49.8
E12M59.099	2	4	10.8					E12M61.053	1	2	17.6	E12M62.119	1	5	63.6
E12M59.093	1	1	47.4									E12M62.106	1	1	37.4
E12M59.077	1	5	20.0												
E12M59.047	1	4	34.8												
E13M47.M3	1	4	36.8	E13M48.492	2	4	32.8	E13M49.M4	2	5	78.7	E13M50.M1	2	5	8.0
E13M47.M1	2	1	15.7	E13M48.248	2	5	20.0	E13M49.M3	1	4	68.8	E13M50.189	2	5	33.4
E13M47.311	1	2	57.8	E13M48.230	2	4	13.4	E13M49.M2	2	4	68.3	E13M50.173	1	1	97.8

E13M47.263	2	4	55.0	E13M48.196	1	4	49.8	E13M49.316	2	5	45.9	E13M50.160	1	5	68.7
E13M47.259	1	4	56.2	E13M48.115	1	5	17.3	E13M49.309	1	4	29.7	E13M50.157	2	1	16.4
E13M47.187	1	4	50.5	E13M48.079	2	3	25.2	E13M49.188	1	5	34.0	E13M50.134	1	5	8.0
E13M47.176	2	3	33.4					E13M49.110	1	3	13.4	E13M50.116	1	5	48.3
E13M47.174	2	2	7.0					E13M49.082	1	1	62.4	E13M50.099	1	1	63.8
E13M47.155	2	1	20.1					E13M49.082	2	1	49.3	E13M50.085	1	4	18.4
E13M47.126	1	3	0.0									E13M50.084	2	4	18.4
E13M47.111	2	1	60.2												
E13M47.100	2	5	20.0												
E13M59.245	2	5	27.5	E13M60.313	1	2	0.0	E13M61.519	2	3	34.7	E13M62.391	1	5	39.5
E13M59.213	2	3	52.4	E13M60.203	2	1	20.1	E13M61.435	1	1	43.4	E13M62.328	2	3	34.7
E13M59.193	2	1	65.8	E13M60.201	1	1	20.1	E13M61.311	2	3	34.7	E13M62.292	1	1	94.2
E13M59.116	1	1	71.8	E13M60.179	2	2	53.2	E13M61.299	1	1	7.8	E13M62.261	2	2	22.4
				E13M60.178	1	2	53.2	E13M61.294	2	1	7.8	E13M62.238	2	5	14.3
				E13M60.168	1	5	54.2	E13M61.196	2	3	5.0	E13M62.088	1	5	47.6
				E13M60.150	2	3	39.0	E13M61.194	1	5	20.0	E13M62.075	2	1	64.4
				E13M60.116	1	3	34.7	E13M61.162	1	4	10.8	E13M62.070	1	4	42.4
				E13M60.083	2	2	17.0	E13M61.150	2	4	42.4	E13M62.052	1	3	37.4
				E13M60.051	2	5	53.4	E13M61.138	2	2	36.6				
				E13M60.048	1	2	16.3								
E14M47.M1	1	4	10.8	E14M48.468	2	5	47.6	E14M49.259	1	4	10.2	E14M50.468	2	2	30.4
E14M47.464	1	3	33.4	E14M48.395	1	5	13.2	E14M49.210	1	4	36.8	E14M50.437	2	5	0.0
E14M47.419	2	3	32.4	E14M48.332	1	2	65.6	E14M49.140	2	2	62.2	E14M50.428	1	5	0.0
E14M47.325	2	5	59.5	E14M48.313	1	2	61.6	E14M49.056	1	2	16.3	E14M50.358	1	3	34.7
E14M47.322	1	5	59.5	E14M48.304	1	3	47.1					E14M50.337	2	3	35.6
E14M47.275	2	4	40.0	E14M48.291	1	5	36.6					E14M50.264	1	3	33.4

(continued)

Table 4 (continued)

Name	P	C#	cM	Name	P	C#	cM	Name	P	C#	cM	Name	P	C#	cM
E14M47.271	2	5	78.7	E14M48.286	1	1	91.4					E14M50.206	1	5	36.6
E14M47.260	2	5	10.8	E14M48.275	2	4	10.8					E14M50.204	2	4	55.0
E14M47.252	2	2	51.8	E14M48.193	2	5	36.6					E14M50.203	1	5	78.7
E14M47.219	1	1	37.4	E14M48.184	1	1	41.9					E14M50.197	2	4	0.0
E14M47.184	1	2	56.4	E14M48.146	1	5	39.5					E14M50.192	2	5	78.7
E14M47.179	1	1	56.3	E14M48.094	2	2	36.6					E14M50.175	2	4	15.2
E14M47.167	1	1	7.8	E14M48.086	2	4	13.4					E14M50.128	1	1	65.3
E14M47.155	2	4	0.9	E14M48.082	1	1	20.1					E14M50.118	2	5	35.3
E14M47.130	1	4	10.2	E14M48.073	2	3	54.5					E14M50.112	2	5	78.7
E14M47.127	2	2	19.3	E14M48.062	1	1	96.6					E14M50.110	1	5	7.5
E14M47.110	2	1	4.5									E14M50.106	2	3	26.7
E14M47.076	2	3	4.1									E14M50.096	2	5	76.6
E14M47.067	2	5	24.2									E14M50.090	1	5	47.6
E14M47.057	1	5	65.8									E14M50.083	2	5	36.6
E14M59.507	1	1	100.0	E14M60.M2	1	3	33.4	E14M61.409	2	3	34.7	E14M62.496	1	5	85.2
E14M59.497	2	1	100.6	E14M60.M1	2	1	43.4	E14M61.394	2	3	9.8	E14M62.458	2	3	7.8
E14M59.316	2	5	27.5	E14M60.490	1	5	70.6	E14M61.329	2	3	34.7	E14M62.388	1	5	62.6
E14M59.280	2	1	81.6	E14M60.457	2	5	71.4	E14M61.237	2	2	53.2	E14M62.333	2	4	32.8
E14M59.274	1	4	10.2	E14M60.408	2	1	65.3	E14M61.183	2	2	38.6	E14M62.315	2	1	42.5
E14M59.268	2	3	5.2	E14M60.317	2	1	43.4	E14M61.173	1	3	8.4	E14M62.293	2	3	39.0
E14M59.240	1	1	28.5	E14M60.275	1	1	42.5	E14M61.109	1	4	36.8	E14M62.254	2	3	3.0
E14M59.238	2	1	29.2	E14M60.255	1	3	3.0	E14M61.068	1	4	32.8	E14M62.198	2	1	44.3
E14M59.212	1	1	81.6	E14M60.250	1	3	35.6	E14M61.063	1	3	33.4	E14M62.152	2	3	34.7

Marker	C#	P	cM	Marker	C#	P	cM	Marker	C#	P	cM	Marker	C#	P	cM
E14M59.200	1	4	40.0	E14M60.225	1	2	13.3	E14M61.060	1	4	66.8	E14M62.146	1	5	36.6
E14M59.183	1	2	38.6	E14M60.208	2	3	34.7	E14M61.052	2	5	48.3	E14M62.099	2	5	34.0
E14M59.159	1	1	37.4	E14M60.169	1	5	90.6					E14M62.092	2	4	10.8
E14M59.123	2	4	34.8	E14M60.164	2	1	4.5					E14M62.061	2	3	34.7
E14M59.118	1	1	12.9	E14M60.162	1	1	4.5								
E14M59.096	1	5	44.3	E14M60.151	1	2	54.3								
E14M59.086	2	3	33.9	E14M60.143	1	2	24.8								
E14M59.084	2	4	9.6	E14M60.142	2	2	54.3								
E14M59.072	2	4	32.8	E14M60.139	2	2	56.9								
				E14M60.119	1	5	80.6								
				E14M60.107	1	2	41.0								
				E14M60.095	1	5	49.6								
				E14M60.084	2	5	14.3								
				E14M60.076	1	3	37.4								
				E14M60.073	1	1	20.9								
				E14M60.055	1	5	36.6								

For the 32 primer combinations we have screened Table 4 has the full marker names, the parental origin (P) of the allele (1 = Landsberg, 2 = Columbia), the chromosome number (C#) and the chromosomal location relative to the RFLP markers (cM). In total, 32 blocks of AFLP markers are shown. Each block has the markers from a single primer combination, all 32 primer combinations are listed. The table allows the identification of primer combinations to target specific chromosomal regions.

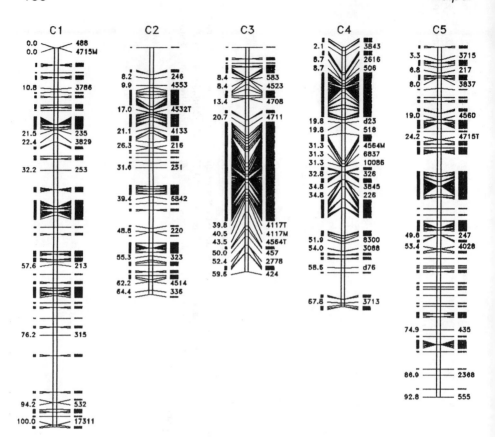

Fig. 2. High density AFLP map: The five linkage groups of the *Arabidopsis* genome are shown. Because of the large number of AFLP markers, their names are shown in very small face. The RFLP marker names are in large face, their centiMorgan position is also given. They may serve as reference points for navigation in the map. The details of the AFLP markers are given in **Table 4**.

corresponding AFLP marker **Table 4**, one may deduce primer combinations optimal for targeting a specific region of the *Arabidopsis* genome.

3.4. Description of the Map

The map shown in **Fig. 2** represents only the first part of our current mapping effort, comprising 32 primer combinations. It covers a total of 387 centiMorgans. The distribution of the markers in the map is given in **Table 5**. It is obvious from **Fig. 2** and **Table 5**, that the highest density of AFLP markers is observed in linkage group C3. The density of AFLP markers detected with the AFLP technology depends on the density of restriction fragments through the genome.

Table 5
Distribution of AFLP Markers over the Different Chromosomes

Chromosome	cM	# Markers
C1	100.6	84
C2	65.6	60
C3	59.6	92
C4	68.8	74
C5	92.8	90

Thus, assuming an equal distribution of polymorphisms over the genome, the density of markers may be taken as a display of the distribution of recombinations over the chromosomes. Dense marker areas thus would coincide with the presence of the centromere. Clearly, there is a pronounced clustering phenomenon in linkage group three where there is a region of 9 cM containing 60 markers, and no RFLP markers.

4. Notes

1. In *Arabidopsis*, *Eco*RI cuts in the order of 30,000 times (extrapolated on the finding, that the average number of fragments amplified with a total of five selective nucleotides is about 60). If each *Eco*RI site is considered to represent an independent locus on the genome, a maximum of 30,000 loci can be screened. Since each *Eco*RI site is bounded by two *Mse*I sites, with this single enzyme combination, a total of close to 60,000 restriction fragments can be detected using all 1024 possible permutations of five selective nucleotides in the primer extensions.
2. The Columbia X Landsberg *erecta* recombinant inbred lines are listed in the *Arabidopsis* Seed List *(2)*, and they are made available through the *Arabidopsis* Stock Centres in Nottingham, UK, and Columbus, OH.
3. Occasionally, highly abundant bands are detected in AFLP fingerprints. These fragments presumably arise from repetitive elements that contain both an *Eco*RI and an *Mse*I site. When amplified, these template molecules consume the majority of the AFLP primers in the PCR reaction, leaving very little primer available for the remaining bands. These fingerprints therefore are usually difficult to evaluate and seldom informative. Using these primer combinations is not recommended.
4. The small genome AFLP primer matrix can also be used on other crops with small genomes. The author has excellent quality fingerprints with rice and cucumber. It is important to note, that primer combinations described here should not be used on yeasts and microorganisms. For such genomes, a special set of primer combinations is needed, with shorter 3' extensions. A specific AFLP kit for microbial fingerprinting is under development. The numbers in **Tables 2** and **3** are approximations only. In general, it is important to note, that the total number of monomorphic and polymorphic bands that can be scored in AFLP finger-

prints very much depends on the quality of the templates and gel images. When the screening was done, the fingerprinting quality was such that all of the polymorphic bandscould not be scored, especially in the higher molecular weight area. Thus, in practice, lower numbers of markers may be scorable than indicated in **Table 3**.

5. Typically, in AFLP fingerprints, both monomorphic and polymorphic loci are detected. The yield of polymorphic loci, or marker bands, depends on the relatedness of the lines used for a screening study.
6. The AFLP reaction is a true multiplex PCR reaction, allowing the simultaneous amplification of up to thousands of restriction fragments. The procedure is not designed for amplification of a single fragment, or even of a few fragments simultaneously. In general, there is a lower limit of fragments that have to be coamplified to warrant reliable and artifact–free amplification. It is prefered not to use primer combinations that target fewer than 10 fragments.
7. When analyzing a population of individuals, preparing templates and preamplifications for all samples simultaneously is advised. Although the AFLP reaction is highly robust and reproducible, occasionally, small differences in intensities of bands are observed between experiments. Presumably slight changes in conditions at either the preamplification or the amplification step may cause minor differences in product concentrations for specific bands. The best comparison of fingerprint patterns can be obtained when all samples are prepared in parallel.
8. For reasons of efficiency, either two *Eco*RI or two *Mse*I primers may be used in a single reaction. The resulting number of fragments will be the sum of the individual fingerprints. One should, however, never use both two different *Eco*RI primers and two different *Mse*I primers in the same reaction. A two-by-two reaction will generate an equivalent of four individual fingerprints. It is recommended to use two *Mse*I primers for some of the E13 and E23 combinations. Individual primer combinations will sometimes give less than 30 bands. When using two *Mse*I primers in a single reaction, it is advised to choose equivalent extensions with respect to G+C content. There is no need to be afraid of problems like the formation of primer dimers. They have never been observed using the optimized protocol.
9. There are some markers in the map that have not been sized, for a number of reasons. These markers have a number in the order of their appearance in the fingerprint, starting from the top of the gel. These markers have yet to be properly sized. All marker information will be made available through the internet and the *Arabidopsis* database.
10. The author has developed an image analysis software package dedicated to the analysis of AFLP images. A full description of this software will be presented elsewhere (manuscript in preparation). Briefly, the software allows the identification and measurement of specific AFLP bands in a pixel image. The origin of the pixel image can be either a phosphorimager, a fluorimager or a sequencer. All these scanner–types provide a dynamic range of sensitivity sufficient for accurate estimation of band intensities. Our software samples a pixel row over the

peaks of bands and the summed pixel information between lane boundaries is used for quantification of the individual bands. As a result, a band can be scored present or absent. With refined quantification procedures, also heterozygosity (corresponding to band intensities 50% of homozygous bands) can be determined (manuscript in preparation).

11. In the analysis of the fingerprints, we encountered several candidates of bi–allelic markers. The majority of the AFLP markers are mono–allelic, resulting from sequence changes in the restriction site or the adjacent nucleotides probed by the primer extensions. Occasionally, however, polymorphism results from insertions or deletions in a restriction fragment. Putative bi–allelic markers are detected with the same primer combination. They usually, but not always, have only a small size difference. Although many of these marker pairs map to a single locus, additional confirmation is needed (sequence information) to exclude the possibility that they are closely linked mono allelic markers in repulsion.

References

1. Vos, P., Hogers, R., Bleeker, M., Reijans, M., Van der Lee, T., Hornes, M., Frijters, A., Pot, J., Peleman, J., Kuiper, M., and Zabeau, M. (1995) AFLP: a new technique for DNA fingerprinting. *Nucl. Acids Res.* **23,** 4407–4414.
2. Anderson, M. and Mulligan, B. (1994) Seed List, June 1994. The Nottingham *Arabidopsis* Stock Centre.
3. Stam, P. and Van Ooijen, J. W. (1995) JoinMap™ version 2.0: software for the calculation of genetic linkage maps. CPRO–DLO, Wageningen.
4. Stam, P. (1993) Construction of integrated genetic linkage maps by means of a new computer package: JoinMap™. *Plant J.* **3,** 739–744

21

Use of Cleaved Amplified Polymorphic Sequences (CAPS) as Genetic Markers in *Arabidopsis thaliana*

Jane Glazebrook, Eliana Drenkard, Daphne Preuss, and Frederick M. Ausubel

1. Introduction

Plant genetic maps can be constructed either with phenotypic or molecular markers. Historically, genetic mapping utilized visible markers, but it is difficult to examine many such markers in a single cross. The recognition that distantly related individuals differ in DNA sequence throughout their genome *(1)* led to the rapid incorporation of DNA markers into mapping strategies. Until recently, the most commonly used DNA markers were restriction fragment length polymorphisms (RFLPs), anonymous single copy number genomic clones that reveal a polymorphism in the length of a restriction fragment, typically by DNA blot hybridization. The RFLP mapping is well suited for determining the genetic location of any newly cloned DNA sequence; the DNA fragment can be used as a hybridization probe (assuming it detects an RFLP) against the DNA filters used to construct the RFLP map. However, many *Arabidopsis* genes are first identified by mutation. Several years ago, mapping such a mutation on to a pre-existing RFLP map was a lengthy procedure requiring the isolation of DNA from individual F_2 plants or F_3 families, and performing DNA blot analysis using each of the RFLP markers as a hybridization probe.

Recently, techniques based on the polymerase chain reaction (PCR) *(2)* have been used to reveal DNA polymorphisms between individuals *(3,4)*, and PCR-generated markers such as random amplified polymorphic DNAs (RAPDs) *(5,6)*, amplified fragment length polymorphisms (AFLPs) *(7)*, cleaved amplified polymorphic sequences (CAPS) *(8)*, and simple sequence length polymorphisms (SSLPs) *(9)* have been developed for *Arabidopsis*.

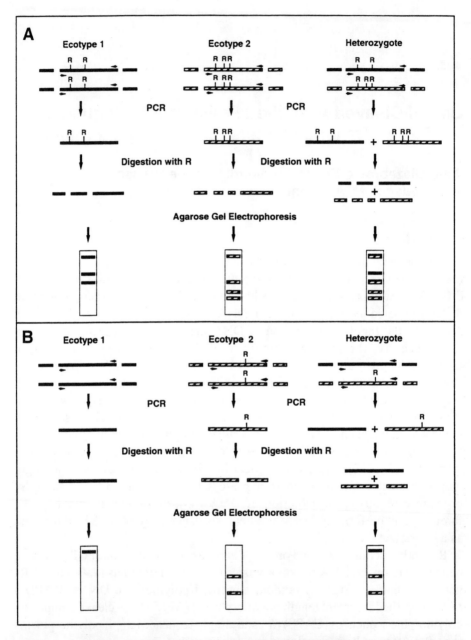

Fig. 1. Assaying CAPS markers by agarose gel electrophoresis. (**A**). A CAPS marker which utilizes a diagnostic restriction endonuclease that cleaves the amplified fragment at either three or two sites. (**B**). A CAPS marker which utilizes a restriction endonuclease that cleaves the amplified fragment at either one site or not at all. In both

This chapter describes the use of cleaved amplified polymorphic sequences (CAPS) for genetic mapping in *Arabidopsis thaliana*. The CAPS method, illustrated in **Fig. 1**, utilizes DNA fragments amplified by the PCR that are digested with a restriction endonuclease to display a restriction site polymorphism *(10–12)*. There are several advantages to the CAPS mapping procedure. First, CAPS markers are codominant, yielding different patterns for plants that are homozygous or heterozygous for the parental alleles. This means that relatively good map positions can be obtained using a fairly small number of F_2 plants, a major advantage for mapping mutations whose phenotypes are difficult to score. Second, because CAPS markers utilize PCR for detection, only a small quantity of DNA is needed to determine a map position; DNA for CAPS mapping can be prepared from portions of F_2 plants using a rapid DNA isolation procedure, rather than from F_3 families. Third, because the difference in size between the cleaved and uncleaved PCR products is relatively large, CAPS markers can be easily assayed using standard agarose gel electrophoresis. Fourth, the CAPS mapping procedure is relatively simple and fast. Assuming that F_2 plants are available, most genes can be mapped by a single investigator in less than 2 wk. Finally, as explained in more detail below, the CAPS mapping procedure is amenable to future automation.

Konieczny and Ausubel *(8)* demonstrated the feasibility of using CAPS markers to map *Arabidopsis* genes by using 18 sets of PCR primers, each of which amplified a single previously mapped and sequenced DNA fragment from both the *Arabidopsis* Columbia and Landsberg *erecta* ecotypes. The amplified products were digested with a panel of restriction endonucleases to identify restriction enzymes that generated ecotype-specific patterns. Using this set of CAPS markers, Konieczny and Ausubel *(8)* showed that an *Arabidopsis* gene could be unambiguously mapped to one of the 10 *Arabidopsis* chromosome arms using only 28 F_2 progeny plants derived from a single cross. Subsequently, many additional CAPS markers have been developed by different investigators. At the time of this writing, a set of 63 CAPS markers has been cataloged by Fred Ausubel and can be accessed via the world wide web browser under:

---

cases, total DNA isolated from ecotype #1, ecotype #2, or from an interspecific hybrid between ecotypes #1 and #2, is used as a template for the polymerase chain reaction using synthetic oligonucleotides as primers (represented by short horizontal arrows) to amplify a DNA fragment. "R" represents the recognition sequence for restriction endonculease R. The amplified fragment is subjected to digestion with restriction enzyme R and the digested fragments are subjected to agarose gel electrophoresis. The rectangular boxes at the bottom of each panel represent an agarose gel and the short horizontal lines represent the patterns of bands observed following electrophoresis and staining with ethidium bromide.

Table 1
CAPS Markers on Chromosome I

Marker	Map position	Fragment size (kb)	Enzyme: ecotype(s)[a], # cuts (size of products in kb)	Primers	Reference # or source
PVV4[b]	1.9 cM on RI map	1.064	*Bsa*AI: C=C24, 2 (0.706, 0.311, 0.047); L, 1 (0.753, 0.311)	5'-GTTTGAAAGTGTAGATGTAACGAC-3' 5'-GGTTGTGTTTTGCTAGCATC-3'	(8)
PAI1[b]	9 cM on RI map	1.4	*Dde*I: C, 1; W, 2 *Sfa*NI: C=L=N, 2; W, 3	5'-GGTACAATTGATCTTCACTATAG-3' 5'-GATCCTAAGGTATTGATATGATG-3'	J. Bender
NCC1[b]	15 cM on RI map	0.97	*Rsa*I: C, 1 (0.87, 0.05); L=C24, 1 (0.92, 0.05)	5'-GTCCTATCTCTACGATGTGGATG-3' 5'-AAGTTATAAGGCATTAGAATCATAATC-3'	(8); T. Altmann, (C24)
PhyA[c]	near m241a	1.3	*Hga*I: C, 1 (1.0, 0.3); L, 0 (1.3)	5'-GATTACTCAACCTCAGTGCG-3' 5'-CGTCATGCAAACTATCAGTGCTCAAACCC-3'	(14)
m59	1.3 cM below g3786, 3.2 cM above g2395	0.78	*Bsh*1236I: C=N, 2 (0.52, 2 x 0.12); L, 1 (0.52, 0.24)	5'-GAATGACATGAACACTTACACC-3' 5'-GTGCATGATATTGATGTACGC-3'	C. Hardtke and T. Berleth
g2395	3.1 cM from pCITN7-31	0.336	*Xba*I: C=N=W, 1 (0.183, 0.153); L, 0 (0.336)	5'-CTCCTTGTCTCCAGGTCCC-3' 5'-GGTTCCATCACCAAGCTCC-3'	C. Hardtke and T. Berleth
m235	44.5 cM on RI map	0.534	*Hind*III: C=N, 1 (0.309, 0.225); L, 0 (0.534)	5'-GAATCTGTTTCGCCTAACGC-3' 5'-AGTCCACAACAATTGCAGCC-3'	C. Hardtke and T. Berleth

[a] C=Col, L=Ler, N=Nd, W=WS.
[b] Marker mapped on RI lines (13).
[c] Marker mapped on RI lines, data not included on RI map.

CAPS as Genetic Markers

http://weeds.mgh.harvard.edu/ausubel.htm/ or http://genome-www.standford.edu/*Arabidopsis*/ (look for *Arabidopsis* information: CAPS marker information and tables for each chromosome) CAPS information can also be obtained via the Gopher under: gopher://genome-gopher.standford.edu/11/*Arabidopsis*/ (look under *Arabidopsis* genetic maps and tables). **Table 1** lists a few of these CAPS markers located on chromosome I and illustrates the format in which the CAPS mapping markers are tabulated. Newly developed CAPS markers can be submitted to this database by sending an e-mail containing the relevant information to Fred Ausubel at ausubel@frodo.mgh.harvard.edu.

Although the number of CAPS markers continues to grow, the usefulness of the current set of CAPS markers is limited by the facts that a variety of different restriction endonucleases, some of which are relatively expensive, are required to display the polymorphisms. A large set of CAPS markers that all utilize the same restriction enzyme would greatly simplify the CAPS procedure. In principle, developing such a set of CAPS markers is feasible because restriction site polymorphisms are abundant in *Arabidopsis* (estimates of differences between Columbia and Landsberg at the nucleotide level range between 1/100 and 1/250 bp) *(15,8)* and are well distributed throughout the genome. For example, this rate of polymorphism translates into 5000–13,000 polymorphic *Alu*I sites between Columbia and Landsberg *erecta* allowing the potential construction of very dense CAPS maps.

2. Materials
2.1. Minipreparation of Arabidopsis DNA

1. Extraction buffer: 100 mM Tris-HCL, 50 mM EDTA, 500 mM NaCl, 10 mM β-mercaptoethanol, pH 8.0. If this solution is prepared without the β-mercaptoethanol, it can be stored indefinitely at room temperature. The β-mercaptoethanol can then be added just before use.
2. 3M Sodium acetate: 3.0M sodium acetate adjusted to pH 5.2 with acetic acid.
3. BTE: 50 mM Tris-HCl, 1.0 mM EDTA, pH 8.0.
4. TE: 10 mM Tris-HCl, 1.0 mM EDTA, pH 8.0.
5. RNaseA: 10 mg/mL DNase free RNaseA dissolved in water and stored in aliquots at –20°C.
6. 20% SDS.
7. 5.0M Potassium acetate.
8. Isopropanol.
9. Ethanol.
10. Liquid nitrogen.
11. Microfuge.
12. Vortex mixer.
13. Water bath or incubator set at 65–70°C.
14. Ice.

15. Mortar and pestle.
16. 1.5-mL microfuge tubes.

2.2. CAPS Reaction

1. 2.5 mM dNTPs: 2.5 mM dATP, 2.5 mM dCTP, 2.5 mM dGTP, 2.5 mM dTTP. Nucleotides should be diluted from commercial molecular biology grade 100 mM solutions, and stored in aliquots at –20°C.
2. Appropriate forward and reverse primers (*see* **Table 1**).
3. Taq polymerase.
4. Taq polymerase 10X buffer (provided by the Taq polymerase supplier).
5. Appropriate restriction enzyme (*see* **Table 1**).
6. 10X concentrated restriction ezyme buffer (recommended or provided by the enzyme supplier).
7. Thermal cycler.
8. Water bath or incubator set at the appropriate temperature for the restriction enzyme used.
9. Tubes that fit the thermal cycler.

3. Methods

3.1. Minipreparation of Arabidopsis DNA

Stategic consideration: How many F_2 plants or F_3 families to use? Konieczny and Ausubel *(8)* calculated that, using codominant markers, linkage between a gene of interest and a marker less than 25 cM away will be detected with 95% confidence using a population of 28 plants that are homozygous at the gene of interest. There are very few, if any (depending on which *Arabidopsis* genetic map is consulted) positions in the genome that are more than 25 cM away from a CAPS marker, so it is reasonable to start a mapping project with a population of 28 plants.

1. Collect about 0.1 g of tissue, place it in a 1.5-mL microfuge tube, and freeze in liquid nitrogen or on dry ice (*see* **Note 1**).
2. Grind the tissue to a fine powder in liquid nitrogen using a mortar and pestle (*see* **Note 2**).
3. Transfer the powder to a 1.5-mL microfuge tube containing 0.5-mL extraction buffer. Vortex 10 s and place on ice until all samples are ground.
4. Add 35 µL 20% SDS and vortex briefly. Incubate at 65–70°C for 10 min.
5. Add 130 µL 5.0M potassium acetate. Shake the tubes vigorously. A white precipitate will form. Incubate on ice for 5 min.
6. Spin the tubes in a microfuge for 5 min at 12,000g.
7. Use a pipet to transfer the supernatant (about 700 µL) into a clean tube. Add 60 µL 3M sodium acetate and 640 µL isopropanol to precipitate the nucleic acids. Mix gently by inversion. Spin the tubes in a microfuge for 5 min at 12,000g.
8. Discard the supernatant. Dissolve as much of the pellet as possible in 200 µL BTE by gently pipeting up and down. The pellet contains some insoluble material.

CAPS as Genetic Markers

9. Spin the tubes in a microfuge for 5 min at 12,000g to pellet the insoluble material.
10. Transfer the supernatant to a clean tube, add 20 µL 3M sodium acetate and 440 µL ethanol. Spin the tubes in a microfuge for 5 min at 12,000g.
11. Remove the supernatant and dissolve the pellet in 100 µL TE containing 0.01 µg/µL RNaseA. Incubate at 37°C for 1 h or 4°C overnight.
12. Add 10 µL 3M sodium acetate and 220 µL ethanol. Spin the tubes in a microfuge for 10 min at 12,000g. Remove the supernatant and wash the pellet with 70% ethanol. Allow the pellet to dry briefly, then dissolve it in 50 µL TE (*see* **Note 3**).

3.2. CAPS Reactions

Strategic consideration: Which CAPS markers to test first? To minimize the number of CAPS markers that must be tested before linkage with the gene of interest is detected, it is best to test a marker near the center of each chromosome first, and then to test markers approx 50 cM away from the previous markers, until linkage is detected. Then any other available CAPS markers in the vicinity can be tested to further define the position of the gene of interest. Workers in many laboratories may wish to take economic considerations into account also. The CAPS mapping requires significant quantities of various restriction enzymes, which vary widely in price. NuSieve agarose is much more expensive than standard agarose. Therefore, another reasonable strategy is to first test markers that depend on inexpensive enzymes, and can be analyzed on 1.5% agarose gels.

1. For each DNA sample to be tested, mix together:
    a. 0.5 µl 2.5 mM dNTPs.
    b. 1.0 µL 20 ng/µL forward primer.
    c. 1.0 µL 20 ng/µL reverse primer.
    d. 1.0 µL 10X Taq buffer (supplied with the enzyme).
    d. 0.1 µL (0.5 U) Taq polymerase.
    e. 5.5 µL water.
    Dispense 9 µL aliquots into microcentrifuge tubes that fit the thermal cycler. To each aliquot, add 1 µL of a DNA sample prepared by the method described above (*see* **Note 4**).
2. Insert the tubes into a thermal cycler, and amplify using the following cycle: 2 min at 95°C, (1 min at 95°C, 1 min at 56°C, 3 min at 72°C) 50 times, 10 min at 72°C (*see* **Note 5**).
3. For each sample to be tested, mix together:
    a. 2 µL appropriate 10X restriction enzyme buffer.
    b. 8 µL water.
    c. 2–10 U appropriate restriction enzyme.
    Add 10 µL to each sample tube and incubate at a suitable temperature for 2 h.
4. Add 2 µL of DNA loading dye. Separate the products on an agarose gel (see **Note 6**).

3.3 Estimation of Genetic Map Distances Using CAPS

If the population used for mapping is composed of F_2 plants that are homozygous for the mutation in the gene of interest, and were derived from a cross between a homozygous mutant of ecotype A crossed to ecotype B, the genetic map distance between a CAPS marker and the gene of interest can be estimated in the following way. Count the number of chromosomes in which the CAPS marker is the ecotype A allele, and the number in which the CAPS marker is the ecotype B allele. (Note that plants in which the CAPS marker is homozygous for the A allele count as two A chromosomes, heterozygotes count as one A chromosome and one B chromosome, and homozygotes for B count as two B chromosomes.) The recombination frequency (r) between the gene of interest and the CAPS marker is then (# of B chromosomes/total chromosomes). In converting recombination frequency to map distance, it is necessary to consider that chromosomes in which two recombination events occured between the markers are counted as having no recombination events, and that this source of error increases with increasing distance between the markers. Futhermore, recombination events can influence the probabilities of a second recombination event occurring in the vicinity, a phenomenon called interference. For *Arabidopsis*, a reasonable estimate of map distance is given by the Kosambi function: $D = 25\ln(100 + 2r/100 - 2r)$, where r is the recombination frequency expressed as a percentage, and D is distance in centiMorgans (cM) *(16)*.

4. Notes

1. The amount of tissue used is not critical. It may vary between 0.05 and 0.2 g. Good results have been obtained using mature leaves, a mixture of leaves, stems, and flowers, and seedlings. Generally, the yield of DNA is considerably higher when samples contain flowers, and when samples consist of seedlings, than it is when only mature leaves are used. If the DNA is to be used only for CAPS, this is not an issue. However, it may be desirable to test a few RFLP markers in addition to CAPS markers (for example, the ARMS markers described in Chapter 22), in which case higher DNA yields are desirable. Tissue may be stored frozen at $-20°C$ for at least 6 mo before processing without adverse effects.
2. If the tissue is not ground to a very fine powder, the DNA yield will be low.
3. This second ethanol precipitation improves the amplification of the DNA in PCR reactions.
4. When working with DNA samples purified on cesium chloride gradients, 8 ng of DNA per reaction gives good results. The amount of DNA in 1 µL of a mini-prep ranges from 10–100 ng, but the PCR reaction works well over this range of DNA concentrations. Konieczny and Ausubel *(8)* reported using oligonucleotide primers at a concentration of 200 ng per 10 µL reaction, but a concentration of 20 ng per reaction gives better amplification, especially when the DNA samples are prepared by miniprep and are of low concentration.

5. If the thermal cycler is not equipped with a heated lid, a small amount of mineral oil must be added to the samples to prevent them from condensing on the lids of the tubes during the reaction. It is still possible to carry out the restriction digests in the same tubes by adding the restriction enzyme mixture under the oil. DNA loading dye may also be added in this manner. In the course of loading the samples into the agarose gel, some of the oil is inevitably transferred into the gel buffer, and onto the electrophoresis apparatus, but is easily removed with soap after the experiment is completed. The amplification program gives good results in standard tubes using water-cooled thermal cyclers or Peltier-effect machines. It requires up to 8 h to run, depending on the ramp time of the machine. Investigators using sophisticated thermal cyclers and thin-walled tubes may find that the incubation times in the cycling reaction can be shortened considerably, and that the product yield can be sufficiently high that it is desirable to cleave only a portion of the product with the restriction enzyme, in order to reduce the amount of enzyme required.
6. Many of the CAPS markers can be successfully scored after electrophoresis in a 1.5% agarose gel. Some markers are based on size differences between fragments < 500 bp long. Resolution of these fragments is greatly improved by using agarose gels consisting of 1% conventional agarose (for strength) and 2% NuSieve GTG agarose (supplied by American Bioanalytical, Natick, MA). Other commercial agarose preparations that can be used at high concentrations should also give good results.

References

1. Botstein, D., White, R. L., Skolnick, M., and Davis, R. W. (1980) Construction of a genetic linkage map in man using restriction fragment length polymorphisms. *Am. J. Hum. Genet.* **32,** 314–331.
2. Mullis, K. B. and Faloona, F. (1987) Specific synthesis of DNA in vitro via a polymerase chain reaction. *Methods Enzymol.* **155,** 355–350.
3. Cox, R. L. and Lehrach, H. (1991) Genome mapping: PCR based meiotic and somatic cell hybrid analysis. *BioEssays* **13,** 193–198.
4. Rafalski, J. A. and Tingey, S. V. (1993) Genetic diagnostics in plant breeding: RAPDS, mcirosatellites and machines. *TIGS* **9,** 275–280.
5. Williams, J. G. K., Kubelik, A. R., Livak, K. J., Rafalski, J. A., and Tingey, S. V. (1990) DNA polymorphisms amplified by arbitrary primers are useful as genetic markers. *Nucl. Acids Res.* **18,** 6531–6535.
6. Reiter, R. S., Williams, J. G. K., Feldman, K. A., Rafalski, J. A., Tingey, S. V., and Scolnik, P. A. (1992) Global and local genome mapping in *Arabidopsis thaliana* by using recombinant inbred lines and random amplified polymorphic DNAs. *Proc. Natl. Acad. Sci. USA* **89,** 1477–1481.
7. Zabeau, M. and Vos, P. (1992) Eur. Patent Appl. 92402629.7 (Publ. 0 534 858A1).
8. Konieczny, A. and Ausubel, F. M. (1993) A procedure for mapping *Arabidopsis* mutations using co-dominant ecotype-specific PCR-based markers. *Plant J.* **4,** 403–410.

9. Bell, C. J. and Ecker, J. R. (1994) Assignment of thirty microsatellite loci to the linkage map of *Arabidopsis. Genomics* **19,** 137–144.
10. Tragoonrung, S., Kanizin, V., Hayes, P. M., and Blake, T. K. (1992) Sequence-tagged-site-facilitated PCR for barley genome mapping. *Theor. Appl. Genet.* **84,** 1002–1008.
11. Weining, S. and Langridge, P. (1991) Identification and mapping of polymorphisms in cereal based on the polymerase chain reaction. *Theor. Appl. Genet.* **82,** 209–216.
12. Williams, M. N. V., Pande, N., Nair, M., Mohan, M., and Bennet, J. (1991) Restriction fragment length polymorphism analysis of polymerase chain reaction products amplified from mapped loci of rice (*Oryza sativa* L.) genomic DNA. *Theoret. Appl. Genet.* **82,** 489–498.
13. Lister, C. and Dean, C. (1993) Recombinant inbred lines for mapping RFLP and phenotypic markers in *Arabidopsis thaliana. Plant J.* **4,** 745–750.
14. Reed, J. W., Nagatani, A., Elich, T. D., Fagan, M., and Chory, J. (1994) Phytochrome A and phytochrome B have overlapping but distinct functions in *Arabidopsis* development. *Plant Physiol.* **104,** 1139–1149.
15. Chang, C., Bowman, J. L., DeJohn, A. W., Lander, E. S., and Meyerowitz, E. M. (1988) Restriction fragment length polymorphism linkage map for *Arabidopsis thaliana. Proc. Natl. Acad. Sci. USA* **85,** 6856–6860
16. Koornneef, M., and Stam, P. (1992) Genetic Analysis in *Methods in* Arabidopsis *Research* (C. Koncz, N.-H. Chua, and J. Schell, eds.), pp. 83–99. World Scientific, Singapore.

22

Mapping Mutations with ARMS

Anton R. Schäffner

1. Introduction

Except the fact that part of T-DNA- or transposon-induced mutations can be directly identified via the T-DNA or transposon tag (*see* Chapters 32–34), non-tagged mutations and mutations induced by chemicals or radiation need to be mapped to chromosomal regions as a first step towards their further analysis and molecular cloning. Besides classical genetic analysis exploiting the segregation of known phenotypic markers with respect to a new mutation (*see* also Chapter 15; *1*) several methods have been developed to achieve mapping by using molecular markers *(2–4)*. These bear the advantage that they are silent by not displaying any phenotypic expression and therefore will not interfere with the mutation itself (*see* also Chapters 20–22). Here, a robust and cheap method for mapping by analyzing restriction fragment length polymorphisms (RFLPs) will be described. A great number of RFLP markers on the *A. thaliana* chromosomes is known *(5–8)* and continuously updated in the *Arabidopsis* databases. RFLPs are codominant markers, thus well suited for mapping as they reveal either homozygosity or heterozygosity at a locus in a Southern experiment using DNA from individuals of a segregating population. However, scoring of several markers might require digestion of DNAs with different restriction enzymes revealing the polymorphisms. To simplify this analysis for an initial rough mapping to chromosomal regions we developed an *Arabidopsis thaliana* RFLP mapping set (ARMS; *3*). The ARMS markers are DNA fragments subcloned into the plasmid vector pBS+/KS (Stratagene, La Jolla, CA) derived from genomic λ- or cosmid RFLP clones *(5)*. All of them detect *Eco*RI polymorphisms and therefore can be scored on a single blot. Furthermore, subcloning was aimed to eliminate all nonpolymorphic signals revealed by the original genomic clones leaving only a single polymorphic band per marker

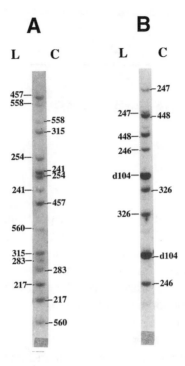

Fig. 1. Autoradiography of combined analysis of ARMS markers. 1 µg of DNA from an F_1 plant heterozygous for Ler (L) and Col (C) had been digested with *Eco*RI and separated on a 0.7% TAE-agarose gel. The bromophenol blue tracking dye migrated approx 20 cm. After transfer to a nylon membrane, the blot was hybridized with eight labeled plasmid DNAs containing eight markers as indicated in **(A)** revealing the Ler (L) and Col (C) polymorphic bands. The probes were stripped off and the membrane anaylzed in the same with five additional markers **(B)**.

and ecotype. Therefore, it is possible to choose and analyze those markers in a single Southern experiment whose signals do not overlap. This greatly accelerates and simplifies RFLP mapping *(3)*. A total of 32 markers had been subcloned, 27 of which reveal polymorphisms for the ecotype crosses Landsberg *erecta* (Ler)/Columbia (Col) and 23 for Ler/Enkheim (En). A selection of 13 or 12 ARMS markers can be combined in two successive Southern experiments without mutual interference of their signals to analyze segregating Ler/Col (**Fig. 1**) or Ler/En populations, respectively. Since these ARMS markers are almost evenly distributed on the five chromosomes and given the known spacing in between them (*3*; **Table 1**) it is sufficient to test about 20 individuals homozygous for the mutant allele. Such a simple experiment allows rough mapping of mutations generated in either Ler, Col, or

Table 1
Chromosomal Position of ARMS Markers and Sizes of Inserts

Polymorphism in mapping population				Marker[a]	Position on chromosome[b]		Size of EcoRI genomic fragment detected (kb)				Sucloned genomic fragments	
Ler/Col	Ler/Ws	Ler/En					Ler	Col	Ws	En	size (kb)	Enzymes for excising insert
	+	+		322A[d]	1	15[d]	10.8	10.8	11.5	13.4	1.1	HindIII, EcoRV
+	+			241A	1	16.0	4.5	5.4	4.9	4.5	4.0	HindIII
+		+		235A	1	39.2	11.6	10.4	11.6	10.4	1.6	HindIII
+	+	+		402A[d]	1	56[d]	9.0	7.8	7.8	7.8	1.6	HindIII, ClaI
+	+	+		254A	1	56.8	6.1	5.1	5.1	5.1	1.6	EcoRI, HindIII
+		+		315B	1	108.6	2.7	7.8	2.7	7.3	1.5	EcoRI, EcoRV
+				453A	1	126.3	7.5	3.0	7.5	7.5	2.2	EcoRI, XhoI
+		+		532B	1	134.6	3.0	10.0	n.t.	10.0	5.0	EcoRI, HindIII
+				132A	1	141.1	5.3	7.2	5.3	5.3	2.6	EcoRI, SpeI, BglI
+	+	+		246B	2	10.8	6.2	2.2	2.2	14.4	2.2	EcoRI
+	+	+		497A	2	13.4	6.5	4.5	6.5	4.4	1.5	EcoRI, HindIII
+				283C	2	58.9	2.6	2.4	n.t.	2.6	0.9	EcoRI, BamHI
	+	+		551C[d]	2	67[d]	3.1	3.1	8.5	6.4	1.6	EcoRI, PstI
+	+	+		336A	2	78.4	8.7	9.9	9.9	9.6	3.8	EcoRI, HindIII
+	+	+		560B1	3	30.8	3.1	1.7	1.7	1.7	1.7	EcoRI
+	+	+		560B2	3	30.8	3.1	1.6	1.6	1.6	1.6	EcoRI
+	+	+		249A	3	72.1	6.6	5.7	5.7	6.3	1.6	EcoRI, HindIII
+		+		457A	3	93.5	12.3	3.9	12.3	13.5	2.5	EcoRI, EcoRV, BglI
+	+			409B[d]	3	95[d]	3.5	3.8	3.8	3.5	1.5	EcoRI, HindIII
+	+	+		424A	3	112.2	5.3	4.6	4.6	n.t.	1.9	EcoRI, HindIII
+	+	+		456A[c,d]	4	25[d]	4.5	3.8	3.8	3.8	3.8	EcoRI

(continued)

Table 1 (continued)

Polymorphism in Mapping Population			Marker[a]	Position on Chromosome[b]		Size of EcoRI genomic fragment detected (kb)				Subcloned genomic fragments	
Ler/Col	Ler/Ws	Ler/En				Ler	Col	Ws	En	size (kb)	Enzymes for excising insert
+	+	+	448A	4	26.7	7.1	8.9	8.9	8.9	2.4	*Eco*RI, *Hin*dIII
+	+	+	518A[c]	4	41.7	10.8	8.1	8.1	8.1	8.1	*Eco*RI
		+	210A[c,d]	4	56[d]	8.7	8.7	n.t.	7.2	8.7	*Eco*RI
+	+	+	326B	4	61.9	3.6	4.3	4.3	4.3	1.0	*Eco*RI, *Hin*dIII
		+	557A[d]	4	86[d]	9.7	9.7	9.7	8.9	3.0	*Hin*dIII, *Bgl*I
+	+	+	d104C	4	95.4	4.8	2.6	2.6	2.6	2.0	*Eco*RI, *Pst*I
+	+		217C	5	10.5	2.1	1.8	1.8	2.1	0.8	*Eco*RI, *Eco*RV
+	+	+	291C	5	37.8	2.2	2.1	2.1	2.1	0.9	*Hin*dIII, *Bam*HI
+	+	+	247A	5	78.4	9.1	12.7	12.7	12.2	1.4	*Hin*dIII, *Bam*HI
+	+	+	558A	5	111.8	11.2	8.8	8.8	8.8	0.9	*Eco*RI, *Sac*I
+	+	+	211A	5	118.9	4.6	4.9	n.t.	4.6	3.2	*Eco*RI, *Hin*dIII, *Bgl*I
+	+	+	pARMS1	1	16.0	4.5	5.4	4.9	4.5	5.6	*Bam*HI, *Xho*I
				1	56.8	6.1	5.1	5.1	5.1		
+			pARMS2	2	58.9	2.6	2.4	n.t.	2.6	1.7	*Bam*HI, *Eco*RV
				5	10.5	2.1	1.8	1.8	2.1		
+		+	pARMS3	3	93.5	12.3	3.9	12.3	13.6	1.6 + 2.0	*Cla*I, *Sac*I
				5	111.8	11.2	8.8	8.8	8.8		
+	+		pARMS4	4	26.7	7.1	8.9	8.9	8.9	3.8	*Xho*I, *Eco*RI
				5	78.4	9.1	12.7	12.7	12.7		
+	+	+	pARMS5	4	61.9	3.6	4.3	4.3	4.3	2.8	*Cla*I, *Pst*I, *Bgl*I
				4	95.4	4.8	2.6	2.6	2.6		

+	+	pARMS6	1	16.0	4.5	5.4	4.9	4.5	7.1	BamHI, XhoI
			1	56.8	6.1	5.1	5.1	5.1		
			1	108.6	2.7	7.8	2.7	7.8		
+		pARMS7	3	93.5	12.3	3.9	12.3	13.6	1.6 + 3.6	ClaI, SacI
			3	30.8	3.1	1.6	1.6	1.5		
			5	111.8	11.2	8.8	8.8	8.3		
+	+	pARMS8	2	10.8	6.2	2.2	2.2	14.4	5.6	XhoI, NotI
			4	26.7	7.1	8.9	8.9	8.9		
			5	78.4	9.1	12.7	12.7	12.7		

[a] For nomenclature of markers see **ref. 3**; the number refers to the original λ-clone *(5)*; d104 is derived from pCITd104 (Meyerowitz lab).
[b] The chromosomal positions are given according to the most recent map based on recombinant inbred lines *(7,8)*.
[c] Smaller, weak hybridizing bands may be observed with these markers.
[d] These markers are not yet mapped on the RI lines *(7,8)*. Therefore the map position was linearly extrapolated in between flanking markers that appear both in the RI map *(8)* and a previously published integrated map *(6)*.
n.t.=not yet tested.

En. A further advantage is that several mutants can be mapped in parallel or successively (re)using the same hybridization solution. Recently, most of the ARMS markers were tested for polymorphisms with another widely used ecotype, Wassilewskija (Ws) (S. Rutherford, personal communication). Twenty-one markers show *Eco*RI polymorphisms for the combination Ler/Ws (**Table 1**). This extends the applicability of the ARMS mapping in a similar way to mutations generated in Ws.

In order to further simplify the handling of ARMS markers and to achieve more uniform signal intensities, several individual marker inserts were combined into a single plasmid. These plasmids were named pARMS (*9*; **Table 1**); e.g., three plasmids pARMS2, pARMS6, and pARMS7 as probes in a single Southern experiment will score eight markers in Ler/Col. All multiple and single marker ARMS clones are available through the DNA stock centers at ABRC in Columbus, OH and at Köln, Germany.

2. Materials

2.1. Digestion, Electrophoretic Separation, and Blotting of Genomic DNA

1. *A. thaliana* DNA of known concentration (*10*; *see* also Chapter 9).
2. *Eco*RI and appropriate incubation buffer.
3. DNase-free RNase A *(11)*.
4. Gel-loading buffer: 20% Ficoll 400, 10 mM EDTA, 0.2 % (w/v) bromophenol blue.
5. TE buffer: 10 mM Tris-HCl, 1 mM EDTA, pH 8.0 *(11)*.
6. Centrifuge for 1.5-mL Eppendorf tubes.
7. Large horizontal electrophoresis chamber allowing a separation of at least 20 cm.
8. TAE-running buffer *(11)* containing 0.5 mg/L ethidium bromide.
9. 0.7% agarose in TAE buffer.
10. Power supply.
11. DNA size marker, e.g., λ-DNA digested with *Bst*EII.
12. UV hand lamp (optional).
13. UV tray and camera.
14. Nylon membrane (Hybond N+, Amersham, Braunschweig, Germany; GeneScreen Plus, DuPont, Bad Homburg, Germany).
15. Pair of scissors or razor blade.
16. Gloves.
17. Vacuum blotting chamber; Pharmacia's VacuGene XL (Uppsala, Sweden) is recommended as its dimension are among the largest that are commercially available.
18. Vacuum pump holding low vacuum of 40 cm H_2O.
19. 0.4M NaOH.
20. 2X SSPE, prepared from a 20X stock solution *(11)*.
21. Plastic wrap.

2.2. Labeling of DNA Probes

1. Plasmid DNA containing marker inserts prepared in mini or large scale by any method described in refs. *11,12* or use of commercial systems.
2. Restriction enzymes according to **Table 1** and appropriate incubation buffers.
3. Gel electrophoresis chamber and TAE-agarose gels.
4. UV tray.
5. Self-made device for spinning and filtering DNA out of an agarose block: puncture a small hole into bottom of 0.5-mL Eppendorf vial and fill in some silanized glass wool (optional).
6. Alternatively to **step 5**, commercial kits for extracting DNA from agarose gels, e.g., QiaQuick Gel Extraction Kit #28704 (Qiagen, Hilden, Germany)
7. Waterbath or heat block for boiling.
8. Ice-water bath.
9. dNTP.
10. Either one of the dNTPs as a labeled α-[^{32}P] dNTP (3000 Ci/mmol) nucleotide.
11. 10X hexanucleotides and reaction buffer (Boehringer Mannheim, Mannheim, Germany, # 1277081): 500 mM Tris-HCl, pH 7.2, 100 mM MgCl$_2$, 1 mM DTT, 2 mg/mL BSA, 62.5 OD/mL random hexanucleotides.
12. Klenow DNA polymerase.
13. Water bath or heat block at 37°C.
14. Centrifuge for 1.5-mL Eppendorf tubes.
15. Stop solution: 10 mM EDTA, 100 mM NaCl, 0.05 mg/mL linear polyacrylamide (*see* **Note 1**).
16. Hand monitor for measuring radioactivity.

2.3. Hybridization

1. (Pre)hybridization solution: 5X SSPE *(11)*, 10% (w/v) dextrane sulfate, 1% (w/v) SDS, 100 µg/mL heat-denatured herring sperm DNA (*see* **Note 2**).
2. Glass roller bottle and hybridization oven.
3. Alternatively to **step 2**, hybridization bag and sealing device.
4. Boiling waterbath or heat block.
5. Ice-water bath.
6. 50-mL polypropylene tube (optional).
7. 2X SSPE, 0.5% SDS *(11)*.
8. 0.5X SSPE, 0.5% SDS *(11)*.
9. Waterbath at 68°C for prewarming 2X SSPE, 0.5% SDS.
10. Waterbath shaker at 68°C with a glass or plastic container if the washing is not done in a roller bottle in hybridization oven.
11. Plastic wrap.

2.4. Evaluation

1. X-ray film and intensifying screen.
2. Alternatively to **step 1**, phosphorimager and plates.

3. Film cassette.
4. Solutions and/or machine for developing X-ray film.
5. 0.4M NaOH.
6. 2X SSPE *(11)*.
7. Plastic wrap.

3. Methods

3.1. Selection of Crosses and F₂ Individuals

According to the occurence of the majority of polymorphisms, mutations generated in Ler background should be crossed with Col, and vice versa. Ws- and En-derived mutation are best analyzed after crossing with Ler. Since the ARMS mapping as a simplified RFLP method depends on the analysis of codominant markers, any recessive or dominant, lethal, or viable mutation can be mapped if basic genetic consideration are taken into account. In all cases, it will be necessary to determine the status of an individual out of a segregating population at the mutant locus as heterozygous or homozygous by a simple genetic analysis. If plants homozygous at the mutant locus or F_3 families representing them are chosen, then a rough mapping will be possible by scoring about 20 individuals with the ARMS markers (*see* **Note 3**).

3.2. Growth of Plant Material and Isolation of DNA

The preparation of DNA is described in detail in Chapter 9 by J. Chory in this book. However, any preparation yielding digestible DNA for Southern analysis will be applicable. The method according to **ref. *10*** or genomic tips from Qiagen (Hilden, Germany) to extract DNA from about 1 g of plant material (F_3 families) grown under sterile conditions in liquid culture is usually used.

3.3. Restriction Enzyme Digestion, Electrophoretic Separation of Genomic DNA, and Transfer to Nylon Membrane

1. 1 µg *A. thaliana* DNA is digested in a reaction volume of 300 µL with 30 U of *Eco*RI at 37°C for at least 8 h (*see* **Note 4**). After phenolization the DNA is precipitated by 2.5 vol of ethanol and redissolved in 25 µL TE-buffer and 5 µL gel-loading buffer (20% Ficoll 400, 10 mM EDTA, 0.2 % (w/v) bromophenol blue; *see* **Note 5**). Always include digests of the parental lines which had been used for the cross. This will serve as an internal reference and verify the polymorphic bands unequivocally. Furthermore, it will exclude or identify spurious bands due to incomplete digestion or "genetic contamination."
2. Separate the digests on a 0.7% TAE-agarose containing 0.5 mg/L ethidium bromide at 1 V/cm in a cold room for about 20 h. The bromophenol blue tracking dye should migrate at least 20 cm and/or it might just start to run out of the gel. All bands with a higher mobility (approx 1.2 kb) will not be detected by any of the ARMS markers.

3. Two lanes with a λ-DNA digest as size marker should be included: approx 500 ng which are visible by UV illumination during (UV hand lamp) and after the gel electrophoretic separation and 1 ng directly adjacent to the *A. thaliana* lanes which will be blotted and might serve as a size marker in the Southern experiment. After the run has been completed a photograph may be taken under UV illumination for record of the run and loading of the lanes.
4. A piece of positively charged nylon membrane is cut according to the dimensions of the gel after excising unused lanes. Also, the region well above 15 kb can be excised to minimize the size of the blot. Always wear gloves when handling the nylon membrane. Use a pencil to label the orientation of lanes and the side of the membrane onto which the DNA is to be transferred. DNA is transferred by vacuum or pressure assisted commercial equipment according to manufacturer protocols (*see* **Note 6**). Alkaline transfer with a vacuum blotting chamber which accommodates gels with dimensions up to 20 cm × 30 cm (Pharmacia VacuGene XL) is routinely used. After the transfer (e.g., about 1 h at 40 cm H_2O with $0.4M$ NaOH), the gel is removed and the membrane is washed and neutralized with 2X SSPE buffer *(11)*.
5. Either procede immediately with the hybridization or dry, wrap in plastic wrap, and store the blot at 4°C (–20°C for longer periods).

3.4. Selection of ARMS Markers

Depending on the particular aim of the experiment, the individual ARMS markers will be chosen. Usually there is not any information about the position of the locus where a particular mutation maps. In this case, a selection covering the whole genome almost evenly will be analyzed first as suggested below for crosses Ler/Col, Ler/Ws, and Ler/En. This will be sufficient to roughly locate the locus. (*see* **Note 7**). If a map location is already known, e.g., by previous analysis of ARMS, SSCP, or CAPS markers or by classical genetics, then only the relevant markers will be chosen for verifying these data or further narrowing the region containing the locus. Refer to the sizes of polymorphic bands given in **Table 1** to know whether different markers can be combined in a single experiment.

3.4.1. Markers for Mapping Mutations in Ler/Col Crosses

Only five multimarker plasmids have to be analyzed *(9)*. Plasmids pARMS2, pARMS6, and pARMS7 can be combined in a first Southern experiment to score eight markers 217C, 241A, 254A, 283C, 315B, 457A, 558A. After striping the blot in a second round two further plasmids, pARMS5 and pARMS8, will analyze five additional markers, 246A, 247A, 326B, 448A, and d104C (**Table 1, Fig. 1**).

3.4.2. Markers for Mapping Mutations in Ler/Ws Crosses

As for Ler/Col crosses, pARMS5 and pARMS8 will analyze markers 246A, 247A, 326B, 448A, and d104C in individuals from a segregating Ler/Ws population (**Table 1**). Markers 217A, 241A, 249A, 402A, and 560B2 (**Table 1**) can be combined in a second hybridization. If necessary, 558A at the bottom of chromosome 5 might be scored in a third experiment. Unfortunately, at present mutations at the bottom of chromosomes 1 and 2 might escape detection since none of the tested ARMS markers showed an *Eco*RI polymorphism for Ler/Ws.

3.4.3. Markers for Mapping Mutations in Ler/En Crosses

Plasmid pARMS5 (harboring markers 326B and d104C) and individual markers 315B, 448A, 457A, and 551A can be combined in one Southern experiment. A second hybridization including individual marker plasmids 254A, 291C, 322A, 497A, 558A, and 560B2 will completely cover the whole genome (**Table 1**).

3.5. Labeling of DNA Probes

Random-primed labeling using radioactively labeled nucloetides can be done using the whole plasmid DNA or isolated fragments as templates *(11)* (*see* **Notes 8** and **9**).

1. Prepare plasmid DNA of individual marker clones or pARMS clones by any miniprep or large scale plasmid isolation method *(11)*. Estimate concentration by gel electrophoresis with a known standard or measure optical density at 260 nm. Check integrity of plasmids by digests given in **Table 1**.
2. For preparation of fragments (*see* **Note 10**) digest approx 200 ng of plasmid DNA (or more for several applications) with the appropriate restriction enzyme(s) (**Table 1**) and separate on a 1% agarose gel. Identify correct band (**Table 1**) under UV illumination through ethidium bromide fluorescence and cut out from the gel with razor blade removing most of the surrounding agarose. To elute the DNA fragment use a self-made device or a filter tip to spin out the DNA into an 1.5-mL Eppendorf tube for 1 min (30–50% recovery). DNA should be concentrated by EtOH precipitation and dissolved in 10 µL of H_2O. Commercial kits are also available for extracting DNA from agarose gels, which give higher yields, e.g., via binding and elution from a spin-column yielding the fragment directly in a volume of 30 µL of H_2O.
3. Either about 200 ng of plasmid DNA diluted into 10 µL H_2O or 10 µL of the eluted fragment are heat-denatured by boiling the tube in a waterbath for 3–4 min (*see* **Notes 10** and **11**). Immediately transfer into an ice-water bath assuring that the tube is fully immersed in ice-water, put also some ice on top of the cap. Leave for at least 3 min.
4. A small scale reaction can be used for random-primed labeling minimizing costs and the production of radioactive waste. The labeling reaction is set up as follows for one individual ARMS marker (*see* **Notes 12** and **13**) :
   a. 0.4 µL 10X hexanucleotides and reaction buffer (Boehringer Mannheim, Mannheim, Germany).

b. 0.6 µL 0.66 m*M* (each) dNTP, omitting the labeled nucleotide (*see* **Note 14**).
c. 1.0 µL α-[^{32}P] dNTP (10 µCi of a 3000 Ci/mmol solution; *see* **Note 14**).
d. 0.2 µL Klenow DNA polymerase (2 U/µL).
Add 1.8 µL of the heat-denatured DNA and mix immediately with the reagents. Some of the water has been condensed at the cap during boiling, however, do not spin the tube, just take the aliquot of denatured DNA from the remaining solution at the bottom of the tube.
5. Tightly close the cap, spin briefly and incubate at 37°C for 30–60 min.
6. To stop the reaction and estimate the incorporation (*see* **Note 15**) add 36 µL of a stop solution and 120 µL of ethanol to precipitate at –70°C for 30 min. Before spinning (15 min/4°C) at full speed in a microcentrifuge estimate the radioactivity of the whole tube in a fixed position in front of a hand monitor. After spinning carefully, remove the supernatant by pipeting into another fresh tube and measure its radioactivity as before. Subtract this value (unincorporated label) from the previous measurement to obtain an estimate for the incorporation into labeled product (*see* **Note 16**). Usually 50–80 % will be incorporated. This small scale reaction will contain sufficient label for up to 25 mL hybridization solution (*see* **Subheading 3.6.**; **Note 17**).
7. Dissolve labeled precipitate in 50 µL H$_2$O and freeze at –70°C or procede to **Subheading 3.6.2.**
8. Since pARMS contain more than one marker, a double-sized reaction should be set up. The DNA template, however, still may be denatured as described (**Subheading 3.5.3.**). For pARMS2 containing markers 217C and 283C the reaction volume should be triplicated (*see* **Note 17**).

3.6. Hydridization

1. To block unspecific binding sites on the membrane it is prehybridized for at least 2 h at 68°C in 5X SSPE, 10% dextrane sulphate, 1 % (w/v) SDS, 100 µg/mL heat-denatured herring sperm DNA *(11,12*; *see* **Notes 2** and **18**). Either seal the membrane into a hybridization bag and incubate in a water bath or put into an cylindrical hybridization glass roller bottle if a hybridization oven is available. Use approximately 1 mL/20 cm^2 of your membrane (*see* **Note 19**).
2. Denature labeled probe(s) by boiling for 5 min (*see* **Note 11**) and immediate chilling in an ice-water bath as described above (**Subheading 3.5.3.**). Again spinning is not necessary; if the tube has to be spun, use a refrigerated microfuge or even denature once more.
3. Add denatured probe directly into prehybridization solution taking into account major differences in labeling efficiency (**Subheading 3.5.6.**). If a λ-DNA marker (1 ng-lane, see **Subheading 3.3.3.**) is to be visualized on the Southern blot, label λ-DNA in the same way as markers and add to the hybridization solution (*see* **Note 20**).
4. Continue incubation in hybridization oven or waterbath at 68°C for at least 15 h.
5. Decant hybridization solution into a 50-mL polypropylene tube if it is to be reused and stored at –20°C (*see* **Note 21**). Otherwise, put to radioactive liquid waste according to the regulations in your department.

6. Rinse membrane once in 2X SSPE *(11)*, 0.5% SDS prewarmed to about 68°C.
7. To remove unspecifically bound probes wash blot twice for about 15 min with 2X SSPE, 0.5% SDS at 68°C followed by two washes with 0.5X SSPE, 05% SDS at the same conditions. All washing solutions should be properly transferred to the liquid radioactive waste. The last washing solution, however, should contain no radioactivy detectable by a hand monitor. If higher amounts are measured, continue washing with fresh 0.5X SSPE, 0.5% SDS (*see* **Note 22**).
8. Remove liquid from membrane by putting in between Whatman papers and leave on the bench for 5 min. However, avoid drying the blot completely which could cause problems if the probes are to be stripped off for another hybridization (**Subheading 3.7.**).
9. Wrap blot into plastic wrap.

3.7. Evaluation of Blot

1. Expose the blot with DNA side facing the X-ray film/intensifying screen or the detection system of a phosphoimager or betacounter. The time for exposure (O/N up to several days) will depend on the sensitivity of the film/intensifying screen or the phosphoimager, on the quality of the blot (e.g., how often it had already been used) and on the specific radioactivity of the probes used for hybridization.
2. Print out result or develop X-ray film.
3. Use the two lanes containing DNA from parental strains and the λ-DNA size marker (if labeled λ-DNA has been included in the hybridization) to identify polymorphic pair of bands corresponding to the individual markers and label.
4. Using these references, score each individual lane as either homozygous for one of the parental backgrounds (A or B) if only one of the respective polymorphic signals is detected or as heterozygous (H) if both bands appear. Collect these data into a table listing markers vs individuals (*see* **Note 23**).
5. If all individuals are homozygous for background A at the locus of the mutation (*see* **Subheading 3.1.**) then scoring A, B, or H will reveal 0, 2, or 1 recombination events between a particular marker and the mutation.
6. Count the number of nonrecombinant and recombinant events for each marker in the population analyzed. Most probably, the mutation will map to a region defined by the marker showing the lowest number of recombinations. If, e.g., 20 plants, i.e., 40 meiotic events had been analyzed ratios of recombinants vs nonrecombinants from 0/40 to 9/31 indicate a highly significant linkage (LOD score 11.9–2.8) with a recombination frequency lower than 0.25 (genetic map distance < 25 cM). Ratios ranging from 10/30 to 13/27 are still indicative of a linkage within less than 35 cM (LOD score 2.3–1.1; *see* **Note 24**). If there is not any tight linkage the first set of experiments with two to three markers per chromosome will reveal at least one or two markers with such a weak linkage. The analysis of another ARMS marker or any other molecular marker from that region will immediately verify or exclude the linkage in this case.
7. Due to the low number of events analyzed, the calculation of recombination frequencies is not accurate at this stage. However, it will give a rough idea to

guide further analyses by defining a region of the published maps from which additional markers for verification and further fine mapping in a larger population are to be choosen. At this level, recombination frequencies p are simply calculated by dividing the number of observed recombination events by the the total number of alleles analyzed (*see* **Note 25**):

p = (number of recombination events observed) (total number of alleles analyzed)$^{-1}$.

8. After evaluation, the probes may be stripped off the membrane by washing the blot twice in 0.4M NaOH for 15 min each. Check the complete removal of radioactivity by a hand monitor. Neutralize the membrane with 2X SSPE buffer *(11)* and blot dry onto Whatman paper. Store the dried membrane wrapped into plastic wrap (–20°C) or start another hybridization (**Subheading 3.6.1.**).
9. To complete, confirm, or narrow the mapping additional ARMS markers from the region identified may now be analyzed on the same blot (**Subheading 3.6.1.**).

4. Notes

1. Linear polyacrylamide is a cheap and efficient carrier for precipitating DNA prepared by polymerizing acrylamide according to **ref. *13***.
2. The (pre)hybridization solution is prepared as follows: 20 mL 2.5 mg/mL sheared herring sperm DNA are boiled for 10 min. Pour DNA solution into beaker containing precooled (ice bath) 250 mL H_2O and 125 mL 20X SSPE *(11)*, add 25 mL 20% SDS, and 50 g dextrane sulphate. Heat this mixture in 68°C waterbath until SDS and dextrane sulfate are completely dissolved. Add H_2O ad 500 mL. Use in hybridization and/or aliquot and freeze at –20°C. Thaw frozen stock in a 68°C waterbath before use.
3. Two examples for selecting appropriate plants are given. First, a recessive, viable mutation generated in Col will allow the immediate identification of mutants (25%) among the F_2 population. If these are chosen, all of them are homozygous for Col at the mutant locus. Second, a recessive lethal mutation generated in Ws was crossed with Ler. With regard to the mutant locus, only heterozygous individuals or plants homozygous for Ler will be recovered from the F_2 population. This has to be revealed in the F_3 generation and the latter may be chosen for the ARMS analysis. Similarly wild-type plant may be selected immediately in the F_2 generation if the mutation is dominant.
4. If DNA preparations contain RNA as it is the case for DNA isolated according to *(10)*, 1 µL of 1 mg/mL DNase-free RNase A *(12)* is added to the reaction.
5. To quickly check for complete digestion and for confirming the amount of DNA, we usually digest 1.3 µg of DNA. After redissolving run 6 µL (1/5th of the digest) on a mini 0.7% agarose gel at 40 V for 1 h. Although not well separated, a characteristic pattern due to repetitive DNA and the organellar DNA content will be visible.
6. An alternative and less expensive method, although more time consuming is the transfer by capillary action overnight *(11)*.
7. If the mapping were not unequivocal, additional individuals and/or markers out of the ARMS collection or others described in this book (*see* Chapter 21) will help clarify the mapping.

8. Instead of random priming, the nick-translation technique is also suitable for preparing labeled DNA probes. Nonradioactive labeling methods can be used as an alternative to radioactive labeling *(11)*.
9. All necessary care for clean and safe handling of radioactive material has to be taken and legislative regulations need to be obeyed. Wear gloves when working with radioactive material *(see* **Note 8**).
10. We usually use the whole plasmids for labeling; however, with some genetic material, spurious bands hybridizing to vector sequences might be observed. As more different markers are combined, increasing amounts of labeled vector are added to the hybridization. If such problems occur, purified fragments give cleaner results.
11. Avoid the heat popping the cap. Use tubes with a special locking device.
12. The volumes given are those used per one labeling reaction of a single ARMS marker; usually several markers will be labeled; therefore a cocktail containing the individual reagents can be prepared on ice, mixed, and distributed into individual tubes (2.2 µL). Start the reaction(s) without delay by adding and mixing the denatured DNA directly into the labeling mix.
13. The reagents used except α-dNTP and dNTP are from Boehringer Mannheim (Mannheim, Germany); the protocol follows their suggestions in a down-scaled version. However, any other reagents may be used according to the manufacturer's suggestions but may be down-scaled in a similar way.
14. Either α-[^{32}P] dATP, α-[^{32}P] dCTP, or α-[^{32}P] dTTP can be used; the respective labeled nucleotide, however, should be omitted from the unlabeled nucleotides.
15. If several independent markers are labeled and are to be used in a combined hybridization, it helps in obtaining uniform signals if approximately similar amounts of labeled products are used for the experiment *(see* **Note 16**).
16. This is a rough estimation since without washing, part of the labeled nucleotides will also be precipitated. Never compare the high radioactivity of the left precipitate itself with the other measurements because there is no more quenching by the liquid.
17. Markers 217C and 283C consistently give weaker signals in the Southern experiment. Thus, it is recommended to use two- to three-fold the amount of radioactively labeled probe if either of them is analyzed together with other ARMS markers. Also 217C and 283C were combined into one pARMS clone, pARMS2.
18. If a membrane has been already used for hybridization, this time might be reduced to 1 h.
19. If more than one membrane is hybridized in one bottle or bag, some additional buffer has to be added to soak the additional membrane.
20. Avoid immediate contact of labeled probe(s) with membrane before it has been diluted with the prehybridization solution.
21. Before reusing a hybridization solution, thaw and then boil for 10 min. Allow to cool down to about 65°C, and add to the prehybridized membrane. Remove prehybridization solution before.

22. If a glass bottle has been used for hybridization, the washing steps can be performed in it.
23. If a marker is not unequivocally scorable, score not detectable (N) in your table.
24. LOD scores indicate the (logarithmic) probability of a linkage at a given recombination frequency p vs the assumption of no linkage (recombination frequency $p = 0.5$) based on the number of observed recombinants:

$$\text{LOD} = \log [p^{\text{number of recombinations}} (1-p)^{\text{number of nonrecombinants}} 0.5^{-(\text{total number of alleles})}].$$

25. If a significantly higher number of alleles has been scored the formula by Kosambi *(14)* is used to calculate more accurate map distances, since it corrects for double crossovers. According to this formula, the genetic map distance x is calculated from the observed recombination frequency p:

$$x = 0.25 \ln [(1 + 2p)(1 - 2p)^{-1}].$$

Acknowledgment

Steven Rutherford, York, UK kindly contributed the information on Ws/Ler and Ws/Col polymorphisms. I thank Christoph Fabri, München, Germany for discussion and technical help. This work was supported by the Deutsche Forschungsgemeinschaft (DFG 454/2-1,2) and by the Fonds der Chemischen Industrie.

References

1. Koornneef, M. and Stam, P. (1992) Genetic analysis, in *Methods in* Arabidopsis *Research* (Koncz, C., Chua, N.-H., and Schell, J., eds,), World Scientific, Singapore, pp. 83–99.
2. Konieczny, A. and Ausubel, F. M. (1993) A procedure for mapping *Arabidopsis* mutations using co-dominant ecotype-specific PCR-based markers. *Plant J.* **4,** 403–410.
3. Fabri, C. O. and Schäffner, A. R. (1994) An *Arabidopsis thaliana* RFLP mappingset to localize mutations to chromosomal regions. *Plant J.* **5,** 149–156.
4. Bell, C. and Ecker, J. R. (1994) Assignment of microsatellite loci to the linkage map of *Arabidopsis. Genomics* **19,** 137–144.
5. Chang, C., Bowman, J. L., DeJohn, A. W., Lander, E. S., and Meyerowitz, E. M. (1988) Restriction fragment length polymorphism linkage map for *Arabidopsis thaliana Proc. Natl. Acad. Sci. USA* **85,** 6856–6860.
6. Hauge, B. M., Hanley, S. M., Cartinhour, S., Cherry, J. M., Goodman, H. M., Koornneef, M., Stam, P., Chang, C., Kempin, S., Medrano, L., and Meyerowitz, E. M. (1993) An integrated genetic/RFLP map of the *Arabidopsis thaliana* genome. *Plant J.* **3,** 745–754.
7. Lister, C. and Dean, C. (1993) Recombinant inbred lines for mapping RFLP and phenotypic markers in *Arabidopsis thaliana. Plant J.* **4,** 745–750.
8. Lister, C. and Dean, C. (1995) RFLP map, in Multinational Coordinated *Arabidopsis thaliana* Genome Research Project. Progress report : Year four, NSF publication 95–43, pp. 40,41.

9. Schäffner, A. R. (1996) pARMS, multiple marker containing plasmids for easy RFLP analysis in *Arabidopsis thaliana*. *Plant Mol. Biol. Rep.* **14,** 11–16.
10. Dellaporta, S. L., Wood, J., and Hicks, J. B. (1983) A plant DNA minipreparation: version II. *Plant Mol. Biol. Rep.* **1,** 19–21.
11. Ausubel, F. A., Brent, R., Kingston, R. E., Moore, D. D., Seidman, J. G., Smith, J. A., and Struhl, K. (eds.) (1995) Current Protocols in Molecular Biology. Greene Publishers and Wiley-Interscience, New York.
12. Sambrook, J., Fritsch, E. F., and Maniatis, T. (1989) *Molecular Cloning. A Laboratory Manual.* Cold Spring Harbor Laborartory, Cold Spring Harbor, New York.
13. Gaillard, C. and Strauss, F. (1990) Ethanol precipitation of DNA with linear polyacrylamide as carrier. *Nucleic Acids Res.* **18,** 378.
14. Kosambi, D. D. (1944) The estimation of map distances from recombination values. *Ann. Eugen.* **12,** 172–175.

23

Mapping Cloned Sequences on YACs

Francis D. Agyare, Gus Lagos, Deval Lashkari, Ronald W. Davis, and Bertrand Lemieux

1. Introduction

The mapping of a large number of cloned sequences on yeast artificial chromosome (YAC) clones is a significant technical challenge. Although yeast colony lifts are suitable for hybridizations with a limited number of probes, this approach can be difficult to scale up to accommodate hundreds or thousands of probes. We have developed a strategy to map *Arabidopsis thaliana* expressed sequence tags (EST) to YACs, which is entirely based on the use of the polymerase chain reaction (PCR) to amplify specific plant genomic sequences. Although we are using these methods to map ESTs *(1,2)*, the same approach can be used to map any sequence onto YAC clones.

The present protocol uses a YAC clone pooling strategy, called *j*–detector pooling *(3)*. The screening process is done in two steps. First, YAC clones are grouped in sets of 36 clones each (called primary pools). DNA samples isolated from these primary pools are arrayed on a 3 × 12 grid which is sampled to generate "super *j*–detector pools" using the *j*–detector key presented in **Fig. 1**. Screening these super *j*–detector pools by PCR allows us to identify which pool of 36 clones contains a given sequence. The PCR reaction products are detected by agarose gel electrophoresis (**Fig. 1**). Second, individual clones within a given primary pool which contain an EST are identified by amplifying EST sequences from a set of "*j*–detector pools" derived from the primary pool (**Fig. 1**). This design has a built-in tolerance for one false negative because each YAC clone is present in five of the 10 pools and each pool shares no more than three YAC clones with other pools such that detection of a PCR product in four pools is sufficient to identify a YAC clone which contains a given EST (**Fig. 1**). The use of super *j*–detector pools in combination with *j*–detectors allows us to screen sets of 36 pools of 36 clones from a given YAC bank with as little as 24 PCR reactions.

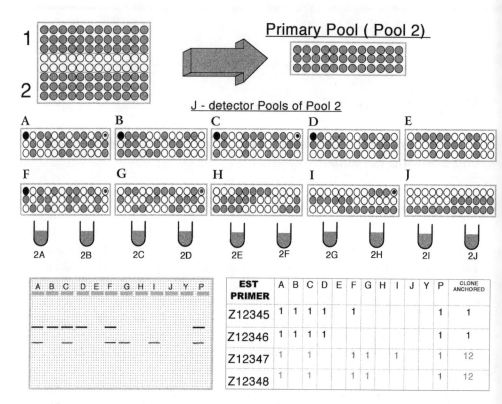

Fig. 1. *J*–detector pooling scheme: This particular example is for the secondary phase of screening. Each 96–well minitube assembly contains two primary pools of 36 clones. A set of 10 *j*–detector pools are assembled by sampling each clone five times. The sampling for each of the 10 *j*–detector pools is represented by a set of 3 × 12 grids, with shaded wells indicating the clones sampled, whereas the pools from which DNA is extracted are shown below the grids. The black circles in the 3 × 12 grids indicate that clone 1 was sampled to generate pools A, B, C, D, and F whereas the gray dot represents the sampling of clone 12 was for pools A, C, F, G, and I. After extraction of DNA from these 10 pools, a PCR screen with a pair of EST amplifying primers is performed and the amplicons analyzed by gel electrophoresis. Yeast DNA (Y) and plant DNA (P) control templates are included in the 11th and 12th lanes. In this example, two hypothetical ESTs (Z12345 and Z12347) mapping to the 1st and the 12th clones are shown. The black bands on the gel were generated by template DNA in clone 1 (indicated by black circles in the key), whereas the gray bands are from clone 12 (gray dots in the key). Also shown is the spreadsheet table on which data is compiled. Note, even ESTs which only yield four amplicons (e.g., Z12346 and Z12348) can be assigned to a clone. This particular detector scheme was conceived by Balding *(3)*.

Table 1
Amino Acid Stock Solutions

Amino Acids	Stock 1,2	For SD++
His	1 g/50 mL	2 mL/L
Arg	1 g/50 mL	2 mL/L
Met	1 g/50 mL	2 mL/L
Tyr	400 mg/100 mL	15 mL/L
Leu	1 g/50 mL	3 mL/L
Ile	1 g/50 mL	3 mL/L
Lys	1 g/50 mL	3 mL/L
Phe	1 g/50 mL	5 mL/L[a]
Glu	1 g/50 mL	10 mL/L[a]
Asp	2 mg/100 mL	10 mL/L[a,b]
Val	3 g/50 mL	5 mL/L
Thr	4 g/50 mL	5 mL/L[a,b]
Ser	8 g/50 mL	5 mL/L

[a] Stored at room temperature.
[b] Add after autoclaving.

We have been screening the CIC and yUP YAC libraries produced by the CNRS/INRA/CEPH collaborative group and J. Ecker (University of Pennsylvania), respectively *(4,5)*. We have also included some clones from the EG and EW YAC libraries *(6,7)* which do not contain repetitive sequences or organelle DNA *(8)*. Note, some of the EG and EW clones have previously been mapped by hybridization to restriction fragment length polymorphism markers (RFLP) *(9)*.

2. Materials
2.1. Yeast Culture

1. Growth medium: YAC clones are grown in standard defined medium (SD) which contains: 0.17% Bacto Yeast Nitrogen base (Difco, E. Molesly, Surrey, UK), 4% dextrose/sucrose, 1% NH_4SO_4. This liquid media is called SD^{++} after it has been supplemented with adenine sulfate to a concentration of 25 mg/L and a series of essential amino acids. **Table 1** lists the stock solutions we use to make SD^{++}. Doubling the concentrations of the amino acids enables the cells to grow faster.
2. YAC Clones: YAC clones are maintained in the AB1380 yeast strain (*Mat–a, can1–100, lys2–1, ade2–1, trp1, ura3, his5[psi +]*). Clones are stored at –80°C in 15% glycerol (Caution! yeast tend to die if stored at a temperature above –55°C). All the *Arabidopsis* YAC banks we are using are available from the *Arabidopsis* Biological Resource Center (ABRC) (http://aims.cps.msu.edu/aims/).
3. Beckman 96-well minitube assemblies.

2.2. DNA Extraction Protocol

1. 10% SDS pH 7.2. Adjust pH to 7.2 by adding concentrated HCl.
2. Ethanol 100% and 70%, keep at –20°C.
3. 7.5M Ammonium acetate.
4. 0.5M EDTA adjusted to pH 8.0 with 10N NaOH.
5. 1M Tris adjusted to pH 7.5 with concentrated HCl.
6. TE 50/10: 50 mM Tris-HCl, pH 7.5, and 10 mM EDTA, pH 8.0.
7. TE 10/1: 10 mM Tris–HCl, pH 7.5, 1 mM EDTA, pH 8.0.
8. High salt TE: 1 mL of 1M Tris–HCl, pH 7.5, 200 µL of 0.5M EDTA, pH 8.0, and 2 mL of 5M NaCl.
9. Distilled phenol, with 8-hydroxyquinoline first equilibrated with equal vol of, 0.5M Tris–HCl, pH 8.0, and later with 0.1M Tris–HCl to 0.1%, pH 8.0 and 0.2 β–mercarptoethanol until pH > 7.6. Keep in obscurity at 4°C.
10. Chloroform and isoamyl alcohol (24:1).
11. Glass beads (size 450–600 microns from Sigma, St. Louis, MO).
12. Ribonuclease A 10 mg/mL, heat treated at 100°C for 20 min, store at –20°C.

2.3. PCR Based Screen

1. Thermal Cycler (Perkin–Elmer, Norwalk, CT system 9600).
2. Thin-walled MicroAmp™ PCR reaction tubes.
3. Taq DNA Polymerase (Perkin Elmer), store at –20°C.
4. 10X PCR Reaction Buffer: 10 mM Tris–HCl, pH 8.3, 50 mM KCl, 1.5 mM MgCl$_2$, 0.1% (w/v) gelatin, store at –20°C.
5. 10X dNTP mix: 2 mM dATP, 2 mM dGTP, 2 mM dCTP, 2 mM dTTP, store at –20°C.
6. 15 mM MgCl$_2$, store at –20°C.
7. Forward and reverse primers with a melting temperature of 65–68°C at 0.2–1.0 µM final concentration (*see* **Note 1**), keep desiccated at –20°C for long term storage (i.e., over 1 mo) and at 4°C when in solution.
8. Ethidium bromide 10 mg/mL, keep in obscurity.
9. Agarose submarine gel apparatus.
10. 6X Loading buffer: 0.25% bromophenol blue, 0.25% xylene cyanol, 40% glycerol, store at 4°C.
11. Agarose.
12. 50X TAE electrophoresis buffer: 242 g/L TRIS base, 57.1 mL glacial acetic acid, 40 mL 0.5M EDTA, pH 8.0.
13. Gel casting tray.
14. Masking tape.
15. Benchtop microcentrifuge.
16. 8- and 12–Channel multichannel pipetes.
17. 13 × 100 mm glass tubes or 50-mL Falcon tube.
18. Savant speedvac SC210A and vacuum pump.
19. Alpha Innotech IS1000 gel documentation system.
20. Sorvall RC-5 centrifuge with SS34 rotor.
21. Beckman Biomek-1000 (optional item).

3. Methods
3.1. Propagation of YAC Clones and DNA Extraction

The YAC clones are grown individually to prevent any bias sampling due to differences in the growth rate of the different clones. Our 96 pools of 36 clones each is represented by 48 deep-well microtiter plates. There are many yeast DNA extraction protocols, however, we decided to use a protocol which can be scaled up for large-scale extractions or down to miniprep volumes and does not require expensive laboratory instrumentation. DNA is isolated from the 96 pools as well as the 36 clones of each pool. To isolate DNA from pools, dispense 150 µL of each clone within the pool into a tube.

1. Dispense 800 µL of SD^{++} into a 96 deep-well minitube assembly plate.
2. Inoculate this medium with 40 µL of cells in 15% glycerol/85% SD^{++} from stock.
3. Incubate at 30°C for about 3 d; at which time the bottom of the well will be saturated with yeast cells. Pink cells indicate that the growth medium contains insufficient adenine.
4. To get more material, concentrate the cells at the bottom of the tube with a low speed centrifugation in a Savant speedvac (SC210A) without vacuum for 3 min and replace the spent growth medium with fresh SD^{++}. Keep growing for as long as needed to obtain enough cells before pooling (usually no more than 1 wk).
5. Transfer one aliquot of 150 µL of the cell mixture and transfer to a 10-mL volume (13 mm × 100) glass test tube. You can collect this volume in the j–detector format manually (*see* **Note 2** and **Fig. 1**) or using an automated workstation.
6. Pellet the cells in a Sorvall centrifuge at 3000g for 5 min.
7. Remove the supernatant and save the pellet.
8. Resuspend the cells in 1.2 mL TE buffer (50/10).
9. Add half a volume of glass beads.
10. Vortex 3X for 40 s each time (total of 2 min).
11. Check under phase contrast (phase 3) 40X objective microscope for rupture of cells by taking 1 µL of vortexed cells and 1 drop of distilled water to a microscope slide. Broken cells will not cause light diffraction and will be dark, whereas unbroken cells will be light. Ensure you have 60% or higher dark cells. Vortexing with glass beads helps to break the cell walls.
12. Add 75 µL of 10% SDS and incubate for 5 min at 65°C. SDS with high temperature solubilizes the cell membranes and the 65°C incubation also helps inactivating contaminating nucleases.
13. Remove cells from tube to 1.5-mL Eppendorf tube by using a 1-mL pipet tip. The glass beads cannot go through the 1-mL pipet tip.
14. Incubate for 5 min at 0°C.
15. Centrifuge in a microcentrifuge for 5 min at room temperature.
16. Transfer 900 µL of supernatant to a new 1.5-mL Eppendorf tube.
17. Add 450 µL of 7.5M NH_3OAC.
18. Incubate 10–20 min at 0°C, to precipitate complex carbohydrates and proteins complexed with SDS.

19. Centrifuge in microcentrifuge for 5 min at room temperature.
20. Transfer 500 µL of supernatant to a new 1.5-mL Eppendorf tube and add 2 vol of ice-cold 95% ethanol to precipitate the DNA.
21. Centrifuge for 10 min in a microfuge at room temperature.
22. Discard the supernatant, keep the pellet.
23. Add 2 vol of ice-cold 70% ethanol to remove trace of salt from the DNA pellet.
24. Centrifuge in a microfuge for 5 min at room temperature.
25. Remove supernatant.
26. Dry the pellet under vacuum (Savant speedvac SC210A) for approx 15 min or in a desiccator for an hour.
27. Resuspend the pellet in 150 µL of TE (10/1).
28. Add 4 µL of DNAse–free RNAse (10 mg/µL) and incubate for 30 min at 37°C.
29. Add 1 vol of equilibrated phenol to extract proteins and the RNAse.
30. Separate the phases by centrifugation in a microfuge for 5 min at room temperature. Phenol will be in the lower phase.
31. Transfer the aqueous (top) phase to a new Eppendorf tube.
32. Add 1 vol chloroform and isoamyl alcohol (24:1) to remove traces of phenol from the aqueous phase.
33. Separate the phases by centrifugation in a microfuge for 5 min at room temperature.
34. Transfer the aqueous (top) phase to a new Eppendorf tube. If there is an intermediate layer, do not transfer it.
35. Add 2 vol of ice-cold 95% ethanol.
36. Centrifuge in a microfuge for 10 min at room temperature.
37. Discard the supernatant, save the pellet.
38. Add 2 vol of ice-cold 70% ethanol.
39. Centrifuge in a microfuge for 5 min at room temperature.
40. Discard the supernatant, save the pellet.
41. Dry pellet in a Savant speedvac (SC210A) for about 15 min.
42. Resuspend pellet in 100–200 µL of water or TE (10/1).
43. DNA can be stored at –20°C or 4°C. The usual yield is 2–4 µg of DNA from 109 cells.

3.2. PCR Based Mapping of Sequences

3.2.1. DNA Amplification

The YAC DNA samples should first be used to amplify a yeast genomic DNA sequence in order to test their suitability for EST mapping (*see* **Note 3**). The EST mapping primers are designed with the PRIMER software package (*see* **Note 1**). The PCR protocol uses 40 cycles in a Perkin–Elmer gene amplification system 9600 and takes about 2 h.

When amplifying many DNA samples it is convenient to make a master PCR mix with all the reagents except template and primers. The following is a master mix for 100 reactions in a 96-well MicroAmp tray assembly: 206 µL 10X PCR reaction buffer, 206 µL 2 mM dNTPs, 186 µL 15 mM MgCl$_2$, 720 µL sterile ddH20, 20 μL *Taq* DNA polymerase. The enzyme is added just prior to dispensing. In order to set-up the amplification reaction, set-up the following steps.

1. Incubate a 96-well format MicroAmp PCR tube/tray assembly at 0°C.
2. Dispense 168 μL of the master mix into eight tubes that will be used as reservoirs for the multichannel pipet.
3. Add 10 μL of the forward and reverse primers to each reservoir tube. This corresponds to 10 pmol of each primer/reaction.
4. With the multichannel pipet mix the solution and dispense 15 μL of the mixture into the MicroAmp assembly tubes.
5. Add 5 μL of template DNA to the MicroAmp assembly and ensure that all the reagents are at the bottom of the tube bycentrifugation. We use the Savant speedvac SC210A, without vacuum, for a 30 s spin.
6. We use the following PCR conditions for our primers (*see* **Note 1**): initial denaturation 95°C for 1 min followed by 40 cycles of: denaturation at 94°C for 30 s, annealing at 61°C for 30 s, and extension at 72°C for 30 s. We include a final extension at 72°C for 5 min and a soak 40°C.

3.2.2. Agarose Gel Analysis of the Products

The classical screening procedure uses agarose gel electrophoresis to analyze PCR reaction products. However, a more easily automated protocol using the ligase chain reaction has been reported *(10)*.

1. To prepare a 2% (w/v), 300 mL agarose gel, dissolve 6 g of agarose in 275 mL of ddH$_2$O.
2. Microwave 6–8 min and add 6 mL of 50*X* TAE buffer and ddH$_2$O to a volume of 300 mL.
3. Cool to about 50°C before adding 8 μL of ethidium bromide 10 mg/mL, mix and pour the gel.
4. Using a multichannel pipet add 5 μL of 3*X* loading dye to the PCR reaction tubes. Touch the sides of the tubes with the pipet tip and spin down/swing in a pendulum fashion later).
5. Load the PCR products in the agarose gel using a 6–12 channel pipet. The first lane corresponds to *j*–detector A, the second to *j*–detector B, and so on. The 11th and 12th lanes correspond to the negative and positive controls.
6. Run at 120 V for 45 min to 1 h or until the bromophenol blue has migrated 2/3 the length of the gel.
7. Take a picture of your gel and note the positions of the PCR products (Alpha Innotech gel documentation system).
8. Using the *j*–detector pooling key (**Fig. 1**), identify which primary pool or individual clone contains the sequence your are amplifying.

Figure 2 presents a sample of the agarose gel electrophoresis data obtained with *j*–detector pool screening. In this particular case, ESTs 5D2, 5D11, and

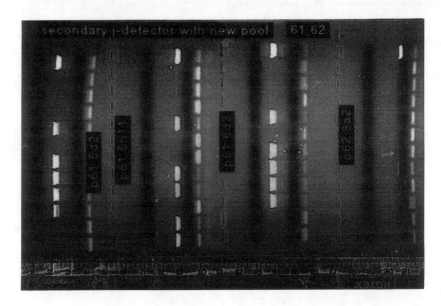

Fig. 2. Agarose gel electrophoresis analysis of j–detector PCR products: PCR products generated from j–detector pools from primary pool 61 and ESTs 5D2, 5D11, and 6D3 each gave five plant DNA products that allow an unambiguous identification of the clones which contain these ESTs. Although primer 8A2 gave PCR products with super j–detector pools which suggested it mapped to primary pool 62, the secondary pool j–detector template failed to give products with this template.

6D3 gave PCR products with super j–detector pools indicated that they mapped to primary pool 61 (data not shown). DNA samples isolated from j–detector pools derived from primary pools 61 were screened with EST 5D2, 5H11, and 6D3. Note, ESTs 5D2, 5H11, and 6D3 all gave five products and could be mapped to individual clones within primary pool 61 (**Fig. 2**). Although primer 8A2 gave PCR products with super j–detector pools which suggested it mapped to primary pool 62 (data not shown), the secondary pool j–detector template failed to give products with it (**Fig. 2**). All of our processed data is listed on the *Arabidopsis thaliana* Genome Center database ATGC (http://cbil.humgen.upenn.edu) and *Arabidopsis thaliana* DataBase AtDB (http://genome–www.stanford.edu/Arabidopsis).

4. Notes

1. Primers are selected using a set of predetermined criteria which are stored in a criteria file for the software package PRIMER, a computer program for automatically selecting PCR primers. PRIMER was developed by Stephen E. Lincoln, Mark J.

Daly, and Eric S. Lander of The Whitehead Institute for Biomedical Research. We use an automated multiplex oligonucleotide synthesis (AMOS) *(11)* to produce primers with a melting temperature of 65–68°C. The sequences of all of the primers we have generated are listed on our homepage (http://science.yorku.ca/units/biology/cm/yorkcore.htm).

2. Growth of yeast cells: 800 µL of SD^{++} medium dispensed in each well of a Beckman minitube assembly plate is inoculated with 72 individual YAC clones such that the first three rows and the last three rows each contain a primary pool of 36 clones. After growth for up to 1 wk with successive rounds of media replenishment every 2 d, these cells are transferred into 13 × 100 mm glass tubes for DNA extraction Change. Each *j*–detector pool will contain 2.7 mL of resuspended cells (i.e., 18 clones × 150 µL/clone) from which DNA is immediately extracted. The pattern shown in **Fig. 1** is used to direct the pipeting operations. These pools can be assembled manually or by an automated laboratory workstation. A p200 pipet tip rack (8 × 12) can be used to assemble the *j*–detector scheme manually. The rack is divided into two with each half representing a primary pool of 36 clones. The first row represents the first 12 clones in pool A, the second row clones 13–24, and the last row clones 25–36. *J*–detector secondary pools are assembled with a multichannel pipet which has tips only for those clones which are sampled in a given secondary pool. This process is facilitated by covering the p200 tip rack with aluminum foil and introducing tips at the positions to be sampled in a given pool such that a direct representation of the pooling scheme is reproduced on the tip rack. The Beckman minitube plates are in the 96 well plate format.

3. The DNA isolated from the YAC clones have to be tested to ensure that there are no contaminants that will hinder the PCR reaction. The results of this PCR amplification can also be used to approximate the DNA concentration of the samples. We use 33–mer oligonucleotides (i.e., SG2: 5'–CCGGTTCTAGAA-CCTTCTCTTTGGAACTTTCAG–3' and SG3: 5'-CCGGTTCTAGACGCT-TCGCTGATTAATTACCCC–3') which give a PCR product of the yeast GAL4 gene, which all clones contain regardless of the *Arabidopsis* DNA insert in the YAC vector.

References

1. Hofte, H., Desprez, T., Amselem, J., Chiapello, H., Caboche, M., Moisan, A., Jourjon, M. –F., Charpenteau, J. –L., Berthomieu, P., Guerrier, D., Giraudat, G., Quigley, F., Thomas, F., Yu, D. –Y., Mache, R., Raynal, M., Cooke, R., Grellet, F., Delseny, M., Parmentier, Y., de Marcillac, G., Gigot, C., Fleck, J., Phillips, G., Axelos, M., Bardet, C., Tremousaygue, D., and Lescure, B. (1993) An inventory of 1,152 expressed sequence tags obtained by partial sequencing of cDNAs from *Arabidopsis thaliana. Plant J.* **4,** 1051–1061.
2. Newman, T., de Bruijn, F. J., Green, P., Keegstra, K., Kende, H., McIntosh, L., Ohlrogge, J., Raikel, N., Somerville, S., Thomashow, M., Retzel, E., and Somerville, C. R. (1994) Genes galore: a summary of methods for accessing

results from large–scale partial sequencing of anonymous *Arabidopsis* cDNA clones. *Plant Physiol.* **106,** 1241–1255.
3. Balding, D. J. (1994) Design and analysis of chromosome physical mapping experiments. *Phil. Trans. R. Soc. London B.* **344,** 329–335.
4. Cresot, F., Fouilloux, E., Dron, M., Lafleuriel, J., Picard, G., Billault, A., Le Paslier,D., Cohen, D., Chaboute, M. –E., Durr, A., Fleck, J., Gigot, C., Camilleri, C., Bellini, C., Caboche, M., andBouchez, D. (1995) The CIC library: a large insert YAC library for genome mapping in *Arabidopsis thaliana. Plant J.* **8,** 763–770.
5. Ecker, J. R. (1990) PFGE and YAC analysis of the *Arabidopsis* genome. *Methods* **1,** 186–194.
6. Grill, E. and Somerville, C. R. (1991) Construction and characterization of a yeast artificial chromosome library of *Arabidopsis* which is suitable for chromosome walking. *Mol. Gen. Genet.* **226,** 484–490.
7. Ward, E. R., and Jen G.C. (1990) Isolation of single–copy–sequence clones from a yeast artificial chromosome library of randomly–shared *Arabidopsis thaliana* DNA. *Plant Mol. Biol.* **14,** 561–568.
8. Schmidt, R., Putterill, J., West, J., Cnops, G., Robson, F., Coupland, G., and Dean, C. (1994) Analysis of clones carrying repeated DNA sequences in two YAC libraries of the *Arabidopsis thaliana* DNA. *Plant J.* **5,** 735–744.
9. Hwang, I., Kohchi, T., Hauge, B., Goodman, H. M., Schmidt, R., Cnops, G., Dean, C., Gibson, S., Iba, K., Lemieux, B., Arondel, V., Danhoff, L., and Somerville, C. R. (1991) Identification and map position of YAC clones comprising one third of the *Arabidopsis* genome. *Plant J.* **1,** 367–374.
10. Kwok, P. –Y., Gremaud, M. F., Nickerson, D. A., Hood, L., and Olson, M. V. (1992) Automatable screening of yeast artificial–chromosome libraries based on the oligonucleotide ligation assay. *Genomics* **13,** 935–941.
11. Lashkari, D., Hunicke–Smith, S. P., Norgen, R. M., Davis, R. W., and Brennan, T. (1995) An Automated multiplex oligonucleotide synthesizer: development of high–throughput, low–cost DNA synthesis. *Proc. Natl. Acad. Sci. USA* **92,** 7912–7915.

V

TRANSIENT AND STABLE TRANSFORMATION

24

Transient Gene Expression in Protoplasts of *Arabidopsis thaliana*

Steffen Abel and Athanasios Theologis

1. Introduction

The transfer of defined DNA segments into plant cells is an established procedure to study regulation of gene expression *(1,2)*. Techniques for direct DNA transfer and transient expression analysis of introduced genes provide a convenient alternative to stable transformation procedures. Such techniques have been utilized extensively in animal cells and in plant cell protoplasts from monocot and dicot species *(2,3)*.

During transient gene expression experiments, the DNA is taken up by the cell without integrating into the host genome, thus avoiding position effects on the expression of the transferred gene. Transient gene expression is usually detected by an early accumulation of RNA and protein followed by a subsequent decline *(4)*. Transient assay systems utilizing plant protoplasts have emerged as a refined method to analyze expression of (mutant) genes within a few hours to days after DNA uptake *(5–11)*. More recently, such assays were used to determine the nuclear localization of various plant proteins *(12–15)*. It is evident that the rapidity of analysis is the major advantage of transient assay systems, particularly in conjunction with the time consuming regeneration of transgenic plants.

Analysis of the expression of mutationally altered genes in plants is greatly facilitated by the use of sensitive and versatile reporter enzymes. Several reporter genes have been used as gene fusion markers to analyze gene expression, such as chloramphenicol acetyl transferase (CAT), neomycin phosphotransferase (NPTII), nopaline synthase (NOS), β-glucuronidase (GUS), luciferase (LUC), or green fluorescence protein (GFP) *(16–19)*.

The methods described in this chapter to prepare and transfect mesophyll protoplasts from *Arabidopsis thaliana* are largely based on previously published procedures by Damm and Willmitzer *(20)* and Damm et al. *(21)*. Protocols are given to monitor GUS reporter gene activity histochemically and spectrophotometrically. Detailed protocols to carry out assays for other reporter enzyme activities have previously been described *(16,18,19)*.

2. Materials
2.1. Plant Material

1. Seeds of *Arabidopsis thaliana* ecotype "Columbia" (Arabidopsis Biological Resource Center at Ohio State University).

2.2. Solutions
2.2.1. Preparation of Protoplasts

1. 500 mM mannitol.
2. 200 mM CaCl$_2$.
3. Protoplasting solution *(20)*: 1% cellulase (Onuzuka R-10, Serva, Heidelberg, Germany), 0.25% Macerozyme R-10 (Serva), 400 mM mannitol, 8 mM CaCl$_2$, 5 mM Mes-KOH, pH 5.6. Dissolve enzymes in buffer with gentle stirring, adjust pH (KOH), remove insoluble material by centrifugation at 3000g for 10 min, filter sterilize supernatant.
4. W5 solution *(22)*: 154 mM NaCl, 125 mM CaCl$_2$, 5 mM KCl, 5 mM glucose, 1.5 mM Mes-KOH, pH 5.6.
5. 0.2% phenosafranine (Serva) in 400 mM mannitol.

2.2.2. Transfection of Protoplasts

1. MaMg solution *(4)*: 400 mM mannitol, 15 mM MgCl$_2$, 5 mM Mes-KOH, pH 5.6.
2. PEG-CMS solution *(4)*: 400 mM mannitol, 100 mM Ca(NO$_3$)$_2$, 40% (w/v) PEG 3350 (Sigma, St. Louis, MO).
3. 400 mM mannitol.
4. K$_3$ medium *(23)*: 400 mM sucrose, 4.3 g/L Gibco-BRL (Gaithersburg, MD) Murashige and Skoog salts, 100 mg/L *myo*-inositol, 250 mg/L xylose, 460 mg/L CaCl$_2$, 10 mg/L thiamine, 1 mg/L pyridoxine, 1 mg/L nicotinic acid, pH 5.8 (KOH), 25 mg/mL chloramphenicol.

2.2.3. Histochemical GUS Assay

1. 400 mM mannitol.
2. GUS assay buffer: 4.3 g/L Murashige and Skoog salts (Gibco-BRL), 300 mM mannitol, 50 mM K-phosphate buffer, pH 6.5, 10 mM EDTA, 0.5 mM K-ferricyanide, 0.5 mM K-ferrocyanide, 2 mM X-Gluc (5-bromo-4-chloro-3-indolyl glucuronide; BioSynth AG, Switzerland). Completely dissolve X-Gluc (FW 444.6) in dimethyl formamide (use 10 µL dimethyl formamide per 1 mg X-Gluc), then add to buffer.
3. Hoechst 33342 (Sigma), 1 mg/mL in sterile water.

2.2.4. Extraction of Protoplasts

1. 400 mM mannitol.
2. Extraction buffer: 250 mM Tris-HCl, pH 7.8, 1 mM EDTA, 0.5 mM PMSF (prepare fresh stock in EtOH).

2.2.5. Colorimetric GUS Assay

1. 100 mM Na-phosphate, pH 7.0.
2. 1M Na$_2$CO$_3$.
3. 10 mM p-Nitrophenyl-β-D-glucuronide (Sigma).

3. Methods
3.1. Growing of Arabidopsis Plants

1. Fill flats with *Arabidopsis* Sunshine mix (or other appropriate soil) and soak with 0.1X Murashige-Skoog salt solution, pH 5.6.
2. Sprinkle *Arabidopsis* seeds onto soil surface. Use a 15-mL Falcon tube with a punctured lid. Do not sow too densely to allow for good development of rosette leaves.
3. Cover flats with a transparent dome and place in a greenhouse at 22–25°C under continuous illumination with fluorescent and incandescent light.
4. After germination, remove domes and grow for 3–4 wk. Subirrigate every 3–4 d with 0.1X Murashige-Skoog salt solution, pH 5.6.

3.2. Preparation of Mesophyll Protoplasts

1. Use a pair of sharp scissors to remove rosette leaves from 3–4 wk old *Arabidopsis* plants. Place leaves immediately into a beaker filled with tap water. Harvest approx 5 g of leaves which will require about 20–50 plants, depending on developmental stage. Avoid unnecessary mechanical damage.
2. Remove traces of soil from leaves by excessive washing with tap water.
3. Rinse leaves excessively with sterile water (*see* **Note 1**).
4. Briefly blot leaves dry on filter paper.
5. Arrange leaves into a bunch using forceps. Cut leaf material with a sharp sterile razor blade into small squares of 5–10 mm^2 (*see* **Note 2**).
6. Transfer leave pieces into a 9-cm Petri dish. Add 10–15 mL of 500 mM mannitol solution. Gently submerge leaf material. Incubate for 1 h at room temperature to allow for preplasmolysis (*see* **Note 3**).
7. Carefully remove the mannitol solution from the cut leaf material. Use a serological pipet.
8. Add 30 mL of protoplasting solution. Gently submerge leaf material.
9. Vacuum-infiltrate mixture until air bubbles emerge from the leaf pieces (approx 2 min at 15 mmHg). Slowly release the vacuum (*see* **Note 4**).
10. Incubate mixture in the dark at 22°C (or at room temperature) for 2.5–3 h with gentle agitation at 50–75 rpm (*see* **Note 5**).
11. Carefully transfer crude protoplast suspension to a funnel containing a folded nylon net (140-μm mesh), on top of a 50-mL Falcon tube. Use a serological pipet (*see* **Note 6**).

12. Rinse remaining macerated leaf tissue with 0.5 vol of 200 mM CaCl$_2$. Pass suspension through the same filter to retain undigested material. Gently mix the combined filtrates.
13. Dispense crude protoplast suspension into 15-mL Falcon tubes. Spin in a swinging bucket rotor for 5 min at 60g (*see* **Note 7**).
14. Remove supernatant with a Pasteur pipet. Do not decant.
15. Gently resuspend pellet in 15 mL of a solution composed of 2 vol 500 mM mannitol and 1 vol 200 mM CaCl$_2$. Spin in a swinging bucket rotor for 5 min at 40g.
16. Remove supernatant with a Pasteur pipet. Do not decant.
17. Gently resuspend pellet in 15 mL of a solution composed of 1 vol 500 mM mannitol and 2 vol 200 mM CaCl$_2$. Spin in a swinging bucket rotor for 5 min at 40g.
18. Remove supernatant with a Pasteur pipet. Do not decant.
19. Gently resuspend protoplast pellet in 5–10 mL of W5 solution.
20. Before starting the transformation procedure, incubate protoplast suspension for at least 30 min on ice. In the meantime, count protoplasts and assess their viability as subsequently described.
21. Gently shake protoplast suspension to ensure good mixing. Remove 50 µL of the suspension and count protoplasts using a hematocytometer. The yield should be around $3-6 \times 10^6$ protoplasts per gram fresh weight of starting leave material.
22. To assess protoplast viability, stain 50 µL of the protoplast suspension with an equal volume of 0.2% phenosafranine. Damaged protoplasts and contaminating cell debris stain red, whereas viable protoplasts exclude the dye and remain green. More than 90% of the protoplasts should be viable.

3.3. Protoplast Transformation

1. Just before starting the transformation procedure, sediment protoplasts by centrifugation for 5 min at 40g.
2. Remove supernatant with a Pasteur pipet. Do not decant.
3. Gently resuspend protoplasts to a density of 5×10^6 protoplasts per mL in MaMg solution.
4. Add 50 µg plasmid DNA (2 mg/mL) and 50 µg sheared salmon sperm carrier DNA (2 mg/mL) in a 15-mL Falcon tube (*see* **Note 8**).
5. Add 0.3 mL of protoplast suspension using a wide bore 1-mL Eppendorf tip (cut tip off with a razor blade). Mix briefly and gently DNA and protoplasts (*see* **Note 9**).
6. Quickly add 0.3 mL of PEG-CMS solution (highly viscous, will sink to the bottom; *see* **Note 10**).
7. Mix carefully to a homogeneous phase. Be patient. Protoplasts will aggregate.
8. Incubate at room temperature for 30 min.
9. Carefully dilute and mix in serial steps transfection mixture with W5 solution over a period of 15–20 min: Add sequentially and mix well but carefully 0.6 mL, 1 mL, 2 mL, and 4 mL of W5 solution. Protoplasts will deaggregate.
10. Sediment protoplasts at 60g for 5 min.
11. Remove supernatant with a Pasteur pipet. Do not decant.
12. Gently resuspend protoplasts in 2 mL of 400 mM mannitol and 0.5 mL W5 solution.

13. Sediment protoplasts at 60g for 5 min.
14. Remove supernatant with a Pasteur pipet. Do not decant.
15. Gently resuspend protoplasts in 3 mL K_3 medium. Use a serological pipet to transfer suspension into a 6-cm Petri dish.
16. Incubate overnight (14–16 h) at 22°C in the dark.

3.4 Histochemical GUS Assay

1. Transfer protoplast suspension into a 15-mL Falcon tube. Use a serological pipet.
2. Add 12 mL 400 mM mannitol and mix carefully.
3. Collect protoplasts by centrifugation at 60g for 5 min.
4. Remove supernatant with a Pasteur pipet. Do not decant.
5. Carefully resuspend protoplasts in 0.5 mL GUS assay buffer (*see* **Note 11**).
6. Incubate at 37°C until suspension turns light blue (10–60 min; *see* **Note 12**).
7. To stain for nuclei, add Hoechst 33342 dye to a final concentration of 20 ng/mL.
8. Inspect protoplasts on a microscope using differential interference contrast optics (Nomarski optics) for GUS staining, and epifluorescence optics for DNA staining (*see* **Fig. 1**).

3.5. Extraction of Protoplasts

1. Transfer protoplast suspension into a 15-mL Falcon tube. Use a serological pipet.
2. Add 12 mL 400 mM mannitol and mix carefully.
3. Collect protoplasts by centrifugation at 60g for 5 min.
4. Remove supernatant with a Pasteur pipet. Do not decant.
5. Add 0.3 mL protoplast extraction buffer.
6. Sonicate for 10 s at 30 W setting (use microtip).
7. Transfer suspension into a 1.5-mL Eppendorf tube and centrifuge at 10,000g for 5 min at 4°C.
8. Transfer supernatant into new 1.5-mL Eppendorf tube and store on ice for enzymatic assays.

3.6. Colorimetric GUS Assay (see Notes 13–16)

1. For each transfection, set up four GUS reactions (1 mL final volume) in glass test tubes: 0.5 mL 100 mM Na-phosphate, pH 7.0, 0.375 mL distilled water, 0.025 mL protoplast extract.
2. Pre-equilibrate at 37°C for 5 min.
3. Start reaction by addition of 0.1 mL 10 mM p-nitrophenyl-β-D-glucuronide and incubation at 37°C.
4. Terminate reactions by addition of 0.4 mL 1M Na_2CO_3 at "zero time" and after 5–60 min.
5. Read absorbance at 415 nm against "zero time" control.
6. Calculate and compare GUS activity of different transfections (1 U of GUS activity equals 1 nanomole/min of p-nitrophenol). Under these conditions, the molar extinction coefficient of p-nitrophenol is approx 14,000. Thus, using a light path of 1 cm (cuvet) and a final assay volume of 1.4 mL, an absorbance at 415 nm of 0.01 represents 1 nanomole of product.

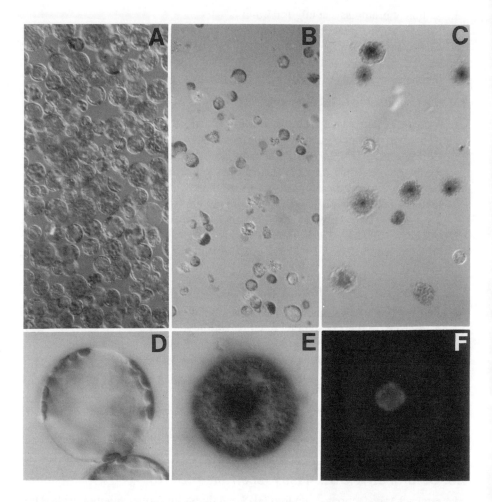

Fig. 1. Histochemical localization of GUS activity in *Arabidopsis* mesophyll protoplasts transfected with translational GUS fusions. Purified protoplasts (A) were mock-transfected (D) or challenged with the following plasmid DNAs: pRTL2-GUS which encodes the nonfused, authentic GUS protein (B); pRTL2-GUS-IAA1 encoding a GUS::IAA1 nuclear fusion protein (C); pRTL2-GUS-PS-IAA6 encoding a GUS::PS-IAA6 nuclear fusion protein (E,F). Transfected protoplasts were cultured for 20 h, stained for GUS activity (B–E) and nuclei (F) as described *(11)*. (*See* color insert following p. 230.)

4. Notes

1. Leaves are usually surface sterilized (brief dipping in 70% ethanol followed by incubation in a 5% NaOCl solution for 30 min; *20*). The ethanol treatment, however, when done too long, can drastically reduce protoplast yield. Bacterial

and fungal contamination can be effectively avoided for transient gene expression experiments when leaves of healthy plants are washed thoroughly with sterile water and all subsequent steps are carried out with care.
2. The use of thin and sharp razor blades to cut leaves is important for good protoplast yield. Do not use blunt or thick tools like scalpels that crunch and unnecessarily damage the tissue. Make sure to cut every leaf since uncut leaves are inaccessible for the protoplasting solution.
3. The preplasmolysis step is optional and appears to have only a marginal effect on protoplast yield.
4. Vacuum infiltration of cut leaves with the protoplasting solution is optional, however, increases protoplast yield and reduces incubation time.
5. The protoplasting solution should turn dark-green and its microscopic examination should reveal abundance of single spherical protoplasts. Incubate longer if protoplasts are low abundant, or not fully spherical. If most of the protoplasts are broken, cancel the experiment and start again.
6. Mesophyll protoplasts are very fragile. From this step on, whenever possible, avoid up-and-down pipeting of protoplasts. Resuspend protoplast pellets with a buffer jet and by very gentle shaking.
7. Observe low centrifugal forces which are essential for high purity of the protoplast preparation. Under these conditions, tissue debris should not sediment.
8. Several parameters affect protoplast transfection, such as PEG concentration, nature of transforming DNA, and other conditions *(21,24,25)*. The protocol of Damm et al. *(21)* to transform *Arabidopsis* mesophyll protoplasts by PEG-mediated transfection will essentially be followed.
9. PEG should be added immediately after DNA addition as protoplasts are a rich source of nucleases (secretion and release by breakage) that may hydrolyze the DNA.
10. It is essential that PEG should be added after and not before DNA. The order of DNA-PEG was found extremely important for the efficiency of DNA uptake *(24,26)*.
11. Protoplasts are stained for GUS activity essentially as described *(17,27)*. The primary product of X-Gluc (5-bromo-4-chloro-3 indolyl glucuronide) hydrolysis is colorless, however, forms the insoluble and highly colored indigo dye upon oxidative dimerization. Indigo formation which is stimulated by atmospheric oxygen can be greatly enhanced by using catalysts such as a K-ferricyanide/ferrocyanide mixture.
12. Incubation time depends on the strength of GUS expression. The time given refers to the vector system developed by Carrington et al. *(12)* to study nuclear localization in plants.
13. Protoplasts are assayed for GUS activity essentially as described *(17,27)*, using the colorimetric test which is straightforward. This test utilizes the chromogenic substrate *p*-Nitrophenyl-β-D-glucuronide. Enzymatic hydrolysis generates the chromophore *p*-Nitrophenol which can be readily detected at 415 nm using a spectrophotometer.
14. The GUS enzyme catalyzes a simple reactions which follows Michaelis-Menten kinetics. In order to obtain reliable quantitative expression data, it is necessary to

measure enzymatic activity under conditions of linear product formation with respect to reaction time and protein concentration. These conditions have to be defined before starting quantitative analysis of gene expression.
15. Based on histochemical GUS staining of transfected *Arabidopsis* protoplasts, we found transformation rates between 20–60%. The GUS activity can be easily monitored using the moderately sensitive colorimetric assay. Otherwise, the more sensitive fluorometric GUS assay is recommended *(17)*.
16. Many plants have nonneglegible levels of endogenous GUS or GUS-like activities that can interfere with the activity originating from the introduced gene, especially if that level is low. Endogenous activities can be effectively suppressed by the addition of 20% methanol to the GUS reaction mixture *(28)*.

References

1. Kuhlemeier C., Green P. J., and Chua N. -H. (1987) Regulation of gene expression in higher plants. *Annu. Rev. Plant Physiol.* **38,** 221–257.
2. Potrykus, I. (1991) Gene transfer to plants. Assessment of published approaches and results. *Annu. Rev. Plant Physiol. Plant Mol. Biol.* **42,** 205–225.
3. Davey, M. R., Rech, E. L., and Mulligan, B. J. (1989) Direct DNA transfer to plant cells. *Plant Mol. Biol.* **13,** 273–285.
4. Negrutiu, I., Shillito, R., Potrykus, I., Biasini, G., and Sala, F. (1987) Hybrid genes in the analysis of transformation conditions. *Plant Mol. Biol.* **8,** 363–373.
5. Walker, J. C., Howard, E. A., Dennis, E. S., and Peacock, J. W. (1987) DNA sequences required for anaerobic expression of the maize alcohol dehydrogenase 1 gene. *Proc. Natl. Acad. Sci. USA* **84,** 6624–6628.
6. Marcotte, W. R. Jr., Bayley, C. C., and Quantrano, R. S. (1988) Regulation of a wheat promoter by abscisic acid in rice protoplasts. *Nature* **335,** 459–457.
7. Lipphardt, S., Brettschneider, R., Kreuzaler, F., Schell, J., and Dengl, J. L. (1988) U.V.-inducible transient expression in parsley protoplasts identifies regulatory cis-elements of a chimeric *Antirrhinum majus* chalcone synthase gene. *EMBO J.* **7,** 4027–4033.
8. Huttly, A. K. and Baulcombe, D. C. (1989) A wheat *a-Amy2* promoter is regulated by gibberellin in transformed aleurone protoplasts. *EMBO J.* **8,** 1907–1913.
9. Skriver, K., Olser, F. L., Rogers, J. C., and Mundy, J. (1991) Cis-acting DNA elements responsive to gibberellin and its antagonist abscisic acid. *Proc. Natl. Acad. Sci. USA* **88,** 7266–7270.
10. Ballas, N., Wong, L. -M., and Theologis, A. (1993) Identification of the auxin-responsive element, *AuxRE*, in the primary indoleacetic acid-inducible gene, *PS-!AA4/5*, of pea *(Pisum sativum). J. Mol. Biol.* **233,** 580–596.
11. Abel, S. and Theologis, A. (1994) Transient transformation of *Arabidopsis* leaf protoplasts: a versatile experimental system to study gene expression. *Plant J.* **5,** 421–427.
12. Carrington, J. C., Freed, D. D., and Leinicke, A. J. (1991) Bipartite signal sequence mediates nuclear translocation of the plant potyviral NIa protein. *Plant Cell* **3,** 953–962.

13. Tinland, B., Koukolikova-Nicola, Z., Hall, M. N., and Hohn, B. (1992) The T-DNA-linked VirD2 protein contains two distinct functional nuclear localization signals. *Proc. Natl. Acad. Sci. USA* **89**, 7442–7446.
14. Howard, E. A., Zupan, J. R., Citovsky, V., and Zambryski, P. C. (1992) The VirD2 protein of A. tumefaciens contains a C-terminal bipartite nuclear localization signal: Implications for nuclear uptake of DNA in plant cells. *Cell* **68**, 109–118.
15. Abel, S. and Theologis, A. (1995) A polymorphic bipartite motif signals nuclear targeting of early auxin-inducible proteins related to PS-IAA4 from pea *(Pisum sativum)*. *Plant J.* **8**, 87–96.
16. Fromm, M., Callis, J., Taylor, L. P., and Valbot, V. (1987) Electroporation of DNA and RNA into plant protoplasts. *Methods Enzymol.* **153**, 351–366.
17. Jefferson, R. A. (1987) Assaying chimeric genes in plants: the GUS gene fusion system. *Plant Mol. Biol. Rep.* **5**, 387–405.
18. Scott, R., Darper, J., Jefferson, R., Dury, G., and Jacob, J. (1988) Analysis of gene organization and expression in plants, in *Plant Genetic Transformation and Gene Expression. A Laboratory Manual.* Blackwell Scientific Publications, pp. 263–339.
19. Niedz, R. P., Sussman, M. R., and Satterlee, J. S. (1995) Green fluorescent protein: an *in vivo* reporter of plant gene expression. *Plant Cell Rep.* **14**, 403–406.
20. Damm, B. and Willmitzer, L. (1988) Regeneration of fertile plants from protoplasts of different *Arabidopsis thaliana* genotypes. *Mol. Gen. Genet.* **213**, 15–20.
21. Damm, B., Schmidt, R., and Willmitzer, L. (1989) Efficient transformation of *Arabidopsis thaliana* using direct gene transfer to protoplasts. *Mol. Gen. Genet.* **217**, 6–12.
22. Menczel, L., Nagy, F., Kiss, Z. R., and Maliga, P. (1981) Streptomycin resistant and sensitive somatic hybrids of *Nicotiana tabacum* + *Nicotiana knightiana*: Correlation of resistance to *N. tabacum* plastids. *Theor. Appl. Genet.* **59**, 191–195.
23. Nagy, J. I. and Maliga, P. (1976) Callus induction and plant regeneration from mesophyll protoplasts of *Nicotiana sylvestris*. *Z. Pflanzenphysiol.* **78**, 453–455.
24. Shillito, R. D., Saul, M. V., Paszkowski, J., Muller, M., and Potrykus, I. (1985) High efficiency direct gene transfer to plants. *Biotechnology* **3**, 1099–1103.
25. Negrutiu, I., Dewulf, J., Pietrzak, M., Botterman, J., Rietveld, E., Wurzer-Figurelli, E. M., De, Y., and Jacobs, M. (1990) Hybrid genes in the analysis of transformation conditions: II. Transient expression vs. stable transformation—analysis of parameters influencing gene expression levels and transformation efficiency. *Physiol. Plant.* **79**, 197–205.
26. Rasmussen, J. O. and Rasmussen, O. S. (1993) PEG mediated DNA uptake and transient GUS expression in carrot, rapeseed and soybean protoplasts. *Plant Sci.* **89**, 199–207.
27. Gallie, D. R., Lucas, W. J., and Walbot, V. (1988) Visualizing mRNA expression in plant protoplasts: factors influencing efficient mRNA uptake and translation. *Plant Cell* **1**, 303–311.
28. Kosugi, S., Ohashi, Y., Nakajima, K., and Arai, Y. (1990) An improved assay for β-glucuronidase in transformed cells: methanol almost completely suppresses a putative endogenous β-glucuronidase activity. *Plant Sci.* **70**, 133–140.

25

Transient Expression of Foreign Genes in Tissues of *Arabidopsis thaliana* by Bombardment-Mediated Transformation

Motoaki Seki, Asako Iida, and Hiromichi Morikawa

1. Introduction

Since the pioneering studies of Klein et al. *(1)* and Christou et al. *(2)*, bombardment-mediated transformation has become a useful method for delivering foreign genes into plant cells *(3)*. The process consists of accelerating particles to a speed at which they can penetrate the surface of the cell and be incorporated into the interior of the cell.

One distinct advantage of bombardment-mediated transformation is the ability to deliver DNA into intact cells and tissues. The transient expression of the introduced genes can be assayed 24–48 h after bombardment, thus providing a quick means to determine the activity of gene constructs in the recipient cells *(4,5)*. Until the advent of biolistics, transient assays for activity of foreign genes and promoters were possible only in protoplasts in which genes could be introduced chemically or by electroporation. It has been reported, however, that foreign genes are not properly regulated in protoplasts *(6,7)*.

Transient expression with the biolistic system is also useful as a tool to optimize bombardment conditions *(8)*. All plant cells can serve as a recipient for bombardment-mediated transformation. The optimal gene delivery conditions, however, vary depending on the type of plant cells or tissues *(8)*. Thus, adjustment of gene delivery conditions depending on the type of the tissues or cells is a vital factor for biolistic transformation *(8)*.

Arimura et al. *(9)* have also reported an *in planta* transient expression assay in *Arabidopsis thaliana* by bombardment-mediated transformation. Recently,

Dekeyser et al. *(7)* reported transient gene expression in intact tissues of rice, maize, wheat, and barley by electroporation. It is not clear, however, whether this method can be adapted for *Arabidopsis* tissues.

In this chapter, we describe a system of transient expression of ß-glucuronidase (GUS) gene in *A. thaliana* leaves and roots by bombardment-mediated transformation *(8)*.

2. Materials

1. *A. thaliana* ecotype C24 seeds.
2. Germination medium (GM): Murashige and Skoog (MS) *(10)* inorganic salts, MS vitamins, 3% sucrose, and 0.6% agarose.
3. Callus regeneration medium (CRM): MS inorganic salts, B-5 *(11)* vitamins, 3% sucrose, 1.0 mg/L 6-benzyladenine (BA), 0.1 mg/L 1-naphthaleneacetic acid (NAA), and 0.8% agar.
4. Callus-inducing medium (CIM): B-5 inorganic salts, B-5 vitamins, 3% sucrose, 0.5 mg/L 2,4-dichlorophenoxyacetic acid (2,4-D), 0.05 mg/L kinetin, and 0.8% agar.
5. Plasmid DNA, pBI221 (Clontech, Palo Alto, CA, USA), in which the *gus* A gene is under the control of the cauliflower mosaic virus (CaMV) 35S promoter and nopaline synthetase (NOS) polyadenylation region (*see* **Note 1**).
6. X-Gluc solution: 1.9 mM 5-bromo-4-chloro-3-indolyl glucuronide (X-gluc, the substrate of GUS; Sigma, St. Louis, MO), 0.5 mM potassium ferricyanide, 0.5 mM potassium ferrocyanide, and 0.3% (v/v) Triton X-100 (*see* **Note 2**). Prepare fresh before use.
7. 3M Sodium acetate, pH 5.2.
8. 70% (v/v) Ethanol.
9. 100% Ethanol.
10. Sterile distilled water.
11. Pneumatic particle acceleration device *(12)* (*see* **Fig. 1**, and **Notes 3** and **4**).
12. Vacuum pump (160-VP-D, Hitachi, Tokyo, Japan).
13. Vortex (LABO-MIXER NS-8, Iuchi Seieido, Osaka, Japan).
14. Binocular microscope (SMZ-10, Nikon, Tokyo, Japan).
15. Funnel with sintered glass (Microfilter 618-4711, Shibata, Tokyo, Japan).
16. Sonicator (Handy Sonic model UR-20P, Tomy Seiko, Tokyo, Japan).
17. Cylindrical plastic container (Agripot, Kirin, Tokyo, Japan).
18. Filter paper (ADVANTEC TOYO No. 2, 55-mm diameter, Toyo Roshi Kaisha, Tokyo, Japan).
19. Gold particle (1.1 µm in diameter, Tokuriki Honten, Tokyo, Japan; *see* **Note 5**).
20. Plastic Petri dishes (5.2-cm id).
21. Polyethylene projectiles (Tokyo Rifle, Tokyo, Japan).
22. Eppendorf tubes.
23. Forceps.
24. Scalpel.

Bombardment-Mediated Transformation

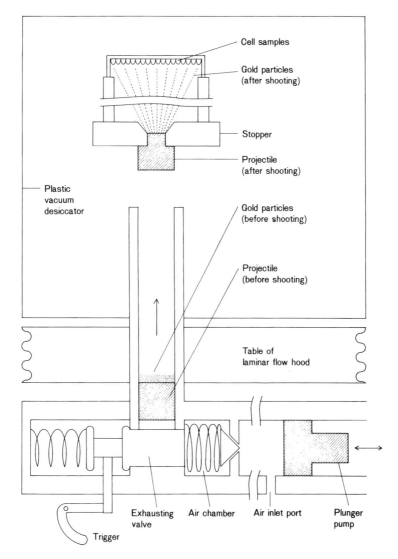

Fig. 1. Design of the pneumatic particle acceleration device.

3. Methods

3.1. Preparation of Leaf and Root Sections

1. Culture sterilized seeds of *A. thaliana* on GM in the cylindrical plastic container (Agripot) for 4–5 wk under 16 h light/8 h dark at 22°C.
2. Prepare leaf sections (1 cm × 0.5 cm) and root pieces (0.5–1.0 cm long) using forceps and scalpel.

3. Place leaf pieces (approx 360 mg fresh weight) in a circle (35-mm diameter) on a plastic Petri dish containing CRM, and root pieces (approx 150 mg fresh weight) on a filter paper. Spread the root sections so that a monolayer forms a disk (35-mm diameter) on the filter paper. Place the filter paper with the root pieces on CIM.
4. Culture for 3 d (*see* **Note 6**).

3.2. Coating Gold Particles with DNA

1. Weight gold particles (5 mg) in an Eppendorf tube.
2. Add 400 µL of 100% ethanol and vortex.
3. Spin in a microcentrifuge at 8000g for 1 s.
4. Discard the supernatant.
5. Add 40 µL of DNA solution (0.5 µg DNA/µL TE buffer) and 4 µL (1/10 vol DNA solution) of 3M sodium acetate and vortex for 5 s.
6. Add 500 µL (12.5 vol DNA solution) of 100% ethanol with vortexing of the tube.
7. Vortex for 5 s and incubate for 15 min at –80°C.
8. Suspend by vortexing.
9. Treat with sonicator 3X 1 s.
10. Place 10 µL of the ethanol suspension of the DNA-coated gold particles on the surface of the projectile and dry. Repeat this step once more (0.8 µg DNA/0.2 mg Au/projectile; *see* **Note 7**) *(13)*.

3.3. Bombardment

1. Introduce a polyethylene projectile with the top surface covered with DNA-coated gold particles from the top of the barrel and place it at the bottom end of the barrel *(8,12)*.
2. Place the filter paper with the root sections on the funnel with sintered glass and remove the medium by vacuum filtration. Turn the Petri dish with leaf sections and the funnel with the root sections, respectively, upside down and place them 10 cm over the stopper.
3. Turn on the vacuum pump. Compress air to 115–200 kg/cm^2 with a plunger pump and accumulate the compressed air in the chamber (*see* **Note 8**).
4. Release the compressed air instantaneously from the chamber to the barrel by triggering the exhaust valve after the pressure in the desiccator is reduced to 60 mmHg (*see* **Note 9**).
5. After bombardment, release the vacuum and remove the bombarded leaf and root sections. Culture the leaf sections and the root sections on the filter paper in the plastic Petri dish containing CRM and CIM under 16 h light/ 8 h dark at 22°C for 24 h. The projectile (stabbed into the aperture of the stopper) can be removed by use of a needle, thus making the device available for the next shot (*see* **Note 10**).

3.4. Assay of Transient ß-Glucuronidase (GUS) Expression

1. Transfer the leaf sections and the filter paper with root sections, to plastic Petri dishes.

2. Add 1.5 mL (leaf) and 800 µL (root) of X-Gluc solution.
3. Incubate the sections for 24 h at 37°C.
4. Add 3 mL of 70% ethanol to the cell-GUS substrate mixture.
5. Count number of GUS-expressing cells (detected as blue-colored spots) under a binocular microscope (*see* **Note 11**).

4. Notes

1. Plasmid DNA can be biolistically delivered in circular or linear form *(12)*.
2. After being bombarded, plant tissues that have high endogenous GUS activity should be incubated with X-Gluc solution containing 20% methanol and $0.1M$ sodium phosphate *(14)*. Nishihara et al. *(14)* reported that the incubation time of bombarded pollen should be less than 12 h.
3. This device has an accelerating pressure of up to more than 220 kg/cm^2 and the following advantages: controllable accelerating pressure, lack of explosion heat, and avoidance of cell damage caused by expanding gas due to a "self-sealing effect" of the projectile. This device is useful for stable transformation as well as transient expression of plant cells and tissues as reported elsewhere *(3)*.
4. Use of various types of particle gun devices have been reported. The devices so far reported can be classified into three types based on the driving forces and the type of macrocarriers: Gas (helium gas)-pressure-driven device using plastic film as a macrocarrier *(15)*, air-gun devices using a plastic projectile as a macrocarrier *(12,16)* and gas-stream-driven devices using no macrocarriers *(17,18)*. Each of these three types seemed to have its own advantages and disadvantages. Helium-gas-driven and air-gun devices need a finely processed plastic film or projectile, respectively, and the running costs for these machines are rather high. Our data indicated that the efficiency of transient expression of the *gus* A gene seemed to be virtually the same using these two machines *(3)*. The cost for construction of gas-stream device is reported to be much less than that for the other two devices *(17)*.
5. Yamashita et al. *(19)* have shown that nuclear delivery of DNA-coated gold particles is a prerequisite for successful expression of the introduced foreign genes. It has also been found that the frequency of nuclear delivery of gold particles changes as a function of the size of gold particles and that, in the case of cultured tobacco cells, 1.1-µm gold particles had more than seven times higher nuclear delivery (and hence about seven times higher gene expression efficiency) than 2.0-µm particles (Morikawa et al., unpublished results).
6. Preculture of the target tissues in the presence of exogenous plant hormones prior to bombardment seems to be an important factor that influences the efficiency of expression of introduced foreign genes. In leaf and root tissues of *A. thaliana (8)*, and cucumber cotyledon tissues *(20)*, preculture for 3 d induced a drastic increase in expression efficiency of the introduced *gus* A gene. One explanation might be that the activation of cellular or nuclear machinery of the plant cells in response to exogenous plant hormone increases transient expression efficiency of the introduced gene. It is conceivable that callus formation itself is inhibitory to biolistic gene delivery because its efficiency decreased with the sections after a longer period of preculture *(8)*.

7. The DNA-coated gold particles must be completely dry for efficient DNA delivery.
8. Adjustment of the accelerating pressure depending on the type of the tissues or cells is a vital factor for biolistic transformation. The accelerating pressure appeared to influence the velocity of the projectile (hence that of the gold particles) as well as the distribution pattern of gold particles (hence the density of the gold particles) on the cell layer *(12)*. Optimal accelerating pressure varies depending on the type of plant cells or tissues, 115–150 kg/cm^2 for leaves of *A. thaliana (8)*, 150–200 kg/cm^2 for roots of *A. thaliana* **(8)**, 150 kg/cm^2 for cultured tobacco cells *(12)*, and 200 kg/cm^2 for shoot primordia of *Haplopappus gracilis (21)*.
9. This accelerates the projectile in the barrel causing it to collide with the stopper, "stabbing" into it and sealing off the aperture, allowing the gold particles to continue through the aperture of the stopper *(12)*.
10. Ten to 15 shots per hour can be performed depending on the tissue type.
11. Norris et al. *(5)* reported fluorometric assay for GUS activity using 4-methyl-umbelliferyl-β-D-glucuronide (MUG) as the substrate. They constructed chimeric genes containing the 5'–flanking regions of three *A. thaliana* polyubiquitin genes, UBQ3, UBQ10, and UBQ11 in front of the coding regions for the GUS, introduced into *A. thaliana* leaves by particle gun, and assayed fluorometrically. Finally, they determined that an intron in the 5'–flanking regions has a quantitative effect on expression of chimeric genes.

References

1. Klein, T. M., Wolf, E. D., Wu, R., and Sanford, J. C. (1987) High velocity microprojectiles for delivering nucleic acids into living cells. *Nature* **327,** 70–73.
2. Christou, P., McCabe, D. E., and Swain, W. F. (1988) Stable transformation of soybean callus by DNA-coated gold particles. *Plant Physiol.* **87,** 671–674.
3. Morikawa, H., Nishihara, M., Seki, M., and Irifune, K. (1994) Bombardment-mediated transformation of plant cells. *J. Plant Res.* **107,** 117–123.
4. Bruce, W. B., Christensen, A. H., Klein, T., Fromm, M., and Quail, P. H. (1989) Photoregulation of a phytochrome gene promoter from oat transferred into rice by particle bombardment. *Proc. Natl. Acad. Sci. USA* **86,** 9692–9696.
5. Norris, S. R., Meyer, S. E., and Callis, J. (1993) The intron of *Arabidopsis thaliana* polyubiquitin genes is conserved in location and is quantitative determinant of chimeric gene expression. *Plant Mol. Biol.* **21,** 895–906.
6. Ingelbrecht, I. L. W., Herman, L. M. F., Dekeyser, R. A., Van Montagu, M., and Depicker, A. G. (1989) Different 3' end regions strongly influence the level of gene expression in plant cells. *Plant Cell* **1,** 671–680.
7. Dekeyser, R. A., Claes, B., De Rycke, R. M. U., Habets, M. E., Van Montagu, M., and Caplan, A. B (1990). Transient gene expression in intact and organized rice tissue. *Plant Cell* **2,** 591–602.
8. Seki, M., Komeda, Y., Iida, A., Yamada, Y., and Morikawa, H. (1991) Transient expression of ß-glucuronidase in *Arabidopsis thaliana* leaves and roots and *Brassica napus* stems using a pneumatic particle gun. *Plant Mol. Biol.* **17,** 259–263.

9. Arimura, G., Kawamura, Y., Irifune, K., Goshima, N., and Morikawa, H. (1996) In planta transient expression assay for the effect of antisense GS cDNA introduced into *Arabidopsis thaliana* leaves by particle bombardment. *Plant Cell Physiol.* **37**, s72.
10. Murashige, T. and Skoog, F. (1962) A revised medium for rapid growth and bioassays with tobacco tissue cultures. *Physiol. Plant.* **15**, 473–497.
11. Gamborg, O. L., Miller, R. A., and Ojima, K. (1968) Nutrient requirements of suspension cultures of soybean root cells. *Exp. Cell Res.* **50**, 151–158.
12. Iida, A., Seki, M., Kamada, M., Yamada, Y., and Morikawa, H. (1990) Gene delivery into cultured plant cells by DNA-coated gold particles accelerated by a pneumatic particle gun. *Theor. Appl. Genet.* **80**, 813–816.
13. Morikawa, H., Iida, A., and Yamada, Y. (1989) Transient expression of foreign genes in plant cells and tissues obtained by a simple biolistic device (particle-gun). *Appl. Microbiol. Biotechnol.* **31**, 320–322.
14. Nishihara, M., Ito, M., Tanaka, I., Kyo, M., Ono, K., Irifune, K., and Morikawa, H. (1993) Expression of the ß-glucuronidase gene in pollen of lily *(Lilium longiflorum)*, tobacco *(Nicotiana tabacum)*, *Nicotiana rustica*, and peony *(Paeonia lactiflora)* by particle bombardment. *Plant Physiol.* **102**, 357–361.
15. Ye, G. N., Daniell, H., and Sanford, J. C. (1990) Optimization of delivery of foreign DNA into higher-plant chloroplasts. *Plant Mol. Biol.* **15**, 809–819.
16. Oard, J. H., Paige, D. F., Simmonds, J. A., and Gradziel, T. M. (1990) Transient gene expression in maize, rice, and wheat cells using an airgun apparatus. *Plant Physiol.* **92**, 334–339.
17. Finer, J. J., Vain, P., Jones, M. W., and McMullen, M. D. (1992) Development of the particle inflow gun for DNA delivery to plant cells. *Plant Cell Rep.* **11**, 323–328.
18. Takeuchi, Y., Dotson, M., and Keen, N. T. (1992) Plant transformation: a simple particle bombardment device based on flowing helium. *Plant Mol. Biol.* **18**, 835–839.
19. Yamashita, T., Iida, A., and Morikawa, H. (1991) Evidence that more than 90% of ß-glucuronidase-expressing cells after particle bombardment directly receive the foreign gene in their nucleus. *Plant Physiol.* **97**, 829–831.
20. Kodama, H., Irifune, K., Kamada, H., and Morikawa, H. (1993) Transgenic roots produced by introducing Ri-*rol* genes into cucumber cotyledons by particle bombardment. *Transgenic Res.* **2**, 147–152.
21. Jin, Y., Tanaka, T., Seki, M., Kondo, K., Tanaka, R., and Morikawa, H. (1993) Transient expression of ß-glucuronidase gene in shoot primordia of *Haplopappus gracilis* by use of a pneumatic particle gun. *Plant Tissue Culture Lett.* **10 (3)**, 271–274.

26

Root Transformation by *Agrobacterium tumefaciens*

Annette C. Vergunst, Ellen C. de Waal, and Paul J. J. Hooykaas

1. Introduction

Genetic transformation and clonal propagation are techniques that play an important role in the identification and characterization of plant genes and their products. The joint efforts to develop *Arabidopsis thaliana* as a model for genetic and molecular analysis of higher plants have produced methods for in vitro propagation *(1)*, regeneration and transformation *(2–15)* using either *Agrobacterium* or direct gene transfer.

Agrobacterium tumefaciens, which belongs to the family of the Rhizobiaceae, is the etiological agent of the crown gall disease. This plant pathogenic soil bacterium is commonly used for the genetic transformation of plants *(16)*, because of its precise and efficient mode of DNA delivery and its ability to use different tissue types, which may display variation in competence for regeneration and transformation, as targets. *Agrobacterium* will deliver any segment of DNA that is surrounded by two imperfect direct repeats, called border repeats to plant cells. In the natural situation, these border repeats enclose a region of oncogenic DNA, called the T(transferred)-DNA that is present on the Ti(tumor inducing)-plasmid of the bacterium. After its transfer to the plant cell nucleus, the T-DNA is randomly integrated into the plant genome. The binary vector strategy *(17)* was developed to provide a system to easily introduce genes between the border repeats in *Escherichia coli* and use the loaded binary vector in transformation experiments after its introduction into an *Agrobacterium* strain containing a helper plasmid. In this way, any gene of interest linked to a selectable marker gene can be used to produce transgenic plants.

For *Arabidopsis*, several *Agrobacterium*-mediated transfer methods were developed using different ecotypes, bacterial strains, selectable marker genes, and tissue types *(2–5,7–12,14,15)*.

From: *Methods in Molecular Biology, Vol. 82: Arabidopsis Protocols*
Edited by: J. Martinez-Zapater and J. Salinas © Humana Press Inc., Totowa, NJ

A frequently followed method is that in which root explants are used to genetically transform *Arabidopsis* (15). Root cells of *Arabidopsis* are highly competent for regeneration. Sangwan and coworkers (18) have shown that the competent cells for transformation are present in dedifferentiating pericycle, but only after preculture treatment with phytohormones (15). Furthermore, root material can easily be obtained in large quantities, which will allow the isolation of many individual transgenic plants. An other advantage of this approach is the possibility to transform *Arabidopsis* roots that are maintained in sustained cultures (1). These cultures make it also possible to clonally propagate roots from mutants that are affected in plant development or from heterozygous plants. Such roots can then be used directly for transformation, but also for the regeneration of a large number of plants with the desired genetic characteristics. *Arabidopsis* somatic tissues display systemic endopolyploidy (19). As a consequence of using these somatic tissues for transformation transgenic polyploid offspring can be obtained. The percentage of polyploid transformants seems to be low when root transformation is used compared to leaf and direct gene transfer methods (20). Different selectable marker genes such as neomycin phosphotransferase (*nptII*, conferring resistance to kanamycin [15]), *bar* (conferring resistance to phosphinothricin [21]), *csr1-1* (encoding an acetolactate synthase resistant to chlorsulfuron [22]), hygromycin phosphotransferase (*hpt* [2,23]), dihydrofolate reductase (*dhfr*, conferring resistance to methotrexate [9]), and negative selective markers such as *codA*, encoding cytosine deaminase (24), and the HSV thymidine kinase gene (25) have been described in connection with the root transformation approach.

Our interest in gene targeting, which occurs only at very low frequencies in plants (10^{-4}–10^{-5} [26,27]), made us decide to optimize the transformation efficiency of *Arabidopsis* root explants. The protocol by Valvekens et al. (15) was modified and adapted to our local conditions. In this chapter, we describe in great detail a highly efficient transformation protocol for ecotype C24 using the *nptII* gene as plant selectable marker. Usually we obtain a transformation frequency of 2–3 putative transgenic calli per root explant (i.e., a 3–5 mm cutting through a single root), although variation between experiments occurs. This frequency is sufficiently high to perform gene targeting experiments. The regeneration efficiency (i.e., the number of calli yielding shoots) reaches 80–95%. Rooting takes place in approx 95% of the shoots. Four to five weeks after the cocultivation period, rooted shoots can be transferred to the greenhouse for seed set, making the overall tissue culture time quite short. Over 95% of the transformants are diploid. Between 40–60% of the primary transformants segregate in a 3:1 (resistant:sensitive) ratio in their offspring, indicating T-DNA insertion(s) at one locus. Of these 3:1 segregating primary transformants, approx 40% contains one insert. The other 60% have T-DNA repeat structures.

For transformation the *Agrobacterium* octopine strain MOG101 *(28)* gave optimal results. Both transient and stable transformation frequencies were determined using a binary vector (pPG1 *[29]*) that harbors chimeric p35S-*gusA* (intron) *(30)*, pnos-*nptII*, and p35S-*hpt* genes. This chapter describes the successive steps in the transformation protocol. The growth of the donor material, followed by culture of the roots, cocultivation with *Agrobacterium*, the counterselection of bacteria, and the selection and regeneration of transgenic shoots are described. This is followed by rooting of the shoots, seed set, and analysis of transformants. The different factors influencing the transformation efficiency, such as growth temperature and light conditions, addition of acetosyringone, and other factors that might help to improve the system will be described. In Section 4., we give additional information on the transformation efficiency of the ecotypes Landsberg *erecta* and Columbia with the described method, selectable markers, and their selection scheme, and additional information of what other researchers have found to be helpful in optimizing the protocol in their hands.

2. Materials

2.1. Tissue Culture Media

Use demineralized (MilliQ) water for preparing media and stock solutions. Stock solutions are filter sterilized through a 0.22-µm filter, except solutions dissolved in dimethyl sulfoxide (DMSO, *see* **Note 1**).

2.1.1. Antibiotic and Acetosyringone Stock Solutions (see **Note 2**)

1. Timentin (ticarcillin/potassium clavanulate, purchased from Duchefa, The Netherlands) is used to counterselect the bacteria after cocultivation. Prepare a stock of 100 mg/mL in H_2O. Filter sterilize and store 1 mL aliquots in Eppendorf tubes at –20°C.
2. Kanamycin (Sigma, St. Louis, MO). Prepare as a 100 mg/mL filter sterilized stock in H_2O. The solution can be used up to several months when stored at 4°C or kept at –20°C for longer time periods.
3. Acetosyringone (3,5-dimethoxy-4-hydroxyacetophenon; purchased from Aldrich Chemie, Germany). Prepare acetosyringone as a 0.2M stock in DMSO (*see* **Note 1**). Store at –20°C in 1 mL aliquots.

2.1.2. Phytohormone Stock Solutions

1. 2,4-D (2,4-dichlorophenoxyacetic acid): 10 mg/mL in DMSO (*see* **Note 1**).
2. Kinetin (6-furfurylamino purine): 5 mg/mL in DMSO.
3. 2-iP (6-(dimethylallylamino)-purine): 20 mg/mL in DMSO.
4. IAA (indole-3-acetic acid): 1.5 mg/mL in DMSO.
5. IBA (indole-3-butyric acid): 1 mg/mL. Dissolve powder in a few drops of 0.1N KOH. Adjust with H_2O. Filter sterilize.

Store all stock solutions in small aliquots at −20°C.
The final concentration of the hormones in the media is shown in **Table 1**.

2.1.3. Composition of the Culture Media

The composition of the media is shown in **Table 1**.

1. Prepare liquid growth medium (LGM): dissolve all ingredients (*see* **Table 1**) for LGM, based on B5 macronutrients, micronutrients, and vitamins *(32)*, in 950 mL H_2O. Adjust to pH 5.7 with $1N$ KOH. Then adjust the volume to 1 L. Autoclave the medium at 110°C for 20 min (*see* **Note 5**).
2. Prepare callus induction medium (CIM): dissolve all components (*see* **Table 1**), except agar and hormones, in 950 mL H_2O. Adjust the pH of the medium to 5.7 and adjust the volume to 1 L. Dispense into bottles, which already contain 0.8% agar (*see* **Note 3**). Add hormones (*see* **Table 1**) and acetosyringone (*see* **Subheading 2.1.1.**) after autoclaving the medium at 110°C for 20 min and cooling down to 60–65°C. Make CIM plates both without (for preculture, *see* **Subheading 3.3.1., step 1**) and with acetosyringone (for cocultivation, *see* **Subheading 3.3.3., step 6**). The final concentration of acetosyringone is 20 μM.
3. Prepare shoot induction medium (SIM) as indicated for CIM. After sterilization and cooling down of the medium, add the appropriate hormones as indicated in **Table 1** and the antibiotics kanamycin to a final concentration of 50 mg/L and timentin to a concentration of 100 mg/L (*see* **Note 2**).
4. Rooting medium (RM) consists of half strength MS macro *(33)* and B5 micro salts *(32)*. Dissolve the components as shown in **Table 1** in 950 mL H_2O. Adjust the pH to 5.8 with $1N$ KOH. Adjust to 1 L. Dispense into bottles containing 0.7% agar and autoclave at 120°C for 20 min. Prepare plates with 1 mg/L IBA and without IBA (*see* **Subheading 3.3.5.**).
5. Prepare germination medium (GM) (half strength MS salts) in a similar way as RM. Add kanamycin after autoclaving the medium, just prior to pouring the plates, to a final concentration of 40 mg/L.

For preparation of GM-agar plates without IBA 100 × 25 mm Labtek Petri dishes (Nunc) are used. For the other agar plates, Greiner 94 × 16 mm Petri dishes with cams are used. Pouring of the agar plates is performed under sterile conditions in a laminar flow cabinet.

2.2. Seed Sterilization and Growth of Donor Plants

1. Seeds of *Arabidopsis thaliana* C24.
2. 1.5-mL Eppendorf tubes.
3. 70% ethanol.
4. Micropipet (P1000) and tips.
5. 1% hypochlorite solution containing 0.1% Tween-20.
6. Demineralized H_2O.
7. LGM in 250-mL Erlenmeyer flasks.

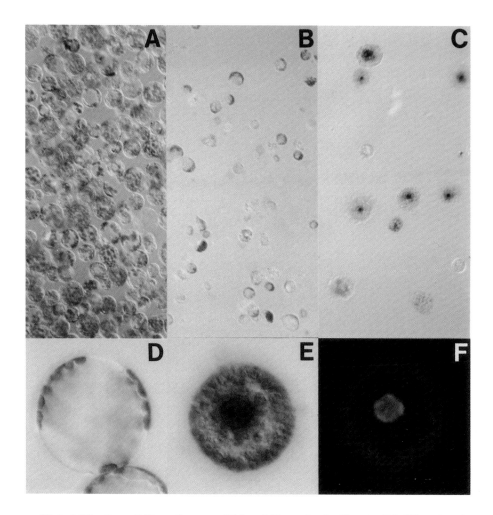

Plate 1 (Fig. 1; *see* full caption on p. 214 and discussion in Chapter 24). Histochemical localization of GUS activity in *Arabidopsis* mesophyll protoplasts transfected with translational GUS fusions. Purified protoplasts (**A**) were mock-transfected (**D**) or challenged with the following plasmid DNAs: pRTL2-GUS which encodes the nonfused, authentic GUS protein (**B**); pRTL2-GUS-IAA1 encoding a GUS::IAA1 nuclear fusion protein (**C**); pRTL2-GUS-PS-IAA6 encoding a GUS::PS-IAA6 nuclear fusion protein (**E,F**). Transfected protoplasts were cultured for 20 h, stained for GUS activity (B–E) and nuclei (F) as described *(11)*.

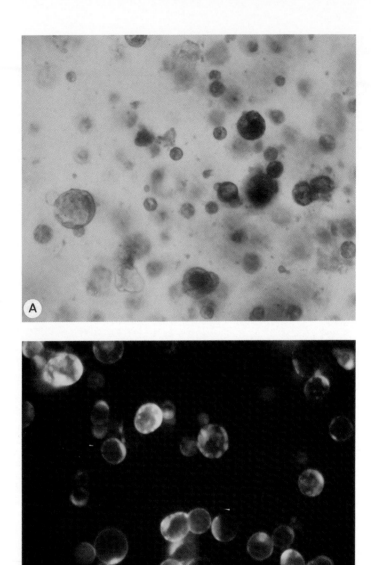

Plate 2 (Fig. 1; *see* full caption on p. 270 and discussion in Chapter 29). GUS and GFP reporter gene expression assays with *Arabidopsis* protoplasts. **(A)** GUS staining of PEG-transformed protoplasts derived from roots of *Arabidopsis* ecotype Columbia after incubation with X-gluc for 6 h at room temperature. **(B)** Leaf mesophyll protoplasts from *Arabidopsis* ecotype Columbia transformed with pCK–GFPs65c exhibit green fluorescence when illuminated with blue light. The chloroplasts emit red fluorescence, whereas the yellow flourescence results from overlapping red and green areas.

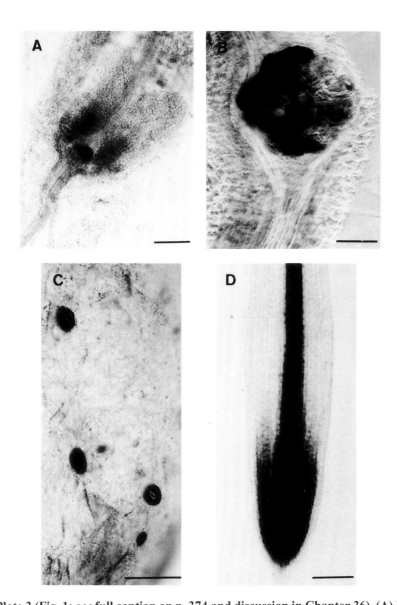

Plate 3 (Fig. 1; *see* full caption on p. 374 and discussion in Chapter 36). (A) WISH on an *Arabidopsis thaliana* seedling hybridized with an antisense *cdc2a* probe detected by gold-labeled antibodies. The signal is visible as a black precipitate resulting from the silver amplification reaction. **(B)** *In situ* localization of *cdc2a* mRNA on a vibroslice of an *Arabidopsis* root infected with a root-knot nematode. The blue precipitate resulted from the histochemical reaction with the substrates X-phosphate and NBT. **(C)** Cotyledon hybridized with an antisense *rha1* probe. Hybrids were visualized by silver amplification of specifically bound gold-labeled antibodies seen as a dark precipitate. **(D)** Chicory root hybridized with an antisense nitrate reductase probe detected by AP-labeled antibodies. A strong expression is visible in the vascular cylinder and in the root meristem. The sample was cleared in CLP for better visualization of the signal.

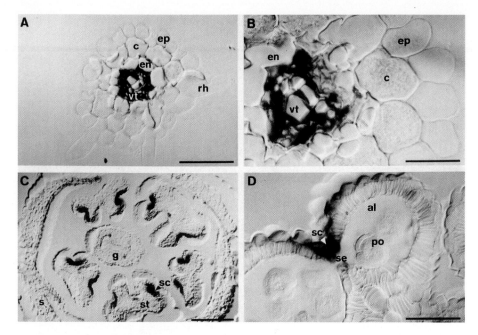

Plate 4 (Fig. 3. *see* full caption on p. 405 and discussion in Chapter 38). X-Gluc reactions on *Arabidopsis* tissue. **(A)** Transverse section through a 2-wk-old root of a *pRPS18A-gus* transformed line *(15)*. **(B)** Detail of the vascular tissue; main GUS activity in vascular tissue. **(C)** Transverse section through a young (stage 9) flower of a line transformed with a *gus* gene fused to a stomium-specific tobacco promoter (provided by T. Beals and P. Sanders, Plant Molecular Biology Laboratory, University of California Los Angeles, CA). **(D)** Detail of mature anther. Main GUS activity located at the stomium, the site of anther dehiscence. Abbreviations: al, anther locule; c, cortex; en, endodermis; ep, epidermis; g: gynoecium; po, pollen; rh, root hair; s, sepal; sc, stomium cells; se: septum; st, stamen; vt vascular tissue. Visualization with Normasky interference microscopy. Bar = 50 µm (A), 20 µm (B,D), and 100 µm (C).

Table 1
Composition of the Root Transformation Culture Media

Component	Concentration				
	LGM	CIM *(15)*	SIM *(15)*	RM *(31)*	GM
Macronutrients[a]					
(final concentration in mg/L)					
NH_4NO_3				825	825
KNO_3	2500	2500	2500	950	950
$MgSO_4 \cdot 7H_2O$	250	250	250	185	185
KH_2PO_4				85	85
$(NH_4)_2SO_4$	134	134	134		
$NaH_2PO_4 H_2O$	150	150	150		
$CaCl_2 \cdot 2H_2O$	150	150	150	220	220
Micronutrients[b]					
(final concentration in mg/L)					
FeNaEDTA	36.7	36.7	36.7	36.7	36.7
H_3BO_3	3	3	3	3	3.1
$MnSO_4 \cdot 4H_2O$	13.2	13.2	13.2	13.2	11.15
$ZnSO_4 \cdot 7H_2O$	2	2	2	2	4.3
$Na_2MoO_4 \cdot 2H_2O$	0.25	0.25	0.25	0.25	0.125
$CuSO_4 \cdot 5H_2O$	0.025	0.025	0.025	0.025	0.0125
$CoCl_2 \cdot 6H_2O$	0.025	0.025	0.025	0.025	0.0125
KI	0.75	0.75	0.75	0.75	0.415
Vitamins, amino acids, buffers[c]					
(final concentration in mg/L)					
nicotinic acid	1	1	1		0.5
pyridoxine-HCl	1	1	1		0.5
thiamine-HCl	10	10	10		0.1
glycine					2
myo-inositol	100	100	100		100
MES	500	500	500	1000	500
Carbohydrates					
(final concentration in g/L)					
glucose	20	20	20		
sucrose				10	20
agar (final concentration in g/L; see **Note 3**)					
Daishin agar		8	8	7	7

(continued)

Table 1 (continued)

Component	Concentration				
	LGM	CIM *(15)*	SIM *(15)*	RM *(31)*	GM
Phytohormones (final concentration in mg/L; see **Subheading 2.1.2.**)					
2,4-D		0.5			
kinetin		0.05			
2-iP			5		
IAA			0.15		
IBA				1	

[a]The macronutrients are made up as 10X concentrated stock solutions. We prepare a seperate stock for $CaCl_2 \cdot 2H_2O$ to prevent precipitation (*see* **Note 4**).

[b]Micronutrients are prepared as 1000X stock solutions (*see* **Note 4**). FeNaEDTA (ethylenediamine tetra acetic acid ferric monosodium salt) is made up separately as a 100X concentrated stock solution.

[c]Vitamins plus glycine are prepared as a 100X stock. Myo-inositol is prepared separately as a 100X stock. MES (2-morpholino-ethanesulfonic acid monohydrate) is added as a powder (*see* **Note 4**).

2.3. Transformation Procedure

2.3.1. Preparation and Preculture of Roots

1. 10-d-old C24 cultures.
2. Empty Petri dishes (Greiner 94 × 16 mm).
3. Scalpels.
4. Forceps.
5. CIM plates without acetosyringone.
6. Urgopore (Chenove, France) tape.

2.3.2. Growth and Preparation of Agrobacterium Strains

1. LC-medium (pH 7.0): Weigh 10 g Bacto-tryptone, 5 g yeast extract, and 8 g NaCl. Adjust with H_2O to a volume of 1 L. For agar plates, 1.5% agar (Difco Bacto agar) is added. Sterilize by autoclaving at 120°C for 20 min.
2. Antibiotic stocks:
    a. Rifampicin (10 mg/mL in methanol, do not sterilize). Final concentration: 20 mg/L.
    b. Spectinomycin (125 mg/mL in H_2O, filter sterilized). Final concentration: 250 mg/L.
    c. Kanamycin (50 mg/mL in H_2O, filter sterilized). Final concentration: 100 mg/L. Store the stock solutions at −20°C in 1 mL aliquots.
3. Bacteria on a fresh LC-agar plate with antibiotics (*see* **Notes 6** and **7**). *Agrobacterium* stocks are stored at −80°C in LC medium containing 14% glycerol.
4. Liquid LC containing antibiotics.

A. tumefaciens Root Transformation

5. 100 mL flask with cotton wool plug.
6. 28°C growth incubator (shaking).
7. 1.5-mL Eppendorf tubes.
8. Micropipet (P1000) and tips.
9. Overnight grown *Agrobacterium* cultures.
10. LGM.
11. Eppendorf centrifuge.

2.3.3. Cocultivation of Root Explants with Agrobacterium

1. Empty Petri dishes (Nunc 100 × 25 mm).
2. LGM.
3. Bacteria in LGM.
4. Scalpel.
5. Forceps.
6. Filter paper.
7. CIM-agar plates containing 20 μM acetosyringone.
8. Urgopore tape.

2.3.4. Selection of Transgenic Calli and Shoots

1. Forceps.
2. Sieve (*see* **Note 8**).
3. Empty Petri dishes (Greiner 94 × 16 mm).
4. LGM.
5. Filter paper.
6. SIM-agar plates with 50 mg/L kanamycin and 100 mg/L timentin.
7. Urgopore tape.

2.3.5. Rooting of Regenerated Shoots

1. Forceps.
2. Scalpel.
3. RM-agar plates containing 1 mg/L IBA.
4. Urgopore tape.
5. RM medium without IBA.

2.4. Analysis of Transformants

1. T2 seeds.
2. 70% Ethanol.
3. 1% Hypochlorite solution with 0.1% Tween-20.
4. Demineralized H_2O.
5. 0.1% Agarose in H_2O.
6. Micropipet (P1000) with tips.
7. Solidified GM containing 40 mg/L kanamycin.
8. Parafilm or Urgopore tape.

3. Methods

Perform all steps under sterile conditions in a laminar or down flow (biohazard) cabinet with sterile equipment and sterile instruments. The growth rooms have a 16-h photoperiod and a relative humidity of 50%. The varying light and temperature conditions are mentioned below (*see* **Note 9**).

3.1. Tissue Culture Media

1. After sterilization of the media and cooling down to 60–65°C, add antibiotics and/or hormones (*see* **Subheading 2.1.3.**). Pour 25 mL of agar medium in Petri dishes.
2. Allow the agar medium to solidify and dry for 45–60 min with the lids of the Petri dishes slightly opened (*see* **Note 3**). Store the plates at 4°C in a closed plastic bag or wrapped in aluminium foil.
3. 50 mL of LGM is dispensed into sterile, 250-mL erlenmeyer flasks capped with aluminium foil.

3.2. Seed Sterilization and Growth of Donor Plants

1. Put approx 3 mg C24 seeds in an Eppendorf tube (*see* **Note 10**).
2. Submerge seeds for 1 min in 70% ethanol. The seeds will sink to the bottom of the tube when left standing.
3. Remove ethanol with a P1000 pipet tip.
4. Add 1 mL of a 1% hypochlorite solution containing 0.1% Tween-20. Make sure all seeds are in contact with the solution. Avoid bubbles. Let stand for 10 min.
5. Remove solution and rinse seeds three times with 1 mL sterile water. Leave small volume of water.
6. Add seeds, using the P1000 pipet, to 50 mL LGM in a 250-mL Erlenmeyer flask capped with aluminium.
7. Put flask on a rotary shaker (80/100 rpm) in growth room (21°C, 1000–1500 lx, *see* **Notes 9** and **11**).

3.3. Transformation Procedure

For all stages in the protocol, Urgopore gas diffusible medical tape is used (*see* **Note 12**).

3.3.1. Preparation and Preculture of Roots

1. Use roots from 10-d-old cultures (*see* **Note 13**). Separate roots from hypocotyls, cotyledons, and leaves (*see* **Note 14**). Keep the roots in LGM while dissecting them. Place the roots on CIM plates, without drying the roots on filter paper. Be careful that all roots are in good contact with the agar-medium. Vegetatively sustained root cultures *(1)* can also be used (*see* **Note 15**).
2. Incubate roots for 3 d in growth room at 25°C with 1000–1500 lx.

3.3.2. Growth and Preparation of Agrobacterium Strains

1. Streak bacteria (*see* **Notes 6** and **7**) on a fresh LC-agar plate with rifampicin, spectinomycin, and kanamycin and grow the bacteria for 2–3 d before preparing liquid cultures at 29°C.
2. Inoculate *Agrobacterium* strain(s) in 10 mL liquid LC containing antibiotics in a 100-mL flask, plugged with cotton wool, from a fresh LC-plate 1 d before the initiation of cocultivation with root explants. Grow the culture overnight at 29°C in a shaking waterbath or rotary shaker (200 rpm).
3. Measure the OD_{600} of the overnight *Agrobacterium* culture. Calculate the desired volume of overnight culture to obtain an OD_{600} of 0.1 in 20 mL LGM and pellet the bacterial cells in an Eppendorf tube (2 min 12,000g). Remove the supernatant and resuspend the pellet in 1 mL LGM. (Optionally 1 mL of an *Agrobacterium* overnight culture can be used.)

3.3.3. Cocultivation of Root Explants with Agrobacterium

1. Transfer roots after the 3-d incubation period to a Petri dish with 19 mL LGM (optional: a sieve can be used; *see* **Note 8**).
2. Add bacteria (from **Subheading 3.3.2., step 3**). The final OD_{600} should be 0.1.
3. Incubate bacteria and roots, while shaking from time to time, for 2 min.
4. Collect roots and place them in the lid of the Petri dish. Cut the roots in pieces of 3–5 mm (one root piece is named explant; *see* **Note 16**). Avoid desiccation of the explants at this stage.
5. Dry the root explants on two layers of sterile filter paper (*see* **Note 17**).
6. Place the root explants on CIM-agar plates containing 20 µM acetosyringone (*see* **Note 18**). Make sure the explants are in close contact with the medium.
7. Cocultivate the root explants and agrobacteria for 2 d in a growth room (25°C, 1000–1500 lx; *see* **Note 9**).

3.3.4. Selection of Transgenic Calli and Shoots

1. Collect the root explants with forceps (bacteria will have overgrown the explants) and transfer them to a sieve (*see* **Note 8**) that is placed in a Petri dish with LGM. Wash the root explants carefully by shaking the sieve. Repeat the washing step in fresh LGM.
2. Blot root explants dry on two layers of sterile filter paper (*see* **Note 17**).
3. Transfer the explants to SIM-agar plates with 50 mg/L kanamycin and 100 mg/L timentin (*see* **Note 19**).
4. Transfer the plates to 3000–4000 lx at 25°C (*see* **Note 9**).
5. Transfer the explants to fresh SIM with timentin and kanamycin every 10–14 d for efficient regeneration.
6. Green, putatively transformed calli (usually they start appearing 1–2 wk after cocultivation) are transferred to fresh SIM agar-plates with kanamycin and timentin. The calli need subculturing every 2–3 wk (*see* **Note 20**).
7. Shoots will start appearing within 2–3 wk.

3.3.5. Rooting of Regenerated Shoots

1. Transfer small shoots to plates with RM containing 1 mg/L IBA. Make sure no callus is left on the shoot (*see* **Note 21**). Put the plates at 21°C with 3000–4000 lx.
2. After 1 wk transfer shoots to Petri dishes containing RM medium without IBA. Roots will soon develop. As soon as a number of roots have grown into the agar, transfer independent transformants (T1) to soil in the greenhouse (*see* **Note 22**).
3. Seeds can be collected 6–8 wk after transfer to the greenhouse.
4. Harvest the seeds and store at 4°C (*see* **Note 23**).

3.4. Analysis of Transformants

1. Sterilize about 100 T2 seeds (*see* **Subheading 3.2., steps 2–5**) to test stable Mendelian segregation of the transgene in the progeny.
2. Remove water from the last washing step and replace with 400 µL 0.1% agarose, in which the seeds will be evenly distributed.
3. Spread seeds in 0.1% agarose on solidified GM containing 40 mg/L kanamycin using a P1000 pipet tip.
4. Place dishes for 4 d at 4°C before transfer to the growth room (21°C, 3000–4000 lx; *see* **Note 11**). Sensitivity to kanamycin results in bleached cotyledons and inhibition of leaf and root growth. Estimate the ratio of sensitive versus resistant seedlings (*see* **Note 24**).
5. Test the ploidy level of the transformants (*see* **Note 25**), since polyploidy is disadvantageous in further genetic analyses.
6. To further analyze the transformants perform Southern blotting or PCR. Pool about 20 T2 plants per T1 transformant, grown in soil, to extract chromosomal DNA and estimate the number and mode of T-DNA integration events by Southern hybridization (*see* **Note 26**).

4. Notes

1. DMSO (sterile filtered, purchased from Sigma) and sterile Eppendorf tubes are used to dissolve the hormones, acetosyringone, or antibiotics. After dissolving, do not sterilize. Do not flame DMSO! DMSO is toxic, so handle with care and do not breathe fume.
2. Besides using the kanamycin-resistance gene for selection, we also tested the *bar*-gene *(34)*, *csr1-1* gene *(22)* and *hpt*-gene *(29)* although we did not optimize the system as for kanamycin (*see* **Note 20**). The *bar*-gene encodes a phosphinothricin acetyltransferase (PAT), that can detoxify phosphinothricin (PPT). The PPT inhibits glutamine synthetase leading to elevated ammonium levels and cell death. We use 20 mg/L phosphinothricin (Duchefa, The Netherlands). The stock solution is 50 mg/mL in H_2O, filter sterilized. The transformation efficiency is similar to that when using kanamycin selection. Chlorsulfuron (Glean, DuPont) inhibits acetolactate synthase (ALS). A single base pair mutation (*csr1-1* allele) confers resistance to chlorsulfuron, a sulfonylurea herbicide. Selection for the mutated ALS-gene is performed on 5 µg/L chlorsulfuron. The stock is 1 µg/µL in H_2O. The selection is not as efficient as kanamycin selection in our hands, but transgenic plants can be obtained.

Hygromycin (Calbiochem, La Jolla, CA) selection is performed on medium containing 20 mg/L. Hygromycin is very toxic and must be handled with care according to safety standards.

Otherwise, we have detected an interaction between agents that are used to inhibit bacterial growth and those for selection of transgenic plant material. We previously counterselected bacteria with 500 mg/L carbenicillin (Duchefa, stock 250 mg/mL in H_2O, stored at $-20°C$) and 100 mg/L vancomycin HCl (Duchefa, stock 100 mg/mL in H_2O. stored at 4°C), while selecting for kanamycin-resistant calli with success. In later stages, we were not able to use this combination of antibiotics (hardly any calli were obtained), but we found timentin to be a very good alternative for counterselection. However, PPT selection is still most efficient in combination with vancomycin and carbenicillin. The apparent transformation frequency is lower in combination with timentin. It is known that carbenicillin has auxin-like properties when broken down in vitro *(35)*. Because we did not alter the protocol, this sudden decrease in transformation frequency can hardly be assigned to an interaction with the phytohormones in the medium as described recently by Lin et al. *(36)*. Furthermore, there seems to be no interaction of vancomycin and/or carbenicillin with PPT selection in our hands.

The above mentioned results do not imply that these problems will occur in every laboratory or with different ecotypes. Unknown factors might have an effect or the production process of the antibiotics might play a role. It is clearly meant as a suggestion in case transformation frequencies are unexpectedly low.

The bactericidal compound cefotaxime is thought to decrease the regeneration efficiency as was also found for *Antirrhinum majus (35)*.
3. Agar is purchased from Brunschwig Chemie (Amsterdam, The Netherlands). Often the choice of agar influences the regeneration and transformation efficiency. In our hands, Daishin agar gives good results. It is very important to take good notice of the way the media are autoclaved. We never autoclave more than 2 L of medium in a pressure cooker at 110°C. Extension of the overall sterilization time (e.g., by sterilizing more medium) can decrease the consistency of the agar after solidifying. Soft agar makes it very tedious to spread the root explants. The agar should feel quite solid when touching with a tip of a pair of tweezers. Agar plates are dried in a laminar flow cabinet with the lid open for 45–60 min. With agar plates prepared in this way, very good regeneration is obtained. Both the consistency of the agar and the lack of accumulation of condensation (by drying the plates and using gas diffusible tape; *see* **Note 12**) yields a high percentage of healthy shoots, without vitrified appearance. Furthermore, *Agrobacterium* is counterselected more efficiently on plates that are prepared in this way.

The RM and GM medium contain 0.7% agar. The agar is more soft than in CIM and SIM plates to allow root growth in the medium.
4. The macro-, micro-, and vitamin stock solutions are not sterilized and are stored at 4°C in the dark. They can be stored up to several months.
5. Media containing glucose are sterilized for 20 min at 110°C.

6. We used *Agrobacterium* strain MOG101 *(28)* for optimization of root explant transformation. This strain has a nopaline C58 chromosomal background with a rifampicine-resistance marker and contains an octopine pTiB6-derived helper plasmid. The complete Tr- and Tl-region have been replaced by a spectinomycin-resistance marker. The octopine strain LBA4404 *(17)*, the supervirulent strain EHA105 *(28)*, and nopaline strain MOG301 *(28)* yielded somewhat lower transformation frequencies than MOG101.

   The pBinl9 *(37)* derived binary vector that was used to optimize the transformation protocol (pPG1 *[29]*) has a bacterial kanamycin resistance gene. Between the T-DNA borders chimeric pnos-*nptII*, p35S-*gusA* (intron), and p35S-*hpt* genes are present. The bacterial strains and binary vectors used define the antibiotics needed.

7. It is recommended to include controls in each experiment. We use a bacterial strain lacking the *nptII* gene on the T-DNA as a negative control. Roots infected with this strain and placed on kanamycin containing medium should not yield transformants. This control allows to test the effectiveness of the selection procedure. A few roots infected with a strain harboring the selectable marker gene are grown on medium without kanamycin, but with timentin to test the regeneration efficiency and the medium.

8. We use stainless steel sieves with 100 mesh screens (purchased from Sigma). Optionally, sieves can be made by cutting 50-mL Greiner tubes, heating the cut surface, and pressing against a nylon membrane. After contamination with bacteria, the sieves are sterilized at 120°C for 20 min in a beaker together with the liquid LGM from **Subheading 3.3.3., step 3**. The sieves are then cleaned with H_2O (do not use soap or any other detergent!), allowed to dry, and sterilized again for further use.

9. We have good results with Philips TL 83HF light tubes which emit light in which the orange/red part of the spectrum is well represented. In the different steps of the protocol, the light intensities and growth temperature are varied. Growth of the plants for root material takes place for 10 d on a rotary shaker at 21°C with a light intensity of 1000–1500 lx. During and after cocultivation (*see* **Subheading 3.3.3., step 7**) the temperature is increased to 25°C. Comparison of transformation frequencies at 21°C and 25°C revealed a large difference. At 21°C the frequency (number of calli per explant) was 6–10-fold lower and the regeneration frequency (number of calli yielding shoots) dropped to 25%. When the roots are incubated at 25°C with higher light intensities (3000–4000 lx) after cocultivation (*see* **Subheading 3.3.4., step 4**) the regeneration efficiency reaches frequencies up to 80–95%. In case only one incubation temperature is available, a temperature of 25°C is advised, especially during cocultivation.

10. Approximately 3 mg seeds is sufficient for 500–800 explants after cocultivation (*see* **Note 20** for transformation frequencies). Some losses are obtained during the washing steps. It is advisable to initiate more cultures, even if low numbers of transformants are needed, due to possible contamination.

11. Stratification of seeds before transfer to the growth room can be desired. Especially when freshly harvested seeds are used, the germination frequency can be low without prior cold treatment. Flasks or agar plates containing the sterilized

seeds are kept at 4°C for 3–4 d. The seeds need to be imbibed during the cold treatment. Stratification of the seeds synchronizes the germination time. Seedlings sown in flasks for transformation can be used l0 d after transfer to the growth room to collect roots (*see* **Subheading 3.3.1., step 1**)

12. Urgopore gas-diffusible tape (Chenove, France) is used at all stages to increase gas exchange. By using parafilm, condensation can accumulate which may cause formation of vitrified regenerants. Furthermore, an increase in concentration of ethylene might negatively influence the transformation and regeneration process. Gas diffusible tape may prevent possible negative effects of ethylene action.

13. The age of the root material that is used for transformation is of great importance. Roots grown for 10 d under our growth conditions yielded significantly higher transformation frequencies than 14- or 21-d-old root material. If other light and/or temperature conditions are used it could be that the donor plants should be older or younger for optimal transformation efficiency.

14. Forceps with a curved tip (e.g., Medicon 07.72.40, Lamèris, The Netherlands) come in very handy in all steps of the transformation process. After 10 d of growth, a clump of seedlings will have formed. Seedlings are easily pulled out in a Petri dish by pulling the leaves with forceps while holding the clump back with a scalpel. Then it is easy to cut the roots just below the hypocotyl. The roots are kept in LGM. Furthermore, we use a black background underneath the Petri dish to make the roots clearly visible.

15. Czako et al. (*1*) have developed a technique to grow sustained root cultures of *Arabidopsis*, which can also be transformed with *Agrobacterium*. This can be especially useful for maintaining lines with a hemi- or heterozygous genotype or with a mutant genotype affecting growth. These cultures can also be used in the herein described transformation protocol. Every 2–3 wk the root cultures are treated with 0.05 mg/L IAA for 2 d and transferred to fresh ARC (*1*) medium. For transformation with *Agrobacterium,* we treated the root cultures with IAA 10 d before initiating the preculture of the roots (*see* **Subheading 3.3.1.**) and transferred to fresh ARC after 2 d. The transformation frequency was about two-fold lower compared to roots collected from 10-d-old seedlings. As the cultures grew older (> 6 mo) the transformation frequency becomes lower and less reproducible in our hands.

16. The roots are arranged in a bundle of approx 5 cm in length. Transverse cuttings are made every 2–3 mm. Again, the roots are arranged in a bundle and the cuttings are repeated. The root explants are further wounded by tapping with a scalpel several times. This will finally result in root explants, which mainly vary in length between 3–5 mm. It is not necessary to wound the roots prior to the 2 min incubation period with bacteria.

17. Take about 15–20 explants each time with forceps having a curved point. While putting the roots on filter paper, spread the explants directly when still moist. This way the explants will separate easily and can be transferred to a plate containing CIM medium. The explants will stick to the tip of the forceps and can be transferred by tipping them on the agar plate. Doing this, the explants can be easily spread on the agar. It is important that the roots are dried, although they

should not dry out! This especially accounts for the transfer to SIM medium (*see* **Subheading 3.3.4., step 3**). If roots are kept too moist it is possible to get *Agrobacterium* overgrowth in later stages. Furthermore, regeneration seems to improve when roots are desiccated a little.

18. We found that addition of acetosyringone during cocultivation could improve the transformation frequency 1.3–2-fold. Sheikholeslam and Weeks *(14)* found an increase in transformation frequency using leaf explants. They cultured the agrobacteria with acetosyringone prior to cocultivation, whereas our experience with root explants has shown that addition of acetosyringone during cocultivation is more efficient.

19. Good results are obtained if the explants are not touching each other (100–150 per Petri dish). In this way, growth and recovery (and counting, if desired) of independent transformants can be enhanced. When transformation frequencies are within the range of 2–3 calli per root explant, it is difficult to separate independent transformation events. It may occur that two or more independently transformed cells give rise to one chimeric callus. Then the number of calli that are counted will even be an underestimation. Transient expression studies (performing a histochemical X-Gluc staining 5 d after cocultivation) showed indeed that several transformed cells can be located at the wounded side of the explants. For that reason, we continue with one putatively transformed shoot per callus.

20. Using the described protocol, we are usually able to obtain a transformation frequency for ecotype C24 of 2–3 kanamycin-resistant calli per root explant. Of these calli, approx 80–95% will develop shoots that can be rooted (95%). In the greenhouse between 1 and 5% of the transformants may be lost during adaptation or may be sterile.

   Besides C24, we tested the ecotypes Landsberg *erecta* and Columbia for their transformation efficiency using kanamycin selection. For Landsberg *erecta Agrobacterium* strain LBA4404 *(17)* yielded transformation frequencies similar to C24 transformed with MOG101. The regeneration efficiency of Landsberg (10%) was less efficient under the conditions optimized for C24. Ecotype Columbia gave lower transformation frequencies with MOG101 than C24, although transformed shoots can readily be obtained. Transformation of other ecotypes has been described by others, for example ecotype Bensheim and Wassilewskija *(2)*, Nossen *(9)* or RLD *(25)*. Already a number of optimized root transformation protocols were described. Mandal et al. *(11)* for instance described the use of 5-azacytidine to improve the regeneration efficiency. Clarke et al. *(5)* mentioned the use of silverthiosulphate, which inhibits ethylene action, to increase transformation efficiencies and the use of an agarose overlay to easily spread the roots. The protocol described in this chapter does not require any addition of silverthiosulphate or 5-azacytidine under our local growth conditions. Also, the use of an agarose overlay only decreased the transformation frequency. In other laboratories though, with differing tissue culture conditions, or different ecotypes or selectable markers used, addition of these compounds might aid in optimizing transformation frequencies. But also under less optimal conditions transformants will be obtained! We encountered severe difficulties with transformation when painting or welding activities were going on elsewhere in the building. Especially, when roots during the cocultivation phase came in contact with gasses produced by painting or welding the transformation frequency decreased drastically.

21. The percentage of shoots that will root will be as high as 95%, provided shoots with a healthy appearance are taken. Vitrified shoots develop roots at a lower frequency. Furthermore, make sure to remove all callus. Callus material left on the base of the shoot can inhibit rooting. The shoots are cultured on RM with IBA for 1 wk. Longer incubation on IBA can give the plants a bushy phenotype. Rooting is normally performed on medium without antibiotics and/or selective compounds. The presence of these compounds can decrease the rooting efficiency. *Agrobacterium* growth is very rarely found after transfer of the shoots to RM plates.

22. For transfer to the greenhouse (60–70% relative humidity, 19°C, >4000 lx, Philips high pressure lamps HPI-T 400W and SON-T AGRO lamps 400W alternating with Philips 60W light bulbs), plants are used which already developed a few roots. We use normal potting soil, mixed with a little sand, in recycling pots. After transfer to soil, a high humidity is maintained by covering the plants with plastic for at least 1 wk. The plants are also covered with filter paper for about 1 wk. The plants can adapt to the greenhouse conditions slowly by first making holes in the plastic foil and subsequently removing the foil. Depending on the health of the transformants, the adaptation period may have to be extended. Soil is kept humid after transfer of the plants. When the plants start bolting, Aracons (Beta Tech, Belgium) are put over the plants. This system makes it possible to grow individual transformants separately, avoiding the risk of cross pollination or seed loss. *See* Chapter 27 on leaf transformation, **Note 20**, for solving problems with the black fly. In vitro seed set can also be obtained in Greiner pots (100–53 mm), but much less seeds are obtained per primary transformant than in vivo in the greenhouse (*see* Chapter 27 on leaf transformation, **Note 19**).

23. Seeds are collected when the siliques have dried. Using the Aracon system, the seeds can be easily collected by pulling the plant through the Aracon base.

24. A segregation ratio of 3 resistant to 1 sensitive seedling predicts T-DNA integration at one chromosomal locus (the χ-square test is used to statistically analyze the data). Using the described protocol 40–60% of the transformants segregate 3:1 in their offspring.

25. For genetic analysis and the isolation of recessive mutants (e.g., by T-DNA tagging approaches), diploid transgenic lines are desired. Polyploidy will hamper further analysis. The level of tetraploid plants in our population of transgenic lines is low. Less than 5% of the plants were tetraploid. Tetraploid plants can be recognized by the size of the seeds (larger compared to diploids), leaves, and late flowering phenotype. It is also possible to count chloroplasts in the guard cells of stomatal cells *(38)*. Therefore, epidermal strips of the leaf can be examined using fluorescence microscopy. Tetraploids contain more (> 12) chloroplasts per two guard cells than diploids *(6–8)*. Altmann et al. *(20)* described an easy method to distinguish between diploid and tetraploid plants by measuring the pollen size.

26. A 3:1 segregation pattern in the offspring of a transformant does not necessarily resemble one insert. Often T-DNA repeats are formed. By isolating chromosomal DNA of 20 pooled T2 plants and performing Southern analysis we found that approx 40% of the 3:1 segregating transformants indeed contain one insert. Less than 1% represents a not transformed plant (=escape).

Acknowledgments

This work was financially supported by the European Union (BIOT CT90-0207). The authors would like to thank Saskia Rueb for helpful discussions and critically reading the manuscript.

References

1. Czakó, M., Wilson, J., Yu, X., and Márton, L. (1993) Sustained root culture for generation and vegetative propagation of transgenic *Arabidopsis thaliana*. *Plant Cell Reps.* **12**, 603–606.
2. Akama, K., Shiraishi, H., Ohta, S., Nakamura, K., Okada, K., and Shimura, Y. (1992) Efficient transformation of *Arabidopsis thaliana*: comparison of the efficiencies with various organs, plant ecotypes and *Agrobacterium* strains. *Plant Cell Reps.* **12**, 7–11.
3. Bechtold, N., Ellis, J., and Pelletier, G. (1993) In planta *Agrobacterium* mediated gene transfer by infiltration of adult *Arabidopsis thaliana* plants. *Life Sciences* **316**, 1194–1199.
4. Chang, S. S., Park, S. K., Kim, B. C., Kang, B. J., Kim, D. U., and Nam, H. G. (1994) Stable genetic transformation of *Arabidopsis thaliana* by *Agrobacterium* inoculation in planta. *Plant J.* **5**, 551–558.
5. Clarke, M. C., Wei, W., and Lindsey, K. (1992) High-frequency transformation of *Arabidopsis thaliana* by *Agrobacterium tumefaciens*. *Plant Mol. Biol. Reporter* **10**, 178–189.
6. Damm, B., Schmidt, R., and Willmitzer, L. (1989) Efficient transformation of *Arabidopsis thaliana* using direct gene transfer to protoplasts. *Mol. Gen. Genet.* **217**, 6–12.
7. Feldmann, K. A. and Marks, M. D. (1987) *Agrobacterium*-mediated transformation of germinating seeds of *Arabidopsis thaliana*: A non-tissue culture approach. *Mol. Gen. Genet.* **208**, 1–9.
8. Huang, H. and Ma, H. (1992) An improved procedure for transforming *Arabidopsis thaliana* (Landsberg *erecta*) root explant. *Plant Mol. Biol. Reporter* **10**, 372–383.
9. Kemper, E., Grevelding, C., Schell, J., and Masterson, R. (1992) Improved method for the transformation of *Arabidopsis thaliana* with chimeric dihydrofolate reductase constructs which confer methotrexate resistance. *Plant Cell Reps.* **11**, 118–121.
10. Lloyd, A. M., Barnason, R. A., Rogers, S. G., Byrne, M. C., Fraley, R. T., and Horsch, R. B. (1986) Transformation of *Arabidopsis thaliana* with *Agrobacterium tumefaciens*. *Science* **234**, 464–466.
11. Mandal, A., Lang, V., Orczyk, W., and Palva, E. T. (1993) Improved efficiency for T-DNA-mediated transformation and plasmid rescue in *Arabidopsis thaliana*. *TAG* **86**, 621–628.

12. Marton, L. and Browse, J. (1991) Facile transformation of *Arabidopsis*. *Plant Cell Reps.* **10,** 235–239.
13. Sangwan, R. S., Bourgeois, Y., Dubois, F., and Sangwan-Norreel, B. S. (1992) In vitro regeneration of *Arabidopsis thaliana* from cultured zygotic embryos and analysis of regenerants. *J. Plant Physiol.* **140,** 588–595.
14. Sheikholeslam, S. N. and Weeks, D. P. (1987) Acetosyringone promotes high efficiency transformation of *Arabidopsis thaliana* explants by *Agrobacterium tumefaciens*. *Plant Mol. Biol.* **8,** 291–298.
15. Valvekens, D., Van Montagu, M., and Van Lijsebettens, M. (1988) *Agrobacterium tumefaciens*-mediated transformation of *Arabidopsis thaliana* root explants by using kanamycin selection. *Proc. Natl. Acad. Sci. USA* **85,** 5536–5540.
16. Hooykaas, P. J. J. and Schilperoort, R. A. (1992) *Agrobacterium* and plant genetic engineering. *Plant Mol. Biol.* **19,** 15–38.
17. Hoekema, A., Hirsch, P. R., Hooykaas, P. J. J., and Schilperoort, R. A. (1983) A binary plant vector strategy based on separation of vir- and T-region of the *Agrobacterium tumefaciens* Ti plasmid. *Nature* **303,** 179–180.
18. Sangwan, R. S., Bourgeois, Y., Brown, S., Vasseur, G., and Sangwan-Norreel, B. (1992) Characterization of competent cells and early events of *Agrobacterium*-mediated genetic transformation in *Arabidopsis thaliana*. *Planta* **188,** 439–456.
19. Galbraith, D. W., Harkins, K. R., and Knapp, S. (1991) Systemic endoploidy in *Arabidopsis thaliana*. *Plant Physiol.* **96,** 985–989.
20. Altmann, T., Damm, B., Frommer, W. B., Martin, T., Morris, P. C., Schweizer, D., Willmitzer, L., and Schmidt, R. (1994) Easy determination of ploidy level in *Arabidopsis thaliana* plants by means of pollen size measurement. *Plant Cell Reps.* **13,** 652–656.
21. De Block, M., Botterman, J., Vandewiele, M., Dockx, J., Thoen, C., Gosselé, V., Rao Movva, N., Thompson, C., Van Montagu, M., and Leemans, J. (1987) Engineering herbicide resistance in plants by expression of a detoxifying enzyme. *EMBO J.* **6,** 2513–2518.
22. Haughn, G. W., Smith, J., Mazur, B., and Somerville, C. (1988) Transformation with a mutant *Arabidopsis* acetolactate synthase gene renders tobacco resistant to sulfonylurea herbicides. *Mol. Gen. Genet.* **211,** 266–271.
23. Van den Elzen, P. J. M., Townsend, J., Lee, K. Y., and Bedbrook, J. R. (1985) A chimeric hygromycin resistance gene as a selectable marker in plant cells. *Plant Mol. Biol.* **5,** 299–302.
24. Perera, R. J., Linard, C. G., and Signer, E. R. (1993) Cytosine deaminase as a negative selective marker for *Arabidopsis*. *Plant Mol. Biol.* **23,** 793–799.
25. Czakó, M., and Márton, L. (1994) The herpes simplex virus thymidine kinase gene as a conditional negative-selection marker gene in *Arabidopsis thaliana*. *Plant Physiol.* **104,** 1067–1071.
26. Halfter, U., Morris, P. C., and Willmitzer, L. (1992) Gene targeting in *Arabidopsis thaliana*. *Mol. Gen. Genet.* **231,** 186–193.
27. Offringa, R., de Groot, M. J. A., Haagsman, H. J., Does, M. P., van den Elzen, P. J. M., and Hooykaas, P. J. J. (1990) Extrachromosomal homologous recombination

and gene targeting in plant cells after *Agrobacterium* mediated transformation. *EMBO J.* **9,** 3077–3084.
28. Hood E. E., Gelvin, S. B., Melchers, L. S., and Hoekema, A. (1993) New *Agrobacterium* helper plasmids for gene transfer to plants. *Transgenic Research* **2,** 208–218.
29. Van der Graaff, E. and Hooykaas, P. J. J. (1996). Improvements in the transformation of *Arabidopsis thaliana* C24 leaf-discs by *Agrobacterium tumefaciens*. *Plant Cell Reps.* **15,** 572–577.
30. Vancanneyt, G., Schmidt, R., O'Connor-Sanchez, A., Willmitzer, L., and Rocha-Sosa, M., (1990) Construction of an intron-containing marker gene: Splicing of the intron in transgenic plants and its use in monitoring early events in *Agrobacterium*-mediated plant transformation. *Mol. Gen. Genet.* **220,** 245–250.
31. Masson, J. and Paszkowski, J. (1992) The culture response of *Arabidopsis thaliana* protoplasts is determined by the growth conditions of donor plants. *Plant J.* **2,** 829–833
32. Gamborg, O. L., Miller, R. A., and Ojima, K. (1968) Nutrient requirement of suspension cultures of soybean. *Exp. Cell Res.* **50,** 151–158
33. Murashige, T. and Skoog, F. (1962) A revised medium for rapid growth and bio-assays with tobacco tissue cultures. *Plant Physiol.* **15,** 473–497.
34. Becker, D., Kemper, E., Schell, J., and Masterson, R. (1992) New plant binary vectors with selectable markers located proximal to the left T-DNA border. *Plant Mol. Biol.* **20,** 1195–1197.
35. Holford, P. and Newbury, H. J. (1992) The effects of antibiotics and their breakdown products on the in vitro growth of *Antirrhinum majus*. *Plant Cell Reps.* **11,** 93–96.
36. Lin, J. -J., Assad-Garcia, N., and Kuo, J. (1995) Plant hormone effects of antibiotics on the transformation efficiency of plant tissues by *Agrobacterium tumefaciens* cells. *Plant Sci.* **109,** 171–177.
37. Bevan, M. (1984) Binary *Agrobacterium* vectors for plant transformation. *Nucl. Acid Res.* **12,** 8711–8721.
38. Bingham, E. T. (1968) Stomatal chloroplasts in alfalfa at four ploidy levels. *Crop Sci.* **8,** 509,510.

27

Transformation of *Arabidopsis thaliana* C24 Leaf Discs by *Agrobacterium tumefaciens*

Eric van der Graaff and Paul J. J. Hooykaas

1. Introduction

Transgenic plants have become a very important tool for the study of gene function. The *Agrobacterium* binary transformation system *(1)* allows the precise transfer of a defined piece of DNA (T-DNA) to the plant genome *(2,3)* of many plant species. In this way, any desired DNA sequence can be introduced into the plant genome. For the small crucifer *Arabidopsis thaliana,* many transformation protocols are available using different tissues as source explant to be transformed with *Agrobacterium (3–11)*.

Although many of these protocols render transformants harboring low numbers of independent T-DNA integration loci, these loci are often comprised of multiple T-DNA inserts in complex structures *(12)*. Such multiple T-DNA structures could lead to T-DNA rearrangements resulting in truncated T-DNA copies and furthermore lead to transcriptional silencing of the genes present on the T-DNA by methylation, a process which has been found to be correlated with the presence of multiple T-DNA inserts *(13)*. Another problem with *Agrobacterium* T-DNA transformation is the occurrence of somaclonal variation. This is the occurrence of changes in phenotype or genotype that are not linked to the T-DNA insertion, but probably caused by extended maintenance of plant tissues in tissue culture conditions *(14,15)*.

In our research, we are studying the role of phytohormones in *Arabidopsis* morphogenesis. One approach is to isolate transgenic *Arabidopsis* plants in which the *Agrobacterium* tumor–inducing genes are expressed to manipulate the endogenous phytohormone levels. Another approach is to isolate phenocopies of these plants after gene tagging with a specialized activator T-DNA tagging vector *(16–18)*. For these purposes, an efficient transformation protocol is

Table 1
Composition of Tissue Culture Medium[a]

Compound	AGM	C1	C2	AGM/IBA	1/2MS
Macrosalts	1/2MS	MS	MS	1/2MS	1/2MS
Microsalts	1/2MS	MS	MS	1/2MS	1/2MS
Vitamins		B5	B5		MS
FeNaEDTA, mg/L	38.5	38.5	38.5	38.5	38.5
Myo–inositol, mg/L		100	100		100
Sucrose, g/L	10	30	30	10	10
MES, g/L	0.5	0.5	0.5	0.5	
PVP, g/L		0.5	0.5		
Plant agar[b], g/L		8	8		8
Daichin agar[c], g/L	8			8	
pH	5.7	5.7	5.7	5.7	5.7
NAA, mg/L		1.0	0.2		
BAP, mg/L		0.1	1.0		
IBA, mg/L				1.0	

[a]1/2 MS, half strength MS macros and micros; MES, 2–(N–morpholino) ethane sulfonic acid; PVP, polyvinyl pyrrolidone.
[b]Plant tissue culture agar (Imperial Laboratories, Ritmeester, The Netherlands).
[c]Daichin agar (Brunschwig Chemie, Amsterdam, The Netherlands).

required resulting in transformants harboring a low number of T-DNA inserts and which is accompanied by only low levels of somaclonal variation.

The leaf transformation protocol reported by van Lijsebettens et al. *(6)* was chosen because in this protocol no 2,4–D, which supposedly causes somaclonal variation *(19,20)* is used and the regeneration is at least as fast as reported in other transformation protocols. This procedure gives rise to transformants with preferentially low numbers of T-DNA inserts. After adapting this protocol in our laboratory, improvements were made on the original van Lijsebettens protocol resulting in an average transformation frequency of 1.6 shoots per leaf explant 4 wk after *Agrobacterium* infection *(18)*.

For transformation of explants, the frequency of transformation can be expressed in various ways. In general, it is expressed as the percentage of explants generating transgenic shoots. By expressing the transformation frequency as the number of transgenic shoots regenerated per (leaf) explant after a fixed regeneration period following the bacterial infection, a reliable transformation frequency can be obtained allowing comparisons between different protocols.

Genetic analysis of transformants obtained using our transformation protocol revealed that mainly (71.7%) single-locus transformants were generated, of which a high percentage (68%) contained one T-DNA insert. Furthermore on a total of 27 transgenic lines with 51 T-DNA inserts, only 8 T-DNA repeat structures were encountered *(18)*, indicating that this transformation protocol is very suited for standard transgenic plants studies in which ingeneral at least 10 independent transgenic lines harboring the same T-DNA insert are included. Given the high transformation frequencies reached using this protocol, small populations of individual transformants for gene–tagging studies can also be generated.

2. Materials
2.1. Plant Tissue Culture Media and Chemicals

Prepare all media and stock solutions in demineralized H_2O and filter sterilize all stock solutions (antibiotics and hormones) using a 0.22-µm membrane.

2.1.1. Antibiotics

1. Vancomycin: (Duchefa, Haarlem, The Netherlands) Prepare as 200 mg/mL stock in H_2O. This solution can be stored for at least 6 mo at –20°C.
2. Augmentin: (Beecham Farma, Amstelveen, The Netherlands) Always prepare fresh as 200 mg/mL stock in H_2O.
3. Kanamycin: (Sigma, St. Louis, MO) Prepare as 50 mg/mL stock in H_2O. This stock can be stored for several months at 4°C in the dark.
4. Hygromycin: (Boehringer, Mannheim, Germany) Prepare as 50 mg/mL stock by dissolving powder in H_2O. Adjust pH to 7.0. Take care to adjust pH with $0.1N$ HCl in single drops to prevent the pH to drop below 6.0, because at acidic pH (below 5.7), the hygromycin antibiotic can become inactivated (*see* **Note 1**). This stock can be stored for several months at 4°C in the dark.

2.1.2. Hormones

1. α-naphthalene acetic acid (NAA): Prepare as 200 mg/L stock by dissolving powder in a few drops of $0.1N$ HCl and adding H_2O to final volume. The stock can be stored for several months at 4°C in the dark.
2. Indole–3–butyric acid (IBA): Prepare as 200 mg/L stock as described for NAA. The stock can be stored for several months at 4°C in the dark.
3. Benzyl amino purine (BAP): Prepare as 50 mg/L stock by dissolving powder in a few drops of $0.1N$ KOH and adding H_2O to final volume. The stock can be stored for several months at 4°C in the dark.

2.1.3. Media

Sterilize all media on 120°C for 20 min and add the stock solutions (antibiotics and hormones) to the media listed below after autoclaving and cooling to approx 65°C.

1. Macro- and micro-salt solutions and vitamins.
    a. MS *(21)* macrosalts (final concentrations in mg/L): 720 NH_4NO_3 950 KNO_3, 185 $MgSO_4 \cdot 7H_2O$, 68 KH_2PO_4, and 220 $CaCl_2 \cdot 2H_2O$.
    b. MS *(21)* microsalts (final concentrations in mg/L): 0.83 KI, 6.2 H_3BO_4, 16.9 $MnSO_4 \cdot 2H_2O$, 8.6 $ZnSO_4 \cdot 7H_2O$, 0.25 $Na_2MoO_4 \cdot 2H_2O$, 0.025 $CuSO_4 \cdot 5H_2O$, and 0.025 $CoCl_2 \cdot 6H_2O$.
    c. Vitamins (final concentrations in mg/L):

	MS *(21)*	B5 *(22)*
Nicotinic Acid	0.5	1
Pyridoxine–HCl	0.5	1
Thiamine–HCl	0.1	10
Glycine	2	0

2. Tissue culture medium (*see* **Table 1**).

2.2. Preparation of Source Plants

1. Seeds of *Arabidopsis thaliana* ecotype Columbia C24 (*see* **Note 2**).
2. 1.5-mL Eppendorf tubes.
3. 1% sodiumhypochlorite solution (commercial bleach) supplemented with 0.05% Tween-20 (Janssen Chimica, Beerse, Belgium).
4. Demineralized H_2O.
5. Demineralized H_2O with 0.1% agarose autoclaved for 20 min at 120°C (*see* **Note 3**).
6. Glass jars (80 × 95 mm) with AGM medium.
7. Gas-diffusable tape (Urgopore, S. A. Sterilco, Brussel, Belgium).
8. Micropipet (p1000) with tips.
9. Source plants grown under short day and blue light conditions (*see* **Note 4**).

2.3. Callus Induction

1. Plates with 25 mL solid C1 medium (Greiner 94 × 16 mm).
2. C1 liquid medium.
3. Empty petridishes (Greiner 94 × 16 mm).
4. Scalpels (long slender).
5. Forceps (long, slender with small tip).
6. Parafilm.

2.4. Bacterial Strains and Growth Conditions

1. Bacterial strains (*see* **Note 5**) on fresh LB (Luria Broth medium: 10 g/L Tryptone, 5 g/L Yeast extract, and 8 g/L NaCl autoclaved 20 min at 120°C) solid medium (1.8% Bacto agar).
2. Liquid LB medium.
3. Acetosyringone: 20 m*M* stock in DMSO.
4. Appropriate stocks of bacterial antibiotics (*see* **Note 6**).
5. 100 mL Erlenmeyer flask with cotton plug.
6. Bacterial growth incubator (28°C).

2.5. Bacterial Infection

1. Empty Petri dishes (Nunc 100 × 25 mm).
2. Filter paper (sterile).
3. Liquid C1.
4. Overnight grown bacterial cultures.
5. Disposable 25 mL pipets.
6. Forceps.

2.6. Washing

1. Glass beakers (250 mL minimum size) with aluminium cap.
2. Shaker (optional).
3. Liquid C1 with antibiotics vancomycin and augmentin (*see* **Note 7**).
4. Filter paper (sterile).
5. Disposable pipets.
6. Forceps.
7. Solid C2 plates (Greiner 94 × 16 mm).

2.7. Regeneration and Seed Set

1. Plates with 25 mL solid C2 medium (Greiner 94 × 16 mm).
2. Forceps.
3. Forceps with sharp, fine tip.
4. Scalpel.
5. IBA rooting plates (15 mL of AGM/IBA: Nunc 60 × 20 mm).
6. Greiner jars (8 mL of AGM: Greiner 53 × 100mm).
7. High petridishes (Nunc 100 × 25 mm) with 25 mL AGM medium.

2.8. Genetic Analysis

1. 1/2MS plates (25 mL: Greiner 94 × 16 mm) with appropriate antibiotic (25 mg/L kanamycin or 20 mg/L hygromycin) to screen for segregation of the number of independent T-DNA loci.
2. Standard protocols for genomic DNA isolation and Southern blot analysis (*see* Chapter 9).

3. Methods
3.1. Plant Tissue Culture Media and Chemicals

1. Autoclave all media for 20 min at 120°C without antibiotics and/or hormones.
2. Add the appropriate antibiotics and/or hormones after the medium has cooled down till approx 60–65°C.
3. Fill the agar containing media and allow to solidify and dry in the laminar flowhood with the lids of the Petri dishes and jars slightly opened.
4. If desired, store media at 4°C in the dark for 4 wk maximum except the C2 plates including the antibiotic augmentin. Prepare this freshly at most 1 d in advance.

5. For the growth of the source plants, pour 50 mL of medium in glass jars (80 × 95 mm), after which the medium is dried for 1 h.
6. For the C1 and C2 solid plates, pour 25 mL of medium in standard Petri dishes (Greiner 94 × 16 mm) and dry the plates for 1 h.
7. For root induction, pour 15 mL of IBA/AGM medium in small Petri dishes (Nunc 60 × 20 mm) and dry the plates for 30 min.
8. For in vitro seed set, transfer 8 mL of AGM medium to plastic jars (Greiner 53 × 100 mm) and allow to dry for 30 min.
9. For root growth after the root induction phase, but prior to transfer to soil, pour 25 mL of AGM medium in high petridishes (Nunc 100 × 25mm) and dry the plates for 1 h.

3.2. Preparation of Source Plants

The *Arabidopsis* seeds will sediment to the bottom of the Eppendorf tubes within 10–20 s, so no centrifuges are needed during the surface sterilization of the seeds.

1. Fill Eppendorf tube with the amount of *Arabidopsis thaliana* seeds (stored at 4°C) needed (1 seed weighs 20 µg).
2. Soak the seeds for 1 min in 70% ethanol.
3. Remove the ethanol using the micropipet and replace by the sodiumhypochlorite/Tween-20 solution.
4. Remove this solution after 5 min and replace with H_2O.
5. Repeat the H_2O wash step at least three times and finally suspend the seeds in H_2O/agarose.
6. Place the seeds in the glass jars by pipeting, placing 20 seeds per jar.
7. Close the jars by fitting the plastic lids loosely on top of the jars and by sealing the remaining gap by gas–permeable tape (Urgopore) to allow efficient aeration.
8. Incubate the jars for a minimum of 4 d at 4°C in the dark to allow equal and efficient germination.
9. Grow the seedlings for 3 wk under short day and blue light conditions (*see* **Note 4**).

3.3. Callus Induction

1. Take the plantlets out of the jars one by one each time closing the lid to prevent desiccation of the plantlets.
2. Cut off the largest leaves (2 or 3) from the plantlet and directly transfer these to liquid C1 medium in a standard Petri dish (Greiner 94 × 16 mm). Take care to remove the complete petiole (*see* **Note 8**).
3. Collect all leaves from one jar in one dish with liquid C1.
4. After all leaves have been collected in liquid C1 medium, cut the leaves into two halves resulting in two leaf explants per initial leaf.
5. Take the leaves out of the liquid medium with a scalpel, transfer to an empty Petri dish and position with a second scalpel. Cut the leaves longitudinally, splitting the main vascular bundle along the length (*see* **Note 9**) and subsequently transfer the leaf explants to solid C1 medium.

A. tumefaciens Leaf Disc Transformation

6. Transfer all leaves from one jar to two C1 plates.
7. Seal the plates with parafilm and incubate for 5 d under white light conditions (*see* **Notes 4** and **10**).

3.4. Bacterial Strains and Growth Conditions

1. Inoculate 1 d prior to the bacterial infection of the leaf parts the *Agrobacterium tumefaciens* strains in 5 mL of LB medium, pH 5.6, supplemented with acetosyringone and appropriate antibiotics (for stable maintenance of the binary vector).
2. Grow cultures overnight at 28°C.

3.5. Bacterial Infection

At this stage, the leaf parts should have expanded in size, and the site where the leaf explants were cut should have become thicker.

1. Dilute the 5 mL bacterial culture grown overnight in 20 mL of liquid medium in a high petridish (Nunc 100 × 25 mm; *see* **Note 11**).
2. Transfer the leaf explants from two randomly picked Petri dishes (*see* **Note 12**) to this 25 mL of diluted bacterial suspension and incubate for 5–10 min with gentle shaking.
3. Remove the leaf explants using a forceps, briefly blot dry on filter paper, and transfer leaf explants to fresh solid C1 plates.
4. Press the leaf explants (gently) into the medium, taking care that the cut edge with the vascular tissue is completely in contact with the medium (*see* **Note 13**).
5. Seal the plates with parafilm and incubate for 2 d under white light conditions (*see* **Notes 5** and **10**).
6. When the leaf explants of more than two dishes have to be transformed with the same bacterial strain, use for each two dishes a new bacterial culture (*see* **Note 14**).

3.6. Washing

The leaf explants at this stage should be slightly overgrown with bacteria.

1. Collect the leaf explants in a sterile glass beaker (250 mL size) for washing.
2. Add at least 50 mL of liquid C1 medium supplemented (prepare fresh) with vancomycin (750 mg/L) and augmentin (200 mg/L), place aluminium cap back on beaker and shake vigorously.
3. Remove the medium using disposable pipets. This medium should be very turbid because of agar pieces and bacteria.
4. Add fresh liquid C1 medium supplemented with antibiotics and again shake vigorously.
5. Remove the liquid medium, which now should almost be completely clear of debris, leaving the leaf parts covered with a small amount of medium.
6. If the removed medium is still turbid, repeat the wash step.
7. When several different transformations are carried out during one experiment, the glass beakers can be placed on a rotary shaker during the wash step while other leaf parts are being handled.

8. Blot the leaf parts in groups of 5–10 pieces to complete dryness and transfer to solid C2 medium supplemented with 750 mg/L vancomycin, 200 mg/L augmentin, and either 50 mg/L kanamycin or 20 mg/L hygromycin (*see* **Note 15**).
9. Press the leaf parts in the medium, making sure that the cut edge, which now clearly is thickening due to callus formation, is in full contact with the selective medium (*see* **Note 13**).
10. Use the explants from two C1 plates to fill two C2 plates, even though the explants have increased in size compared to the initial source explants.
11. Seal the plates with gas-diffusable tape and incubate under white light conditions (*see* **Note 4**).

3.7. Regeneration and Seed Set

At 7 d after the bacterial infection, individual calli can be observed. These calli (3–7 per leaf part) expand in size and form a complete callus ridge on the initial cut side 14 d after bacterial infection (*see* **Note 16**). The first regenerated shoots can be removed for root induction 3 wk after the bacterial infection.

1. Transfer the leaf parts to fresh C2 plates every 2 wk. Lower the vancomycin concentration gradually from 750 mg/L, to 500 mg/L, 400 mg/L, and 250 mg/L on subsequent transfers: the augmentin concentration is maintained at 200 mg/L.
2. Remove organized shoots with three or more leaves from the leaf explants. Use either a knife or a sharp, pinned forceps to clear the shoots from any attached callus (*see* **Note 17**).
3. Transfer these shoots to root-induction plates, and incubate for 1 wk under white light conditions (*see* **Notes 4** and **18**).
4. Remove the callus surrounding the regenerated shoot from the leaf explant (discarding roughly 1/3 of the complete callus ridge) to ensure the regeneration of individual shoots.
5. Transfer the shoots after root induction to either Greiner jars for root proliferation and in vitro seed set or to high Petri dishes with AGM medium for root formation under white light conditions (*see* **Note 4**).
6. For in vitro seed set, it is important that the Greiner jars are free of condensation. During seed set and ripening, the jars can be incubated opened in laminar flowhoods for several hours to allow condensation to disappear. Collect the seeds as soon as they have ripened (*see* **Note 19**).
7. For in vivo seed set, the transgenic plantlets must have formed a root of at least 2 cm before they can be transferred to soil. Seed set and seed isolation is facilitated using the Aracon seed set system (*see* **Note 20**).
8. Transfer the leaf parts to fresh C2 plates for a maximum of 8 wk after bacterial infection. After this period, the leaf parts either are completeley cleared from callus, or will only regenerate aberrant shoots.

3.8. Genetic Analysis

1. Sow seedlings (50–100) on 1/2MS plates (25 mL with appropriate antibiotic selection).

2. Determine the ratio of resistant vs sensitive seedlings.
3. Calculate the number of independently segregating T-DNA loci (*see* **Note 21**).
4. Analyze plant lines by Southern blotting, taking care that the probes and restriction enzymes used to digest the genomic plant DNA allow the detection of right and left border T-DNA fragments as well as internal T-DNA fragments resulting from T-DNA repeat structures (*see* **Note 22**).

4. Notes

1. In general, all chemicals and especially antibiotics are toxic, so great care must be taken when preparing the antibiotic stock solutions. Always avoid breathing dust and direct contact with the skin. The antibiotic hygromycin is extremely toxic andgreat care must be taken not only when preparing the stock solution (50 mg/mL), but also when filling out of this stock solution in the selection medium. Hygromycin is very basic and the pH of the stock solution should be adjusted to pH 7.0. Because hygromycin can be inactivated at acidic pH (below 5.7), adjust the pH with $0.1N$ HCl in small drops under thorough stirring of the solution.
2. Only the *Arabidopsis thaliana* ecotype C24 has been used in this leaf transformation protocol so far.
3. The use of H_2O/agarose to suspend the seeds allows efficient and fast spreading of *Arabidopsis* seeds by pipeting using micropipets. The *Arabidopsis* seeds are too small to be transferred using a small paint brush (when seeds are dry), but alternatively could be handled with a forceps having a small tip. However, handling in this way is very tedious. The use of yellow micropipet tips allows the handling of the seeds one by one (seeds cannot enter the tip itself), whereas the use of blue micropipet tips allows the spreading of multiple seeds per drop of H_2O/agarose (seeds can enter the blue tip).
4. Source plants grown in the AGM medium, which lacks vitamines *(23)*, under a 12 h light (3000 lx) 20°C/12 h dark, 16°C regime (short day) using Duro–Test® 5500k-91 tubes (Philips, The Netherlands) which provide a light spectrum containing relatively more blue light in growth cabinets (Polar 5,0 KU2) were found to generate more transformants than plants of the same age grown on half strength MS medium (1/2MS) under our standard white light and day/night conditions (16 h light 3000 lx/8 h night regime) at 21°C. Although the leaves from plants grown under the short day/blue light conditions are much smaller compared to those from plants grown under white light, these smaller leaves show a two-fold higher regeneration capacity. Although the regenerating callus is formed exclusively on the cut edge where the main vascular bundle is situated, the transformation frequency cannot be increased using older plantlets. Because the leaves are larger, the wounded area is increased, but the transformation frequency decreases for older (and larger) leaves.
5. Optimal transformation frequencies have been obtained using the GV2260 bacterial strain, which was also used in the original van Lijsebettens protocol. The strain LBA1115 (=MOG101), which is a similar octopine-type helper strain, renders comparable transformation frequencies. The use of the LBA4404

bacterial strain *(1)* or the strain EHA105, which carries a nononcogenic derivative of the supervirulent helper plasmid pTiBo542 *(24)*, did not result in higher transformation frequencies. The EHA105 strain furthermore resulted in frequent bacterial overgrowth.

6. The bacterial antibiotic selection used depends on the bacterial strain and the binary vector used. All antibiotic stocks are filter sterilized and are added to the LB medium after autoclaving and cooling down of the LB medium.

7. The antibiotics vancomycin and augmentin can be omitted from the liquid C1 medium used for washing because this washing procedure is very thorough. Alternatively, lower antibiotic concentrations or only one of the antibiotics can be used during this wash step. Initially, the antibiotic cefotaxim was used to kill off the bacteria, but cefotaxim was found to inhibit the regeneration of the transgenic shoots, resulting in lower transformation frequencies and shoots of lower quality.

8. Initially, pieces of the petioles still attached to the leaves were also used for transformation. The petiole pieces show a good regenerative capacity, but since these petioles contain cells of mainly higher ploidy level compared to the leaf tissues of the same age (investigated by flow cytometric analysis), the proportion of tetraploid transgenic shoots could be increased by including petiole tissue.

9. Initially, transversely cut leaf explants were used. However, regenerating calli were only observed where the main vascuar bundle was wounded (cut). Therefore, subsequently leaves were cut longitudinally, splitting this main vascular bundle along its length. This cutting does not have to be very exact, because the position of this vascular bundle can only be estimated using the (small) leaves from the source plants grown under short day/blue light conditions. Nevertheless almost all leaf explants cut in this way regenerate 3–7 calli along the complete wounded area. Given the fragility of the small leaves used as source explant, simply pressing the scalpel blade on the leaf tissue suffices to cut the leaf into two leaf halves. In this way, the scalpel blade remains sharp for a longer period and does not have to be changed more than once per two jars.

10. These white light conditions are the standard growth conditions for *Arabidopsis* plants and tissues in our laboratory. Incubation during the callus induction phase and bacterial infection at higher temperatures (25°C) resulted in the formation of larger calli, but shoot regeneration is not increased. However, due to the regeneration of larger calli, it turned out to be more tedious to remove all callus from the base of the shoots.

11. Some bacterial strains like LBA4404 tend to form slimy clusters when the fully grown bacterial cultures are not shaken anymore. Such cultures were removed from the growth device only shortly before the bacterial infection.

12. When using several different bacterial strains (and/or binary vectors), this random selection of plates allows better comparisons within one transformation experiment. Occasionally (some of) the source plants from one particular jar appear to be of lower quality compared to those from other jars used in the same experiment. In general, two dishes (one original jar with source plants) are used for

transformation with one bacterial strain. This should render enough transformants per strain used to allow thorough analysis of the transformants harboring the same construct (10 diploid, single locus, single T-DNA insert lines).
13. From this stage on, the leaf explants not only have expanded in size compared to the initial leaf explants, but also have become more rigid. The leaf parts tend to start curling, pushing part of the regenerating calli out of the medium. Because the calli will regenerate only at or very near the original cut site, this wound site should always be in good contact with the medium both for nutritional uptake and efficient exposure to the antibiotic selection, preventing the regeneration of shoots escaping this selection.
14. To ensure equal circumstances for each independent transformation (different bacterial strain/binary vector) within one experiment, fresh bacterial dilutions are used for each two Petri dishes.
15. The results with hygromycin during the optimization of this protocol are based on the hygromycin supplied by Boehringer (as powder). The hygromycin used now is purchased from Duchefa. This hygromycin is delivered as a concentrated hygromycin solution in DMSO. This solution can simply be diluted with H_2O, but the pH should still be adjusted as described above in **Note 1**. This new hygromycin has not been used in this transformation protocol, yet, but selection on seedling stage also works well on 20 mg/L hygromycin.
16. When independent regenerated calli have to be maintained these should be removed before 10 d after the bacterial infection, because otherwise the individual calli tend to form one large callus ridge. The separation of individual calli does not lead to a marked increase in the regeneration of shoots. In both cases (separated calli and calli still attached to the leaf explants) the percentage of calli regenerating shoots is approx 30%.
17. Take care to remove all pieces of callus from the base of the regenerated shoots, because root formation is hampered by callus present at the shoot base.
18. At this stage, regenerated shoots of poor quality tend to die very rapidly. In our hands, on an average 30% of the regenerated shoots died. The shoot quality is dependent on the growth conditions present in the plant growth facilities, but can also be influenced in a negative way by, for instance, painting and welding activities in the building in which these plant-growth facilities are situated. This may result not only in a lower frequency of regeneration but also in a lower shoot quality.
19. In case in vivo seed set is possible, this is preferable over in vitro seed set. The in vitro seed set nevertheless can be important in cases in which the regenerants show poor or no root formation. Root formation not only is important for survival of the regenerant in soil but also for seed production when maintained in vitro. Several of the regenerants harboring the T–cyt gene of *Agrobacterium* were not able to form roots. However, cutting off the flower stalks formed on these shoots and cultering these separated flower stalks in the Greiner jars resulted in root formation and (low) seed production. Once the seeds have ripened, they must be collected as soon as possible. Seed germination ranges from 70–100%. Seed isolation several months after seed ripening resulted in a much lower seed

germination and a lower seed quality, reflected by increasing proportions of aberrant seedlings.
20. In vivo seed set using the Aracon system is preferable, but can only be used when good root formation has taken place on the regenerants. High humidity must be prevented to exclude the possibility of fungal infections on the flower stalks. Such infections could lead to the association of fungi with the seeds, which are not killed by the seed surface sterilization procedure. Another problem might be the presence of eggs or larvae of insects in the soil used for growth of the plants. These bugs really like fresh *Arabidopsis* plants. One major problem we encountered was the larval stage of the black fly. These slug-like larvae tend to eat the leaves of *Arabidopsis* plants, but these attacks can easily be prevented by covering the soil surface with sand, preventing the larvae from reaching the leaves. The root system will not be eaten by these larvae.
21. In general 50–100 seeds are sown for segregation analysis. In most cases, simple ratio determination of the number of resistant seedlings vs the number of sensitive seedlings will give the number of individual loci. However, in some cases statistical analysis is needed to determine the number of loci. Using this transformation protocol, 71.7% of the transformants segregated for one T-DNA locus, whereas 18.5% and 1.4% of the transformants segregated for two and three T-DNA loci, respectively. The remaining 8.3% of the transformants showed aberrant segregation making it unable to determine the number of T-DNA loci *(18)*.
22. Using Southern blot analysis, we determined that 68% of the transformants segregating for one T-DNA locus actually contained one T-DNA insert. The average number of T-DNA loci in 27 transgenic lines analyzed was 1.9 T-DNA insert per plant. In these 27 lines, a total of 8 T-DNA repeat structures were encountered *(18)*.

References

1. Hoekema, A., Hirsch, P. R., Hooykaas, P. J. J., and Schilperoort, R. A. (1983) A binary plant vector strategy based on separation of vir– and T–region of the *Agrobacterium tumefaciens* Ti–plasmid. *Nature* **303**, 179–180.
2. Hooykaas, P. J. J. and Schilperoort, R. A. (1992) *Agrobacterium* and plant genetic engeneering *Plant. Mol. Biol.* **19**, 15–38.
3. Hooykaas, P. J. J. and Beijersbergen, A. G. M. (1994) The virulence system of *Agrobacterium tumefaciens. Ann. Rev. Phytopathol.* **32**, 157–179.
4. Feldmann, K. A. and Marks, M. D. (1987) *Agrobacterium*–mediated transformation of germinating seeds of *Arabidopsis thaliana*: a none tissue cultue approach. *Mol. Gen. Genet.* **208**, 1–9.
5. Valvekens, D., Van Montagu, M., and Van Lijsebettens, M. (1988) *Agrobacterium tumefaciens*–mediated transformation of *Arabidopsis thaliana* root explants by using kanamycin selection. *Proc. Natl. Acad. Sci. USA* **85**, 5536–5540.
6. Van Lijsebettens, M., Vanderhaeghen, R., and Van Montagu, M. (1991) Insertional mutagenesis in *Arabidopsis thaliana:* isolation of a T–DNA–linked mutation that alters leaf morphology. *Theor. Appl. Genet.* **81**, 277–284.

7. Schmidt, R. and Willmitzer, L. (1988) High efficiency *Agrobacterium tumefaciens*–mediated transformation of *Arabidopsis thaliana* leaf and cotyledon explants. *Plant Cell Rep.* **7**, 583–586.
8. Sangwan, R.S., Bourgeois, Y., Brown, S., Vasseur, G., and Sangwan-Norreel, B. (1992) Characterization of competent cells and early events of *Agrobacterium*–mediated genetic transformation of *Arabidopsis thaliana*. *Planta* **188**, 439–456.
9. Akama, K., Shiraishi, H., Ohta, S., Nakamura, K., Okada, K., and Shimura, Y. (1992) Efficient transformation of *Arabidopsis thaliana*: comparison of the efficiencies with various organs, plant ecotypes and *Agrobacterium* strains. *Plant Cell Rep.* **12**, 7–11.
10. Bechtold, N., Ellis, J., and Pelletier, G. (1993) In *Planta Agrobacterium*–mediated gene transfer by infiltration of adult *Arabidopsis thaliana* plants. *C.R. Acad. Sci. Paris* **316**, 1194–1199.
11. Chang, S. S., Park, S. K., Kim, B. C., Kang, B. J., Kim, D. U., and Nam, H. G. (1994) Stable genetic transformation of *Arabidopsis thaliana* by *Agrobacterium* inoculation in *Planta*. *Plant J.* **5**, 551–558.
12. Castle, L. A., Errampalli, D., Atherton, T. L., Franzmann, L. H., Yoon, E. S., and Meinke, D. W. (1993) Genetic and molecular characterization of embryonic mutants identified following seed transformation in *Arabidopsis*. *Mol. Gen Genet.* **241**, 504–514.
13. Matzke, A. J. M., Neuhuber, F., Park, Y. D., Ambros, P. F., and Matzke, M. A. (1994) Homology–dependent gene silencing in transgenic plants: epistatic silencing loci contain multiple copies of methylated transgenes. *Mol. Gen. Genet.* **244**, 219–229.
14. Walden, R., Hayashi, H., and Schell, J. (1991) T–DNA as gene tag. *Plant J.* **1**, 281–288.
15. Koncz, C., Németh, K., Rédei, G. P., and Schell, J. (1992) T–DNA insertional mutagenesis in *Arabidopsis*. *Plant Mol. Biol.* **20**, 963–976.
16. Walden, R. and Schell, J. (1994) Activation T–DNA tagging–a silver bullet approach to isolating plant genes. *Agro–Food–Industry Hi–Tech* 9–12.
17. Walden, R., Fritze, K., Hayashi, H., Miklashevichs, E., Harling, H., and Schell, J. (1994) Activation tagging: a means of isolating genes implicated as playing a role in plant growth and development. *Plant Mol. Biol.* **26**, 1521–1528.
18. Van der Graaff, E. and Hooykaas, P. J. J. (1996) Improvements in the transformation of *Arabidopsis thaliana* C24 leaf–discs by *Agrobacterium tumefaciens*. *Plant Cell Rep.* **15**, 572–577.
19. Pavlica, M., Papes, D., and Nagy, B. (1991) 2,4–Dichlorophenoxyacetic acid causes chromatin and chromosome abnormalities in plant cells and mutation in cultured mammalian cells. *Mutation Res.* **263**, 77–81.
20. Sangwan, R. S., Bourgeois, Y., and Sangwan-Norreel, B. (1991) Genetic transformation of *Arabidopsis thaliana* zygotic embryos and identification of critical parameters influencing transformation. *Mol. Gen. Genet.* **230**, 475–485.
21. Murashige, T. and Skoog, F. (1962) A revised medium for rapid growth and bioassays with tobacco tissue cultures. *Plant Physiol.* **15**, 473–497.

22. Gamborg, O. L., Miller, R. A., and Ojima, K. (1968) Nutrients requirements of suspension cultures of soybean root cells. *Exp. Cell Res.* **50,** 151–158.
23. Masson, J. and Paszkowski, J. (1992) The culture response of *Arabidopsis thaliana* protoplasts is determined by the growth conditions of donor plants. *Plant J.* **2,** 829–833.
24. Hood, E. E., Gelvin, S. B., Melchers, L. S., and Hoekema, A. (1993) New *Agrobacterium* helper plasmids for gene transfer to plants. *Trans. Res.* **2,** 208–218.

28

In Planta Agrobacterium-Mediated Transformation of Adult Arabidopsis thaliana Plants by Vacuum Infiltration

Nicole Bechtold and Georges Pelletier

1. Introduction

Plant genetic transformation was initiated and developed in the 80s thanks to the convergence of constant progress in the protocol of regeneration from tissue culture, molecular techniques leading to well-expressed marker genes after transfer in plant cells, and the diversification of DNA delivery methods.

Arabidopsis thaliana can be transformed through direct DNA uptake in protoplasts or after cocultivation of leaf or root explants with *Agrobacteria* (*see* Chapters 26 and 27 in the book). It is also possible to transform this species by directly applying *Agrobacteria* to the plant and recovering transformants in the progeny.

The first *in planta* method was described by Feldmann and Marks in 1987 *(1)* and consisted of the imbibition of seeds with *Agrobacteria*. Another whole plant transformation procedure is that of Chang, Park, and Nam *(2)* in which young inflorescences are cut off and the wounded surfaces are inoculated with *Agrobacterium*.

These procedures offer two main advantages: tissue culture and the resulting somaclonal variations are avoided and only a short time is required in order to obtain entire transformed individuals. However, the mean frequency of transformants in the progeny of such inoculated plants is relatively low and very variable.

The infiltration method proposed later by Bechtold et al. *(3)* was based on the assumption that the stage at which the T-DNA transfer takes place with these methods is late in the development of the plant, either at the end of gametogenesis or at the zygote stage. This assumption was deduced from the observa-

Table 1
Infiltration Medium and In Vitro Culture Medium

Composants	Infiltration medium (mg/L) (see **Note 1**)	In Vitro culture medium (mg/L) (see **Note 2**)
Macroelements		
NH_4NO_3	1650	0
KNO_3	1900	506
$CaCl_2, 2H_2O$	440	0
$Ca(NO_3)_2 4H_2O$	0	472
$MgSO_4, 7H_2O$	370	493
KH_2PO_4	170	340
Microelements		
H_3BO_3	6.3	4.3
$MnSO_4, 4H_2O$	22.3	0
$MnCl_2, 4H_2O$	0	2.8
$ZnSO_4, 7H_2O$	8.6	0.29
KI	0.83	0
$Na_2MoO_4, 2H_2O$	0.25	0.05
$CuSO_4, 5H_2O$	0.025	0.13
$CoCl_2, 6H_2O$	0.025	0.0025
NaCl	0	0.58
Morel and Wetmore vitamins		
Myo-inositol	0	100
Calcium panthothenate	0	1
Niacine	0	1
Pyridoxine	0	1
Thiamine HCl	0	1
Biotine	0	0.01
Other Compounds		
6-Benzylaminopurine	0.010	0
Sucrose	50,000	10,000
MES[a]	0	700
Agar BIOMAR	0	7000
Ammonium iron (III) citrate		50
pH	5.8	5.8

[a]MES, 4-morpholineethanesulfonic acid.

tion that transformants are hemizygous for T-DNA insertions. The following protocol makes use of adult plants that are infiltrated with *Agrobacterium* at the reproduction stage. Each treated plant gives, on average, 10 transformants that can be selected from the progeny after 4–6 wk.

2. Materials

2.1. Greenhouse Materials

1. 22 × 16 cm Aluminium alimentary trays (Bourgeat, Les Abrets, France).
2. Net pots diameter = 5.5 cm, 28 × 38 cm transport tray (TEKU, Lohne/Oldb, Germany).
3. 45 × 33 × 3.5 cm incubator for seed trays (BHR, St. Germain du Plain, France).
4. 40-Well multipot trays (TEKU, Lohne/Oldb, Germany).
5. Perforated plastic wrap (1000 holes/m_2).
6. Subirrigation potting mix (WOGEGAL, St. Pierre-des-Corps, France).
7. 0.5 mm Sieved sand.
8. Perlite.
9. Birlane CE40 compost disinfection treatment (chlorfenvinphos, Agrishell, Tassin-La-Demi-Lune, France).
10. Hypnol (nicotine) plant louse treatment (CP Jardin, Bavay, France).
11. FINAL™ (phosphinothricin) transformed plant selection (Procida, Marseille, France).

2.2. Laboratory Materials

1. Rotary shaker (Infords, Orsay, France).
2. Vacuum oil pump (Alcatel/CIT, Paris, France).
3. Dessicator (Nalgene, Rochester, NY, 10 L volume).
4. *Agrobacterium* culture medium (Luria-Bertani [LB] medium): 10 g/L Bacto-tryptone, 5 g/L Bacto-Yeast extract, 10 g/L NaCl, pH 7.0, pH adjusted with NaOH 1*M*. The medium is sterilized by autoclaving 20 min at 115°C.
5. Infiltration medium: *see* **Table 1**.
6. In vitro culture medium: *see* **Table 1**.
7. Sterilization solution: Dissolve 1 tablet of Bayrochlore (contains sodium dichlorocyanate and releases 1.5 g of active chlorine. Bayrol GMBH, München 70, Germany) in 40 mL of distilled water and add some drops of 1% Tween. Take 5 mL in 45 mL ethanol 95%.

2.3. Plant Materials

A. thaliana (L.) Heyn., ecotype Wassilevskija (Ws-0) was used to perfect the infiltration protocol. Ecotypes: Columbia (Col-0), Nossen (No-0), and Landsberg *erecta* may also be used with good efficiency.

2.4. Agrobacterium *Strains and Vectors*

We used the *Agrobacterium* strain MP5-1 for most of the experiments *(4)*. This strain carries the binary vector pGKB5, which was constructed for T-DNA insertional mutagenesis. This plasmid is very stable in *Agrobacterium* under nonselective conditions and confers resistance to kanamycin. It was introduced into the helper strain C58C1(pMP90) *(5)*, which contains a disarmed C58 Ti plasmid, to produce strain MP5-1. The T-DNA contains a promoterless

GUS reporter gene fused to the right border, and kanamycin and Basta resistance genes as plant selection markers.

Other binary vectors and helper strains could also be used (strains ABI, ASE, GV3101; vectors pBin19, pOCA18, pCGN, pDE1001). However, strain C58C1(pMP90) gave the best results in our conditions. Commonly used binary vectors confer resistance to kanamycin, and selection of transformants has then to be done in vitro under sterile conditions.

3. Methods
3.1. Growth Conditions of the Plant Material Before Infiltration

1. Prepare some plastic trays (20 × 30 cm) with compost. Wet and treat them against *Sciridae* larvae by spraying with a commercial preparation of chlorfenvinphos.
2. Sow 50 carefully separated seeds on the surface of the compost.
3. Place the trays at 4°C for 64 h for stratification.
4. Place the trays in the greenhouse (16 h day photoperiod, 15°C night and 20–25°C day temperature with additional artificial light (105 µE/m^2/s) (*see* **Note 3**) and subirrigate with a layer of tap water under the trays, until germination. Afterward water moderately for 4–6 wk.

3.2. Agrobacterium *Culture and Preparation*

1. Prepare a preculture by inoculating 10 mL of LB medium containing the appropriate antibiotics with 100 µL of a last culture or a glycerol, or with a colony taken from a Petri dish (*see* **Note 4**).
2. Place at 28°C with good aeration for one night.
3. Inoculate a 2-L Erlenmeyer flask containing 1 L of LB medium and the appropriate antibiotics with the 10 mL of preculture.
4. Grow at 28°C with good aeration until the OD (600 nm) reaches at least 0.8 (*see* **Note 5**).
5. Spin the culture at 8000*g* for 7 min. Gently resuspend the bacteria with 1/3 of the initial volume of infiltration medium (*see* **Note 6**).

3.3. Infiltration

1. Carefully remove the 4–6-wk-old plants from the soil with the roots intact (*see* **Note 7**).
2. Briefly rinse the roots in water to eliminate any adhering soil particles.
3. Put 25–50 plants in an aluminium tray (22 × 16 cm) and stack a second (perforated) tray inside the first one to hold the plants in place. Put 300 mL of fresh bacterial suspension into the trays. The same suspension can be used several times (*see* **Note 8**).
4. Place the trays in a 10-L vacuum chamber and apply 10^4 Pa (0.1 atm) of vacuum pressure for 20 min. Gently break the vacuum and remove the trays (*see* **Note 9**).
5. During the infiltration fill a plastic tray (28 × 38 cm) with compost, treat, and water.

6. Immediately replant the infiltrated plants (T0) in the trays; 30 plants/tray. Cover with a perforated plastic wrap or a seed tray incubator and place some water underneath (*see* **Note 10**).
7. Remove the cover 3–4 d later. Water the plants moderately until maturity (4–6 wk) and then let the plant dry progressively (*see* **Note 11**).
8. Harvest the seeds from 30 plants in bulk. Let the siliques dry at 27°C for 2 d, then thresh and clean the seeds (*see* **Note 12**).

3.4. Screening of Transformants
3.4.1. In Greenhouse

1. Sow each bulk of seeds in a 55 × 36 tray containing perlite and a top layer of fine sand, previously wet with water containing an appropriate herbicide. The nature of the herbicide should vary depending on the selection marker used; preferably phosphinothricin (final [Basta] 5–10 mg/L) or glyphosate (Roundup, 50 mg/L; *see* **Note 13**).
2. Synchronize germination at 4°C for 64 h.
3. Place the trays in the greenhouse. Subirrigate permanently with water containing the herbicide, as described previously, for 4 wk. Transformants (T_1), in the Basta selection, (normal green cotyledons and first leaves formed) can be observed after 2 wk. Untransformed plantlets are blocked just after germination (no expansion of cotyledons which rapidly turn yellow; *see* **Note 14**).
4. Water and treat prepared pots (5.5-cm diameter) containing compost.
5. Transfer resistant plantlets into individual pots when they are sufficiently developed (four to five leaf stage) and cover to facilitate rooting.
6. Water moderately and alternatively with tap water and a nutrient solution until the flowering stage. At this time, care must be taken to individualize plants, to prevent cross-pollination and/or seed contaminations. Progressively reduce watering while plants finish producing flowers (*see* **Note 15**).
7. When siliques are dry, harvest, and clean the T_2 seeds from each T_1 plant. Generally, enough seeds are obtained for most experiments without further propagation. The in vitro segregation of the T-DNA selectable markers and Southern blotting experiments allow the estimation of the number of loci and the number of copies of T-DNA (*see* **Note 16**).

3.4.2. In vitro

1. Divide the seeds into 1.5-mL microtubes; 100 mL of seeds/tube.
2. Add 1 mL of the sterilization solution, close the tubes, and mix.
3. Lay the tubes down in a laminar flow cabinet for 8 min to disperse the seeds into the solution (*see* **Note 17**).
4. Remove the solution with a pipet and rinse twice with 1 mL of pure 95% ethanol. Remove as much of the ethanol as possible. Let the seeds dry in the flow cabinet for one night.
5. Sow no more than 500 seeds in sterile conditions on a 10-cm Petri dish containing the selective medium. Close the dishes with only two pieces of adhesive tape to prevent high levels of humidity.

6. Place the dishes at 4°C for 64 h.
7. Transfer the dishes into a growth chamber (16-h day length; 20°C).
8. Tranformants (green-rooted plants) may be scored 10 d later for kanamycin selection.
9. Plant the resistant plantlets into individual pots when they are sufficiently developed (four to five leaf stage). Transfer to a growth chamber and cover to facilitate rooting.
10. Continue the process as in **Subheading 3.4.1.**, **steps 6** and **7**.

4. Notes

1. The microelements and 6-benzylaminopurine (BA) are made up as concentrated stock solutions (at 1000X and 1 mg/L, respectively) and stored at 4°C. The pH is adjusted with KOH and the medium is sterilized by autoclaving at 115°C for 20 min.
2. 5 mL of a filter-sterilized ammoniacal iron citrate stock solution (1%) and for kanamycin selection 1 mL of a filter-sterilized kanamycin stock solution (100 mg/mL) must be added after autoclaving at 115°C for 20 min. Macroelements, microelements, vitamins, and MES are made up as concentrated stock solutions and stored at room temperature or at 4°C for the microelements and the vitamins ($1M$ KNO_3, $1M$ KH_2PO_4, $1M$ $MgSO_4$ $7H_2O$, $1M$ $Ca(NO_3)_2$, $4H_2O$, microelements 1000X, vitamins 500X, MES 14%). Agar is added in each bottle before autoclaving. The pH is adjusted with KOH. The medium is sterilized by autoclaving at 115°C for 20 min.
3. Plants must be as vigorous as possible. A better development is observed when they are grown during the rosette stage under relatively short days (13 h). It is also preferable to avoid etiolation by providing sufficient lighting. The optimal stage for infiltration is when the plants have the first siliques formed and the secondary floral stems are appearing.
4. It is convenient to maintain (for 1 mo) a sample of the last culture at 4°C as this allows the inoculation of the next culture without necessity of a preculture preparation.
5. For the MP5-1 strain, the culture needs 15 h to reach the desired OD, but it may take longer for other strains.
6. It is convenient to centrifuge the culture in a GSA (Sorvall, Wilmington, DE) rotor with 250-mL tubes. To resuspend the bacteria pellet, it is just necessary to add a small volume of infiltration medium and to shake the tube. A better transformation frequency is obtained with a suspension that is 3 times more concentrated than the original suspension.
7. It is also possible to leave the plants in the tray and to infiltrate only the leaves and the stems *(6)*. In this case, the number of transformants is generally lower, but may be sufficient if only a few are needed.
8. Plants must be entirely immersed into the suspension. It is possible to use other trays depending on the size of plants and of the volume of the vacuum chamber. Avoid infiltrating more than 50 plants at once otherwise the transformation frequency decreases dramatically.

9. Twenty minutes are needed to obtain a complete infiltration of the plants with the suspension. The principal cause of plant death after the treatment is where the drop in pressure is too rapid (because of the power of the pump), or when the vacuum is broken too suddenly. Nevertheless, the pump must be sufficiently powerful to degas 300 mL of liquid in the vacuum chamber.
10. Replanting must be done immediately after the treatment. Do not let the plants desiccate after the treatment and before replanting (do not treat too many plants at once). Be careful to put only the roots in the soil. If leaves or the base of the rosette is in the soil, it usually causes the deterioration of the plant. Replanting must be performed using chirurgical gloves on a table covered with a plastic wrap. This permits the elimination of all things which have been in contact with the *Agrobacterium* as they may be destroyed afterwards. Avoid high illumination and water condensation while the plants are recovering in greenhouse. Do not completely cover the plants.
11. Following the treatment, the leaves dry rapidly but the floral stems become erect and continue to flower. If the plants do not continue flowering, it may be due to the vacuum conditions or to a high temperature in the greenhouse after infiltration.
12. To prevent contamination when an in vitro selection of transformants is done, avoid harvesting soil particles with the seeds.
13. Perlite is useful to lighten the trays. The size of sand particles has to be small enough to allow a constant imbibition of the seeds sown on the surface. The selection with herbicide is more efficient (at avoiding the selection of sensitive plants) with a minimal medium of tap water and sand, instead of compost. Take care to homogenize the seed sowing on the trays to allow the even development of the transformants. It is advisable to supply the plants with a nutrient solution when the transformants reach the two-leaf stage.
14. A continuous presence of the herbicide is necessary because the germination of all the seeds can take 4 wk. To prevent the drying of the sand, a constant watering is required.
15. Plants must be staked and may be individualized with perforated transparencies rolled up around the pot.
16. One can expect to obtain > 50% of transformants with a T-DNA insertion at a single mendelian locus. Seventy percent of the T-DNA insertions are in tandem.
17. Do not sterilize more than 10 tubes at once to prevent a too long contact of the seeds with the sterilization solution (no more than 8 min).

References

1. Feldmann, K. A. and Marks, M. D. (1987) *Agrobacterium*-mediated transformation of germinating seeds of *Arabidopsis thaliana*: a non-tissue culture approach. *Mol. Gen. Genet.* **208**, 1–9.
2. Chang, S. S., Park, S. K., and Nam, H. G. (1994) Transformation of *Arabidopsis* by *Agrobacterium* inoculation on wounds. *Plant J.* **5(4)**, 551–558.
3. Bechtold, N., Ellis, J., and Pelletier, G. (1993) In planta *Agrobacterium* mediated gene transfer by infiltration of adult *Arabidopsis thaliana* plants. *C. R. Acad. Sci. Paris, Life Sci.* **316**, 1194–1199.

4. Bouchez, D., Camilleri, C., and Caboche, M. (1993) A binary vector based on Basta resistance for *in planta* transformation of *Arabidopsis thaliana*. *C. R. Acad. Sci. Paris, Life Sci.* **316,** 1188–1193.
5. Koncz, C. and Schell, J. (1986) The promoter of TL-DNA gene 5 controls the tissue-specific expression of chimaeric genes carried by a novel type of *Agrobacterium* binary vector. *Mol. Gen. Genet.* **204,** 383–396.
6. Bent, A. F., Kunkel, B. N., Dahlbeck, D., Brown, K. L., Schmidt, R., Giraudat, J., Leung, J., and Staskawicz, B. J. (1994) RPS2 of *Arabidopsis thaliana:* a leucine-rich repeat class of plant disease resistance genes. *Science* **265,** 1856–1860.

29

PEG–Mediated Protoplast Transformation with Naked DNA

Jaideep Mathur and Csaba Koncz

1. Introduction

Direct introduction of DNA into plant protoplasts facilitates a rapid analysis of transient gene expression, as well as the generation of stably transformed transgenic plants. Transient gene expression assays performed after DNA transformation permit a comparative analysis of *cis*–acting regulatory sequences and their function in transcriptional control of plant genes by signaling pathways mediating cellular responses to different environmental and hormonal stimuli *(1)*. There are a number of methods for introducing DNA into plant protoplasts, but the most commonly used technique is the polyethylene glycol (PEG)–mediated DNA uptake. The PEG–mediated transformation is simple and efficient, allowing a simultaneous processing of many samples, and yields a transformed cell population with high survival and division rates *(2)*. The method utilizes inexpensive supplies and equipments, and helps to overcome a hurdle of host range limitations of *Agrobacterium*–mediated transformation. The PEG–mediated DNA transfer can be readily adapted to a wide range of plant species and tissue sources.

In *Arabidopsis thaliana,* several methods of direct gene transfer to leaf mesophyll *(3–5)* and root-derived protoplasts *(6)* have been reported. They are all derived from a PEG–mediated direct gene transfer technique established originally for tobacco protoplasts by Negrutiu et al. *(7)*. This chapter describes a method for PEG–mediated transformation of protoplasts derived from leaves, roots, and cell suspensions of *A. thaliana*. Leaf mesophyll protoplasts are able to regenerate after embedding into alginate, but their yield is relatively low. In comparison, cell suspensions provide an unlimited source of rapidly dividing protoplasts that can be obtained within 2–3 h and show a transient expression of introduced genes within 24 h. Root-derived protoplasts also feature a high division and regeneration capability. A low autofluorescence of cell suspension and root-derived protoplasts is of particular importance, when light emitting enzymes, such as the green fluorescence protein (GFP) or luciferases, are being used as reporter proteins in nondestructive in vivo gene expression assays *(8–9)*. In leaf protoplasts, the red fluorescence of chloroplasts is a deterrent for effective monitoring of the *GFP* reporter gene activity. Nonetheless, the different protoplast systems provide a choice of material according to the specific questions addressed in the experiments.

2. Materials

2.1. Materials and Equipment

Items listed in Chapter 6 are required along with the following:

1. Fluorimeter.
2. Fluorescence microscope with FITC filters.
3. Diapositive films (Kodak Ektachrome 320T; Kodak Panther 1600, Braunschweig, Germany).

2.2. Media and Solutions

The pH is adjusted to 5.8 with $1M$ KOH or $1N$ HCl. The protoplast medium (PM), and the enzyme solution are filter sterilized. Growth regulator stocks (1 mg/mL) are filter sterilized and added separately to sterilized media. Solutions 14, 16, 17, and 18 should be filter sterilized and stored at $-20°C$.

1. Basal medium (BM): MS medium *(5)* (pH 5.8) containing B5 vitamins *(6)*, with or without gelling agents (0.8% agar or 0.2% gelrite), and 3% sucrose, if not stated otherwise.
2. 0.5 BM Medium: consisting of half concentration of MS macroelements *(5)*, B5 vitamins, and 3% sucrose with and without gelling agents (0.8% agar or 0.2% gelrite, pH 5.8).
3. MSAR I medium *(7)*: BM medium containing 2.0 mg/L indole–3–acetic acid (IAA), 0.5 mg/L 2,4–dichloro–phenoxyacetic acid (2,4–D), 0.5 mg/L 6–(γ,γ–dimethylallylamino)–purine riboside (IPAR) (pH 5.8).
4. MSAR II medium *(7)*: BM medium containing 2.0 mg/L IPAR, 0.05 mg/L α–naphtaleneacetic acid (NAA) with 0.2% gelrite (pH 5.8).
5. MSAR III medium *(7)*: BM medium containing 1.0 mg/L (IAA), 0.2 mg/L indole–3–butyric acid (IBA), 0.2 mg/L 6–furfurylaminopurine (kinetin), 0.2% gelrite (pH 5.8).
6. Protoplast medium (PM): 0.5X BM medium with $0.45M$ sucrose or $0.45M$ mannitol (pH 5.8).
7. Enzyme solution: 1.0% Cellulase (Onozuka R–10; Serva), 0.25% Macerozyme (R10; Serva) dissolved in PM medium.

PEG-Mediated Protoplast Transformation

8. $0.45M$ Mannitol and $0.45M$ sucrose solutions (pH 5.8).
9. Sodium alginate solution: 1% (w/v) solution in BM medium containing $0.45M$ sucrose.
10. Calcium agar plates: 20 mM calcium chloride, $0.45M$ sucrose and 1% agar.
11. $0.5M$ MaMg solution: $0.5M$ mannitol, 15 mM MgCl$_2$·6H$_2$O, 0.2% MES (morpholinoethane sulfonic acid, pH 5.8).
12. PEG solution (PEG 1450): 40 g PEG in 100 mL MaMg solution.
13. GUS extraction buffer: $0.1M$ potassium phosphate (pH 7.8), 2 mM Na$_2$EDTA, 2 mM dithiothreitol, 5% glycerol.
14. GUS substrate: 1 mM crystalline 4–methylumbelliferyl–β–D–glucuronide (Sigma M 9130, St. Louis, MO) in GUS extraction buffer.
15. GUS stop buffer: $0.2M$ Na$_2$CO$_3$ (store at 4°C).
16. MU standard: 1 mM 4–methylumbelliferone in GUS stop buffer. Prepare dilutions for calibration in stop buffer.
17. X–Gluc solution for tissue staining: for preparing 100 mL solution, dissolve 100 mg X–Gluc in approx 50 μL N,N–dimethylformamide and add 98 mL ($0.1M$) potassium phosphate buffer (pH 7.0), 1 mL of potassium ferricyanide (5 mM), 1 mL of potassium ferrocyanide (5 mM), and 0.1 mL Triton X–100.
18. X–gluc solution for staining of protoplasts: For 100 mL solution add to 100 mg X–gluc dissolved in 50 μL of N,N,–dimethylformamide 100 mL of CaCl$_2$ (125 mM) solution containing $0.45M$ mannitol.
19. Hygromycin B solution: 1 g dissolved in 20 mL PBS may be obtained from Boehringer Mannheim (Mannheim, Germany, cat. no. 843555). Dilute to 15 mg/mL concentration in sterile water. Store at –20°C.

3. Methods

3.1. Isolation of Protoplasts

Methods for the isolation of protoplasts from leaf mesophyll tissue, root and cell suspension cultures are described in Chapter 6.

3.2. PEG–Mediated Transformation of Protoplasts

1. After washing with $0.45M$ mannitol (*see* Chapter 6), resuspend the protoplast pellet in 1 mL of MaMg solution.
2. Count the number of protoplasts and adjust the protoplast density to approx 1×10^6 cells/mL (*see* **Note 1**).
3. Place the protoplast suspension on ice for 35 min (*see* **Note 2**).
4. Centrifuge the protoplasts at 60g for 5 min.
5. Resuspend the protoplast pellet in 0.3 mL of MaMg solution and carefully transfer them as a single droplet in the middle of a 9-cm, glass Petri dish (*see* **Notes 3** and **4**).
6. Add slowly 25–35 μg of plasmid DNA dissolved in water into the drop of protoplast suspension. Shake the Petri dish very gently to mix the DNA well with the protoplast suspension (*see* **Notes 5** and **6**).
7. After 5 min add 0.3 mL of PEG solution at the circumference of the drop of protoplast suspension.

8. Tilt the Petri dish gently to allow mixing of the PEG solution with the protoplasts. (Alternatively, carefully mix the drops of PEG solution surrounding the drop of protoplast suspension by a sterile micropipet tip.)
9. After 10 min add from the sides of the protoplast droplet 1 mL of $0.45M$ mannitol solution.
10. Add at 2 min intervals 2 mL of $0.45M$ mannitol solution with gentle shaking until about 12 mL are present in the Petri dish (*see* **Note 7**).
11. Collect the protoplasts in a 12 mL centrifuge tube and centrifuge them at $60g$ for 5 min.
12. Assay for transient gene expression, using reporter (GUS or GFP) gene constructs, after 24–48 h (*see* **Subheading 3.3.**) or start embedding of the protoplasts into alginate, in order to select for stably transformed cells and process them further to plant regeneration (*see* **Subheading 3.4. and 3.5.**).

3.3. Transient Gene Expression Assays with GUS and GFP Reporters

3.3.1. Application of the GUS Reporter Gene

Qualitative and quantitative assays of GUS reporter enzyme activity are carried out as described by Jefferson *(13)*.

1. Transient expression of the *GUS* reporter gene may be assayed 24–48 h after the PEG–mediated DNA uptake. For this, collect the protoplasts by centrifugation at $60g$ for 5 min.
2. Resuspend the protoplast pellet in approx 1 mL of X–gluc solution for protoplasts and incubate them at room temperature for 6–12 h.
3. Take a drop of protoplast suspension (e.g., 10 µL) and count the number of cells using a hemocytometer and an inverted microscope. Transformed protoplasts appear blue (*see* **Notes 8** and **9, Fig. 1A**).
4. The ratio between the total number of cells plated and the number of GUS expressing cells indicates the relative transformation efficiency expressed in percentage.

3.3.2. The Use of Green Fluorescent Protein as Reporter

The mGFP4 gene construct described by Haseloff and Amos *(14)*, as well as its mutated derivative (Reichel et al., *17*) work effectively in all *Arabidopsis* protoplast preparations (*see* **Note 10**).

Fig. 1. *(opposite page)* GUS and GFP reporter gene expression assays with *Arabidopsis* protoplasts. **(A)** GUS staining of PEG–transformed protoplasts derived from roots of *Arabidopsis* ecotype Columbia after incubation with X–gluc for 6 h at room temperature. **(B)** Leaf mesophyll protoplasts from *Arabidopsis* ecotype Columbia transformed with pCK–GFPs65c exhibit green fluorescence when illuminated with blue light. The chloroplasts emit red fluorescence, whereas the yellow fluorescence results from overlapping red and green areas. (*See* color insert following p. 208.)

PEG-Mediated Protoplast Transformation

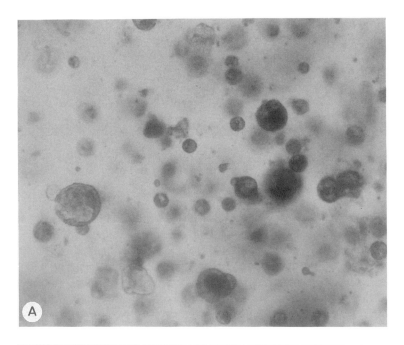

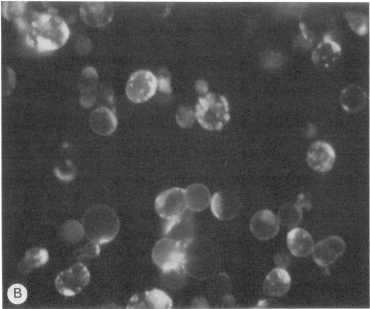

Fig. 1

1. For assaying the transient expression of GFP reporter gene, take an aliquot of 200 µL from the transformed protoplasts and transfer into a hemocytometer (*see* **Note 11**).
2. Observe and count the number of green fluorescent cells using a fluorescence microscope equipped with fluorescein isothiocyanate (FITC) filters (*see* **Fig. 1B** and **Notes 12** and **13**).
3. Take pictures in bright field and UV/blue light of the same field for records, using Kodak 320T and P–1600 slide films, respectively (*see* **Note 14**).

3.4. Selection of Stable Transformants

1. For embedding of the cells into alginate after PEG–mediated DNA uptake, adjust the density of protoplasts to $3–5 \times 10^5$ cells/mL and create alginate gel–drops of 250–500 µL on calcium agar plates (*see* **Notes 15** and **16**).
2. After 45 min remove the alginate drops containing the embedded protoplasts and place them in 55-mm Petri dishes containing 5 mL of PM medium.
3. Culture the protoplasts in low light (500–700 lx) conditions at 25°C.
4. After 7 and 14 d remove 2.5 mL of the PM medium and add 2.5 mL of fresh PM medium containing appropriate antibiotic(s) for selection (*see* **Notes 17** and **18**).
5. The frequency of stable transformation is determined at different time points, following the application of relevant antibiotic selection (*see* **Note 18**). When using GFP, the transformation efficiency can also be determined without applying an antibiotic selection (*see* **Note 19**). Take out small aliquots or sections of growing cells or tissues respectively in a Petri dish. Count the total number of cells in a field using white light and the number of cells showing GFP fluorescence under UV/blue light after transformation with the GFP reporter gene. This ratio expressed on a percentage basis indicates the relative transformation frequency (*see* **Notes 19** and **20**). Similarly, when using GUS reporter constructs, small aliquots from dividing cell cultures or microcalli are incubated for 6–12 h in X–gluc solution supplemented with appropriate osmoticum, and the ratio of GUS–stained cells vs the total cell number is determined.
6. Proceed for regeneration of stable transformants after applying a proper antibiotic selection (**Subheading 3.5.**). Remove 1 mL of PM medium and add 1 mL of MSAR I medium containing antibiotic(s) on d 21, 28, and 35 (*see* **Note 21**).

3.5. Plant Regeneration from Protoplast–Derived Transformed Calli

1. Remove the liquid medium from the Petri dishes using a pipet and transfer the alginate beads carrying microcalli, after dividing them into four to five pieces, into MSAR II medium containing the selective antibiotics in Petri dishes (*see* **Notes 21** and **22**).
2. Place the Petri dishes in an illuminated culture chamber (3000 lx) set for 16 h of light and 8 h of dark at 25°C.
3. Transfer green calli and regenerating shoots to MSAR II medium containing the selective antibiotics and grow the shoots in glass jars until they attain a size of about 2 cm (*see* **Note 23**).

4. Transfer the shoots into MSAR III medium to induce root formation for 3–6 d, then place the plantlets in culture tubes containing 0.5 BM agar medium with only 0.5% sucrose for flowering and seed setting (*see* **Note 24**).

4. Notes

1. If the number of protoplasts exceeds 100 cells/square in the hemocytometer, dilute the protoplast suspension.
2. Do not leave the protoplasts, especially those obtained from cell suspensions, too long (> 60 min) on ice.
3. A glass test tube can also be used for PEG–mediated DNA uptake. However, we found the use of Petri dishes better because the addition of PEG solution, the consequent clumping, and the subsequent restoration of protoplasts can easily be monitored under an inverted microscope.
4. The use of glass Petri dishes is recommended for this step, because the PEG–treated protoplasts tend to adhere to the surface of plastic Petri dish and therefore often get damaged during the process of reconstitution following the dilution of PEG.
5. The use of carrier DNA has been recommended in a number of protocols and suggested to increase the efficiency of transformation. However, we found no significant increase in the transformation rates by applying carrier DNA. On the other hand, unnecessary clumping of protoplasts occurs in the presence of excessive amounts of carrier DNA. The washing step that removes the enzymes seems to have a greater bearing on the transformation efficiency, because protoplast samples that have not been washed very well always yield lower transformation, division, and survival rates.
6. The DNA should be sterile (ethanol-precipitated and dissolved in sterile water). Do not incubate DNA for too long with the protoplasts because it may result in lower transformation efficiency *(16)* owing to nuclease digestion.
7. It is worthwhile to include DNA untreated controls that are processed through the same PEG treatment and washes as the DNA–treated protoplasts. The PEG–treated protoplasts will clump up and look shrunken. However, they should not burst. If they appear to be damaged, the PEG concentration of 20% is probably too high and should be reduced. Following dilution of the PEG, the protoplasts should declump and regain their former shape.
8. Use the microscope with a fully open diaphragm and at low magnifications, since at either higher magnifications or low light, an illusion of a blue tinge in cell suspension and root derived protoplasts may lead to erroneous counts.
9. The indigo–colored reaction product of the β–glucuronidase enzyme and the substrate X–gluc (5–bromo–4–chloro–3–indolyl glucuronide) is cytotoxic and may sometimes lead to a collapse of transformed protoplasts. This may cause errors in counting of the transformed cells because the dye diffuses into the vicinity of the collapsed protoplast.
10. In the modified GFP, available from J. Haseloff, the cryptic splice sites were removed. This GFP is detectable in both UV (excitation filter 340–380 nm, reflection short pass 400 nm, long pass 430 nm) and blue light (450–490 nm,

short pass 510 nm, long pass 520 nm), whereas the GFP protein translated from the modified pCK–GFPs65c gene construct (Reichel et al., *17*) works optimally in blue light with no visible green fluorescence in UV light.
11. Optimal fluorescence is observable only after 48 h, although some transformed protoplasts start exhibiting GFP fluorescence 24 h after transformation.
12. A filter set specifically designed for the observation of GFP (GFP–41014; HQ GFP–LP 41015) is available from the Chroma Technology (Houston, Texas). However, the commonly available FITC filter set is also adequate for most purposes.
13. When viewed on a Leitz Aristoplan fluorescence microscope (Leitz, Bensheim, Germany), using the filter sets described in **Note 10**, the *Arabidopsis* cell suspension and root-derived protoplasts exhibit a faint blue autofluorescence, whereas dead protoplasts display an intense yellow–orange fluorescence. Under blue light, the chloroplasts emit a bright red fluorescence. In certain cases, an overlapping of the green and red fluorescence results in yellow fluorescence in the pictures taken (*see* **Fig. 1B**).
14. Some photobleaching occurs in the UV light and therefore, when using the original GFP construct from J. Haseloff, it is advisable to take pictures first in normal light, then under blue light, and finally in UV light. The use of a fast film is recommended for the same reason. The problem of photobleaching is not encountered with the improved pCK–GFPs65c GFP construct.
15. A certain proportion of protoplasts will invariably break during and following the PEG treatment. The debris of dead cells is detrimental for a continued liquid culture of surviving protoplasts. Therefore, embedding of the PEG–treated protoplasts is necessary for obtaining stable transformants.
16. Larger drops may be prepared and later cut into smaller pieces. However, we find it easier to use small drops which avoid the problem of spreading.
17. Depending on the quality of protoplast preparations, up to 75% of the protoplasts survive and 20–40% of cells will undergo divisions during the first 5–7 d of culture.
18. After 14 d of culture, the dividing cells should form colonies of 4–16 cells. In the case of leaf mesophyll protoplasts, this time may be longer by another 7 d. Selection pressure should be applied at this stage to inhibit the growth and development of untransformed calli. We use 15 µg/mL hygromycin (Boehringer Mannheim) when selecting for constructs carrying the *hpt* (hygromycin phosphotransferase) selectable marker gene under the Cauliflower Mosaic Virus (CaMV) 35S promoter.
19. Although epifluorescence illumination adapted for use with an inverted microscope would be ideal for viewing of transformed cell clumps and regenerating shoots, in the absence of such a lighting system, the following method may be used: Remove a single alginate drop carrying protoplasts from the culture plate and place in a sterile Petri dish. After sealing the Petri dish with parafilm, invert the plate, and observe the cells in the alginate gel directly under the fluorescence microscope. The same procedure can be adopted for microcalli and regenerating structures.
20. In certain leaf pieces, the green fluorescence is entirely masked by the chlorophyll pigment. In case of doubts about the transgenic nature of regenerated plant, a

small portion of leaves may be used to make protoplasts in 1 mL of enzyme solution and the protoplasts may directly be observed as they are released. Alternatively, the sample tissue may be fixed in formali:acetic acid:70% ethanol (FAA 5:5:90) for an hour and then observed after washing out the fixative with few rinses of distilled water.
21. By d 35, microcalli are visible and in some, the differentiation of roots and somatic embryo-like structures may already be observed.
22. Alternative approaches involve either depolymerisation of the alginate gel in 20 mM sodium citrate solution containing an appropriate osmoticum *(4),* or transferring the microcalli from the alginate gel into regeneration medium *(9)*. Since we use a lower concentration of alginate than earlier reported *(3–4)*, the regeneration of calli consisting of more than 64 cells is not hindered by the alginate embedding.
23. Care must be taken to remove the dead cells from the regenerating cultures.
24. The culture tubes are capped with loose cotton to facilitate proper aeration necessary for seed setting. Take care not to place the tubes close to the light source because this will cause moisture condensation inside the tubes and result in a low pollination rate. Rooted plants can be transferred to soil and, after proper hardening, will flower and set seed.

References

1. Morgan, M. K. and Ow, D. W. (1995) Polyethylene glycol–mediated transformation of tobacco leaf mesophyl protoplasts: an experiment in the study of Cre–Lox recombination, in *Methods in Plant Molecular Biology: A Laboratory Course Manual* (Maliga, P., Klessig, D. F., Cashmore, A. R., Gruissem, W., and Varner, J. E. eds.,), Cold Spring Harbor Laboratory Press, Cold Spring Harbor, NY, pp. 1–17.
2. Potrykus, I. (1991) Gene transfer to plants: assessment of published approaches and results. *Ann. Rev. Plant Physiol. Plant Mol. Biol.* **42**, 205–255.
3. Morris, P. C. and Altmann, T. (1994) Tissue culture and transformation, in *Arabidopsis* (Meyerowitz, E. M. and Somerville, C. R., eds.), Cold Spring Harbor Laboratory Press, Cold Spring Harbor, NY, pp. 173–222.
4. Damm, B., Schmidt, R., and Willmitzer, L. (1989) Efficient transformation of *Arabidopsis thaliana* using direct gene transfer to protoplasts. *Mol. Gen. Genet.* **217**, 6–12.
5. Karesch, H., Bilang, R., Mittelsten–Scheid, O., and Potrykus, I. (1991) Direct gene transfer to protoplasts of *Arabidopsis thaliana. Plant Cell Rep.* **9**, 575–578.
6. Mathur, J., Szabados, L., and Koncz, C. (1995) A simple method for isolation, liquid culture, transformation and regenerationof *Arabidopsis thaliana* protoplasts. *Plant Cell Rep.* **14**, 221–226.
7. Negrutiu, I., Shillito, R., Potrykus, I., Biasini, G., and Sala, F. (1987) Hybrid genes in the analysis of transformation conditions. I. Setting up a simple method for direct gene transfer in plant protoplasts. *Plant Mol. Biol.* **8**, 363–373.
8. Chalfie, M. Tu, Y. Euskirchen, G., Ward, W. W., and Prasher, D. C. (1994) Green fluoresence protein as a marker for gene expression. *Science* **263**, 802–805.

9. Heim, R., Prasher, D. C., and Tsien, R. Y. (1994) Wavelength mutations and posttranslational autoxidation of green fluorescent protein. *Proc. Natl. Acad. Sci. USA* **91,** 12501–12504.
10. Murashige, T. and Skoog, F. (1962) A revised medium for rapid growth and bioassay with tobacco tissue cultures. *Physiol. Plant.* **15,** 473–497.
11. Gamborg, O. L., Miller, R. A., and Ojima, K. (1968) Nutrient requirements of suspensions cultures of soybean root cells. *Exp. Cell Res.* **50,** 151–158.
12. Koncz, C., Schell, J., and Rédei, G. P.(1992) T–DNA transformation and insertional mutagenesis, in *Methods in Arabidopsis Research* (Koncz, C., Chua, N.–H., and Schell, J., eds.), World Scientific, Singapore, pp. 224–273.
13. Jefferson R. A. (1987) Assaying chimeric genes in plants: the GUS gene fusion system. *Plant Mol. Biol. Rep.* **5,** 387–405.
14. Haseloff, J. and Amos, B. (1995) GFP in plants. *Trends Genet.* **11,** 328–329.
15. Mason, J. and Paszkowski, J. (1992) The culture response of *Arabidopsis thaliana* protoplasts is determined by the growth conditions of donor plants. *Plant J.* **2,** 829–833.
16. Rasmussen, J. O. and Rasmussen, O. S. (1993) PEG–mediated uptake and transient GUS expression in carrot, rapeseed and soybean protoplasts. *Plant Sci.* **89,** 199–207.
17. Reichel, C., Mathur, J., Langenkemper, C., Eckes, P., Koncz, C., Rei§. B., Schell, J., and Maas, C.(1996) Enhanced green fluorescence by the expression of an *Aequorea victoria* GFP mutant in mono– and dicotyledonousplant cells. *Proc. Nat. Acad.Sci. USA* **93,** 5888–5893.

VI

GENE CLONING STRATEGIES

30

Cloning Genes of *Arabidopsis thaliana* by Chromosome Walking

Jeffrey Leung and Jérôme Giraudat

1. Introduction

Chromosome walking is a versatile technique applicable to cloning virtually any gene of interest identifiable by mutations *(1)*. The principle of this approach consists of using a molecular probe (for instance, a restriction fragment length polymorphism [RFLP] marker) mapping near the target locus to identify genomic clones, particularly those constructed in high capacity vectors. The extremities of the cloned insert are recovered and are used in turn as new probes to identify overlapping clones extending towards the gene. Repeated rounds (or steps) of such screening would eventually yield an overlapping set of clones (contig) bridging the initial RFLP marker and the gene of interest. *Arabidopsis thaliana* is particularly suited for isolating specific genes by this approach because of the small size of the haploid content of its genome and the paucity of repetitive DNA *(2)*.

Prior to walking, the genetic map position of the target mutation must be determined accurately by constructing a "high resolution map" by segregation analysis of the mutation with tightly linked phenotypic and molecular markers (for an example, see **ref. 3**). Candidate RFLP and phenotypic markers for this purpose can be identified by consulting several independently generated maps *(4–7)*. Updates of the RFLP map with the use of inbred *Arabidopsis* RI lines *(8,9)* are available via world wide web and Gopher servers (see **Table 1** footnotes). An important caveat is that the accuracy of these RFLP maps is largely limited by the size of the mapping population and the distances between particular markers being tested. Thus, these global maps should never be mistaken as replacements for high resolution mapping to define the precise order of phenotypic and molecular markers in the target gene region.

Table 1
High Capacity Genomic Libraries for Chromosome Walking[a]

Libraries	Average inserts in kb	Clones	Vector (ecotype)	Ref.
YACs				
CIC[b]	420	1152	YAC4 (Columbia)	*17*
Yup[c]	250	2300	YAC4 (?)	*16*
EG1	160	2300	YAC41 (Columbia)	*14*
abi1	120	2100	YAC41 (Landsberg; abi1 muant)	*14*
S	50–100	9600	YAC45 (Landsberg)	*14*
U	100	6700	YAC45 (Landsberg; abi2 mutant)	*14*
EW	150	2200	YAC 3 (Columbia)	*15*
Other libraries				
P1	80	10080	pAd10*Sac*BII (Columbia)	*19*
BAC[d]	100	3948	pBeloBAC11 (Columbia)	*18*

[a]The availability of these genomic libraries can be inquired at the *Arabidopsis* Biological Resource Center at Ohio State, 1735 Neil Avenue Columbus, OH 43210. Orders can be made by fax: 614-292-0603 or by e-mail: dna@genesys.cps.mus.edu
ADDITIONAL INFORMATION ACCESSIBLE ELECTRONICALLY:
[b] CIC YAC library, and up-dated RFLP maps:
http://nasc.nott.ac.uK:8300/
[c]Yup YAC library and physical maps: http://cbil.humgen.upenn.edu/~atgc/ATGCUP.html
[d]BAC library: http://www.tree.caltech.edu

The physical distance of the walk necessary—between the closest RFLP marker and the target gene—can be extrapolated from the high resolution genetic map. From the cloning of the chromosome 4 of *Arabidopsis (10)*, and several localized walks to isolate loci of interest *(11–13)*, the average ratio of physical to genetic distance is about 200 kb/cM but can vary from 30 to >550 kb/cM *(3,10)*. Several genomic banks of *Arabidopsis* for walking purposes have been constructed in yeast artificial chromosomes (YACs) *(14–17)*, and more recently in bacterial artificial chromosomes (BACs) *(18)* and P1 phages *(19)* (**Table 1**). Regardless of the type of vector used to propagate the bank, the theoretical principles of chromosomal walking described above are identical. YACs will probably remain indispensable as they have already been used extensively and practical knowledge on their handling has been completely worked out *(1,16)*. Also, independent YAC banks are available (representing 30 *Arabidopsis* genomes in equivalence) and this redundancy is important in that unstable or chimeric YAC clones *(17,20)* encountered in one bank may be circumvented by screening another. The map positions of many of these YACs are known from the current genome efforts and there is a high probability that the target mutation may already be covered by an existing contig *(10,21)*.

Finally, the insert sizes in YACs are generally two- to four-fold larger than those in BACs and P1, and thus effectively reduce the number of walking steps. There are at present 462 mapped RFLP markers *(9)*. Assuming an average distance of < 250 kb between adjacent RFLP markers, this means that most genes could be readily cloned in a single-step walk, particularly if mega-YACs *(17)* are used (**Table 1**).

Here we outline the basic techniques of yeast colony screening with probes such as RFLP markers to identify positive YAC clones (**Subheading 3.1.**). To verify and to estimate the sizes of the YACs (i.e., the physical distance walked), pulsed-field gel electrophoresis as a routine method is described (**Subheading 3.2.**). For the next step in the walk, we outline the rapid isolation of the ends of the YAC insert (as new hybridization probes) by inverse polymerase chain reaction (**Subheading 3.3.**) to identify overlapping YAC clones extending towards the target gene. The last section, in two parts (**Subheading 3.4.1. and 3.4.2.**), describes subcloning of the YAC insert. Because of space constraints, we can only present what we consider the simplest strategy, consisting of constructing a total genomic bank of the yeast harboring the YAC in a cosmid and then screening for *Arabidopsis*-specific DNA. Although the vector to be used for subcloning is a matter of personal choice, we present *cos*Pneo *(22,23)* as a means here to illustrate some general and desirable features that the practitioners may want to look for in a vector if a large amount of subcloning should become unavoidable (**Fig. 1**). These cosmid subclones are of critical importance as they form the basis for the final stages of pin-pointing the gene by methods such as functional tests in transgenic plants and DNA sequencing.

2. Materials

2.1. Culture Media and Reagents for Yeast Colony Screen

1. YPD: 10 g Bacto® yeast extract (Difco, Detroit, MI), 20 g Bacto peptone (Difco), 20 g dextrose (D-glucose, Prolabo, Paris, France), made up to 1 L.
2. YPDA: YPD supplemented with 40 mg/L of adenine hemisulfate.
3. SD: 1.7 g Bacto yeast nitrogen base (Difco), 5 g $(NH_4)_2SO_4$, 20 g dextrose, made up to 1 L.
4. Hybond-N® membrane (Amersham, Arlington Heights, IL).
5. Whatman® 3MM filters (Whatman, Maidstone, England).
6. L-amino acid mix in powder (mg, Sigma, St. Louis, MO): arginine (800), aspartic acid (4000), histidine (800), leucine (800), lysine (1200), methionine (800), phenylalanine (2000), threonine (8000), and tyrosine (1200).
7. Bacto Agar (Difco).
8. 10N NaOH.
9. Replica plater for 8 × 12 microtiter dishes (Sigma).
10. 12 × 12 cm square Petri plates (Greiner, Frickenhausen, Germany).

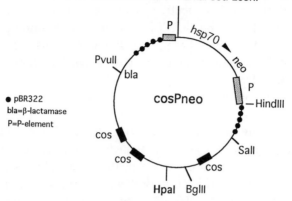

Fig. 1. Restriction map of cosPneo (redrawn and modified from ref. 23). This vector, 9.5 kb in size, was originally designed for cloning and transforming large pieces of DNA into *Drosophila*, but has been used in our laboratory for subcloning YAC clones of *Arabidopsis* DNA *(22)*. This cosmid can accommodate up to 43 kb of foreign DNA and can be easily isolated from *E. coli* (DH5α) in large quantities by simple minipreps because it replicates as a high copy number plasmid. The vector also contains multiple *cos* sites (black boxes), which allow the cloning of a larger range of insert sizes as well as significantly increasing the yield of recombinant clones over background (never observed in our hands). The polylinker cloning sequence (shown as the series of elevated restriction sites) can be exploited for easy isolation of specific ends of the cloned *Arabidopsis* DNA. Note that not all of these restriction sites are unique (e.g., *Hin*dIII and *Sal*I). The cloning site (*Bam*HI) for inserting *Arabidopsis* DNA and the opposing site (*Hpa*I) for preparing arms are bolded. The origin of the various parts of the vector is shown on the left of the map. The selective marker, *neo*, driven by the *hsp70* promoter is irrelevant here since it is only expressed in *Drosophila*.

11. SOE: $1M$ sorbitol, $0.1M$ sodium citrate, $0.01M$ ethylenedinitrilo tetraacetic acid disodium salt dihydrate (EDTA), pH 7.0.
12. ß-Mercaptoethanol (Merck, Darmstadt, Germany).
13. NDS: $0.5M$ EDTA, pH 8.0, 1% (w/v) sodium *N*-lauroylsarcosine, 10 mM Tris-HCl, pH 8.0.
14. Zymolyase 20-T (ICN, Costa Mesa, CA).
15. 10 mg/mL Proteinase K (Boehringer Mannheim, Meylan, France) in H_2O.
16. 0.1X SSC, 0.1% (w/v) sodium dodecyl sulfate (SDS) (20X SSC/L: 175 g NaCl, 27.6 g sodium citrate, pH 7.0, made up to 1 L).
17. Sodium phosphate stock: 134 g of $Na_2HPO_4·12H_2O$, 4 mL of 85% H_3PO_4, pH 7.2, made up to 1 L.
18. Church and Gilbert buffer: 50 mL sodium phosphate stock, 35 mL 20% (w/v) SDS, 15 mL H_2O, 200 µL $0.5M$ EDTA.
19. Deprobing solution: 2 mM Tris-HCl, pH 8.2, 0.1% (w/v) SDS.

2.2. Pulsed Field Gel-Electrophoresis

1. 1% (w/v) Low gelling temperature agarose (NuSieve® GTG®, FMC, Rockland, ME).
2. 125 mM EDTA, pH 8.0.
3. Clinical centrifuge (Heraeus Sepatech Minifuge RF 4350g or equivalent, Heraeus, Les Ulis, France).
4. 50 mM EDTA, pH 8.0.
5. Normal low electroendosmosis (EEO) agarose (Boehringer Mannheim).
6. ß-Mercaptoethanol.
7. 30 mg/mL zymolyase 20-T dissolved in 10 mM sodium phosphate buffer, pH 7.5, 50% (v/v) glycerol (stable for at least 1 yr at –20°C).
8. Plug mold (BioRad, Hercules, CA).
9. Scalpel.
10. Thin-blade spatula.
11. LET: 0.5M EDTA, pH 8.0, 10 mM Tris-HCl, pH 8.0.
12. NDS: 0.5M EDTA, pH 8.0, 1% (w/v) sodium N-lauroylsarcosine, 10 mM Tris-HCl, pH 8.0.
13. 5X TBE: 54 g Tris base, 27.5 g boric acid, 2 mL 0.5M EDTA, made up to 1 L.
14. Pulsed-field electrophoresis apparatus (CHEF-DR®II, BioRad).
15. Agarose gel tray (13 × 12.5 cm, BioRad).
16. Well formers (BioRad).
17. Ethidium bromide (10 mg/mL stock in water, Sigma).
18. Microbiological culture tubes (Falcon® Los Angeles, CA 15 mL or equivalent).

2.3. End Probes from YAC Insert

1. 1M sorbitol, 20 mM EDTA.
2. ß-Mercaptoethanol.
3. 30 mg/mL Zymolyase 20-T in 10 mM sodium phosphate buffer, pH 7.5, 50% (v/v) glycerol.
4. 0.1M EDTA, 0.15M NaCl.
5. 20% (w/v) SDS.
6. 5M Potassium acetate (KOAc).
7. Eppendorf centrifuge.
8. TE: 10 mM Tris-HCl, 1 mM EDTA, pH 8.0.
9. 4.4M LiCl.
10. 95% Ethanol.
11. 70% Ethanol.
12. Restriction enzymes AluI and TaqI and 10X buffers (New England BioLabs, Montigny-le-Bretonneux, France).
13. Phenol-chloroform (1:1 [v/v]).
14. Chloroform.
15. 7.5M Ammonium acetate (NH$_4$OAc), pH 7.5.
16. Isopropanol.

17. T4 DNA ligase and 10X T4 DNA ligase buffer (Pharmacia Biotech, St. Quentin-Yvelines, France).
19. 10 mM ATP (Boehringer Mannheim).
20. *Taq* DNA polymerase and 10X buffer (Appligene, Pleasanton, CA).
21. Polymerase chain reaction (PCR) deoxyribonucleotide mix with all four dNTP (Pharmacia Biotech) each at 1.25 mM.
22. PCR machine (Hybaid®, Woolbridge, NJ).
23. PCR primers:
    U44: 5' - GTC GAA CGC CCG ATC TCA AG - 3'
    U45: 5' - CCT AAT GCA GGA GTC GCA TA - 3'
    C47: 5'- GTA GCC AAG TTG GTT TAA GG - 3'
    C101: 5' - GTG CCT GAC TGC GTT AGC AA -3'
24. "GAT" deoxyribonucleotide mix of 0.5 mM dGTP, 0.5 mM dATP, 0.5 mM dTTP.
25. [α^{32}P]-dCTP (3000 Ci/mmol) (Amersham, Arlington Heights, IL).
26. 16-mL Polypropylene tubes.
27. 1-mL Disposable syringe.
28. G-50 Sephadex in TE.
29. Sterile glass wool.

2.4. Subcloning YACs

2.4.1. Isolating Total Genomic DNA from Yeast

1. YPDA.
2. Sterile 250-mL flask for yeast culture.
3. 0.1M Sorbitol, 20 mM EDTA.
4. Zymolyase T-20 (30 mg/mL).
5. ß-Mercaptoethanol.
6. 10 mg/mL of proteinase K in sterile water.
7. 20% (w/v) SDS.
8. 5M Potassium acetate.
9. Phenol-chloroform (1:1 [v/v]) equilibrated with 10 mM Tris-HCl, pH 8.0.
10. Isopropanol at room temperature.
11. TE: 10 mM Tris-HCl, pH 8.0, 1 mM EDTA.
12. 50X TAE: 242 g Tris base, 57.1 mL glacial acetic acid, 100 mL 0.5M EDTA, made up to 1 L.
13. λ DNA (cI857 Sam7; Boehringer Mannheim) cut with *Sal*I or with *Hin*dIII as markers.
14. Pasteur pipets.
15. Eppendorf microcentrifuge.
16. Waterbaths set at 37, 55, and 65°C.
17. 1.5-mL Eppendorf tubes.
18. Minigel electrophoresis system.
19. Dilution buffer for *Mbo*I: 10 mM Tris-HCl, pH 7.5, 50 mM KCl, 0.1 mM EDTA, 500 µg/mL bovine serum albumin (BSA), 50% (v/v) glycerol (sterilized by filtering through a 0.22 µm membrane, for example, Syrfil® MF from Costar, Cambridge, MA).

20. 10X BSA: 1 mg/mL.
21. *Mbo*I restriction enzyme and 10X buffer (New England BioLabs).
22. 0.5M EDTA.
23. 7.5M NH$_4$OAc.
24. Normal low EEO agarose (Boehringer Mannheim).
25. 15% (w/v) Ficoll (Type 40, Pharmacia Biotech), 100 mM EDTA.
26. Ethidium bromide (10 mg/mL) in H$_2$O.
27. Chloroform.
28. 70% Ethanol.
29. 80% Ethanol.
30. 1.25M NaCl, made in either distilled H$_2$O or TE.
31. 5.0M NaCl, made in either distilled H$_2$O or TE.
32. Beckman tubes (14 × 89 mm) for gradient (Beckman, Palo Alto, CA).
33. Gradient former (Hoefer SG15 or equivalent; Hoefer Scientific Instruments, San Francisco, CA).
34. Beckman SW41 rotor.
35. Beckman ultracentrifuge.
36. Clinical centrifuge (Heraeus Sepatech Minifuge RF).
37. 50-mL Sterile polypropylene tubes (or equivalents such as Corex® tubes).
38. 16-mL Sterile polypropylene tubes (or equivalents).

2.4.2. Preparing Cosmid Arms and Construction of Genomic Bank

1. *Hpa*I and 10X buffer (New England BioLabs).
2. Calf intestinal phosphatase (CIP; Boehringer Mannheim).
3. Water baths set at 37, 50, and 65°C.
4. Phenol-chloroform (1:1 [v/v]).
5. Chloroform.
6. 7.5M NH$_4$OA$_C$.
7. 80% Ethanol.
8. Isopropanol.
9. T4 DNA ligase and 10X buffer (Pharmacia Biotech).
10. 20 mM Dithiothreitol (DTT).
11. 10 mM ATP.
12. 40 mM Spermine (Sigma).
13. 10X BSA: 1 mg/mL.
14. λ DNA (*c*I857 Sam7; Boehringer Mannheim).
15. *Mbo*I, *Bam*HI, and 10X buffers from suppliers (New England BioLabs).
16. TE: 10 mM Tris-HCl, pH 8.0, 1 mM EDTA.
17. Cosmid (*cos*Pneo, *23*).
18. Normal low EEO agarose (Boehringer Mannheim).
19. 50 mM ATP.
20. 20 mM DTT.
21. 20% (w/v) maltose stock.
22. 1M MgSO$_4$ stock.

23. λ packaging extract (Stratagene, La Jolla, CA).
24. SB: 32 g Bacto tryptone, 20 g Bacto yeast extract, 5 g NaCl, made up to 1 L.
25. SB ampicillin plates: SB plus 15 g Bacto agar and 50 mg of ampicillin per L; we use 132-mm Petri plates.
26. *Escherichia coli* strain DH5α[*sup*E44 Δ*lac*U169(φ80 *lac*ZΔM15) *had*R17 *rec*A1 *gyr*A96 *thi*-1 *rel*A11].
27. Eppendorf centrifuge.
28. SM: 5.8 g NaCl, 2 g MgSO$_4$·7H$_2$O, 50 mM Tris-HCl, pH 7.5, 0.01% (w/v) gelatin, in 1 L.

3. Methods
3.1. Replica Plating and Colony Hybridization of Yeast

1. Prepare SD medium in 1 L of distilled H$_2$O. Adjust pH to between 6.0 and 6.5 with a few drops of 10N NaOH and add 870 mg of amino acid mix and Bacto agar to 1.5%. Sterilize in a kitchen stovetop pressure cooker for approx 15 min (*see* **Note 1**).
2. Pour Petri plates after cooling the media to approx 50–60°C (*see* **Note 2**).
3. Replica plate yeast colonies with a 96-pin replicator from stock plates or microtiter dishes onto these SD (-trp, -ura) plates. Incubate at 30°C for 2–4 d to regenerate colonies. Repeat the transfer if necessary to achieve homogeneous colony size (*see* **Note 3**).
4. Prepare a set of YPD-agar plates. Place a sheet of Amersham Hybond-N membrane cut to size onto each YPD-agar plate. Replica plate yeast colonies onto the Amersham Hybond-N membranes (*see* **Note 4**).
5. Allow colonies to grow on the membranes placed on YPD until the colonies are slightly pinkish in color which usually takes between 12 and 20 h at 30°C or overnight at room temperature (*see* **Note 5**).
6. Transfer the Hybond-N membranes with the yeast colonies on top onto a sheet of Whatman 3MM saturated with SOE supplemented with 1% ß-mercaptoethanol (*see* **Note 6**).
7. After 5 min, transfer the membranes onto fresh sheets of Whatman 3MM saturated with SOE, 1% ß-mercaptoethanol, 1.5 mg/mL zymolyase 20-T. Incubate at 37°C for 4 h to overnight (*see* **Note 7**).
8. Transfer the membranes onto Whatman 3MM filters pre-saturated with NDS for 5 min. Transfer them again onto another set of Whatman 3MM filters saturated with NDS supplemented with 1 mg/mL of proteinase K. Incubate the filters at 50°C for approx 5 h to overnight. Next day, air dry the membranes with the colonies.
9. Protect the membranes by placing them between sheets of dry Whatman 3MM. Wrap the entire ensemble with aluminum foil and dry autoclave it for 2.5 min (at 120°C and 15 lbs/in.^{2}) to denature and fix the DNA onto the membranes (*see* **Note 8**).
10. Wash the filters with 0.1X SSC, 0.1% SDS, pH 8.0 at room temperature until the wash solution is clear. Rub the filters gently with gloved hands if necessary.
11. Prehybridize the membranes in Church and Gilbert buffer at 65°C for 5 min to one hour and then hybridize the membranes overnight in fresh buffer containing the radiolabeled probe (*see* **Note 9**).

12. After hybridization, wash filters in a shaking waterbath as follows: twice in 0.1X SSC, 0.1% SDS, 10 min each, at room temperature and then again twice, 20 min each, in the same washing solution between 55 and 68°C depending on the desired stringency. Expose the filters for 1 h to overnight (*see* **Note 10**).
13. To deprobe the filters, wash them in dehybridization solution at 68°C for 15–20 min. Air dry and store them at room temperature in the dark until needed (*see* **Note 11**).

3.2. Analysis of YAC Clones by Pulsed–Field Gel Electrophoresis

1. Grow 2 mL of the positive yeast colonies in round bottom 15-mL tubes (e.g., Falcon) to roughly 10^8 cells/mL in YPD with or without adenine at 30°C. This usually takes approx 24–36 h starting from single colonies (*see* **Note 12**).
2. While washing the cells as described below, prepare 1% low gelling temperature agarose (LGT) in 125 mM EDTA, pH 8.0. Keep the LGT molten at 38°C until needed.
3. Pellet yeast cells by centrifugation at 3000–4000g for 5 min at room temperature. Pour off the supernatant and resuspend the cell pellet in equal volume of 50 mM EDTA, pH 8.0. Repeat this washing step twice for a total of three washes. Drain off as much of the liquid as possible by inverting the tube and placing the tube briefly upside down on a clean piece of paper towel.
4. Resuspend cells by adding a small volume (e.g., 50 µL) of 50 mM EDTA, pH 8.0 to facilitate resuspending the cells. Equilibrate the temperature of the tubes for approx 3–5 min in a 38°C bath. Add 2 µL of zymolyase, 4.6 µL ß-mercaptoethanol and then 150–180 µL of LGT. Mix the content quickly by tapping with fingers. Transfer the embedded cells to a mold to cast plugs. Solidify the LGT by chilling the mold on ice or at 4°C for about 10–30 min.
5. Transfer the plugs from the mould into a small volume of LET with 7.5% ß-mercaptoethanol. Incubate at 37°C from 6 h to overnight (*see* **Note 13**).
6. Wash the plugs three times in 5 mL of NDS buffer at room temperature for 15 min each. Then transfer to approx 2–3 mL of NDS with final concentration of 1–2 mg/mL proteinase K. Incubate at 50°C from 6 h to overnight (*see* **Note 14**).
7. Cool the plates to room temperature. Pipet off the NDS and wash the plugs in 50 mM EDTA, pH 8.0 by slow shaking. Change the EDTA solution every 15–30 min for a total of three times. These plugs can be stored in EDTA at 4°C or can be used directly for CHEF gel analysis (*see* **Note 15**).
8. To prepare CHEF gels, first cool the 0.5X TBE running buffer in the gel chamber to 12–14°C 20–30 min before electrophoresis. Cast a 1.0% agarose gel in 0.5X TBE.
9. Cut a 2–3 mm strip from the yeast plugs and insert it into the gel slots.
10. Seal the slots with 1% LGT or normal agarose. Run the gel (*see* **Note 16**). After the run the chromosomes can be visualized by staining the gel in water or 0.5X TBE with 0.5 µg/mL ethidium bromide for 30–45 min.
11. The DNA from the gel can then be transferred to membranes by standard technique of Southern blotting *(24)* and hybridized with RFLP probes to confirm the results obtained by colony screens.

3.3. Preparing End Probes by Inverse Polymerase Chain Reaction

1. Grow up 5 mL of yeast in YPDA at 30°C.
2. Centrifuge at 4000g for 5 min to harvest cells.
3. Resuspend the cells in equal volume of $1M$ sorbitol, 20 mM EDTA and collect cells again by centrifugation. Repeat this washing step once and then resuspend cells in 150 µL of $1M$ sorbitol, 20 mM EDTA.
4. Add 5 µL of zymolyase, 11.5 µL of ß-mercaptoethanol, and incubate for about 2 h at 37°C.
5. Centrifuge at 1200g for 5 min at room temperature to harvest spheroplasts. Remove the supernatant gently with a pipet.
6. Add 0.5 mL of $0.1M$ EDTA and $0.15M$ NaCl to resuspend the spheroplasts. Pipet up and down to break up the pellet. Add 25 µL of 20% SDS and incubate at 65°C for 20 min.
7. Add 200 µL of $5M$ KOAc. Invert several times to mix and set on ice for at least 30 min. Centrifuge for 3 min at room temperature in an Eppendorf centrifuge at maximum speed (12,000g).
8. Transfer supernatant into a clean 1.5-mL Eppendorf tube. Fill the tube with 95% ethanol and mix by inversion. Precipitate the nucleic acids by centrifugation in an Eppendorf microfuge for 10 s at maximum speed at room temperature. Discard supernatant.
9. Resuspend the pellet in 250 µL of TE, and if necessary, by occasionally heating it at 65°C to accelerate the process. Add equal volume of $4.4M$ LiCl. Incubate the tube on ice for 30 min. Centrifuge for 5 min to precipitate out the RNA.
10. Transfer the supernatant to clean 1.5-mL tube and precipitate the DNA by filling the tube with 95% ethanol. Chill the tubes at –20°C for a few min and then precipitate the DNA by centrifugation in an Eppendorf centrifuge for 5 min at maximum speed. Discard supernatant.
11. Rinse the pellet two times with 0.5 mL 70% ethanol and dry the pellet briefly *in vacuo*. Resuspend the DNA in 100 µL of TE.
12. Remove 10 µL or about 0.5 µg of DNA for restriction digest (*see* **Note 17**). Digest the DNA in a 50-µL volume with *Alu*I at 37°C for 1 h. Set up another restriction digest using *Taq*I as the enzyme and digest for 1 h at 65°C.
13. Extract the digests once with phenol and chloroform (1:1 [v/v]). Then, extract once with chloroform. Add half volume of $7.5M$ NH$_4$OAc and equal final volume of isopropanol. Chill the tubes at –20°C for 20 min and centrifuge at 4°C for 10 min at maximum speed in an Eppendorf microcentrifuge. Wash the pellet twice with 300 µL ice-cold 70% ethanol with 2 min centrifugation between washes. Desiccate to dry the pellets.
14. Resuspend the pellets in 30 µL of TE. Remove 10 µL (100–150 ng) of the digested DNA and add 5 µL of 10X T4 DNA ligase buffer, 5 µL of 10 mM ATP, 30 µL of water, and 5 U of the T4 DNA ligase (*see* **Note 18**). Leave the ligation at room temperature for a few hours to overnight.
15. In a 0.5-mL tube, add 5–25 µL of the ligations, 10 µL of 10X Taq polymerase buffer, 1 µM (or about 0.5 µg) each of the PCR primers (*see* **Note 19**), 16 µL of

PCR deoxyribonucleotide mix. Complete the volume to 100 µL with clean H_2O. Add about 1 U of Taq DNA polymerase. Layer on top a drop of clean mineral oil with a P1000 Pipetman if necessary. Carry out the PCR reactions in an automatic apparatus according to the supplier's instructions (*see* **Note 20**).

16. After the PCR reaction, put the tubes at –70°C for a few min to freeze the aqueous phase of the PCR reaction. Remove all of the oil on top with a clean pipet.
17. Prepare a 2–4% gel by mixing 3:1 (w/w) of LGT and normal agarose in 1X TBE. Analyze 5–10 µL of the amplified products by electrophoresis (*see* **Note 21**). Quantify the amount of PCR product by gel and then dilute in clean H_2O to obtain approx 2 ng/µL.
18. To make hybridization probe, remove 10 µL of the amplified DNA, and add 1µ*M* each of the same two specific primers used in the previous PCR reactions (*see* **Note 19**).
19. Add 1 µL of the GAT deoxyribonucleotide mix.
20. Add 5 µL of 10X Taq DNA polymerase buffer. Bulk up to 40 µL with clean distilled water.
21. Add 10 µL of [$\alpha^{32}P$]-dCTP and 0.5 U of Taq DNA polymerase.
22. Top with a drop of mineral oil using P1000 Pipetman if required depending on the type of PCR apparatus employed.
23. Carry out PCR amplification for three cycles.
24. Freeze the aqueous phase of the PCR reaction by chilling the tube at –70°C for a few min and remove the oil with a pipet.
25. Make a small column with a 1-mL disposable syringe *(24)*: Plug the bottom with a bit of sterile glass wool and then fill it with 1-mL of G-50 sephadex in TE. Place the syringe inside a 16-mL polypropylene tube, and centrifuge at 100*g* in clinical centrifuge for 2 min to remove the TE.
26. Transfer the syringe to a clean 16-mL polypropylene tube and load the radioactive sample. Centrifuge at 100*g* for 2 min.
27. Add 50 µL of TE to the syringe and centrifuge for 2 min at 100*g*.
28. Collect the eluate and remove a small amount (e.g., 1 µL) to estimate the specific activity of the probe by Cerenkov counting, which should be on average 10^9 cpm/µg of DNA.
29. Boil the eluate for 10 min and use as hybridization probe (*see* **Note 22**).

3.4. Subcloning YACs into Cosmid

3.4.1. Isolating Large Genomic DNA From Yeast

1. Grow 100 mL of yeast containing the desired YAC in YPDA for approx 40–48 h (*see* **Note 23**).
2. Transfer cells to 50-mL polypropylene tubes and collect cells by centrifugation at 2500*g* in a clinical centrifuge (Heraeus Sepatech Minifuge RF) for 10 min.
3. Resuspend the cells in 5 mL of 0.1*M* sorbitol, 20 m*M* EDTA. Repeat the washing and centrifugation steps in the same sorbitol-EDTA solution two more times.
4. Add 100 µL of zymolyase T-20 (30 mg/mL stock), 100 µL ß-mercaptoethanol. Incubate the cells at 37°C for 1.5–2.0 h. Centrifuge at 1200*g* for 5 min to collect spheroplasts.

5. Resuspend pellet in a final volume of 5 mL of 0.1M sorbitol, 20 mM EDTA. Add 0.5 mL of 10 mg/mL proteinase K and 0.25 mL of 20% SDS. Incubate at 55°C for 45–60 min.
6. Add 2 mL of 5M KOAc. Slowly mix by tapping with finger or by inversion of the tube and leave the tube on ice for 30 min.
7. Centrifuge 20 min at 3000g. Remove supernatant carefully with P1000 Pipetman with a cut off pipet tip to provide a wider opening of approx 3 mm in diameter.
8. Add phenol-chloroform and gently rock the tube horizontally (*see* **Note 24**). Repeat the extraction step if necessary either by replacing the organic phase with a fresh aliquot or gently transferring the aqueous phase to a fresh tube and then adding phenol-chloroform. Finally, wash the aqueous phase with equal volume of chloroform.
9. Transfer the aqueous phase to a clean Corex or polypropylene tube. Without mixing, gently layer an equal volume of isopropanol on top of the aqueous phase. Tilt the tube at a slight incline and rotate it in one direction for a few min and then in the other direction to slowly partition out the high molecular weight DNA. Change the isopropanol every 5–10 min and continue the extraction until almost all of the aqueous phase has disappeared (*see* **Note 25**).
10. Retrieve the ball of DNA with a flame-sealed Pasteur pipet with a hook at one end (formed by flaming).
11. Wash the ball of DNA by dipping the pipet up and down several times in a series of 5 mL of 70% ethanol. (*see* **Note 26**).
12. Get rid of most of the ethanol by touching the ball of DNA on the inside wall of a clean Eppendorf tube several times. Slightly dry the ball of DNA in air, but not completely. Gently shake the ball of DNA loose in 1.0 mL of TE (*see* **Note 27**). Leave it for 30 min to overnight at 4°C until the DNA is completely resuspended.
13. Check 3–5 µL of the DNA for potential degradation by electrophoresis on a 0.3% normal low EEO agarose minigel made in 1X TAE. Use uncut λ DNA (e.g., 100 ng) as control. Run the gel slowly, for example, 2–3 V/cm (*see* **Note 28**).
14. Set up a test partial digest by mixing: 25 µL of the genomic DNA, 10 µL of BSA, 10 µL 10X *Mbo*I buffer, 55 µL of H$_2$O. Prewarm the mix for 5 min or so at 37°C (*see* **Note 29**).
15. Add diluted *Mbo*I enzyme (usually 0.05–0.005 U). Remove 20 µL at 5, 10, 20, 30, and 45 min after addition of enzyme. Add 5 µL of 15% Ficoll, 100 mM EDTA to stop the digests. Heat all samples at 65°C for 5–10 min and run a 0.3% agarose minigel in 1X TAE at 1–2 V/cm overnight (*see* **Note 30**).
16. For the large-scale digest, mix 750 µL of genomic DNA, 300 µL BSA, 300 µL 10X *Mbo*1 buffer, and 1650 µL H$_2$O for a total of 3 mL in a 16-mL polypropylene tube. Prewarm the tube for 15 min at 37°C.
17. Add the appropriate units of *Mbo*I diluted exactly to the same final concentration as in the test digest so that the ratios of the various volumes remain identical. Digest the DNA, but reduce the duration by 50% as that determined in the initial small-scale digests which yielded most of the fragments between 35 and 50 kb (*see* **Note 31**).
18. Add EDTA to a final concentration of 20 mM.

19. Gently extract the digested DNA with phenol-chloroform. Extract once with chloroform.
20. Add 0.1 vol of 7.5M NH$_4$OA$_C$, mix by tapping with finger and add equal vol of isopropanol. Incubate at –20°C for 20 min and then centrifuge at 12,000g for 20 min at 4°C (*see* **Note 32**).
21. Rinse the pellet once with 80% ethanol and centrifuge briefly at 4°C. Remove as much ethanol as possible with a drawn-out Pasteur pipet. Add 100 µL of TE to the pellet and leave overnight at 4°C (*see* **Note 33**). Remove a small aliquot (e.g., 1 µL) for analytical gel electrophoresis in a 0.3% agarose in 1X TAE against λ controls as above to verify that there has been no degradation during storage.
22. Make an 11-mL continuous NaCl gradient of 1.25–5.0M in Beckman tubes (**Fig. 2**).
23. Heat the DNA samples at 65°C and then load on top of the gradient with a P1000 Pipetman.
24. Centrifuge in a SW41 rotor at 39,000g for 3.5 h or 35,000g for 6 h (*see* **Note 34**).
25. Collect 0.5-mL fractions with a P1000 Pipetman and a series of numbered 2-mL Eppendorf tubes (*see* **Note 35**).
26. Add an equal volume of distilled H$_2$O to each fraction, then fill the tube with isopropanol. Cool the tubes at –20°C for about 20 min, then centrifuge for about 15 min in an Eppendorf centrifuge (*see* **Note 36**).
27. Gently resuspend the pellets in 300 µL of TE, add 30 µL of 7.5M NH$_4$OA$_C$, and then equal volume of isopropanol. Precipitate the DNA as above.
28. Wash the pellet twice with ice-cold 80% ethanol and briefly centrifuge for 2 min in between washes.
29. Remove the supernatant *carefully* either with a Pipetman or a drawn-out Pasteur pipet (*see* **Note 37**). Briefly desiccate the pellets. Resuspend in 10 µL TE. Leave overnight, if necessary, at 4°C.
30. Check an aliquot (1 µL) from every third fraction on a 0.3% mini-agarose gel made in 1X TAE. Collect all fractions containing DNA fragments between 35 and 50 kb for cloning (*see* **Note 38**).

3.4.2. Ligation of Size-Fractionated DNA into Cosmid

1. Mix 40 µL of *cos*Pneo vector DNA (1 mg/mL; **Fig. 1**), 20 µL of 10X *Hpa*I buffer and 140 µL H$_2$O. Add 20 U of *Hpa*I and digest at 37°C for 2–4 h (*see* **Note 39**).
2. Add 10 U of CIP and incubate at 50°C for approx 45 min (*see* **Note 40**).
3. Phenol-chloroform extract once (*see* **Note 41**). Chloroform extract once. Add 0.1 vol of 7.5M NH$_4$OA$_C$ and equal volume of isopropanol. Incubate at –20°C for 20 min and centrifuge in the cold room for 15 min.
4. Carefully rinse pellet with 0.5 mL of 80% ethanol (*see* **Note 37**). Dry the pellet and resuspend the pellet in 80 µL of TE (*see* **Note 42**).
5. If the CIP has worked well, then digest the *cos*Pneo with *Bam*HI by mixing: 75 µL DNA, 20 µL 10X *Bam*HI buffer, 20 µL 40 mM spermine, 20 µL BSA, and 65 µL H$_2$O. Add 20 U of *Bam*HI, digest at 37°C for 2 h (*see* **Note 43**).
6. Remove a small aliquot for gel analysis. If the digest is complete, phenol-chloroform extract once; chloroform extract once. Add 0.1 vol of 7.5M NH$_4$OA$_C$

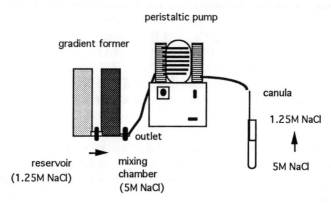

Fig. 2. Making continuous salt gradients to size fractionate partially digested total yeast genomic DNA: The basic gradient former consists of two chambers, the reservoir and mixing chamber, which are connected at the base by a small passage and can be closed or opened by regulating a valve. The $1.25M$ NaCl (5.5 mL) is first added to the reservoir and the valve is opened and then closed immediately to allow the solution to fill the passage; place a small magnetic stir bar to the mixing chamber and then add 5.5 mL of the $5.0M$ NaCl. The gradient is made by first starting a stirring motor. The outlet is then opened and simultaneously, the peristaltic pump is activated and the passage between the two chambers is opened completely by regulating the valve. The canula is placed at the bottom of the receiving centrifuge tubes. When the gradient solution exits (we aim for a flow rate of approx 1 drop/s), slowly raise the canula, or lower the tube, but at all times keeping the tip of the canula in contact with the wall of the tube and just slightly above the surface of the gradient to minimize disturbance. Some workers prefer reversing the order of the two salt solutions, that is, $1.25M$ NaCl in the mixing chamber and the $5.0M$ NaCl in the reservoir. In this case, the canula is left at the bottom of the centrifuge tubes throughout the procedure and then removed slowly after the continuous gradient is made. We found this second method more convenient, especially if a large number of gradients are needed, but the resolution in our hands is less sharp in comparison with the first method above.

and then equal vol of isopropanol. Precipitate the DNA by centrifugation in an Eppendorf centrifuge for 10 min. Rinse pellets twice in 80% ethanol with 2 min centrifugation between washes. Remove ethanol (*see* **Note 37**) and dry pellet (*see* **Note 44**).

7. Mix in an Eppendorf tube on ice: 1 µL of 10X T4 DNA ligase buffer, 1 µL of 50 mM ATP (*see* **Note 45**), 1 µL 20 mM DTT, 1 µL BSA, 1 µL H$_2$O, 1–2 U (1 µL) of T4 DNA ligase.
8. Divide the mix into two, 3-µL aliquots. To each aliquot add 1 µL of vector arms (1 µg) and approx 0.5 µg of the size-fractionated insert DNA in 1 µL. Mix carefully with a Pipetman 5–6 times. Incubate the ligation at 16°C overnight (*see* **Note 46**).

9. Package the ligation mix with λ packaging extracts according to supplier's instructions. After packaging reaction is completed, add SM to 0.5–1 mL and store at 4°C.
10. Grow a 10-mL culture of DH5α in SB supplemented with 0.2% maltose and 10 mM MgSO$_4$ for approx 10–12 h at 37°C.
11. Test titer of the packaged cosmids by infecting 0.1 mL of the bacteria with 1 and 5 μL of the SM from **step 9**. Incubate at 37°C for 15 min (with slow shaking if possible). Then add 0.9 mL of SB. Incubate at 37°C (slow shaking if possible) for 45 min.
12. Spread 100, 200, 300, and 400 μL of the cells onto SB plates containing no more than 50 μg/mL of ampicillin. Incubate the plates overnight at 37°C and calculate the number of colonies.
13. For large-scale plating, infect 0.2 mL of an overnight bacterial culture with enough of the packaged ligation in SM to give approx 5000–10,000 colonies/132-mm Petri plates. Incubate at 37°C for 15 min as usual for the infection step, then add 1–2 mL of SB and incubate further at 37°C for 45 min (*see* **Note 47**).
14. Centrifuge the tubes in Eppendorf microfuge at 12,000g for 30 s, resuspend the cells in 0.5 mL of SB for each 132-mm Petri plate. Spread the cells onto SB-agar plates containing 40–50 μg/mL ampicillin. These colonies are ready to be screened for YAC-specific DNA (*see* **Note 48**).

4. Notes

1. Add the various required amino acids in a mortar and grind to a fine powder to ensure homogeneity. This amino acid mix is stable for years and is designed for selecting YAC vectors carrying tryptophan and uracil as markers propagated in the yeast host AB1380 (genotype is MAT α, Ψ$^+$ *ura 3 trp1 ade2-1 can1-100 lys2-1 his5*). We routinely use a pressure cooker to sterilize the media, which minimizes the degradation of the amino acids and caramelization of dextrose. If autoclave is used and the media are caramelized as a result, it is advisable to prepare a 10X stock of dextrose and autoclave this separately. There is normally more than sufficient amounts of the various amino acids to withstand the usual autoclaving at 15 lb/in.^{2} for 15 min. If the stability of the amino acids is a concern, one can prepare liquid amino acid stocks, filter sterilized by passage through a 0.22-μm filter (for example, Syrfil MF). The following can be used as a base for variations to suit personal convenience. They are prepared as mg/100 mL: add the powder and resuspend in 70 mL of distilled water and add dropwise 10N NaOH until all of the amino acids have dissolved. Complete with distilled water to 100 mL. Mix 1: met (240), leu (720), phe (600), thr (2400), ser (4500), his (240). Mix 2: glu (1200), asp (1200). Mix 3: arg (240), lys (360), tyr (1800). Add 8.3 mL of each mix to 975 mL of autoclaved SD after cooling to 60°C, the pH should be automatically adjusted to around 6.0–6.5 by these additions. These liquid stocks are stable at room temperature for a few months. Normally these stock amino acid solutions are colorless, but if they become tainted, fresh stocks should be prepared.

2. Although the sizes and shapes of the Petri plates are personal choices, we use 12 × 12 cm square plates. One liter of the SD medium is sufficient for about 12 plates.
3. Homogeneous colony sizes are essential to obtain unambiguous signals in the subsequent screen.
4. We cut sheets of Amersham Hybond-N membrane measuring 8.5 × 11.5 cm, which are labeled at the top right corners (to orientate and to distinguish the individual filters). We replicate four plates of yeast colonies onto a single Hybond-N membrane by plating each set of the 96 colonies slightly offset horizontally and vertically from those of the previous set. The limiting factor concerning the number of colonies that can fit onto the Hybond-N membrane here is the diameters of the pins of the replicator.
5. We deliberately leave out the adenine in the YPD so that the yeast colonies will eventually turn red as they age. The ideal colonies for screening are when they have turned slightly pinkish in color, and in our case, are about 1–2 mm in diameter (this will be influenced by the diameter of the pins of the replicator and should not be used as the sole criterion). If the colonies are dark red, which is a visual cue that they have been overgrown, their cell walls become more resistant to zymolyase treatment. The results of such colony screens are usually high background. It is our experience to better restart from replica plating the cells onto the Hybond-N membranes again rather than continue on with the screen.
6. Sterile techniques are not critical throughout this step. The Whatman filters should not be overly saturated with SOE and 1% ß-mercaptoethanol or else the colonies will diffuse. The addition of the ß-mercaptoethanol is used to weaken the cell wall. The Whatman 3MM is cut to the same dimension as the Hybond-N membranes and wetted by placing them slowly in a 12 × 12 cm Petri plate containing 5 mL of the solution. After the filter is fully saturated, the excess liquid is poured off. The Petri plates for these steps are recovered by removing the media from old plates and can be reused many times.
7. The zymolyase is directly dissolved in SOE, then ß-mercaptoethanol is added to a final concentration of 1%. We usually put the plates inside sealed plastic bags (e.g., the same packaging bags for the Petri plates) or boxes to suppress the odor of ß-mercaptoethanol as much as possible and as reassurance that the membranes do not dry accidentally. Usually, a zymolyase digest of 4 h at 37°C is sufficient.
8. We have compared other more traditional procedures for denaturing and fixing the DNA onto the membranes such as lysing the spheroplasts chemically and then fixing the DNA either by UV cross-linking or by baking the membranes at 80°C *in vacuo*. None of these other procedures in our hands produced hybridization signals with the consistency and rapidity as those from simply autoclaving. This technique also allows a large number of membranes to be processed simultaneously. For traditional chemical lysis of yeast spheroplasts and fixation of DNA onto membrane, *see* **ref. *16***).
9. The Church and Gilbert buffer used here differs from that of the original *(25)* by the omission of BSA. We found no influence of the BSA on the severity of background in our colony screens, so we consider this as optional; its presence,

however, has been reported to decrease the kinetics of hybridization *(15)*. The vessels for prehybridization and hybridization are a matter of personal choice. If a large number of membranes are to be screened (e.g., 20–30), we find it convenient to do the hybridization in the 12 × 12 cm Petri plates (*see* **Note 6**). The Hybond-N membranes with the lysed colonies are placed one at a time into the Petri plate containing 10–30 mL of Church and Gilbert buffer (either with or without radiolabeled probe) and the Petri plate is swirled to ensure even distribution of the buffer between each addition of the membranes. Finally, a piece of used X-ray film, cut to the same size as the Petri plate is placed on top of the stack of membranes to ensure that all of them stay properly submerged during hybridization. The Petri plates are placed inside a plastic container and the assembly is floated in a water bath set at 65°C. There is no need for agitation of the membranes during prehybridization or hybridization. The hybridization signals are sufficiently clear such that we consider simultaneous hybridization of duplicate sets of colonies as unnecessary.

10. If these membranes are to be rescreened, they should be kept moist during the exposure, for example, by putting them on a sheet of Whatman 3MM wetted with 0.1X SSC, 0.1% SDS. Once the membranes are dried, it becomes very difficult to strip off the old probe. Under the screening conditions that we have described, a strong positive signal can be detected routinely within a few hours, however, an overnight exposure would be required to observe the background necessary to realign the colonies for identification of the positives. Note that if the same set of filters is to be rescreened, then the same overexposed film can be reused for all subsequent realignment with the filters to identify the positive colonies.

11. If these filters are not deprobed, they can be stored moist after hybridization at –20°C (to prevent fungal growth on the filters). If these filters are deprobed after each round of screening, the hybridization signals will get progressively weaker. We have used the same set of filters for up to 10 rounds of screening. The alternative, to prolong the longevity of these filters, is to dehybridize them once every three to four screens.

12. We use disposable polypropylene tubes with a round bottom (Falcon). Avoid using conical tubes, since mixing the reagents in the subsequent steps is more difficult. Also, avoid tubes with tightly fitting caps as this slows down the growth of the yeast.

13. We remove the plugs from the mold by loosening them first around the edges with a clean scalpel. Using a thin blade spatula, the plugs are then nudged out into a 2.5–5.0-cm Petri plate containing 2–3 mL of LET. For a visual cue of the zymolyase activity, the red pigments of the yeast cells should leach out and the plugs will turn white within a few hours.

14. During the proteinase K digest, 0.1 mg/mL RNaseA can also be added.

15. For long term storage, we keep the plugs in $0.5M$ EDTA at 4°C. They are stable for about a year before the chromosomal bands are noticeably less sharp upon analysis by pulsed field gel electrophoresis.

16. We routinely start with 1% agarose gels. For rapid analysis of a small number of positives, an alternative could be a minigel without preformed slots. The small

pieces of plugs can be adhered at one end of the gel by capillary action, and a drop of agarose can be used to secure the pieces of plugs. The gel is held in place by the flow of the circulating TBE buffer coming from one end which pushes it against a rubberband stretched across two pins at the other end. We use the clamped homogeneous electric field system from BioRad (CHEF-DR gel chamber driven by the model 200/0.2 power supply and the Pulsewave®760 Switch). The pulse conditions for optimal resolution depends on many parameters, the most important being the pulse rates (described in the user's manuals). As a routine, for gel dimension of 13 × 12.5 × 0.5 cm and assuming the YACs are at least 100 kb, we use a time ramp of between 30 and 90 s for 18–20 h. For a 1% minigel of dimensions of 10 × 7 × 0.5 cm, a time ramp of between 20 and 40 s for 6–8 h can be used. We use the constant conditions of 200 V, 0.5X TBE as running buffer maintained at between 12–14°C and the pulse ratio of 1.0 between forward and reverse times. For estimating the sizes of the YAC clones that are at least 250 kb, the intact chromosomes of yeast can be used. We have initially used the chromosomes of the yeast strain *YNN295* as standards, which are available from BioRad. The sizes (in kilobases supplied by BioRad) of the chromosomes (in Roman numerals) are: 2200 (XII), 1600 (IV), 1125 (XV, VII), 1020 (XVI), 945 (XIII), 850 (II), 800 (XIV), 770 (X), 700* (XI), 630 (V), 580 (VIII), 460 (IX), 370* (III), 290 (VI), and 245* (I). Those chromosomes marked with (*) are similar in sizes to those in the strain *A1380* (and thus can be used routinely as rough internal size markers if desired); however, the other chromosomes seem slightly different in mobility between the two yeast strains. For additional size markers, a ladder consisting of λ concatamers can be made *(24)*.

17. We assume that the total recovery of the yeast genomic DNA is approx 5 µg.
18. From theoretical considerations *(26)*, we estimate that there should be 0.6 or more units of T4 DNA ligase/µL of ligation.
19. Use the primer pair U44 and U45 for amplifying insert end linked to the URA3 marker on the vector arm and the primer pair C47 and C101 for the insert end linked to the CEN4 arm. These primer pairs can be used to amplify both *Alu*I and *Taq*I digested and recircularized YAC DNA. Additional IPCR primer sequences for the YAC banks constructed by Grill and Somerville *(14)* can be found in **ref. 1**. An alternative to IPCR for amplifying YAC ends is by "vectorette" *(16)*.
20. The PCR amplification procedures are fairly standard and we have found it unnecessary to clean up the ligation mixes before performing the reactions. For each positive YAC clone, we thus perform four IPCR reactions in total (URA3- and CEN4-linked insert ends for *Alu*I digest and then the same for the *Taq*I digest) to ensure that at least one of the religated products will yield an amplified fragment of several hundred base pairs. In extremely rare occasions (in our hands, approx 5% of the cases) where neither digests yield amplified fragments of more than 100 bp, there are two solutions. When analyzing the PCR products by gel electrophoresis, it is our experience that in many cases, there will be several coamplified fragments; the larger ones correspond to amplified products resulting from incomplete *Alu*I or *Taq*I digested and religated YAC DNA. These higher

molecular weight bands can be purified from the gel and reamplified by PCR. The other solution is to purposely under digest the yeast DNA containing the YAC clone with *Alu*I or *Taq*I, and then recircularize the DNA for IPCR amplification. These bands can be gel purified and then used directly as hybridization probes. It should be mentioned that if too much input templates are used for the amplification reactions, extra faint bands of yeast DNA accompanied by a smeary background in the gel could be seen. In this situation, simply lower the input DNA for the PCR, or gel purify the real amplification products, which are usually much more abundant, and reamplify to dilute out the contaminating yeast DNA (for obvious reasons, PCR products contaminated with yeast DNA should not be used directly as hybridization probes for yeast colony screens). For the PCR cycles, many variations will work but we have generally used the following parameters: 94°C for 1 min, 55°C for 1 min, and 72°C for 1 min for 30 cycles. Some newer types of PCR machines do not require an oil overlay in the samples. If no amplification products are obtained despite several attempts, it is definitely worthwhile to check the Taq polymerase. We find the enzymes to be extremely variable in quality depending on the supplier and particular batches. We use 1–10 ng of pBluescript plasmid (better yet, one with a small cloned insert) and T7 and T3 primers as a control with each new batch of enzyme.

21. Specific bands corresponding to the PCR products can be cut out of the gel and purified by various means such as GeneClean® or its homemade version *(27)*. The usual yield is about 500 ng or more amplified YAC-specific DNA product, which is sufficient for making about 10 hybridization probes.

22. These end probes can sometimes give high background when used directly to screen YAC banks if the size ratio between the insert and coamplified adjacent vector sequences is low (e.g., amplified *Arabidopsis* DNA is < 200 bp). The other possibility is that there is yeast contaminating DNA in the PCR product (from using too much input religated DNA). In these situations, we simply used these probes to screen a λ genomic bank of *Arabidopsis* which we always keep on hand and, in turn, used these λ clones as probes to screen the YAC banks. The same plates of λ genomic bank of *Arabidopsis* (and the corresponding set of filters for screening) can be used for up to 6 mo if stored at 4°C. The end probes prepared by IPCR, however, usually pose no problems for mapping by restriction fragment length polymorphism to monitor the chromosome walk. To search for polymorphisms between ecotypes, we routinely digest the genomic DNA with 6–7 different "6-cutter" and 6–7 "4-cutter" restriction enzymes.

23. The culture is grown to saturation, but not overgrown to the point where the cells are red even in the presence of added adenine. Also the culture is intended for extraction and cloning of the total genomic DNA along with the YAC. Some practitioners prefer purifying the YAC DNA first. However, this option requires making agarose plugs and multiple runs of pulsed-field gel electrophoresis to obtain sufficient amount of YAC DNA which is likely to be still contaminated with yeast genomic DNA. Because of the often small quantity of YAC DNA recovered, it is difficult to monitor the extent of partial digestion without resorting

to labeling a small amount of it as internal control. Furthermore, the resulting subclones could contain chimeras of YAC and yeast DNA because of difficulties in applying existing techniques to purify size-fractionated inserts with such small amounts of DNA. Cloning the entire yeast genome is much less time consuming, easier to handle, and it is highly reproducible. Rough estimate suggests that if the YAC insert is only 150 kb, it would still represent 1% of the total yeast genome (14,000 kb) and thus can be easily retrieved by standard screening of the *E. coli* colonies. Another advantage is that the yeast DNA can act as a carrier to protect the YAC from shearing during isolation, as well as serving as a visible marker to monitor the extent of digest and efficiency of size-fractionation procedures.

24. The tube is held in a horizontal position and gently rolled back and forth. This maximizes the surface of contact between the organic and aqueous phases necessary for efficient extraction, but much gentler than mixing by inversion in order to avoid excessive shearing of the DNA.
25. This method of slow extraction takes a long time, but ensures that the highest possible molecular weight genomic DNA is recovered (*see* **Note 27**). Alternatively, if this technique is too tedious, mix the tubes gently by inversion although there is always the attendant risk of shearing the DNA.
26. We place several tubes containing 5 mL of 70% ethanol and wash the ball of DNA by dipping it up and down several times in each tube successively.
27. When the ball of DNA starts to dry in the air, it collapses slowly onto the tip of the Pasteur pipet. Never let it dry completely, as most of the DNA would not come off the tip. If the DNA adheres strongly to the pipet, we simply break off the tip and leave it in the TE. The tiny amount of residual ethanol poses no problem later with restriction digest analysis. We avoid using centrifugation to precipitate the DNA, as high molecular weight DNA is very difficult to resuspend without imposing damage. The sizes of the DNA fragments recovered with this method routinely range from approx 150–600 kb as determined by pulsed-field gel electrophoresis and contain little RNA contaminant. Small DNA fragments also do not precipitate well by this gentle isopropanol extraction as no salt, such as sodium acetate, is added during this step which is carried out at room temperature.
28. We prefer using TAE as the running buffer because it gives better resolution with high molecular weight DNA. The important point here is that the starting genomic DNA should be at least 150–200 kb before proceeding to cosmid cloning. For those isolating high molecular DNA for the first time, it is worthwhile to more precisely evaluate the size range of the DNA by pulsed-field gel electrophoresis against yeast chromosome standards (*see* **Note 16**) as well as in a normal 0.3% agarose gel to become familiar with its mobility pattern vis-à-vis the λ marker. Once the DNA isolation procedure becomes familiar, the latter normal agarose gel electrophoretic analysis alone suffices as a routine check. With electrophoretic run in a 0.3% normal agarose gel at 2–3 V/cm, we routinely see good separation of the isolated yeast genomic DNA and the λ marker after 3 h, which based on our experience, indicates that the size of the starting DNA is at least 150 kb.

29. The quantitation of the genomic DNA is never really accurate either by absorbance at 260 nm (because of residual RNA contaminants or ribonucleotides from degraded RNA) or by gel electrophoresis (large molecular weight DNA binds more ethidium bromide than smaller fragments which gives a false impression of the amounts based on relative fluorescence). We prefer to use volume as the reference point for scaling up and down the restriction digestion. For the 10X *Mbo*I buffer, consult the catalog of the supplier for the recipe if there is not one already provided along with the enzyme.
30. The number of units of *Mbo*I suggested in this protocol should be considered as starting points, since the absolute amount of genomic DNA is not accurate and the activity of the enzyme varies from batch to batch. If none of the timepoints gave partial *Mbo*I products in the range of 35–50 kb, the partial digest should be repeated by varying the amount of the enzyme. We do not include dyes such as bromophenol blue or xylene cyanol in the loading buffer (except for size markers), as the color can sometimes obscure the fluorescence of the ethidium bromide. When photographing the gel, underexpose the film; overexposure creates problems in interpreting the extent of digestion extrapolated from the relative intensity of the stained DNA. As size markers, we include undigested λ (50 kb), λ digested with *Sal*I (35 and 15 kb) and digested with *Hin*dIII (23, 9.4, 6.6, 4.4, 2.3, 2.0 kb, and two other small bands of approx 0.56 and approx 0.125 kb, which can be ignored). If desired, the RNA contaminants in the high molecular weight genomic DNA preparation can be removed by digesting with 20 µg/mL of RNaseA at 37°C for 30 min before embarking on the *Mbo*I partial digestion.
31. For a theoretical treatment of partial digests, *see* **ref. 28**. Briefly, the intensity of ethidium bromide staining does not accurately reflect the number of molecules in that range because larger DNA fragments bind more ethidium bromide. A 50% reduction in the time for the large-scale digest ensures that the DNA is not overdigested; moreover, it does not distort the final representation of the genome complexity in the library. An alternative is to remove 1 mL aliquots at 25, 50, and 100% relative to the "optimal" digest duration determined previously in the small-scale trial. These fractions can then be pooled together after phenol extraction.
32. For convenience, the DNA can be divided into several Eppendorf tubes and precipitate DNA in a microfuge.
33. Residual ethanol does not matter, as long as most of it is removed.
34. The tubes can accommodate a maximum of 12.5 mL. If necessary, fill the tubes almost to the top with TE or water, leaving approx 1–2 mm from the brim. We also prefer using the non-Ultra Clear Beckman centrifuge tubes which are more sturdy.
35. Here, the middle fractions are the most important. The pipet tip should be marked to indicate 0.5 mL, because the increase in NaCl in the gradient makes pipeting increasingly inaccurate toward higher salt density.
36. As an alternative to reduce the amount of unnecessary work, one can simply analyze the middle fractions of the gradient which are usually around fractions 12–16 (out of the total of 24 fractions). The rest of the fractions can be stored at

−20°C. Another possibility is that every third of the 24 fractions can be analyzed right from the start.
37. There is not much DNA and the pellets are not visible if they are clean. All manipulations should be done slowly because pellets are easily dislodged from the tube. We use 80% ethanol (instead of the usual 70%) for rinses because the DNA pellets appear to adhere slightly better to the wall of the test tube.
38. All fractions between 35 and 50 kb can be collected by precipitation including those that had not been actually tested by gel electrophoresis. If necessary, small amounts of DNA from these fractions can be used for test ligations and packaging efficiency to optimize the conditions. Do not pool these fractions. Include size standards in the gel (*see* **Note 30**). This analysis also provides a crude estimate of the amount of size-fractionated DNA needed for ligation later.
39. The *cos*Pneo is used as an example here because of its ease of manipulation. However, if only a limited chromosomal region represented by the YAC contig is needed to be subcloned and the main goal is to identify the target gene by plant transformation (or complementation), then a cosmid with transferred-DNA (T-DNA) borders might be a better choice. Recently such a vector (called 04541) has been constructed by C. Lister and C. Dean and they have generously made this available to the *Arabidopsis* community. Although high complexity banks can be readily made in this vector (J. Leung and J. Giraudat, unpublished data; for example of its use, *see* **ref. *13***), its large size (30 kb, and thus the maximum cloning capacity is only 20 kb) and low copy number may pose difficulties for those with little experience with genomic bank construction.
40. This higher temperature rather than the more usual 37°C is used to prevent potential activity of contaminating nucleases.
41. Back extract with a small volume of TE if necessary if the interphase is thick with proteins.
42. The efficiency of the CIP treatment can be tested by mixing 5 µL of vector DNA (2.5 µg), 1 µL 10X T4 DNA ligase buffer, 1 µL of 20 mM DTT, 1 µL 10 mM ATP, and 2 µL H$_2$O. Remove 1–2 µL as control. Add approx 1–2 U of T4 DNA ligase to the rest and incubate at 16°C overnight. Next day, check both control and ligated samples by running a 0.5–0.6% agarose gel.
43. If digest is incomplete, add more enzymes and incubate at 37°C for another 1–2 h. Care should be taken to avoid unnecessary prolonged digestion in case the ends of the vector might be damaged by nonspecific nucleases which cannot be detected by gel electrophoresis. This can be very time costly for subsequent experiments.
44. A test of the ligation efficiency should be done here, particularly if the quality of the various enzymes is not known and small damages at the ends of the vector or inserts are not detectable by gel analysis. Cut up some λ DNA as test inserts by mixing: 20 µL of λ DNA (5 µg), 5 µL of BSA, 5 µL 10X *Mbo*I buffer, and 20 µL H$_2$O. Add 10 U of *Mbo*I, digest for 1–3 h at 37°C. Phenol-chloroform extract as before; ethanol precipitate the DNA. Redissolve the pellet in 10 µL TE (0.5 µg/µL). Mix 5 µL *cos*Pneo (5 µg), 5 µL of test insert (2.5 µg), 2 µL of 10X T4 DNA

ligase buffer, 2 μL of 20 m*M* DTT, 2 μL 10 m*M* ATP, and 4 μL of H₂O. Remove 4–5 μL as control. To the rest of the mix, add 1–2 U of T4 DNA ligase. To test vector–vector ligation, mix: 5 μL *cos*Pneo (5 μg), 2 μL of 10X T4 DNA ligase buffer, 2 μL of 20 m*M* DTT, 2 μL 10 m*M* ATP, and 6 μL of H₂O. Remove 4 μL as control. To the rest add 1–2 U of ligase. For all of the ligation tests, incubate at 16°C overnight. Next day, heat the ligation and control mixes at 65°C for a few min, and then analyze the ligation in a 0.3% agarose gel made in 1X TAE.

45. A higher concentration of ATP is used to suppress blunt-end ligation at the *Hpa*I site that may not be completely dephosphorylated.

46. We aim for a 5–10 molar excess of the vector. The ligation can be done in 0.5-mL Eppendorf tubes or in flame-sealed capillaries. One of the two, 3-μL aliquots of the mix can be used as a control without insert (by adding TE instead of DNA). However, if you have enough familiarity with bank construction, one aliquot can be used to ligate smaller size-fractionated inserts (approx 30 kb), whereas the other maximizing for 40 kb inserts (in this case, we ligate size-fractionated DNA in the > 50 kb range). With all of the test ligations, usually one is fairly confident that the reaction will proceed as planned. If there is residual doubt, one can remove 1 μL of the ligation for analysis by gel electrophoresis. If successful, one should see a 9.5-kb band corresponding to the cosmid arms religated at the *Bam*HI site (recall that the arms are in molar excess) and a high molecular weight smear corresponding to arms ligated to size-fractionated yeast DNA. Unused portions of the size-fractionated genomic DNA can be preserved by adding 2 vol of 95% ethanol and storing at –70°C. When needed, the content of the tube can be rapidly dried by desiccation and the DNA rehydrated again with sterile distilled H₂O.

47. From our experience, the total packaged cosmids should routinely yield anywhere between $1-2 \times 10^6$ colonies. We avoid doing the infection for the entire bank in a single tube, which always carries the risk of contaminating the whole bank by mishaps and the annoyance of readjusting the time of incubation because of the large volumes. Also, instead of plating out all the infected bacteria (which sometimes takes several hundred plates), one can simply plate out a small aliquot of the bank (e.g., 50,000 colonies) for routine screening and store the rest of the infected *E. coli* cells in 1–2-mL aliquots at –80°C (*see* **Note 48**) as a back-up in case of unstable cosmids.

48. We find that the colony sizes are more homogeneous on rich media such as SB than those plated on LB. This leads to less distortion in representation of the cloned DNA if these colonies are to be stored for future use. Also avoid using ampicillin at concentrations higher than 50 μg/mL, as this tends to favor the growth of cosmids with deletions during propagation. For a quick check on the quality of the bank, pick 10–20 colonies and grow them in 3 mL of SB with 50 μg/mL ampicillin. Analyze these cosmids by "minipreps" and restriction digests. A reasonable bank is that all of the cosmids tested should yield different and complex restriction profiles and no empty vectors. Note that the vector bands could differ from each other by about 650 bp depending on the *cos* sites selected by the λ terminase during in vitro packaging. For storing the genomic bank, the

cells are recovered by washing each 132-mm plate with 3–5 mL of SB (under a bacterial hood to avoid contamination). The bacteria from all the plates are then pooled into one sterile flask or graduated cylinder (if the volume is large, we then harvest the cells by centrifugation and resuspend them in a smaller volume of SB). Add an equal volume of 65% glycerol, $0.1M$ $MgSO_4$, $0.25M$ Tris-HCl, pH 8.0, or on ice, with dimethyl sulfoxide (DMSO) to a final concentration of 7% (DMSO generates heat), swirl to mix and aliquot 1–2 mL into Eppendorf tubes or other convenient storage vials. The cells are flash frozen in liquid nitrogen and stored at –70 to –80°C. Keep one or two tubes for routine use and the rest of the aliquots for long-term storage. To use the library, a small amount of the frozen cells is quickly scraped from the tube with a sterile loop and deposited in a small amount of SB containing 50 µg/mL ampicillin. Make serial dilutions in SB with 50 µg/mL ampicillin to determine titer and plate to a density of about 5000–10,000 colonies per 132-mm Petri plate. When screening these colonies, they should be grown to no more than 1–2 mm in diameter, because overgrown colonies tend to smear during lifting with membrane and the original colonies are sometimes very difficult to regenerate afterward. To further prevent smearing of the colonies, the plates should be cooled at 4°C for 30 min to an hour before doing the lift. Better yet, for almost no smearing, membranes are deposited onto the colonies when they are just barely visible. The plates with the membranes are returned to 37°C for further growth (2 h) and are then transferred to 4°C for another 2 h. The membranes are then removed and placed colony side up (could be invisible) onto fresh SB agar plates supplemented with 50 µg/mL ampicillin for further growth at 37°C (along with the original plates to regenerate the colonies). The membranes are then air-dried for about 30 min, sandwiched between sheets of Whatman 3MM filters, and wrapped in aluminum foil. The colonies are lysed by autoclaving in a dry cycle for 2.5 min at 120°C at 15 lb/in.^{2} (or chemical lysis, *see* **ref. *24***). The lysed colonies are then ready for hybridization (there is no further need to fix the DNA on the membrane after autoclaving). The ends corresponding to the *Arabidopsis* inserts in the positive cosmids can then be isolated and used further as hybridization probes, for example, to search for overlapping clones or as RFLP markers. Some cosmid vectors are equipped with prokaryotic RNA promoters, such as those from bacteriophage T3 and T7, flanking the cloning site to make end-specific RNA probe by in vitro transcription with corresponding polymerases and radioactive ribonucleotides. In those cases in which there is no such provisions in the cosmid, it can be digested with a restriction enzyme which cuts within the polylinker cloning site (e.g., the *Eco*RI site in *cos*Pneo), and also most likely in several places within the *Arabidopsis* insert. The digested DNA is diluted and recircularized by ligation with T4 DNA ligase followed by retransformation into *E. coli*. The recircularized cosmid recovered this way would retain only one extreme end of the original segment of the *Arabidopsis* insert, which can then be excised, purified by gel electrophoresis, and used as a probe to rescreen the cosmid bank (or an *Arabidopsis* genomic bank if desired) for overlapping clones. The opposite extreme end of the insert can be recovered in a similar fashion by choosing another enzyme (e.g., *Xba*I) which cuts in the polylinker cloning site on the other side flanking the *Arabidopsis* insert. Another possibility is to choose a restriction enzyme that does not cut at all within the vector (including the polylinker), then both extremities of the insert can be recovered at the same time. These methods obviate the need for a detailed restriction map or for subcloning pieces of the insert in order to identify those corresponding to the extreme ends.

References

1. Gibson, S. I. and Somerville, C. (1992) Chromosome walking in *Arabidopsis thaliana* using yeast artificial chromosomes, in *Methods in* Arabidopsis *Research* (Koncz, C., Chua, N.-H., and Schell, J., eds.), World Scientific Publishing, Singapore, pp. 119–143.
2. Meyerowitz, E. M. (1994) Structure and organization of the *Arabidopsis thaliana* nuclear genome, in *Arabidopsis* (Meyerowitz, E. M. and Somerville, C., eds.), Cold Spring Harbor, Cold Spring Harbor Laboratory Press, NY, pp. 21–36.
3. Giraudat, J., Hauge, B. M., Valon, C., Smalle, J., Parcy, F., and Goodman, H. M. (1992) Isolation of the *Arabidopsis* ABI3 gene by positional cloning. *Plant Cell* **4,** 1251–1261.
4. Reiter, R. S., Williams, J. G. K., Feldmann, K. A., Rafalski, J. A., Tingey, S. V., and Scolnik, P. A. (1992) Global and local genome mapping in *Arabidopsis thaliana* by using recombinant inbred lines and random amplified polymorphic DNAs. *Proc. Natl. Acad. Sci. USA* **89,** 1477–1481.
5. Nam, H.-G., Giraudat, J., den Boer, B., Moonan, F., Loos, W .D. B., Hauge, B. M., and Goodman, H. M. (1989) Restriction fragment length polymorphism linkage map of *Arabidopsis thaliana. Plant Cell* **1,** 699–705.
6. Chang, C., Bowman, J. L., DeJohn, A. W., Lander, E. S., and Meyerowitz, E. M. (1988) Restriction fragment length polymorphism linkage map for *Arabidopsis thaliana. Proc. Natl. Acad. Sci. USA* **85,** 6856–6860.
7. Hauge, B. M., Hanley, S. M., Cartinhour, S., Cherry, J. M., Goodman, H. M., Koornneef, M., Stam, P., Chang, C., Kempin, S., Medrano, L., and Meyerowitz, E. M. (1993) An integrated genetic/RFLP map of the *Arabidopsis thaliana* genome. *Plant J.* **3,** 745–754.
8. Lister, C.,and Dean, C. (1993) Recombinant inbred lines for mapping RFLP and phenotypic markers in *Arabidopsis thaliana. Plant J.* **4,** 745–750.
9. Lister, C. and Dean, C. (1995) Latest Lister and Dean RI Map. *Weeds World* **2,** 11–18.
10. Schmidt, R., West, J., Love, K., Lenehan, Z., Lister, C., Thompson, H., Bouchez, D., and Dean, C. (1995) Physical map and organization of *Arabidopsis thaliana* chromosome 4. *Science* **270,** 480–483.
11. Vijayraghavan, U., Siddiqi, I., and Meyerowitz, E. (1995) Isolation of an 800 kb contiguous DNA fragment encompassing a 3.5-cM region of chromosome 1 in *Arabidopsis* using YAC clones. *Genome* **38,** 817–823.

12. Putterill, J., Robson, F., Lee, K., and Coupland, G. (1993) Chromosome walking with YAC clones in *Arabidopsis*: isolation of 1700 kb of contiguous DNA on chromosome 5, including a 300 kb region containing the flowering-time gene CO. *Mol. Gen. Genet.* **239,** 145–157.
13. Bent, A. F., Kunkel, B. N., Dahlbeck, D., Brown, K. L., Schmidt, R., Giraudat, J., Leung, J., and Staskawicz, B. J. (1994) RPS2 of *Arabidopsis thaliana:* a leucine-rich repeat class of plant disease resistance genes. *Science* **265,** 1856–1860.
14. Grill, E. and Somerville, C. (1991) Construction and characterization of a yeast artificial chromosome library of *Arabidopsis* which is suitable for chromosome walking. *Mol. Gen. Genet.* **226,** 484–490.
15. Ward, E. R. and Jen, G. C. (1990) Isolation of single-copy-sequence clones from a yeast artificial chromosome library of randomly-sheared *Arabidopsis thaliana* DNA. *Plant Mol. Biol.* **14,** 561–568.
16. Matallana, E., Bell, C. J., Dunn, P. J., Lu, M., and Ecker, J. R. (1992) Genetic and physical linkage of the *Arabidopsis* genome: methods for anchoring yeast artificial chromosomes, in *Methods in* Arabidopsis *Research* (Koncz, C., Chua, N.-H., and Schell, J., eds.), World Scientific Publishing, Singapore, pp. 144–169.
17. Creusot, F., Fouilloux, E., Dron, M., Lafeuriel, J., Picard, G., Billault, A., Paslier, D. L., Cohen, D., Chaboute, M., Durr, A., Fleck, J., Gigot, C., Camilleri, C., Bellini, C., Caboche, M., and Bouchez, D. (1995) The CIC library: a large insert DNA library for genome mapping in *Arabidopsis thaliana*. *Plant J.* **8,** 763–770.
18. Choi, S., Creelman, R. A., Mullet, J. E., and Wing, R. A. (1995) Construction and characterization of a bacterial artificial chromosome library of *Arabidopsis thaliana*. *Plant Mol. Biol. Rep.* **13,** 124–128.
19. Liu, Y., Mitsukawa, N., Vazquez-Tello, A., and Whittier, R. F. (1995) Generation of a high-quality P1 library of *Arabidopsis* suitable for chromosome walking. *Plant J.* **7,** 351–358.
20. Schmidt, R., Putterill, J., West, J., Cnops, G., Robson, F., Coupland, G., and Dean, C. (1994) Analysis of clones carrying repeated DNA sequences in two YAC libraries of *Arabidopsis thaliana* DNA. *Plant J.* **5,** 735–744.
21. Hwang, I., Kohchi, T., Hauge, B., Goodman, H. M., Schmidt, R., Cnops, G., Dean, C., Gibson, S., Iba, K., Lemieux, B., Arondel, V., Danhoff, L., and Somerville, C. (1991) Identification and map position of YAC clones comprising one-third of the *Arabidopsis* genome. *Plant J.* **1,** 367–374.
22. Leung, J., Bouvier-Durand, M., Morris, P.-C., Guerrier, D., Chefdor, F., and Giraudat, J. (1994) *Arabidopsis* ABA response gene *ABI1*: features of a calcium modulated protein phosphatase. *Science* **264,** 1448–1452.
23. Steller, H. and Pirrotta, V. (1985) A transposable P vector that confers selectable G418 resistance to *Drosophila* larvae. *EMBO J.* **4,** 167–171.
24. Sambrook, J., Fritsch, E. F., and Maniatis, T. (1989) *Molecular Cloning: A Laboratory Manual,* 2nd ed. Cold Spring Harbor Laboratory, Cold Spring Harbor, NY.
25. Church, G. M. and Gilbert, W. (1984) Genomic sequencing. *Proc. Natl. Acad. Sci. USA* **81,** 1991–1995.

26. Collins, F. S. and Weissman, S. M. (1984) Directional cloning of DNA fragments at a large distance from an initial probe: a circularization method. *Proc. Natl. Acad. Sci. USA* **81,** 6812–6816.
27. Boyle, J. S. and Lew, A. M. (1995) An inexpensive alternative to glassmilk for DNA purification. *Trends Genet.* **11,** 8.
28. Seed, B., Parker, R. C., and Davidson, N. (1982) Representation of DNA sequences in recombinant DNA libraries prepared by restriction enzyme partial digestion. *Gene* **19,** 201–209.

31

Chromosome Landing Using an AFLP™-Based Strategy

Ann Van Gysel, Gerda Cnops, Peter Breyne, Marc Van Montagu, and Maria Teresa Cervera

1. Introduction

Chromosome walking and, more recently, chromosome landing experiments have been successfully used in higher plants to identify genes correlated with specific genetic loci *(1,2*; for a review, see **ref. 3**). In *Arabidopsis thaliana,* these strategies are mainly used for mapping recessive mutations (*see* **Note 1**) as are described here. Both approaches require a rough map position of the target locus to start with. To this end, the mutant is crossed with a wild-type plant of another genetic background. The resultant F1 plants are self-fertilized and the F2 progeny is used to determine the initial map position relative to molecular markers. DNA markers such as cleaved amplified polymorphic sequences (CAPS), simple sequence repeats (SSRs), and *Arabidopsis* restriction fragment length polymorphism (RFLP) mapping sets (ARMS) were identified between the *Arabidopsis* ecotypes Columbia and Landsberg *erecta* but can also be used in other ecotypes (*see* **refs. 4–6** and Chapters 21 and 22). These codominant markers allow rapid and efficient rough mapping.

The target locus is subsequently fine mapped relative to molecular and genetic markers in order to identify flanking DNA markers. To simplify screening for recombinants, it can be useful to make additional crosses with T-DNA lines *(7,8)* lines containing embryo-lethal mutants *(9)* or visual marker lines (http://nasc.nott.ac.uk/). These markers are linked in repulsion to the target locus and can only be scored in the nonmutant progeny and specific recombinants. Alternatively, molecular markers can be used to score both parental ecotypes. Chromosome walking from these flanking markers allows the isolation of overlapping genomic clones covering the target locus. The walk continues until clones that actually cover the target locus are isolated.

Currently, chromosome landing by the identification of tightly linked molecular markers is the preferred method for gene isolation. This approach requires the generation of a sufficient number of recombinants and the rapid identification of large numbers of markers. The recently developed amplified fragment length polymorphism (AFLP) analysis allows such rapid screening for polymorphisms and identification of linked DNA markers (*see* **ref. 10** and Chapter 19). To obtain tightly linked AFLP markers, different approaches can be used (*see* **Note 2**). Our strategy is based on the selection of recombinants at both sides of the target locus using easily scorable markers. In order to obtain sufficient recombinants, we recommend screening of 1000 F2 individuals (or F3 families). Subsequently, this pool of recombinants is used for the identification of AFLP markers tightly linked to the target locus.

Recently, Thomas et al. *(11)* reported the identification of two AFLP markers on opposite sides of the *Cf-9* gene in tomato, separated by only 15.5 kbp of intervening DNA. In potato, a high resolution map around the *R1* locus was constructed based on AFLP markers *(12)*. Two flanking AFLP markers closely linked to the *Gro1* locus were identified in potato *(13)*. The AFLP technology has been used for the identification of several tightly linked AFLP markers to the *TORNADO-1 (TRN1)* locus in *Arabidopsis thaliana (14)*. These studies illustrate the strength of the AFLP technology and demonstrate how it can be exploited for gene isolation by positional cloning. We describe an AFLP-based strategy for the identification of markers linked tightly to a recessive mutation.

2. Materials
2.1. Genomic DNA Extraction

1. Micromax® centrifuge (IEC, Needham Heights, MA).
2. Thermomixer 5436 (Eppendorf, Hamburg, Germany).
3. Liquid nitrogen.
4. DNA extraction buffer: $0.1M$ Tris-HCl, pH 8.0, $0.5M$ NaCl, $0.05M$ ethylenediaminetetraacetic acid (EDTA), $0.01M$ β-mercaptoethanol.
5. Cetyltrimethylammonium bromide (CTAB) buffer: $0.2M$ Tris-HCl, pH 7.5, $2M$ NaCl, $0.05M$ EDTA, 2% (w/v) CTAB.
6. Sodium dodecyl sulfate (SDS).
7. Isopropanol.
8. 96% Ethanol.
9. 70% Ethanol.
10. TE buffer: 10 mM Tris-HCl, pH 7.4, 1 mM EDTA, pH 8.0.
11. Chloroform/isoamylalcohol (24/1).
12. RNAse (10 mg/mL; Pharmacia, Uppsala, Sweden).

2.2. AFLP Analysis

1. PE9600 cycler (Perkin Elmer, Norwalk, CT; *see* **Note 3**).
2. Vortex (Heidolph Elektro GmbH & CoKG, Kelheim, Germany).
3. Magnetic particle concentrator MPC-M (Dynal, Oslo, Norway).
4. Gel casting system Sequi-Gen GT system 38 × 50 cm (BioRad, Hercules, CA).
5. Autoradiograms (Fuji, Tokyo, Japan).
6. Filter paper 3MM (Whatman, Maidstone, UK).
7. γ-^{33}ATP (Amersham, Aylesbury, UK).
8. Dyna beads M-280 (Dynal).
9. STEX buffer: 100 mM NaCl, 10 mM Tris-HCl, 1 mM EDTA, 0.1% Triton X-100, pH 8.0.
10. TE 10/0.1: 10 mM Tris-HCl, pH 8.0, 0.1 mM EDTA.
11. Formamide loading buffer: 10 mM EDTA, 98% formamide, 0.06% (w/v) bromophenol blue, 0.06% xylene cyanol.
12. Denaturing polyacrylamide gels: 4.5% acrylamide/bisacrylamide (19/1), 7.5M urea, 1 × TBE (10 × TBE for 1 L: 108 g Tris, 55 g boric acid, 40 mL 0.5M ETDA, pH 8.0).
13. Restriction enzymes: *Eco*RI (Eurogentec, Seraing, Belgium) and *Mse*I (New England Biolabs, Beverly, MA).
14. *Mse*I and biotinylated *Eco*RI adaptors (Genset, Paris, France).
15. *Eco*RI, *Mse*I primers (made on a 392 DNA/RNA synthesizer; Applied Biosystems, Foster City, CA).

2.3. Isolation and Cloning of AFLP Markers

1. PE9600 cycler (Perkin Elmer; *see* **Note 3**).
2. Thermomixer 5436 (Eppendorf).
3. Scalpel and needle.
4. High salt buffer: 20% ethanol, 1M LiCl, 10 mM Tris-HCl, pH 7.5.
5. Taq Polymerase (Beckman, Fullerton, CA).
6. 5M NaCl.
7. 96% Ethanol.
8. pGEM-T vector system II (Pharmacia).
9. Luria Bertani (LB) medium (for 1 L: 10 g bacto-tryptone, 5 g bacto-yeast extract, 10 g NaCl, pH 7.5).
10. Plates with LB medium, 100 µg/mL ampicillin (Amp), 12 µg/mL isopropyl–ß–D–thiogalactopyranoside (IPTG; Sigma, St. Louis, MO), 30 µg/mL 5-bromo-4-chloro-3-indolyl-ß-D-galactoside (X-Gal; Duchefa, Haarlem, The Netherlands).
11. Plasmid midi preparation kit (Qiagen, Chatsworth, CA).
12. JM109 competent cells (Promega, Madison, WI).

3. Methods

3.1. Genomic DNA Extraction

DNA is prepared from individual recombinants, the homozygous mutant and both parental ecotypes. The DNA should be sufficiently pure to allow complete digestion. We obtained good quality DNA using the minipreparation method described by Dellaporta et al. *(15)* modified as described.

1. Grind 2- to 3-wk-old leaves (100 mg) in Eppendorf tubes in liquid nitrogen.
2. Add 1 mL DNA extraction buffer and 50 µL 10% SDS.
3. Incubate for 30 min at 65°C and mix regularly.
4. Spin the cell debris in a microfuge for 10 min.
5. Transfer the supernatant (be careful not to transfer cell debris) to a new Eppendorf tube and add an equal volume of isopropanol; mix.
6. Incubate for approx 10 min on ice for DNA precipitation, centrifuge for 10 min in a microfuge and remove supernatant.
7. Dissolve DNA pellet in 400 µL TE (put at 65°C if DNA dissolves slowly).
8. Add 1 µL 10 mg/mL RNAse (optional) and incubate for 10 min at 37°C.
9. Add 400 µL CTAB buffer and incubate for 15 min at 65°C; mix.
10. Add 800 µL chloroform/isoamylalcohol and mix (this step can be repeated).
11. Centrifuge for 5 min in a microfuge.
12. Transfer the aqueous phase to a fresh 2.5-mL Eppendorf tube, add 1.4 mL ethanol (96%), and incubate at room temperature for 15 min.
13. Centrifuge for 10 min in a microfuge, discard the supernatant, and wash pellet with 70% ethanol.
14. Centrifuge and remove ethanol, leaving pellet as dry as possible.
15. Dissolve the DNA in 200 µL H_2O.

3.2. Identification of AFLP Markers Closely Linked to the Target Locus

AFLP markers linked to the target locus in coupling (markers in P1) or in repulsion (markers in P2) can be identified (**Fig. 1A**). To limit the number of AFLP reactions, the DNA of the recombinants is grouped in pools. Recombinants between the locus of interest (TL) and flanking markers C1 or C2 are selected from the recombinant F2 individuals or F3 populations to identify markers in coupling (**Fig. 1A**, left panel). Recombinants at the left of the target locus and homozygous for marker C2 are pooled in group G1 and recombinants at the right and homozygous for C1 are pooled in group G2. To identify markers linked in repulsion, recombinants that lost the mutation and one of the flanking markers (C1 or C2) are selected and pooled in group G3 and G4, respectively (**Fig. 1A**, right panel; *see* **Note 4**). The polymerase chain reaction (PCR) technology limits the size of the pools to maximum 20 individuals; it is possible that a specific marker only present in one individual will not be detected in larger pools *(16,17)*.

We performed the AFLP technique essentially as described by Vos et al. *(16)* outlined in Chapter 19, with minor modifications. *Arabidopsis* has a genome size of approx 120,000 kb, hence a restriction digest with *Eco*RI and *Mse*I generates approx 80,000 *Eco*RI-*Mse*I fragments. Double-stranded adapt-

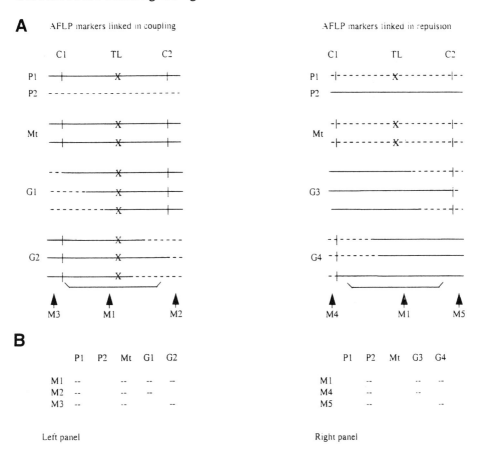

Fig. 1 (**A**) The left and right panels show the identification of AFLP markers linked in coupling and repulsion, respectively. G1 and G2 each show three possible genotypes of recombination events between C1 and TL and C2 and TL, respectively, in homozygous F_2 plants. G3 and G4 represent possible genotypes of F_2 recombinants that lost the mutation but are homozygous for C1 (G4) or C2 (G3), respectively. P1, mutant parental genotype; P2, nonmutant parental genotype; Mt, homozygous mutant; TL, target locus; C1 and C2, flanking markers linked in coupling to the TL. (**B**) Marker M1 shows a band pattern that identifies AFLP marker closely linked to the TL (left panel, in coupling; right panel, in repulsion). Markers M2 and M3 are absent in G2 or G3, respectively, and, therefore, situated further away from the locus than marker M1. Markers M4 and M5 are absent in G4 and G3, respectively, and thus not necessarily tightly linked to the locus.

ers (*Eco*RI adapter is biotinylated) are ligated to the *Eco*RI-*Mse*I fragments. Using streptavidin-coated magnetic beads, the DNA sequence complexity of the DNA mixture is reduced by selecting the biotinylated DNA fragments.

1. Mix the templates gently with 100 μg Dyna beads for 30 min at room temperature to optimize the binding.
2. Collect the beads with the MPC–M magnet and wash them three times with 200 μL STEX buffer. Between washes, the beads are resuspended by vortexing and transferred to a fresh tube.
3. Resuspend the beads with the coupled DNA fragments in 200 μL TE 10/0.1.

The selective AFLP-PCR amplification is carried out using AFLP primers with two (*Eco*RI primer) and three (*Mse*I primer) additional nucleotides at the 3' ends (the *Eco*RI primer is end labeled with γ-^{33}ATP). Approximately 80 fragments ranging from 100–500 bp are amplified and subsequently denatured by adding formamide loading buffer and heating for 3 min at 95°C. The fragments are loaded on denaturing polyacrylamide gels. After electrophoresis, the gels are dried on 3MM filter paper and autoradiographed.

The pooled recombinants and parental ecotypes are screened using different selective AFLP primers to identify markers linked to the target locus (**Fig. 1**). AFLP markers linked in coupling and restricted to the genomic region defined by the markers C1 and C2, will be present in both pools of recombinants (G1 and G2), the homozygous mutant (Mt), and the mutant parental ecotype (P1), but absent in the other parental ecotype (P2) (M1; **Fig. 1**, left panel). Markers within the C1–C2 region and linked in repulsion will be present in both pools of recombinants (G3 and G4) and the nonmutant parental ecotype (P2), but absent in the homozygous mutant and the mutant parental ecotype (M1; **Fig. 1**, right panel). The AFLP markers M2, M3, M4, and M5 are only present in one of the two pools of recombinants and, therefore, not as closely linked to the TL as marker M1. The AFLP primer combinations that produce an M1 AFLP pattern (**Fig. 1B**) are used on the individual recombinants to map the markers correctly around the TL identifying tightly linked markers (*see* **Notes 5** and **6**).

Due to the extensive coverage of the *Arabidopsis* genome with yeast artificial clones (YACs) and cosmid contigs, the possibility exists that contigs are available around the TL. If the mutation is in the same ecotype as the genomic libraries (predominantly Columbia ecotype), these clones can be pooled and used immediately in an AFLP analysis to identify genomic clones containing closely linked AFLP markers. This approach allows fast access to YAC or bacterial artificial clones (BACs) containing the gene. When using an ecotype other than Columbia, this strategy is not necessarily successful, since not all identified AFLP markers will be conserved between the different ecotypes.

3.3. Isolation and Cloning of AFLP Markers

To prevent contamination with closely or comigrating bands, a less dense gel pattern should be generated prior to excision of the AFLP marker. Addition of one extra base to the 3' end of each primer increases selectivity and, therefore, fewer bands are amplified. To reamplify the markers with primers containing one additional nucleotide each, 16 different combinations of primers (+1/+1) need to be tested. If genomic BAC or YAC clones are available in the area of the TL, it is advisable to isolate the AFLP marker from these clones since they contain considerably lower amounts of plant-specific DNA. Therefore, probably no additional bands will be contaminating the isolated AFLP marker, which allows direct sequencing of the AFLP fragment without prior cloning.

1. Determine the correct position of the AFLP marker on the dried polyacrylamide gel by puncturing the corners of the gel and the film with a needle. After exposure, align the autoradiogram on the gel and excise the AFLP bands with a scalpel: Both gel and the 3MM paper are excised.
2. Add 400 μL high salt buffer to the gel fragment and incubate for 1 h at 65°C.
3. Remove the supernatant (containing the eluted DNA) and precipitate the DNA in 0.15M NaCl and 2.5 vol ethanol for 30 min at −70°C.
4. Collect the DNA by centrifugation and resuspend in 20 μL H_2O.
5. Reamplify 1/4 of the resuspended DNA using the following PCR conditions: 35 cycles of 30 s at 95°C, 30 s at 45°C, and 1 min at 72°C.
6. Clone the PCR product in the pGEM-T vector (*see* **Note 7**) in a 3:1 vector/insert molar ratio: 1 μL T4 DNA ligase buffer (10X), 1 μL pGEM-T vector (75 ng), 1 μL PCR product (25 ng), 1 μL T4 DNA ligase (1 U/μL), and 6 μL H_2O.
7. Incubate for 3 h at 15°C.
8. Heat the reaction for 10 min at 70–72°C and allow to cool to room temperature.
9. Gently mix 2 μL of the ligation mixture with a 50-μL aliquot of freshly thawed JM109 competent cells (*see* **Note 7**).
10. Heat shock the cells for 45–50 s at 42°C.
11. Return the tubes on ice for 2 min.
12. Add 1.4 mL LB medium to the tubes.
13. Gently invert the tubes to mix and incubate for 1 h at 37°C.
14. Mix the tube and spread 50 μL of cells onto LB/Amp/IPTG/X-Gal plates.
15. Incubate for 24 h at 37°C (*see* **Note 8**).
16. Prepare the recombinant plasmid for each cloned PCR fragment from approx 10 colonies using the plasmid midi/preparation kit.

It is advisable to sequence the cloned PCR fragments of approx 10 colonies to be confident of cloning the correct AFLP band. When the AFLP marker is isolated without contamination of nearby migrating bands, the sequences of all 10 inserts will be identical. However, comigration of the AFLP marker with underlying bands cannot always be prevented. The sequence of 10 clones will reveal which of the fragments is most abundant and hence, most probably represents the AFLP marker (*see* **Note 9**).

The cloned AFLP markers are used to isolate corresponding genomic clones. If the marker is isolated from YAC or BAC clones and sufficiently pure, a random multiprime probe can be made directly from the AFLP marker without prior cloning and used to screen genomic libraries (*see* **Subheading 3.3.**, ref. *1*, and **Note 9**).

4. Notes

1. With minor modifications, the described approaches can also be used to map dominant mutations.
2. An alternative approach for the identification of tightly linked AFLP markers to a genetic locus has been reported by Liscum and Oeller *(18)*. They suggested screening of at least 250 F_2 individuals for AFLP markers linked in repulsion. Linked polymorphic bands should be present in the nonmutant parental ecotype and absent in most of the mutant F_2 individuals. Screening for absence of bands in the F_2 population allows one to distinguish heterozygotes from homozygotes. However the individual plants cannot be pooled during the initial screening for informative AFLP primer combinations which makes the AFLP work quite labor intensive (*see* **Subheading 3.2.**).
3. Other thermocyclers can be used with equal success (e.g., PCH-3; Techne, Cambridge, UK), although it is advisable to monitor the quality and reproducibility of the amplification reactions in time.
4. One must be aware that the identification of AFLP markers linked in repulsion requires screening of three-fold more F_2 plants compared to screening for markers linked in coupling.
5. Depending on how far the markers C1 and C2 are located from the TL, it can be necessary to redefine the pools based on more closely identified flanking AFLP markers.
6. Screening of a 3-centiMorgan region around the *TRN1* locus with approx 200 *Sac*I/*Mse*I (+2/+2) and 100 *Eco*RI/*Mse*I (+2/+3) AFLP primer combinations revealed respectively 12 and 5 closely linked AFLP markers *(14)*.
7. The pGEM-T vector system II contains the pGEM-T vector, that has a multicloning site between an SP6 and T7 promoter, allowing cloning of PCR fragments. The presence of the *lacZ* sequences allows blue/white colony screening for recombinants. The vector contains a β-lactamase gene rendering ampicillin resistance. JM109 high-efficiency competent cells are provided with the system and recommended for efficient transformation, although other host strains may be used. Selection for transformants should be on LB/Amp/IPTG/X-Gal plates.
8. The characteristics of the PCR fragments cloned into the pGEM-T vector can significantly affect the ratio of blue/white colonies. PCR fragments (usually a multiple of 3-bp length and without stop codons) that are cloned in-frame with the *lacZ* gene can result into an increased amount of blue colonies. In this case, it can be useful to screen both blue and white colonies, since blue colonies can represent the cloning of a PCR fragment in the other orientation compared to white colonies.
9. Cloning can be omitted and direct sequencing of the PCR fragment is feasible if the AFLP marker is not contaminated with nearby or comigrating bands (including underlying nonlabeled fragments). This is likely to be the case when

the bands are isolated from YACs and BACs or from total plant DNA reactions if the band migrates at a sufficient distance from other fragments. Amicon Micropure™ separators (Beverly, MA) enable very efficient purification for direct sequencing of PCR fragments without subcloning.

Acknowledgments

The authors thank Tom Gerats for critical reading of the manuscript and Martine De Cock for assistance with its final preparation. This work was supported by a grant from Vlaams Actieprogramma Biotechnologie (ETC 002) and in part by the European Communities BIOTECH Programme, as part of the Project of Technological Priority 1993–1996. Peter Breyne and Maria Teresa Cervera are indebted to the Vlaams Instituut voor de Bevordering van het Wetenschappelijk-Technologisch Onderzoek in de Industrie for a postdoctoral fellowship and to the European Union for a fellowship from the Human Capital and Mobility program (41AS8694), respectively.

References

1. Meyer, K., Leube, M. P., and Grill, E. (1994) A protein phosphate 2C involved in ABA signal transduction in *Arabidopsis thaliana*. *Science* **264**, 1452–1455.
2. Mindrinos, M., Katagiri, F., Yu, G.-L., and Ausubel, F. M. (1994) The A. thaliana disease resistance gene *RPS2* encodes a protein containing a nucleotide-binding site and leucine-rich repeats. *Cell* **78**, 1089–1099.
3. Tanksley, S. D., Ganal, M. W., and Martin, G. B. (1995) Chromosome landing: a paradigm for map-based gene cloning in plants with large genomes. *Trends Genet.* **11**, 63–68.
4. Konieczny, A. and Ausubel, F. M. (1993) A procedure for mapping *Arabidopsis* mutations using co-dominant ecotype-specific PCR-based markers. *Plant J.* **4**, 403–410.
5. Morgante, M. and Olivieri, A. M. (1993) PCR-amplified microsatellites as markers in plant genetics. *Plant J.* **3**, 175–182.
6. Fabri, C. O. and Schäffner, A. R. (1994) An *Arabidopsis thaliana* RFLP mapping set to localize mutations to chromosomal regions. *Plant J.* **5**, 149–156.
7. Liu, Y.-G., Mitsukawa, N., Oosumi, T., and Whittier, R. F. (1995) Efficient isolation and mapping of *Arabidopsis thaliana* T-DNA insert junctions by thermal asymmetric interlaced PCR. *Plant J.* **8**, 457–463.
8. Van Lijsebettens, M., Wang, X., Cnops, G., Boerjan, W., Desnos, T., Höfte, H., and Van Montagu, M. (1996) Transgenic *Arabidopsis* tester lines with dominant marker genes. *Mol. Gen. Genet.* **251**, 356–372.
9. Franzmann, L. H., Yoon, E. S., and Meinke, D. W. (1995) Saturating the genetic map of *Arabidopsis thaliana* with embryonic mutations. *Plant J.* **7**, 341–350.
10. Vos, P., Hogers, R., Bleeker, M., Reijans, M., van de Lee, T., Hornes, M., Frijters, A., Pot, J., Peleman, J., Kuiper, M., and Zabeau, M. (1995) AFLP: a new technique for DNA fingerprinting. *Nucleic Acids Res.* **23**, 4407–4414.

11. Thomas, C. M., Vos, P., Zabeau, M., Jones, D. A., Norcott, K. A., Chadwick, B. P., and Jones, D. J. G. (1995) Identification of amplified restriction fragment polymorphism (AFLP) marker tightly linked to the tomato *Cf-9* gene for resistance to *Cladosporium fulvum*. *Plant J.* **8,** 785–794.
12. Meksem, K., Leister, D., Peleman, J., Zabeau, M., Salamini, F., and Gebhardt, C. (1995) A high-resolution map of the vicinity of the *R1* locus on chromosome V of potato based on RFLP and AFLP markers. *Mol. Gen. Genet.* **249,** 74–81.
13. Ballvora, A., Hesselbach, J., Niewöhner, J., Leister, D., Salamini, F., and Gebhardt, C. (1995) Marker enrichment and high-resolution map of the segment of potato chromosome VII harbouring the nematode resistance gene *Gro1*. *Mol. Gen. Genet.* **249,** 82–90.
14. Cnops, G., den Boer, V., Gerats, A., Van Montagu, M., and Van Lijsebettens, M. (1996) Chromosome landing at the Arabidopsis *TORNADO1* locus using and AFLP-based strategty. *Mol. Gen. Genet.* **253,** 32–41.
15. Dellaporta, S. L., Wood, J., and Hicks, J. B. (1983) A plant DNA minipreparation: version II. *Plant Mol. Biol. Rep.* **1,** 19–21.
16. Michelmore, R. W., Paran, I., and Kesseli, R .V. (1991) Identification of markers linked to disease-resistance genes by bulked segregant analysis: a rapid method to detect markers in specific genomic regions by using segregating populations. *Proc. Natl. Acad. Sci. USA* **88,** 9828–9832.
17. Williams, J. G. K., Reiter, R. S., Young, R. M., and Scolnick, P. A. (1993) Genetic mapping of mutations using phenotypic pools and mapped RAPD markers. *Nucleic Acids Res.* **21,** 2697–2702.
18. Liscum, M. and Oeller, P. (1996) AFLP: not only for fingerprinting, but for positional cloning. http://carnegiedpb.stanford.edu/methods/aflp.html.

32

Transposon Tagging with *Ac/Ds* in *Arabidopsis*

Deborah Long and George Coupland

1. Introduction

Transposon tagging provides an effective method of isolating plant genes by identifying a mutation tagged by a DNA insertion, and then using the DNA tag to isolate the mutant gene (reviewed in **ref. *1***). The first examples of this approach in plants involved transposon tagging in maize and *Antirrhinum* using endogenous transposons. More recently, maize transposons were shown to be active in a wide range of plant species, and in some of these have facilitated the use of transposon tagging *(2–5)*. In *Arabidopsis, Activator/Dissociation (Ac/Ds)* have been used to isolate recessive loss–of–function mutations *(6–8)*, dominant gain of function mutations *(9)* and to isolate genes based on their patterns of expression *(10,11)*.

In *Arabidopsis, Activator* has been used as a single component system in which the transposon carries the transposase gene required for its own transposition *(12,13)*. This system has the advantage of simplicity; no genetic crosses are required to initiate transposition. More often a two component system is used. The two components are comprised of a transposase gene that can not itself transpose, and of a nonautonomous element, *Ds*, that will transpose in the presence of the transposase gene but not in its absence. *Ds* transposition is activated by crossing plants carrying the transposase gene to those carrying *Ds* and activating transposition in the progeny *(14–17;* **Fig. 1**). The two–component system is more flexible than the use of an autonomous element. For example, the same *Ds* element can be activated by transposase genes engineered to be expressed in different ways. The *Ds* element can be constructed to carry marker genes such as those conferring resistance to antibiotics, so that their inheritance can be followed, or to carry modified β–glucuronidase genes to monitor the expression of plant genes close to the

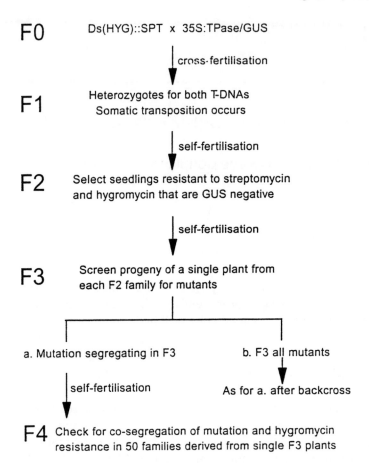

Fig. 1. A genetic strategy used to isolate mutations with the *Ac/Ds* system in *Arabidopsis*, and to determine whether a copy of the transposon is genetically linked to the mutation. This strategy was used and described in **refs. 6** and **7**.

transposon *(18)*. The *Ds* element is also stable until the transposase gene is introduced by crossing, so that lines carrying *Ds* at known positions can be established without the possibility of the element changing position.

Marker genes are used to monitor transposition and to select plants carrying transposed elements. The first type of marker used is an excision marker. The transposon is inserted within the excision marker gene, usually within the untranslated leader, inactivating the gene. Excision of the transposon from this position causes the gene to be activated, and therefore excision can be scored by monitoring expression of the marker gene. The most commonly used marker gene for this purpose is one encoding streptomycin phosphotransferase (SPT) that encodes resistance to the antibiotic streptomycin *(19)*, but one encoding neomycin phosphotransferase (NPT) that encodes resistance to kanamycin is also used *(12,20)*. The second type of marker gene used is within the *Ds* transposon. Typically, this is another resistance gene, conferring resistance to hygromycin *(14,21)*, methotrexate *(16)*, or chlorosulfuron *(17)*. In addition to the marker genes associated with the transposons, the T-DNAs also carry genes enabling their inheritance to be followed. Most constructs contain kanamycin resistance genes to permit selection of transformed plants by the method of Valvekens *(22)*, although the availability of vacuum infiltration transformation *(23)* allows a wider variety of selectable markers to be used. In two component systems, it is often an advantage to be able to identify plants in subsequent generations that do not carry the transposase T-DNA, because any transposed *Ds* elements will be stable in the absence of the transposase gene. Transposase T-DNAs are therefore usually engineered to carry genes such as β–glucuronidase *(15,17)*, enabling plants to be screened for the absence of the transposase containing T-DNA, or an IAAH gene *(14)*, which allows the population to be selected for those individuals that do not carry transposase.

The transposase genes used to drive *Ds* transposition in transposon tagging strategies are usually expressed from either the CaMV 35S or *Ac* promoters. Deletion of the untranslated leader of the *Ac* transposase gene increases the activity of the gene if it is expressed from the *Ac* promoter *(14)*. Transposase genes expressed from the CaMV 35S promoter give rise to excisions early during plant development *(15,21)*. This has two major consequences. Many progeny of a single plant carry the same transposed element, and therefore, in mutant screens, it is usually necessary to take only one plant carrying a transposed element from each family (F_2 in **Fig. 1**). However, early excisions have the advantage that progeny plants rarely inherit an active excision marker gene from one gamete, and a *Ds* in its original position from the other gamete *(21)*. Therefore, over 90% of F_2 plants selected for the presence of a transposed element actually contain a copy of the transposon at a new position. Use of the *Ac* promoter fused to the transposase gene causes excision much later in plant development, so all of the plants from a single family that are selected for the presence of a transposed element can be used in tagging screens. However, in this case only approx 50% actually carry a transposed element *(14,21)*. In the others, *Ds* excised from the marker gene but did not reinsert (or was not inherited after reinsertion) and the other chromosome carries *Ds* at its original position within the T-DNA. In several plant species, including *Arabidopsis*, the majority of transposed *Ds* elements insert at locations genetically linked to their starting position *(24,25)*. Several mutant screens have been performed using approaches in which plants carrying transposed elements were identified by selecting for activity of the excision marker and for expression of the gene carried by the transposon *(6,7)*. In these screens, the majority of insertions are genetically linked to the starting position, increasing the likelihood of tagging genes in this vicinity. An alternative method has been developed in which screens are performed to identify plants carrying insertions unlinked to the starting position. This method involves selecting for the transposed *Ds* element and selecting against both the T-DNA containing the transposase gene and the T-DNA from which the Ds excised *(11,18)*. This method provides a population of insertions throughout the genome rather than directed to the vicinity of the original insertion.

Most transposon tagging experiments performed so far in *Arabidopsis* have not been targeted toward the isolation of a particular gene. Basically, plants carrying transposed elements have been identified, self–fertilized, and their progeny has been screened for mutants. This has led to the isolation of mutants with a wide range of phenotypes. However, transposon tagging in maize can be directed toward the isolation of particular genes by crossing one line carrying transposons to a second line homozygous for a mutant allele of the gene of interest. Progeny F_1 plants that show the mutant phenotype, are likely to have inherited a transposon tagged allele from the wild–type parent (reviewed in **ref. *1***). The efficiency of the identification of tagged alleles can be increased by using a wild–type line with a transposon close to the gene of interest. This approach is difficult to use in *Arabidopsis* because of the relative difficulty of crossing. However, genes have been targeted by simply self–fertilizing plants carrying a transposon genetically linked to the gene of interest *(8)*.

The effectiveness of applying transposon tagging to *Arabidopsis* is also determined by the proportion of mutants that carry a copy of the transposon at the mutant locus. Typically, approx 50% of mutations are likely to be tagged *(6,7)*, although in one population, the proportion of tagged mutations was as low as 7–11% *(26)*. The origin of the mutations which are not tagged is unclear. However, tagged mutations can be reliably identified by testing for genetic linkage between the mutation and the transposon, and by demonstrating that excision of the transposon can result in phenotypic reversion of the mutation.

In this protocol, we describe the methods used to identify plants carrying a transposed *Ds* by selecting such plants with a combination of hygromycin and streptomycin. We also describe the methods used to determine whether a mutation is caused by *D*s insertion, and how to clone the gene affected by such a mutation.

2. Materials
2.1. Selection of Seedlings on Antibiotic Plates
1. Miracloth (Calbiochem, La Jolla, CA).
2. 250-mL Specimen container (Sterilin, R & L Slaughter, Stoumarket, UK).
3. 70% Ethanol.
4. 0.5% Sodium dodecyl sulfate (SDS), 5% sodium hypochlorite.

5. Germination medium (*see* **ref. 22**) containing 1% glucose and streptomycin (175 mg/L; Sigma, St. Louis, MO) and hygromycin (40 mg/L; Calbiochem; *see* **Note 1**).

2.2. Identification of Mutant Phenotypes

1. Germination medium (*see* **ref. 22**).
2. Dissecting microscope.
3. Parafilm.
4. Plantpak trays (22 × 37 cm).
5. *Arabidopsis* soil mix (1 bag Levington M3, 1 bushel of grit, 4.5 g dimilin, King's Horticulture, Colchester, UK).

2.3. Testing Whether a Mutation Is Caused by Transposon Insertion

1. Germination medium (*see* **ref. 22**) with and without 40 mg/L hygromycin.
2. Plantpak trays (22 × 37 cm).
3. *Arabidopsis* soil mix (1 bag Levington M3, 1 bushel of grit, 4.5 g dimilin, King's Horticulture).

2.4 Gene Cloning

2.4.1. Cetryltrimethylammonium bromide (CTAB) Plant DNA Extraction

1. CTAB buffer: 140 mM sorbitol, 220 mM Tris-HCl, pH 8.0, 22 mM EDTA, 800 mM NaCl, 1% N–lauryl sarcosine, 0.8% CTAB, combine and check that the solution is pH 8.0, then autoclave.
2. Mortar and pestle cooled with liquid nitrogen.
3. 50-mL Plastic tube (Corning).
4. Chloroform.
5. Isopropanol
6. TE: 10 mM Tris HCl, pH 8.0, 1 mM EDTA.
7. 4M Lithium acetate.
8. 100% Ethanol.
9. 3M Sodium acetate.

2.4.2. DNA Digestion and Ligation

1. 3M Sodium acetate.
2. Ethanol cooled to –20°C.
3. 5X Ligase buffer (Boehringer, Mannheim, Germany).
4. Ligase enzyme (Boehringer).
5. Phenol:chloroform (1:1).

2.4.3. Inverse Polymerase Chain Reaction (PCR)

1. 1.25 mM dNTPs.
2. 10X PCR buffer.

3. Taq DNA polymerase.
4. Mineral oil.
5. The sequences of the primers used for amplification are as follows:

D73	CGGGATTTTCCCATCCTACTTTCATCCCTG
B34	ACGGTCGGTACGGGATTTTCCCAT
B35	TATCGTATAACCGATTTTGTTAGTTTTATC
E4	AAACGGTAAACGGAAACGGAAACGGT
D71	CGTTACCGACCGTTTTCATCCCTA
B39	TTCGTTTCCGTCCCGCAAGTTAAATA
B38	GATATACCGGTAACGAAAACGAACG
D75	ACGAACGGGATAAATACGGTAATC
B49	ATCTGTCAACAAACATAAGACACAT
D72	GAATCTGTGAACTAACACGGCTGGG
DL6	TTGCTGCAGCAATAACAGAGTCTAGC

2.4.4. Cloning the PCR Fragments

1. Phenol:chloroform (1:1).
2. $3M$ Sodium acetate.
3. 100% Ethanol.
4. 70% Ethanol.
5. $1.25M$ dNTPs.
6. 10X T_4 DNA polymerase buffer (Boehringer).
7. Bovine serum albumin (BSA) 10 mg/L (Promega, Madison, WI).
8. T_4 DNA polymerase (Boehringer).
9. TAE: $0.04M$ Tris acetate; $0.01M$ EDTA, pH 8.0.
10. Qiaex beads (Qiagen, Chatsworth, CA).
11. pKR vector (*see* **ref. *28***).
12. 10X Blunt end ligation buffer: $0.66M$ Tris HCl, pH 7.5, 50 mM MgCl$_2$, 50 mM dithiothreitol (DTT), 5 mM spermidine, 2 mM ATP, 10 mM hexamino cobalt chloride.

3. Methods

3.1. Selection of Seedlings on Antibiotic Plates

A population of 200–500 seeds per F_2 family derived from the cross between *Ac*- and *Ds*-containing lines (**Fig. 1**) is sown on a 9-cm agar plate to isolate plants that are fully resistant to both the excision marker (streptomycin) and the reinsertion marker (hygromycin).

1. Seeds are placed on a sheet of 7 × 7 cm Miracloth that is folded and then closed with a paper clip.
2. Up to 50 of these packets are soaked in a Sterilin 250 mL specimen container in 70% ethanol for 2 min then in 0.5% SDS, 5% sodium hypochlorite for 15 min. At this point, the procedure should continue in a sterile flow hood.

3. The packets should be washed with copious amounts of sterile water (at least 500 mL/50 packets) until all traces of the detergent are gone, then opened with forceps onto Petri plates and allowed to dry before sowing.
4. Once dry, the seeds can be scraped off the Miracloth with a small spatula into the bottom of the empty Petri dish and then sown onto germination medium (GM) medium *(22)* containing glucose, streptomycin, and hygromycin by placing the agar plate on top and inverting the dishes.
5. The plates are sealed with micropore tape and placed at 4°C for 2–7 d to promote germination and then transferred to a growth room at 20°C with 16 h light/8 h dark.
6. After 2–3 wk of growth, seedlings fully resistant to the antibiotics may be identified as fully green plants with expanded cotyledons. These can be tested for the presence of transposase (*see* **Notes 2** and **3**). The selected F2 plants are then transferred to the greenhouse, grown to maturity, and the seeds of individual plants collected in F3 families.

3.2. Identification of Mutant Phenotypes in F3 Families

The identification of mutant phenotypes in seedlings is best observed on agar plates, whereas phenotypes in older plants are easier to identify and handle when grown on soil. Families in which transposase is no longer present are more easily screened as their segregation ratios are more predictable.

1. Screening on plates requires the sterilization of approx 50 seeds and their subsequent sowing on GM medium as described in **Subheading 3.1., step 4**. After 2–7 d at 4°C, the plates are placed in the growth room and observed by eye and with a dissecting microscope every 2–3 d.
2. To grow plants on soil approx 50 seeds are sown on wet filter paper in Petri dishes and wrapped with parafilm before being placed at 4°C for 2–7 d. They are then sown in Plantpak trays on a mixture of wet *Arabidopsis* mix and covered until the first two leaves appear. From this point on they are observed every 2–3 d.

3.3. Testing Whether a Mutation Is Caused by Transposon Insertion

Two methods are used to determine whether a mutation has been caused by the insertion of a transposon. These are genetic linkage of the mutation to the *Ds* element and the reversion of the mutant phenotype to wild–type in the presence of transposase.

3.3.1. Determining Genetic Linkage Between Ds and Mutations Causing Seedling Phenotypes

1. Collect seeds from each member of the affected F_3 family (**Fig. 1**). If every member of the family shows the mutant phenotype, then this line must be backcrossed to wild–type to obtain a segregating family before linkage analysis can be tested.

2. For a family segregating a seedling phenotype, the genotype of 50 family members can be determined by sowing 50 seeds derived from each of the 50 plants on GM plates as in **Subheading 3.1., step 4**.
3. Once the phenotype becomes apparent and the seedlings have a minimum of two leaves, the seedlings should be transferred to GM plates containing 40 mg/L hygromycin.
4. After approx 1 wk, the plants may be scored for their sensitivity to hygromycin which can be seen as a cessation of growth, a darkening of the tissues, and a curling under of the leaves. All mutant plants should be resistant to hygromycin. Families which segregate mutants should also segregate hygromycin resistance. Any families with only wild–type plants should be sensitive to hygromycin.

3.3.2. Determining Genetic Linkage Between Ds and Mutations Causing Mature Plant Phenotypes

1. F_3 families with mutant phenotypes that are apparent only in older plants are most easily analyzed by sowing 50 seeds from each of 50 family members on both GM plates and soil.
2. The plants on GM should be transferred to GM and 40 mg/L hygromycin once they have two leaves, and scored 1 wk later for their sensitivity to the antibiotic.
3. The plants on soil should be scored for the presence of the phenotype once it becomes apparent and compared with the results of the antibiotic tests to determine linkage (as described in **Subheading 3.3.1, step 4**).

3.3.3. Determining Whether Ds Excision Causes Phenotypic Reversion of the Mutation

1. Reversion analysis requires the isolation of a mutant plant that also contains the transposase gene. The seed stocks required to isolate these plants may already be available as the progeny of sibling F_2 plants (**Fig. 1**). If not, then the mutant line must be crossed to a line carrying transposase.
2. Mutant plants carrying transposase (identified by checking for the presence of either the GUS or IAAH markers) are self–fertilized and the seed collected.
3. Large numbers of seeds are sown on either agar plates or soil, whichever is appropriate to identify for revertant plants showing a wild–type phenotype.
4. Putative revertant plants should be tested for the presence of transposase as a wild–type plant that does not contain the transposase gene must have inherited a revertant allele, rather than be the result of somatic reversion (*see* **Note 4**).
5. The DNA present after *Ds* excision should be sequenced to demonstrate the presence of an alteration in the DNA sequence (*see* **Note 5**).

3.4. Gene Cloning

A tagged gene is isolated by first extracting DNA from the mutant line then using the inverse PCR technique to isolate fragments of genomic DNA from either side of the *Ds* element (*see* **ref. 27**). Nested primers are used to confirm that the fragment obtained is correct.

Ac/Ds Transposon Tagging

3.4.1. CTAB Plant DNA Extraction

1. Grind 2–5 g of frozen leaves to a very fine powder using N_2 cooled mortar and pestle. Transfer to a 50-mL tube containing 25 mL CTAB buffer.
2. Incubate at 65°C with occasional vigorous shaking for 20 min. Add 10 mL of chloroform, shake well, and place on an inverter at room temperature for 20 min.
3. Centrifuge at 2000g for 10 min to resolve the phases.
4. Transfer the aqueous phase to a fresh tube, add 17 mL of isopropanol, mix, and place on ice for 10 min.
5. Centrifuge at 2000g for 5 min to collect the precipitate.
6. Drain the liquid and dry the sides of the tube, but do not dry the pellet.
7. Add 4 mL of TE and dissolve the precipitate by gentle pipeting.
8. Add 4 mL of 4M LiAc, incubate on ice for 20 min.
9. Centrifuge 2000g for 10 min.
10. Transfer supernatant to a fresh tube, add 16 mL of ethanol, and place on ice for 20 min.
11. Centrifuge 2000g for 10 min to collect the precipitate.
12. Drain liquid and dry the sides of the tube, but do not dry the pellet.
13. Dissolve DNA in 3 mL of TE by gentle pipeting.
14. Add 300 µL 3M NaOAc and place in a 15-mL tube.
15. Extract with an equal volume of phenol, then phenol:chloroform, then chloroform. Add 2 vol of ethanol and place on ice for 5 min.
16. Centrifuge at least 5 min to collect the precipitate.
17. Evaporate as much ethanol as possible without drying the pellet out. The pellet will be very gelatinous.
18. Dissolve in a minimum volume of TE, usually at least 100 µL (*see* **Note 6**).

3.4.2. DNA Digestion and Ligation

1. Digest 2 µg of DNA with an appropriate enzyme (*see* **Note 7**).
2. Precipitate the DNA with 0.1 vol of 3M NaOAc and 2.5 vol EtOH at –20°C for 20 min, then spin the pellet at 4°C for 20 min.
3. Dissolve the pellet in 40 µL of water.
4. For the ligation, take 20 µL of the dissolved pellet and add to 60 µL of 5X ligase buffer, 1 µL of ligase, and 220 µL of water.
5. Ligate overnight at 15°C.
6. Extract with an equal volume of phenol:chloroform.
7. Precipitate the DNA with 0.1 vol of 3M NaOAc and 2.5 vol EtOH at –20°C for 20 min, then spin the pellet at 4°C for 20 min.
8. Dissolve the pellet in 10 µL water.

3.4.3. PCR

1. Into a 0.5-mL tube, add 2.5 µL DNA, 56.5 µL water, 16 µL dNTPs (1.25 mM), 10 µL 10X PCR buffer (with magnesium added), 5 µL each of 20 µM primers and 0.5 µL of Taq DNA polymerase. Frequently used primers are shown in **Fig. 2** (*see* **Note 8**).
2. Layer with 100 µL mineral oil.

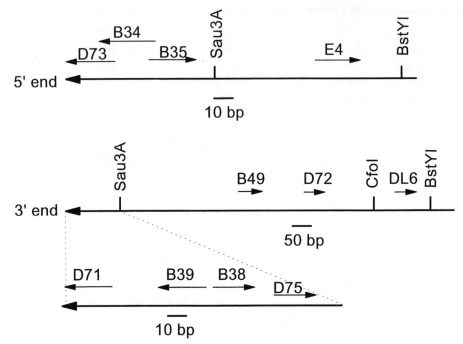

Fig. 2. Primers used to amplify DNA adjacent to *Ds* by IPCR. The 5' end of the transposon is shown at the top of the figure, and is defined as the end from which transcription of the *Ac* transposase gene is initiated. The large arrow head denotes the 11-bp inverted repeat at the very end of the transposon. The primers used are shown as arrows, and arenumbered. Their sequence is presented in **Subheading 2.4.3.** The *Sau*3A and *Bst*YI restriction sites used to cleave the DNA prior to ligation and PCR are shown (*see* **Note 7**). The 3' end is shown in small and large scale. The inverted repeat of the transposon, the primers, and restriction sites are shown as for the 5' end.

3. Run on the following program: 94°C for 4 min; 34 cycles of 94°C for 1 min, 55°C for 2 min, 72°C for 3 min; 72°C for 3 min; then 4°C for up to 24 h.
4. Run out 10 µL of the reaction mixture on a 2% agarose gel.
5. If amplified fragments are detected, then take 1 µL of the reaction mixture and dilute it 1000-fold. Use 2.5 µL of this diluted mixture in the PCR reaction (*see* **Subheading 3.4.3., step 3**) with the appropriate nested primers (*see* **Fig. 2**) and an increased annealing temperature of 65°C (as opposed to the 55°C used previously). A fragment that is smaller by the predicted amount should be detected when the products are analyzed on an agarose gel.

3.4.4. Cloning the PCR Fragments

1. Pool the products of three PCR reactions. Add an equal volume of phenol:chloroform (1:1), mix well, and microcentrifuge for 3 min.

2. Remove the aqueous phase to a fresh tube and add 0.1 vol of $3M$ NaOAc and 2.5 vol EtOH.
3. Precipitate at $-20°C$ for at least 30 min, then microcentrifuge at $4°C$ for 20 min.
4. Wash the pellet with cold 70% EtOH then allow it to air dry.
5. Redissolve in 20 μL water.
6. To blunt end the fragments add to the 20 μL of DNA, 24 μL dNTPs (1.25 mM stock), 5 μL 10X T_4 buffer, 0.4 μL BSA, 1 μL T_4 DNA polymerase in a total volume of 50 μL.
7. Incubate at $37°C$ for 30 min.
8. Heat at $70°C$ for 10 min.
9. Gel purify the fragments by slowly running them on a 1X TAE gel made up with 2% agarose.
10. Cut out the bands and extract them from the gel with Qiaex beads (Qiagen).
11. The isolated fragments can now be ligated into the pKR vector *(28)* after it has been digested with *Eco*RV or into any standard vector which has been digested with an appropriate enzyme.
12. The ligation should be carried out in a final volume of 10 μL. Use a 3:1 molar ratio of insert:vector where the vector is approx 150–300 ng.
13. Mix DNA and water and heat to $45°C$ for 5 min. Then put on ice for 5 min.
14. Add 1 Weiss unit of T_4 DNA ligase and 1 μL of 10X blunt-end ligation buffer.
15. Incubate at $16°C$ overnight.
16. Heat kill the ligase by incubation at $65°C$ for 10 min.
17. Transform *Escherichia coli* with 2 μL of ligation mix, and select ampicillin-resistant transformants.
18. To confirm that the correct fragment has been isolated, PCR may be performed on a 1000-fold dilution of a DNA miniprep of the cloned fragment, (*see* **Subheading 3.4.3.**) to ensure the presence of the original primer sites.
19. Sequencing of the isolated fragments will confirm the presence of some *Ds* sequence at the end of the genomic fragment.
20. The use of the cloned fragment as a probe on a southern blot of wild–type DNA and the tagged mutant line should show a band shift due to the presence of the *Ds* element in the mutated line.

4. Notes

1. Variable results were achieved using hygromycin from different suppliers. A concentration of 40 mg/L works well for that supplied by Calbiochem, but if hygromycin is used from other suppliers, it will be necessary to empirically determine the most effective concentration.
2. F_2 plants selected as carrying a transposed copy of the *Ds* element are usually tested to determine whether they carry a copy of the transposase T-DNA. If the transposase is marked with a β glucuronidase (GUS) gene, then the seedlings are removed form the agar plates and incubated at $37°C$ for 10 min in small Petri plates with a small volume (just enough to wet the roots) of 50 mM NaPO4, pH 7.0, 0.1% Triton X–100, and 0.5 mg/mL 5–Bromo–4–chloro–3–indoyl–beta–D–

glucuronide cyclohexylammonium salt (this must be dissolved in a small volume of dimethyl formamide before addition to the substrate solution, i.e., 50 mg dissolved in 1 mL). The roots of plants carrying the transposase marker will become blue. All of the seedlings should be taken out of the solution and placed onto GM plates without antibiotics until the plants are transplanted to the greenhouse.

3. When a transposase source marked with the IAAH gene is used, then fully green seedlings should be transferred to segmented Petri dishes (multi–well) containing GM supplemented with 0.4 mg/L α–napthaleneacetamide (NAM). The stock solution is made up of 20 mg NAM dissolved in 1 mL of dimethyl sulfoxide. Each seedling should be placed in a separate segment as the phenotype of neighboring plants is affected. The roots of seedlings carrying the IAAH gene will become very hairy. All seedlings may be transplanted directly to the greenhouse.

4. Identification of revertant plants that do not carry the transposase gene is straightforward if the transposase T-DNA carries the GUS gene, and the parental mutant plant was heterozygous for the transposase T-DNA. Mature, wild–type plants are tested for the presence of the GUS gene on the transposase T-DNA. The presence of GUS may be quickly tested by putting a small amount of leaf tissue into an Eppendorf tube with 300 µL of the substrate solution described in **Note 2** and grinding the tissue with an Eppendorf pestle until chlorophyll is apparent in the solution. This should be incubated at least 30 min at 37°C until a blue color is seen in some of the tubes.

5. The presence of revertant alleles is usually confirmed by amplifying the excision site of the transposon and demonstrating that this has a different DNA sequence to that of wild–type. Primers can be designed from the sequence of DNA flanking the *Ds* insertion as described below and the site of excision sequenced. A change in the sequence from wild–type denotes a footprint that signifies an imperfect excision of the *Ds* element upon reversion. This rules out the possibility that the "revertant" is the consequence of the experiment becoming contaminated with a seed of a wild–type plant, or of cross–fertilization of a mutant by a wild–type plant.

6. This method yields approx 20 µg of DNA/gram of rosette leaf tissue. The DNA is quantified on an agarose gel by comparison with known amounts of lambda DNA.

7. The enzyme used is usually *Sau*3A or *Bst*YI or *Bst*YI + *Bcl*I or *Cfo*I (**Fig. 2**). The restriction enzyme used is determined by the pattern of sites in the flanking DNA, however as this is not usually known prior to the IPCR, the appropriate enzyme is determined empirically by testing which produces a final PCR product. Generally *Sau*3A or *Bst*YI are tested first.

8. Combinations of the primers shown in **Fig. 2** are used for PCR. For example, to amplify fragments adjacent to the 3' end, DNA prepared after a *Sau*3A digestion could be used in a PCR with primers B38 and B39 and subsequently reamplified with the nested primers D71 and D75. To amplify fragments adjacent to the 5' end, the same DNA can be used in a PCR with primers D73 and B35 and reamplified with B34 and B35.

References

1. Walbot, V. (1992) Strategies for mutagenesis and gene cloning using transposon tagging and T–DNA insertional mutagenesis. *Ann. Rev. Plant Physiol. Plant Mol. Biol.* **43,** 49–82.
2. Chuck, G., Robbins, T., Nijjar, C., Ralston, E., Chourtney–Gutterson, N., and Dooner, H. K. (1993) Tagging and cloning of a petunia flower color gene with the maize transposable element *Activator*. *Plant Cell* **5,** 371–378.
3. Aarts, M. G. M., Dirkse, W. G., Stiekema, W. J., and Pereira, A. (1993) Transposon tagging of a male sterility gene in *Arabidopsis*. *Nature* **363,** 715–717.
4. Jones, D. A., Thomas, C. M., Hammond Kosack, K. E., Balintkurti, P. J., and Jones, J. D. G. (1994) Isolation of the tomato *Cf–9* gene for resistance to *Cladosporium fulvum* by transposon tagging. *Science* **266,** 789–793.
5. Whitham, S., Dineshkumar, S. P., Choi, D., Hehl, R., Corr, C., and Baker, B. (1994) The product of the tobacco mosaic virus resistance gene *N*, similarity to *TOLL* and the interleukin–1 receptor. *Cell* **78,** 1101–1115.
6. Bancroft, I., Jones, J. D. G., and Dean, C. (1993). Heterologous transposon tagging of the *DRL1* locus in *Arabidopsis*. *Plant Cell* **5,** 631–638.
7. Long, D., Martin, M., Sundberg, E., Swinburne, J., Puangsomlee, P., and Coupland, G. (1993a) The maize transposable element system *Ac/Ds* as a mutagen in *Arabidopsis*: identification of an *albino* mutation induced by *Ds* insertion. *Proc. Natl. Acad. Sci. USA* **90,** 10,370–10,374
8. James, D. W., Jr., Lim, E., Keller, J., Plooy, I., Ralston, E., and Dooner, H. K. (1995). Directed tagging of the *Arabidopsis FATTY ACID ELONGATION1 (FAE1)* gene with the maize transposon *Activator*. *Plant Cell* **7,** 309–319.
9. Wilson, K., Long, D., Swinburne, J., and Coupland, G. (1996) A *Dissociation* insertion causes a semidominant mutation that increases expression of *TINY*, an *Arabidopsis* gene whose product contains an APETALA2 domain. *Plant Cell* **8,** 659–671.
10. Smith, D. L. and Fedoroff, N. V. (1995). *LRP1*, a gene expressed in lateral and adventitious root primordia of *Arabidopsis*. *Plant Cell* **7,** 735–745.
11. Springer, P. S., McCombie, W. R., Sundaresan, V., and Martienssen, R. A. (1995). Gene trap tagging of *Prolifera*, an essential MCM2–3–5–like gene in *Arabidopsis*. *Science* **268,** 877–880.
12. Schmidt, R. and Willmitzer, L. (1989) The maize autonomous element *Activator (Ac)* shows a minimal germinal excision frequency of 0.2–0.5% in transgenic *Arabidopsis thaliana* plants. *Mol. Gen. Genet.* **220,** 17–24.
13. Dean, C., Sjodin, C., Page, T., Jones, J., and Lister, C. (1992) Behaviour of the maize transposable element *Ac* in *Arabidopsis thaliana*. *Plant J.* **2,** 69–82,
14. Bancroft, I., Bhatt, A. M., Sjodin, C., Scofield, S., Jones, J. D. G., and Dean, C. (1992) Development of an efficient two–element transposon tagging system in *Arabidopsis thaliana*. *Mol Gen. Genet.* **233,** 449–461.
15. Swinburne, J., Balcells, L., Scofield, S., Jones, J. D. G., and Coupland, G. (1992) Elevated levels of *Activator* transposase mRNA are associated with high frequencies of *Dissociation* excision in *Arabidopsis*. *Plant Cell* **4,** 583–595.

16. Grevelding, C., Becker, D., Kunze, R., Von Menges, A., Fantes, V., Schell, J., and Masterson, R. (1992) High rates of *Ac/Ds* germinal transposition in *Arabidopsis* suitable for gene isolation by insertional mutagenesis. *Proc. Natl. Acad. Sci. USA* **89,** 6085–6089.
17. Honma, M. A., Baker, B. J., and Waddell, C. S. (1993) High frequency germinal transposition of Ds^{ALS} in *Arabidopsis. Proc. Natl. Acad. Sci. USA* **90,** 6242–6246.
18. Sundaresan, V., Springer, P., Volpe, T., Haward, S., Jones, J. D. G., Dean, C., Ma, H., and Martienssen, R. (1995) Patterns of gene action in plant development revealed by enhancer trap and gene trap transposable elements. *Genes Dev.* **9,** 1797–1810.
19. Jones, J. D. G., Carland, F., Maliga, P., and Dooner, H. (1989) Visual detection of transposition of the maize element *Activator (Ac)* in tobacco seedlings. *Science* **244,** 204–207.
20. Baker, B., Coupland, G., Fedoroff, N., Starlinger, P., and Schell, J. (1987) Phenotypic assay for excision of the maize controlling element *Ac* in tobacco. *EMBO J.* **6,** 1547–1554.
21. Long, D., Swinburne, J., Martin, M., Wilson, K., Sundberg, E., Lee, K., and Coupland, G. (1993b) Analysis of the frequency of inheritance of transposed *Ds* elements in *Arabidopsis* after activation by a CaMV 35S promoter fusion to the *Ac* transposase gene. *Mol. Gen. Genet.* **241,** 627-636.
22. Valvekens, D., Van Montagu, M., and Van Lijsebettens, M. (1988). *Agrobacterium tumefaciens*–mediated transformation of *Arabidopsis* root explants using kanamycin selection. *Proc. Natl. Acad. Sci. USA 85,* 5536–5540.
23. Bechtold, N., Ellis, J., and Pelletier, G. (1993) In planta *Agrobacterium* mediated gene transfer by infiltration of adult *Arabidopsis thaliana* plants. *C R Acad. Sci. Paris, Sci. de la Vie/Life Sci.* **316,** 1194–1199.
24. Bancroft, I. and Dean, C. (1993). Transposition pattern of the maize element *Ds* in *Arabidopsis thaliana. Genetics* **134,** 1221–1229.
25. Keller, J., Lim, E., and Dooner, H. K. (1993) Preferential transposition of *Ac* to linked sites in *Arabidopsis. Theor. Appl. Genet.* **86,** 585–588.
26. Altmann, T., Felix, G., Jessop, A., Kauschmann, A., Uwer, U., Pena–Cortes, H., and Willmitzer, L. (1995) *Ac/Ds* transposon mutagenesis in *Arabidopsis thaliana*: mutant spectrum and frequency of *Ds* insertion mutants. *Mol. Gen. Genet.* **247,** 646–652.
27. Ochman, H., Gerber, A. S., and Hartl, D. L. (1988) Genetic applications of an inverse polymerase chain–reaction. *Genetics* **120,** 621–623.
28. Waye, M. M. Y., Verhoeyen, M. E., Jones, P. T., and Winter, G. (1985) *EcoK* selection vectors for shotgun cloning into M13 and deletion mutagenesis. *Nucl. Acids Res.* **13,** 8561–8571.

33

Transposon Tagging with the *En–I* System

Andy Pereira and Mark G. M. Aarts

1. Introduction

Transposon tagging *(1)* is a method for the isolation of genes, which display a mutant phenotype. Mutants that contain the transposon insert are "tagged" with the transposon and often display the characteristic chimeric or variegated phenotype owing to transposase-mediated excision. With the help of the transposon sequences, the mutant allele and then the corresponding wild type gene can be isolated *(2)*.

Two transposable element systems from maize have been successfully used for transposon tagging in heterologous plants *(3)*. These are the *Activator–Dissociation (Ac–Ds)* and the *Enhancer–Inhibitor (En–I)* two–component transposable element systems. The *En–I (4)* system is genetically and molecularly *(5–8)* equivalent to the *Suppressor–mutator (Spm–dSpm) (9)* system, the names merely refer to their difference in origin. Accordingly we refer to the *En–I* system *(6)* described here for tagging.

The autonomous 8.3-kbp long *En/Spm* element *(7)* codes for two gene products *TnpA* and *TnpD*, which are required for transposition *(10,11)*. The nonautonomous *I/dSpm* components are usually deletion derivatives of the autonomous *En/Spm* components *(12)*, which lack specific gene functions and therefore cannot transpose by themselves. Members of the *En–I(Spm)* transposable element family contain a 13-bp terminal inverted repeat and create a 3-bp target site duplication on insertion. Imprecise excision leaves a footprint behind, which may generate a mutant or new allele.

Subsequent to the molecular isolation and characterization of the maize transposable elements, transposon tagging *(1)* was employed for the isolation of a number of genes in maize *(2)* using established transposable element containing mutants. The property of predominant transpositions to closely

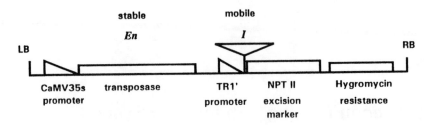

Fig. 1. The "in *cis* two element *En–I* system." The T-DNA contains a hygromycin-resistance marker gene for transformation and segregation analysis and a kanamycin-resistance (NPTII) gene as excision marker. The *En* transposase genes are under control of the CaMV 35S promoter, the mobile *I* element is originally inserted in the NPTII gene, but in the plant it transposes into the genome. LB and RB are left and right T-DNA borders, respectively.

linked sites by members of the *Ac–Ds* and *En–I(Spm)* systems, was later employed in improved tagging strategies *(2)* to isolate new mutants of specific genes using linked transposons.

The autonomous *Ac* *(13)* and *En* *(14)* elements were shown to transpose actively in tobacco. In contrast to other plants *(3)*, the *Ac–Ds* system did not function effectively in *Arabidopsis* *(15,16)* whereas the *En–I* system revealed favorable characteristics for transposon tagging *(17,18)*.

The transposon system described here is based on an "in cis two component *En–I*" construct *(18)*, which has a nonmobile *En*–transposase under control of the CaMV 35S promoter and a mobile *I* element (**Fig. 1**). A transformant obtained by this construct displayed efficient *I* transposition and led to the isolation of a number of tagged genes *(19,20)* in later generations. This transformant contained two T-DNA inserts containing the 35S–*En* element. The TEn5 insert comprising five tandem T-DNA copies is about five times as active as the two-copy TEn2 insert in inducing the transposition of *I* elements *(18)*. These multiple T-DNA inserts yielded many (ranging from 2–10) transposed *I* elements in the genome. Continuous transposition has been observed over several (5–10) generations, and has resulted in the generation of a number of independent lines bearing unique transposition events. The frequency of independent transpositions *(itf)* ranges between 5–30% *(18)* in different progenies, so that plants with about 10 *I* elements contain at least one new transposed *I* element.

The *En* transposase under control of the CaMV 35s promoter exhibits "constitutive" transposition right through development of the plant. The high incidence of independent *I* inserts is due to transpositions late in development giving rise to flowers or inflorescence sectors yielding novel inserts in the progeny. In contrast for *Ac*, transposase overexpression with the CaMV 35S promoter leads to high levels of transposase *(21)* which inhibit later transposition. Also for the *Ac–Ds* system *(22)*, selection for transposition is required, although because of the efficient reinsertion of excised *I* elements, independent transposed *I* elements are easily obtained.

Subheadings 3.1. and **3.2.** describe genetic strategies for random and targeted/directed tagging, and **Subheading 3.3.** gives a detailed protocol on the analysis of mutants using the molecular techniques which are described in **Subheadings 3.4.** and **3.5.**

2. Materials

1. Greenhouse or growth chamber facilities for growing transgenic *Arabidopsis*, as required by national rules, which should include autoclaving of plant waste and exclude dispersal of seed and pollen by insects.
2. The *En–I* transposon system in *Arabidopsis thaliana* is in the ecotype Landsberg *erecta*. The transposase TEn2 locus is on chromosome 1 closely linked to *cer1* and the TEn5 locus on chromosome 2 between markers m220 and m323 *(23)*. *I* element containing lines will be available through the Nottingham Arabidopsis Stock Center.
3. Hygromycin selection media: Murashige and Skoog (MS) medium (Sigma, St. Louis, MO), 0.7% agar, no sucrose, hygromycin 20 mg/L; seeds placed on this media do not need sterilization (resistant seedlings are quickly transferred to soil).
4. DNA extraction buffer *(24)*: $0.3M$ NaCl, 50 mM Tris-HCl pH 7.5, 20 mM ethylenedamine tetra-acetic acid (EDTA), 2% sarkosyl, 0.5% sodium dodecyl sulfate (SDS), $5M$ urea, 5% phenol (equilibrated in $0.5M$ Tris-HCl pH 8.0). The first five ingredients are mixed as a 2X stock solution, and urea and phenol are added before use.
5. DNA probes: *I* element specific, about 280 bp of the left terminus (to the *Sal*I restriction site) *(6)*.
6. *I* Primer sequences *(8)*:

	5'	3'
1st PCR:	ILT1: GAA TTT AGG GAT CCA TTC ATA AGA GTG T	
	ILT2: TTG TGT CGA CAT GGA GGC TTC CCA TCC GGG GA	
2nd PCR:	IRT1: ATT AAA AGC CTC GAG TTC ATC GGG A	
	IRT2: AGG TAG TCG ACT GAT GTG CGC GC	
3rd PCR:	ITIR: GAC ACT CCT TAG ATC TTT TCT TGT AGT G	

7. Equipment: Eppendorf centrifuge, polymerase chain reaction (PCR) machine, temperature controlled incubators.

3. Methods

3.1. Random Tagging with the En–I system

En–I transposon lines containing various numbers of transposed *I* elements *(18)* can be continuously generated by selfing *En*–transposase bearing genotypes. Most of the mutants obtained by this strategy have been represented more than once in the progeny, suggesting a late sector in the parent plant giving rise to a few homozygous progeny.

1. To generate populations containing new inserts, select lines homozygous for the TEn transposase that display 100% hygromycin resistance. Because most individual plants carry new independent inserts, selfed progeny will directly yield plants homozygous for some inserts (*see* **Note 1**).
2. Harvest siliques from different parts of a plant to produce a population of independent insertions. The seeds from the individual siliques may be kept separate, which will yield families diverged for *I* element inserts and aid in recovering lethal mutants (*see* **Note 2**).
3. A systematic screen of general mutants can be done as described for the T-DNA *(25)* and *Ac–Ds* screens *(26)*. Primarily a screen in vitro will reveal form and morphology mutants with the advantage that seedling lethality can also be scored. Greenhouse-based screens at densities up to 4000 plants/m^2 will reveal late phenotypes like sterility (*see* **Note 3**).

3.2. Targeted tagging with Linked I Elements

To select transposon-tagged mutants of specific mapped loci, use of transposons linked to the target gene will enable tagging at a higher frequency than would be expected at random. The efficiency of this step depends on the extent of linked transposition and the frequency of independent transpositions. In *Arabidopsis* linked transposition with the *En–I* system is not as pronounced as for the *Ac–Ds* system *(27)*. Approximately 20% of *I* transpositions are linked to positions approx 20 cM from the donor locus. The major advantage of targeted tagging with the *En–I* system is the ease in producing a large number of new transpositions without selection.

1. As source of *I* transposons at defined positions in the genome, a number of mapped *I* elements as "jumping pads" are being made available through the stock centers. To produce such *I* jumping pads (*see* **Note 4**), a set of lines containing a low number of *I* elements are outcrossed to Landsberg *erecta* and selfed to obtain F$_2$ families segregating for the *En* transposase locus (TEn5 or TEn2).
2. The line containing the target locus linked *I* element and the TEn transposase source (*see* **Note 5**) is confirmed by DNA blot hybridization prior to crossing. Preferably the hemizygous TEn5 locus should be used which confers high transposition with the option of recovering stable tagged mutants in the progeny.
3. Cross to produce an F$_1$ population for directed tagging, represented by:

    a/a; ms/ms; –/– × A/A_linked *I*; Ms/Ms; TEn/–

    where *A* is the target locus. The use of the male sterile *(ms/ms)* female plant will enable the production of large amount of crossed seeds. About 5000 seed per plant can be obtained in this way with a nondeleterious mutant phenotype (*see* **Note 6**). Various modifications of the simplified cross are possible depending primarily on the genotype availability and fertility of the target mutant. Another way of targeted or "directed" tagging with linked transposons can be carried out by producing a selfed population of inserts enriched for transpositions near the target locus. Independent transpositions can be selected by harvesting and keeping individual siliques or inflorescences separate and planting out a small sample from this seed. This strategy is particularly useful for loci which may have a lethal phenotype. Families that display the mutant in progeny can then be analyzed for heterozygosity at the target locus.

3.3. Analysis of a Putative Transposon Tagged Mutant

After isolation of a mutant from the transposon lines following a random or targeted/directed tagging strategy, genetic analysis of the mutants is necessary to judge whether it is tagged with the transposon. For common recessive mutants, which do not affect fertility or viability, the following strategy is applicable.

In the presence of *En* transposase, the mutant should display chimerism or revertant sectors for a cell–autonomous gene action. In the absence of somatic reversion, germinal reversion in the progeny can also be sought for to give the first indication of a tagged mutant *(2)*. For the analysis it is very important to use as few generations as possible avoiding secondary transposition (*see* **Note 7**). An adequate proof for cloning the correct gene can be obtained by correlating the sequence of excision alleles with the plant phenotype.

1. Transplant mutant plants to large (5 cm) pots that will enable longer growth for harvest of leaf material and making necessary crosses. Harvest young leaves in branches or inflorescences with unopened flowers for DNA isolation.
2. Use the mutant for making crosses:
    Cross A: with wild-type, e.g., Landsberg *erecta*.
    Cross B: with tester line containing mutant allele (target locus) in targeted/directed tagging strategy.
    Cross C: with a transposase line if the mutant does not carry the transposase.
3. Perform Southern blot analysis *(28)* using isolated mutant plant DNA cut with an enzyme like *Hin*dIII, *Bgl*II, or *Eco*RI, which do not cut in the *I* element. Hybridize with an *I*-specific probe, revealing new transpositions, and after stripping off the probe, rehybridize with a TEn flanking probe which will indicate transposase genotype.
4. Follow inheritance and segregation of the mutant phenotype in the selfed progeny using DNA blot hybridization with *I* element and TEn flanking probes. Search for revertants, first with small populations of 24–48 plants and, if unsuccessful, increase to 500–1000 progeny. In a directed/targeted tagging strategy, the Cross B results will confirm the targeted mutation.
5. F_2 population segregation analysis: Use an F_2 population of about 50 plants from Cross A to identify an *I* insertion cosegregating with the mutation by DNA blot hybridization (*see* **Note 8**). Preferably use a population without transposase segregating 3:1 for wild-type:mutant. From Cross C, use the F_2 population segregating for transposase to screen for mutant plants with somatic reversion.

6. Isolate plant DNA flanking the identified cosegregating *I* element by IPCR *(28)* preferably from a plant lacking the transposase locus and carrying few *I* elements. For these genotypes, backcross generations (from Cross A) may be necessary.
7. Confirm the cloning of genomic DNA flanking the transposon by hybridizing the *I* element flanking probe to a Southern blot containing DNA from mutant and revertant plants to reveal a "band shift" caused by *I* excision in the revertant allele.
8. Determine the DNA sequence of the cloned IPCR fragments. Design PCR primers flanking the *I* insert; PCR amplify, clone, and sequence the target site sequences from wild-type, revertant, and mutant alleles without inserts. All revertants should have at least one allele with wild-type (or nonmutant) DNA sequence. All mutants should have only alleles featuring frame shifts, aberrant termination, or amino acid exchanges.
9. Isolate genomic and cDNA clones from appropriate λ phage libraries using the IPCR products as probes, and determine their DNA sequence.

3.4. Single Plant DNA Isolation (24)

1. Harvest two to three leaves (<1.5 cm long) up to approx 200 mg of leaf material per plant in an Eppendorf tube, freeze in liquid nitrogen. Alternatively use one to three young inflorescences (three to five flower buds) of older plants.
2. Grind material to a fine powder in the tube with a pestle that fits into the Eppendorf tube. Add 250 µL of DNA extraction buffer and grind once more. Add additional 250 µL of extraction buffer and 500 µL phenol/chloroform, hand vortex, and leave samples at room temperature until batches (18–24) of samples are ready.
3. Centrifuge for 10 min, remove 450 µL supernatant, and add 0.7 vol isopropanol. Mix, keep tubes at room temperature for 5 min, then centrifuge for 5 min in Eppendorf centrifuge at full speed. Wash DNA pellet with 70% ethanol and briefly air dry.
4. Dissolve DNA pellets in 100 µL TE *(28)* containing 10 µg/mL RNase. DNA samples may be stored at 4°C (few months) or at –20°C.

3.5. Isolation of DNA Probes Flanking I Inserts by IPCR

1. Digest 300 ng DNA with 10–15 U *Hin*fI in 100 µL 1X*Hin*fI buffer containing 1mM spermidine (3 h at 37°C). Add 1 µL 2.5 mM dNTPs (mix of dATP, dTTP, dCTP, and dGTP) and 1 U DNA polymerase I Klenow fragment. Incubate for 10 min at room temperature. Phenol/chloroform extract and precipitate DNA with NaAc/isopropanol for 20 min at –20°C *(28)*. Centrifuge DNA for 20 min, wash pellet in 70% ethanol, and dry.
2. Resuspend in 90 µL water and then add 10 µL 10X ligation buffer *(28)*. Add 2.5 U T4 ligase and self–ligate DNA fragments overnight at 14°C. Inactivate ligase by heating the sample at 65°C for 10 min and add 200µL sterile water.
3. To 100 µL ligate mix add 10 µL 10X*Sal*I buffer and 10 U *Sal*I. To another 100 µL ligate mix, add 10 µL 10X*Xba*I buffer and 10 U *Xba*I and incubate for 2 h at 37°C. Phenol/chloroform extract the *Sal*I and *Xba*I digested DNA samples as

well as the remaining ligate mix; and recover the DNA from supernatant by NaAc/ isopropanol precipitation. Wash the DNA pellet with 70% ethanol, dry, and resuspend in 30 µL H_2O.
4. Transfer DNA template into a PCR tube and add 4 µL 10X PCR buffer, 2 µL primer ILT1, 2 µL primer IRT1 (both at 120 ng/µL), 2 µL dNTPs (2.5 mM each). Prepare 10 µL of 1X PCR buffer with 2.5 U *Taq* DNA polymerase.
5. PCR reaction: 5 min at 95°C; hot–start by adding 10 µL *Taq* DNA polymerase mix; set 25 cycles of PCR: 1 min 95°C, 1 min 55°C, and 3 min 72°C; and elongation for 5 min at 72°C.
6. Transfer 2 µL aliquot to a new PCR tube. Add 38 µL 1X PCR buffer, containing 2 µL primer ILT2, 2 µL primer IRT2 (both at 120 ng/µL), and 2 µL dNTPs (2.5 mM each). Second PCR for 25 cycles using the conditions described in **step 5**.
7. To clone IPCR derived *I* element flanking DNA fragments: Dilute 45 µL IPCR mix to 100 µL with water. Phenol/chloroform extract and precipitate with NaAc/isopropanol. Wash pellet in 70% ethanol and dissolve in 18 µL water. Add 2 µL 10X Klenow buffer and 1 U DNA polymerase I Klenow fragment. Keep at room temperature for 10 min, add 1 µL dNTPs and incubate at room temperature for further 10 min. Size separate on a 1.2% TBE–agarose gel, cut out the DNA bands from the gel, elute ,and clone in an appropriate vector (e.g., Bluescript SK$^+$).
8. To obtain probes with very little *I* element sequence, use 2 µL of IPCR mix (or isolated fragment) in a 50 µL PCR reaction containing 2 µL of ITIR primer (at 105 ng/µL), but with annealing at 50°C instead of 55°C. Digest the PCR reaction mix with *Bgl*II (restriction site in the ITIR primer) and clone in the compatible *Bam*HI site of a suitable vector (e.g., Bluescript SK$^+$).

4. Notes

1. In some cases, it might be advantageous to produce several F_1 progeny by outcrossing to wild–type Landsberg *erecta* and screening the F_2s, as this would then yield plants with stable homozygous *I* elements.
2. For convenience, the seeds from a plant or family can be bulked and used to sow a next generation for mutant screening.
3. In bulk planting, even distribution of seeds is important, otherwise local high density patches of seedlings with slower irregular growth hamper the selection of real mutants. Individual seed sowing is advantageous as mutant plant phenotypes are better scored.
4. To map the *I* elements in different lines, the TEn segregating populations are screened with *I* element and transposase locus discriminating probes and *En*-free plants are used for IPCR and isolation of *I* flanking DNA. The cloned *I* flanking DNA fragments are used for RFLP mapping in recombinant inbred lines (RILS) *(23)*. Mapped probes are then used back on the original segregating population blot to identify plants containing the specific *I* element and the TEn locus.
   A number of transposon-tagged mutant alleles have also been isolated (e.g., *cer1, cer6, ap1, gl2, ms2, lfy, abi3*), which can serve as jumping pads for targeted tagging. Most of these mutants display major phenotypes involved in flower and plant

form or color, are easily scorable, and often revertants can be selected and used for crosses.
5. To determine the *En* transposase genotype, a probe from the hygromycin gene reveals the presence of the TEn5 or TEn2 locus. A more discriminatory probe is the DNA flanking the T-DNA insertion, which was obtained by plasmid rescue of the TEn2 and the TEn5 loci. DNA blot hybridization with the corresponding flanking T-DNA probes reveals whether the TEn5 or TEn2 loci are hemi– or homozygous.
6. A plant homozygous for, e.g., an EMS–induced mutation at the target locus a is crossed with a nuclear male sterile plant (available from stock centers) and from the F_2 male sterile *aa* mutant, plants are selected for crosses with transposon plants containing the *En* transposase and the specific target locus linked (<5–10 c*M*) *I* insert. In a large F_1 population heterozygous for the target locus, mutants resulting from *I* transposition into the target locus are screened for. With a transposition frequency of 20%, such an F_1 population of 20,000 plants is expected to carry about 4000 new *I* inserts, enriched for insertion near or at the target locus. Using as many different transposon plants as possible (10–50) will increase the independence of transpositions and thus the tagging frequency.
7. Homozygous mutant plants may not always contain a transposon insert in both mutant alleles. Occasionally, an insert may transpose and leave an excision footprint behind, thus generating a stable mutant allele. For insertions in coding regions, this is rather a rule than exception. Excision of these elements often deletes or duplicates a few basepairs, generating a frameshift that leads to a premature stopcodon.
8. Load equal amounts of DNA per lane to distinguish between homozygous and hemizygous inserts, based on intensity, in comparison to other bands in lane.

References

1. Bingham, P. M., Levis, R., and Rubin, G. M. (1981) Cloning of DNA sequences from the white locus of *D. melanogaster* by a novel and general method. *Cell* **25,** 693–704.
2. Walbot, V. (1992) Strategies for mutagenesis and gene cloning using transposon tagging and T–DNA insertional mutagenesis. *Annu. Rev. Plant Physiol. Plant Mol. Biol. 43,* 49–82.
3. Haring, M. A., Rommens, C. M. T., Nijkamp, H. J. J., and Hille, J. (1991) The use of transgenic plants to understand transposition mechanisms and to develop transposon tagging strategies. *Plant Mol. Biol.* **16,** 449–461.
4. Peterson, P. A. (1953) A mutable *pale green* locus in maize. *Genetics* **38,** 682,683.
5. Peterson, P. A. (1965) A relationship between the *Spm* and *En* control systems in maize. *Am. Nat.* **44,** 391–398.
6. Pereira, A., Schwarz–Sommer, Zs., Gierl, A., Bertram, I., Peterson, P. A., and Saedler, H. (1985) Genetic and molecular analysis of the Enhancer (En) transposable element system of *Zea mays. EMBO J.* **4,** 17–25.
7. Pereira, A., Cuypers, H., Gierl, A., Schwarz–Sommer, Zs., and Saedler, H. (1986) Molecular analysis of the *En/Spm* transposable element of *Zea mays. EMBO J.* **5,** 835–841.

8. Masson, P., Surosky, R., Kingsbury, J., and Fedoroff, N. V. (1987) Genetic and molecular analysis of the *Spm–dependent a–m2* alleles of the maize *a* locus. *Genetics* **117,** 117–137.
9. McClintock, B. (1954) Mutations in maize and chromosomal aberrations in *Neurospora. Carnegie Inst. Wash. Year Book* **54,** 254–260.
10. Frey, M., Reinecke, J., Grant, S., Saedler, H., and Gierl, A. (1990) Excision of the *En/Spm* transposable element of *Zea mays* requires two element–encoded proteins. *EMBO J.* **12,** 4037–4044.
11. Masson, P., Rutherford, G., Banks, J., and Fedoroff, N. (1989) Essential large transcripts of the maize *Spm* transposable element are generated by alternative splicing. *Cell* **58,** 755–765.
12. Gierl, A., Saedler, H., and Peterson, P.A. (1989) Maize transposable elements. *Ann. Rev. Genet.* **23,** 71–85.
13. Baker, B., Schell, J., Lorz, H., and Fedoroff, N. (1986) Transposition of the maize controlling element 'Activator' in tobacco. *Proc. Natl. Acad. Sci. USA* **83,** 4844–4848.
14. Pereira, A. and Saedler, H. (1989) Transpositional behaviour of the maize *En/Spm* element in transgenic tobacco. *EMBO J.* **8,** 1315–1321.
15. Schmidt, R. and Willmitzer, L. (1989) The maize transposable element *Activator (Ac)* shows a minimal germinal excision frequency of 0.2–0.5% in transgenic *Arabidopsis thaliana* plants. *Mol. Gen. Genet.* **220,** 17–24.
16. Dean, C., Sjodin, C., Page, T., Jones, J., and Lister, C. (1992) Behaviour of the maize transposable element *Ac* in *Arabidopsis thaliana. Plant J.* **2,** 69–81.
17. Cardon, G. H., Frey, M., Saedler, H., and Gierl, A. (1993) Mobility of the maize transposable element *En/Spm* in *Arabidopsis thaliana. Plant J.* **3,** 773–784.
18. Aarts, M. G. M., Corzaan, P., Stiekema, W. J., and Pereira, A. (1995) A two–element *Enhancer–Inhibitor* transposon system in *Arabidopsis thaliana. Mol. Gen. Genet.* **247,** 555–564.
19. Aarts, M. G. M., Dirkse, W., Stiekema, W. J., and Pereira, A. (1993) Transposon tagging of a male sterility gene in *Arabidopsis. Nature* **363,** 715–717.
20. Aarts, M. G. M., Keijzer, C. J., Stiekema, W. J., and Pereira, A. (1995) Molecular characterization of the *CER1* gene of Arabidopsis involved inepicuticular wax biosynthesis and pollen fertility. *Plant Cell* **7,** 2115–2127.
21. Scofield, S. R., English, J. J., and Jones, J. D. G. (1993) High level expression of the *Activator* transposase gene inhibits the excision of *Dissociation* in tobacco cotyledons. *Cell* **75,** 507–517.
22. Bancroft, I., Jones, J. D. G., and Dean, C. (1993) Heterologous transposon tagging of the *DRL1* locus in *Arabidopsis. Plant Cell* **5,** 631–638.
23. Lister, C. and Dean, C. (1993) Recombinant inbred lines for mapping RFLP and phenotypic markers in *Arabidopsis thaliana. Plant J.* **4,** 745–750.
24. Shure, M., Wessler, S., and Fedoroff, N. (1983) Molecular identification and isolation of the *Waxy* locus in maize. *Cell* **35,** 225–233.
25. Feldmann, K. A. (1991) T–DNA insertion mutagenesis in *Arabidopsis*–mutational spectrum. *Plant J.* **1,** 71–82.

26. Altmann, T., Schmidt, R., and Willmitzer, L. (1992) Establishment of a gene tagging system in *Arabidopsis thaliana* based on the maize transposable element *Ac*. *Theor. Appl. Genet.* **84,** 371–383.
27. Bancroft, I. and Dean, C. (1993) Transposition pattern of the maize element *Ds* in *Arabidopsis thaliana*. *Genetics* **134,** 1221–1229.
28. Sambrook, J., Fritsch, E. F., and Maniatis, R. (1989) *Molecular Cloning, A Laboratory Manual* (2nd ed.), Cold Spring Harbor Laboratory Press, Cold Spring Habor, NY.

34

Cloning Genes from T-DNA Tagged Mutants

Brian P. Dilkes and Kenneth A. Feldmann

1. Introduction

Insertion mutagenesis has a major advantage over other types of mutagenesis in that it not only causes a disruption of the gene, but the insertion element serves as a vehicle for recovery of the flanking DNA. Two types of insertion mutagens used extensively in plants are T-DNAs and transposons. Each of these mutagens has their own advantages and disadvantages. A few of the advantages of T-DNAs are that they are:

1. Generally single copy;
2. Large and therefore more likely to cause a gene disruption on insertion into introns or untranslated regions;
3. Stable on integration; and
4. Easily engineered to carry scoreable and selectable markers.

The advantages of transposons are:

1. They excise, causing reversions and footprints;
2. They can be in high copy number, meaning that fewer plants need to be screened to isolate a specific mutant;
3. It is easier to generate an allelic series with allele-specific phenotypes; and
4. It is easier to generate large numbers of novel insertion events because they transpose.

The cloning of genes via T-DNA insertion mutagenesis is described in this chapter, whereas transposon tagging is described in Chapters 32 and 33.

More than 20,000 transformed lines of *Arabidopsis* have been generated during the past decade *(1–4)*. These have been generated using various whole plant as well as tissue culture transformation procedures. More than 6000 of the lines generated by Forsthoefel et al. *(3)*, 1500 of the lines generated by Bechtold et al. *(4)* and approx 1000 lines generated by Koncz et al. *(4)* are available through the *Arabidopsis* Biological Resource Centers (ABRC) and it is likely that tens of thousands more will be available over the next several years. Approximately 20% of these transformed lines contain a visible alteration in phenotype but only 40%, depending on the population, are actually the result of a functional T-DNA insert *(5–7)*. Because of the high percentage of untagged mutants, it is important to first show tight linkage between the T-DNA-encoded selectable marker and the phenotype of interest before initiating a gene cloning procedure.

Currently dozens of plant flanking DNAs have been recovered from T-DNA tagged mutants utilizing a variety of protocols. Techniques to isolate the plant flanking DNA include:

1. Plasmid rescue;
2. Isolation from a genomic lambda library;
3. Inverse polymerase chain reaction (IPCR); and
4. Thermal asymmetric interlaced (TAIL)-PCR.

The procedure selected for any specific mutant is dependent on what is known about the insertion pattern, the availability of laboratory equipment, and the experience of the researcher. We describe these procedures and detail the protocols for plasmid rescue and TAIL-PCR in this chapter (*see* **Notes 1** and **2**).

Once plant flanking DNA has been isolated, it must be used to recover a wild-type genomic clone spanning the site of insertion. The putative wild-type gene is isolated either from a lambda or cosmid library. A cosmid library, which allows direct transformation of clones into Agrobacterium and plants via a binary plant transformation vector, has been generated by M. Bennett *(10)* and is available through the ABRC (stock CD4-11).

Finally, it must be shown that the correct gene has been isolated. This can be accomplished by transforming a minimally sized genomic fragment into the mutant to show that it restores the phenotype. If wild-type plants are recovered that cosegregate with the transgene, then a functional unit sufficient to complement the mutant has been identified. Alternatively, several mutant alleles can be sequenced to associate an altered gene with the phenotype. This entire process has been completed for more than two dozen *Arabidopsis* mutants. In this chapter, we describe what has been learned from these experiences.

2. Materials
2.1. Materials for Cosegregation Analysis

1. Kan (Kanamycin) and Murashige and Skoog (MS) media: MS salts (Gibco-BRL, Gaithersburg, MD), 0.5% (w/v) sucrose, 0.8% (w/v) bacto-agar (Difco, E. Molesly, Surrey, UK) in 1 L double-distilled water. Autoclave 15 min, cool to 50–60°C and pour in sterile petri plates (30–35 mL/plate). For Kan media allow

T-DNA Tagged Mutants

media to cool to 50–60°C and add 50 mg/L kanamycin (Sigma, St. Louis, MO); pour as above.
2. Crossing tools: fine tipped forceps, dissecting needles, stereoscope.
3. Sterilization solutions: 50% (v/v) commercial bleach containing 0.05% (v/v) Triton-X 100 (make fresh each time); double-distilled water, sterilized by autoclaving; and 95% ethanol.
4. Laminar flow hood.
5. Whatmann number 1, 90-mm circular filter paper.
6. 4°C Storage for seed stratification.
7. Growth chamber; 24°C, 24 h light.
8. Pots, soil, and flats for transplantation and plant growth.
9. Plastic disposable transfer pipets

2.2. Materials for Plasmid Rescue

1. Genomic plant DNA.
2. $mrcABC^-$, $hsdRMS^-$, rec^- *Escherchia coli* strain (e.g., SURE or DH10B).
3. Restriction endonucleases and appropriate reaction buffers.
4. T4 DNA ligase and ligase reaction buffer.
5. Molecular biology plasticware.
6. $0.1M$ spermidine (store at $-20°C$).
7. Double distilled water, sterilized by autoclaving.
8. Phenol/chloroform 1:1 (v/v).
9. $3M$ sodium acetate (pH 4.5).
10. 100% Ethanol.
11. 80% Ethanol ($-20°C$).
12. SOC: 20 g tryptone (Difco), 5 g yeast extract (Difco), and 0.5 g NaCl in 900 mL double-distilled water. Add 10 mL of $0.25M$ KCl, adjust volume to 975 mL with double-distilled water. Adjust pH to 7.0 with NaOH and autoclave 20 min. Add 5 mL filter-sterilized $2M$ $MgCl_2$ and 20 mL $1M$ filter-sterilized glucose.
13. YT+Amp, Tet, or Kan plates: 16 g tryptone (Difco), 10 g yeast extract (Difco), 5 g NaCl, 1.2% (w/v) bacto-agar (Difco) in 800 mL double-distilled water. Adjust pH to 7.0 with NaOH. Adjust the volume to 1 L with double-distilled water and autoclave for 20 min. Allow media to cool to 55–60°C, add ampicillin (U.S. Biochemical, Cleveland, OH) to a final concentration of 60 mg/mL, tetracycline (Sigma) to 10 mg/mL, or kanamycin (Sigma) to 25 mg/mL, and pour in sterile Petri plates.
14. Microcentrifuge.
15. Vortex.
16. 16°C Water bath: easily made by placing a small water bath in a 4°C cold room or refrigerator.
17. 37°C Incubator.

2.3. Materials for TAIL-PCR

1. DNA extraction buffer: $0.3M$ NaCl, 50 mM Tris-HCl, pH 7.5, 20 mM EDTA, 2% N-lauroylsarcosine, 0.5% (w/v) sodium dodecyl sulfate (SDS), $5M$ urea, and 5% (v/v) phenol.

2. Variable speed drill and microcentrifuge tube bit or disposable hand-held pestle.
3. Phenol/chloroform/isoamyl alcohol 24:25:1 (v/v/v).
4. RNase TE: 20 µg/mL DNase free RNase A, in 10 mM Tris-HCl, pH 7.9, and 1 mM EDTA.
5. dNTP's: 200 µM solution of each nucleotide: ATP; CTP; GTP; TTP. Make in sterile double-distilled water and store in 50 mL aliquots at –20°C.
6. 10X reaction mix: 100 mM Tris-HCl, pH 8.3, 500 mM KCl, 20 mM MgCl$_2$, 0.01% (w/v) gelatin.
7. Arbitrary degenerate (AD) primers sequences taken from Liu et al. *(11)*.
   AD1: NTCGA(G/C)T(A/T)T(G/C)G(A/T)GTT.
   AD2: NGTCGA(G/C)(A/T)GANA(A/T)GAA.
   AD3: (A/T)GTGNAG(A/T)ANCANAGA.
8. Specific primers (*see* **Note 3**).
9. *Taq* polymerase.
10. PCR thermocycler.
11. Loading dye: 0.25% (w/v) bromophenol blue, 0.25% (w/v) xylene cyanol, 50% (w/v) sucrose in electrophoresis buffer.
12. 1.5% (w/v) Agarose gel in electrophoresis buffer.
13. DNA size markers.
14. Liquid nitrogen

3. Methods
3.1. Cosegregation Analysis

Before plant flanking DNA is isolated from a putative insertion mutant, it is essential to establish cosegregation between the insertion element and the mutant phenotype. First, a line heterozygous for one insertion must be established. If the mutant is isolated from a single segregating line, it is simple to test for the number of independent inserts by examining the segregation pattern of the progeny on selective medium. Since the resistance markers are dominant, a single insertion in a heterozygous plant will produce progeny that segregate 3:1 Kan^R:Kan^S, two unlinked inserts will result in a 15:1 Kan^R:Kan^S ratio, whereas two linked inserts will generate a Kan^R:Kan^S ratio between 3:1 and 15:1. A line segregating for three or four unlinked inserts will segregate 63:1 and 255:1 Kan^R:Kan^S, respectively.

If the mutant is homozygous recessive and isolated from a large pooled population, such as those that are available from the ABRC, mutants should be backcrossed to a wild-type (WT) plant. The F$_1$ resulting from the cross, should be 100% resistant on selective media. Resistant-progeny (F$_1$s) are grown to maturity and selfed. The segregation of the F$_2$ population will be dependent on the number of inserts. If more than one insert is detected, F$_3$s should be tested until a line can be isolated that segregates 3:1 Kan^R:Kan^S and displays the proper segregation ratio for the phenotype of interest. One can be >95%

confident of isolating a subline with a single insert from a line segregating for two or three inserts if 20, or 71 Kan^R plants are selfed, respectively. Fewer individuals can be used if you can transplant only Kan^R phenotypically wild-type plants. Therefore, it is best to backcross a line with three to four independent inserts twice before trying to identify a subline with a single T-DNA. Alternatively, one could amplify T-DNA-specific products using TAIL-PCR and demonstrate cosegregation of a particular PCR product, with the mutant phenotype.

There are several ways that cosegregation analyses can be accomplished depending on the expression of the phenotype of the mutant. The following method should allow you to perform an accurate cosegregation analysis under a number of phenotypic contingencies.

3.1.1. Cosegregation Analysis (see **Note 4**)

1. Place approx 1500 seeds from a line known to segregate for your phenotype and containing one T-DNA in a 1.5-mL microcentrifuge tube.
2. Work in a laminar flow hood from this point forward utilizing sterile techniques. Fill the microcentrifuge tube with 50% bleach solution, cap, and flick to fully suspend the seeds. Let stand 4 min, then flick the tube again, and let stand 4 more min. The seeds will settle to the bottom of the tube. Pour off the liquid.
3. Rinse seeds three times with sterile, double-distilled water, mixing the seeds each time, allowing them to settle, and pouring off the liquid.
4. Place a filter paper on top of a 50-mL beaker toward the back of the hood.
5. Fill the microcentrifuge tube with 95% ethanol and quickly pipet the seeds and liquid, using a transfer pipet, onto the filter paper (*see* **Note 5**).
6. After the seeds have dried, plate them at a density of 200 individuals/plate on Kan media.
7. For uniform germination, stratify the seeds by storing at 4°C for 2 d. Transfer plates to 24°C and 24 h light.
8. Transplant phenotypically wild-type Kan^R seedlings, individually, to soil and self (*see* **Note 6**). The number transplanted depends on the degree of linkage desired (*see* **Note 7**).
9. Plate 50 progeny, from each selfed individual from **step 8**, on Kan and MS media in groups of three (*see* **Note 8**). Score and perform a Chi-square analysis for Kan^R:Kan^S and WT:mut respectively. Transplanting 29 or 299 Kan^R phenotypically wild-type seedlings, all of which subsequently segregate properly for the mutant phenotype and 3:1 for kanamycin resistance, is sufficient to provide a >95% confidence in linkage of the mutation and T-DNA within 10 or 1 c*M* respectively (*see* **Note 7**). An unlinked insertion should produce at least one line containing only Kan^R progeny with 11 transplants >95% of the time.

3.2. Plasmid Rescue

Plasmid rescue has been used on the greatest number of mutants because it is inexpensive and rapid *(13–15)*. This procedure is dependent on cassettes within the T-DNA containing a bacterial antibiotic resistance gene and a plasmid origin of replication. Briefly, genomic DNA from the mutant is digested to completion with a restriction endonuclease and the resulting fragments are self-ligated. The ligated DNA is transformed into *E. coli* and plated on ampicillin to select for plasmids with the pBR322 sequences contained in the T-DNA. Amp^R colonies are transferred to kanamycin plates to identify colonies with internal pieces of T-DNA. Amp^R, Kan^S colonies are restriction mapped and tested for plant flanking DNA via Southern analyses of wild-type DNA. The restriction enzymes used depend on the insertion type and border region one is attempting to recover (**Fig. 1A,B**). The T-DNA used by Bechtold et al. *(4)* does not contain a bacterial origin of replication (**Fig. 1C**). Therefore, plant flanking DNAs cannot be recovered by this method. The protocol presented was synthesized from our experiences and others' in which it has been used successfully *(20)*.

3.2.1. Digestion of Genomic DNA with Restriction Endonucleases

1. To 1 mg of genomic DNA isolated from the mutant of interest, add 5 µL of the appropriate 10X reaction buffer, 2 µL of 0.1M spermidine, and water to a final volume of 49 µL. Incubate on ice for 5 min. Add 1 µL of restriction enzyme (10 U) and incubate at 37°C for 2 h (*see* **Note 9**).
2. Bring the volume to 250 µL with water, mix, and add 250 µL of phenol/chloroform. Vortex for 15 s. Separate the phases by centrifugation (15,000g for 10 min) and transfer the aqueous phase (top) to a clean microcentrifuge tube.
3. Add 25 µL 3M sodium acetate (pH 4.5) and invert the tube several times. Add 500 µL of 100% ethanol. Incubate on ice for 1 h and pellet the DNA by centrifugation for 15 min at 15,000g in a microcentrifuge.
4. Decant the supernatant and wash the pellet twice with 80% ethanol (–20°C).
5. Resuspend the pellet in 50 µL water.

3.2.2. Ligation of Fragments

1. In three microcentrifuge tubes add 1, 5, and 10 µL of the DNA from **Subheading 3.2.1., step 5**. To each add 10 µL 10X ligation buffer, 4 U T4 ligase and double-distilled water to a final volume of 100 µL. Incubate at 16°C for 16 h (*see* **Note 10**).
2. Bring the volume to 250 µL with water and add an equal volume of phenol/chloroform. Vortex for 15 s. Separate the phases by centrifugation (15,000g for 10 min) and transfer the aqueous phase to a new microcentrifuge tube.
3. Add 25 µL of 3M sodium acetate (pH 4.5) and 2 vol 100% ethanol. Incubate on ice for 1 h and pellet the DNA by centrifugation (15 min at 15,000g) in a microcentrifuge.
5. Decant the supernatant and wash the pellet with 80% ethanol twice.
6. Resuspend the pellet in 5 µL water.

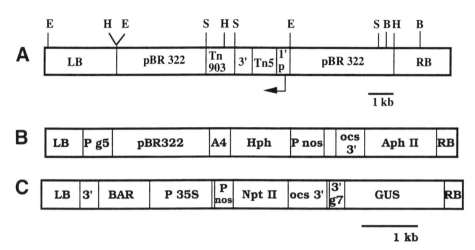

Fig. 1. **(A)** Structure of the 3850:1003 T-DNA in the Feldmann-generated transformants *(1,3,16)*. Components of the T-DNA and their GENBANK accession number or reference include: pBR322, pBR322.vec; Tn903, trn903.ba; 3' OCS, Tipct.ba; Tn5, trn5neo.ba; 1'2' promoter, *(17)*; and right border region, Tipost37.ba. The following ABRC stocks contain this T-DNA structure: CS2361-CS3068, CS3115, CS3116, CS6401-CS6502. Restriction enzyme sites are as follows: E, *Eco*RI; H, *Hin*dIII; S, *Sal*I; and B, *Bam*HI. **(B)** Structure of the pPCVNFHyg T-DNA used in some Koncz-generated transformants *(2,18)*. See Koncz et al. *(18)* for components. *Aph*II confers kanamycin resistance when transcriptionally fused to a plant promoter or in frame with an actively translated mRNA, and Pnos:Hph confers constituitive hygromycin resistance. **(C)** Structure of the pGKB5 T-DNA in the Bechtold-generated infiltration transformants *(4)*. For components see Bouchez et al. *(19)*. Figure adapted from a personal communication from D. Bouchez. P35S:BAR and Pnos:*Npt*II confer constitutive resitance to phophinothricin and kanamycin, respectively. The promotorless GUS gene may be active as either an in-frame translational fusion or a transcriptional fusion at the site of integration.

3.2.3. Transformation of E. coli

1. Transform highly competent *E. coli* cells with 1 µL of each ligation mixture.
2. Allow the transformed cells to recover in SOC medium at 37°C for 1 h.

3.2.4. Analysis of Clones

1. Pick individual colonies with a toothpick onto a grid on YT+Kan, and YT+Amp plates. Use one toothpick for each colony on both plates. Incubate the plates at 37°C overnight.
2. Pick single colonies displaying the appropriate resistance and grow overnight in 3 mL of YT+Amp. Isolate plasmid DNA by standard methods *(21)*.
3. Analyze clones for intermolecular ligation events by digesting plasmids with the same restriction enzyme used in **Subheading 3.2.1.** and size fractionate the

products on an agarose gel. The only product should be a single, linear DNA molecule.
4. If the cloned DNA is the plant DNA flanking an insert, a Southern should demonstrate a difference in restriction patterns in the wild-type and mutant (*see* **Note 11**).

3.3. TAIL-PCR

TAIL-PCR is a method recently developed by the lab of R. Whittier *(24)* that, like IPCR, requires sequence information from only one side of the amplified target. A specific target can be amplified by using an arbitrary degenerate primer and nested specific primers to a known sequence. "Supercycles" of varying temperature conditions are used to generate a population of PCR products from specific and nonspecific priming events. The products are diluted and further amplification, using the nested specific primers, "purifies" the flanking DNA product. Liu et al. *(11)* have described the use of this technique to isolate plant DNAs flanking T-DNA insertion events. Confirmation of the identity of the TAIL-PCR product is accomplished by demonstrating a polymorphism between DNA from wild-type plants and the mutant via Southern analyses. With one set of nested right border primers and three different arbitrary primers, a right border-specific TAIL-PCR product from 96% of the 190 T-DNA lines tested was generated *(11)*.

3.3.1. DNA Isolation

1. Place three leaves in a micro-centrifuge tube and freeze in liquid nitrogen. These can be stored at $-70°C$ or used immediately.
2. Using a chilled bit in a variable speed drill, grind the tissue in the presence of 50 µL DNA isolation buffer.
3. Add 100 µL DNA isolation buffer, mix by inverting.
4. Add 10 µL phenol/chloroform/isoamyl alcohol; mix by inverting.
5. Centrifuge 5 min at 11,000g, and transfer the supernatant to a fresh microcentrifuge tube.
6. Add 2 vol of ethanol; mix by inverting.
7. Centrifuge 5 min at 7000g, discard supernatant.
8. Wash the pellet in 70% ethanol twice, centrifuging at 11,000g for 5 min each time.
9. Dissolve the pellet in 50 µL RNase-TE.
10. Spectrophotometrically quantify DNA according to Sambrook et al. *(21)*.

3.3.2. Primer Design

3.3.2.1. Nonspecific Primers

1. The nonspecific (AD) primers of Liu et al. *(11)* are perfectly suited to cloning plant flanking DNA from any insertion mutagen in *Arabidopsis* (e.g., transposon, or T-DNA).

2. If further nonspecific primer sequences are desired, additional factors should be considered (*see* **Note 12**)

3.3.2.2. SPECIFIC PRIMERS

1. Because there are many insertion types, a set of specific primers for the various T-DNAs must be designed.
2. Specific primers should be designed with approximate melting temperatures of 57–63°C according to the formula: 69.3 + 0.41(%GC content) – 650(primer length)$^{-1}$ *(12,26)*.
3. The first two specific primers should be as close together as possible, optimally, overlapping by only a few nucleotides.
4. The third primer should leave 50–100 nucleotides between it and the end of specific primer 2.
5. The proximity of the end of the border sequence and third primer determines both the likelihood that priming sites will be lost owing to minor truncations, and in the absence of rearrangements the minimum size of a fragment containing plant DNA.
6. The first primer should have a higher annealing temperature than the following two.

3.3.3. TAIL-PCR Amplification

3.3.3.1. PRIMARY CYCLES

1. In a PCR reaction tube, add 2 µL 10X reaction mix, 2 µL dNTPs, 20 ng DNA, 0.8 U *Taq* polymerase, 0.2 µM specific primer 1, 2 µM AD1, 3 µM AD2, or 4 µM AD4; bring the volume to 20 µL with double-distilled water.
2. Denature at 93°C 1 min, followed by 95°C 1 min.
3. 5 cycles of 94°C 30 s, 62°C 60 s, 72°C 150 s.
4. Low stringency cycle consisting of 94°C 30 s, 25°C 180 s, ramp from 25–72°C over 180 s, hold 72° 150 s.
5. 15 TAIL cycles: 94°C 10 s, 68° 60 s, 72°C 150 s, 94°C 10 s, 68° 60 s, 72°C 150 s, 94°C 10 s, 44°C 60 s, 72°C 150 s.
6. Complete elongation for 5 min at 72°C, and hold at 4°C.

3.3.3.2. SECONDARY CYCLES

1. Transfer 1 µL from the primary cycle reaction tube to a fresh tube and add water to final volume of 50 µL; mix by inverting. Save the primary reaction mix.
2. To 1 µL of the dilution add 2 µL 10X reaction mix, 2 µL dNTPs, 0.6 U *Taq* polymerase, 0.2 mM specific primer 2, the same nonspecific primer at 1.5 µM AD1, 2 µM AD2 or 2 µM AD3. Bring the volume to 20 µL with double-distilled water.
3. 12 TAIL cycles of : 94°C 10 s, 64° 60 s, 72°C 150 s, 94°C 10 s, 64° 60 s, 72°C 150 s, 94°C 10 s, 44°C 60 s, 72°C 150 s.
4. Complete elongation for 5 min at 72°C, and hold at 4°C.

3.3.3.3. TERTIARY CYCLES

1. Transfer 1 µL from the secondary cycle reaction tube to a fresh tube and add double-distilled water to a final volume of 10 µL; mix by inverting. Save the secondary cycle reactants.

2. To 1 µL of the 10 fold dilution, add to a final volume of 100 µL: 10 µL 10X Reaction mix, 200 µM dNTPs, 0.6 U *Taq* polymerase, 0.2 µM specific primer 3, the same nonspecific primer at 1.5 µM AD1, 2 µM AD2 or 2 µM AD3, and double-distilled water.
3. Run 20 reduced stringency cycles of 94°C 15 s, 44°C 60 s, 72°C 150 s.
4. Run for 5 min at 72°C.

3.3.4. Confirmation of Insert-Specific Amplification

1. Mix 7 µL of each PCR reaction product with 1 µL loading dye.
2. Load on a 1.5% agarose gel with size marker ranging from 200bp to 3 kb, and electrophorese until the shortest, predicted, plant flanking DNA molecule is 75% down the gel.
3. Tertiary cycle PCR products from insert-specific priming events should be 50–100 nucleotides shorter than corresponding secondary cycle products. Shifted fragments should be isolated from the gel and used as probes for a Southern analysis of the restriction pattern in wild-type and mutant plants.
4. Fragments showing a decrease in size in the tertiary reaction, demonstrating a band shift in Southern analyses should be used to isolate cDNA and genomic clones of the disrupted genetic material, or may be used for direct sequencing *(11,12)*.

4. Notes

1. There are a number of reasons for generating a lambda library from the mutant of interest to recover the plant flanking DNA. The isolation of a series of lambda clones spanning the T-DNA may be the most reliable way to isolate flanking DNA if nothing is known about the T-DNA structure in the mutant. Isolation of plant flanking DNA is accomplished by screening for colonies homologous to the right and/or left border sequences. Probing replicate filters with border probes will help eliminate clones likely to contain concatemeric T-DNA structures. Clones that hybridize to only the left border or right border should be chosen for restriction and Southern analyses. The plant DNA can be subcloned for sequencing, or used directly as a probe for isolating cosmids for complementation.
2. IPCR cloning of plant flanking DNA has only been reported for two T-DNA tagged mutants *(8,9)*. This technique will become more popular as *Arabidopsis* researchers discover the speed with which PCR approaches can be used to recover plant flanking DNAs. Sequence information is required only from one side of the amplified target. Following a restriction digest, the DNA is self-ligated. A known sequence, the T-DNA in this case, now occupies both the 3' and 5' ends of the neighboring unknown sequence. Outfacing primers designed to the T-DNA can now be used to amplify the plant DNA originally flanking the insert. This technique is described in more detail elsewhere in this volume.
3. In TAIL-PCR, annealing temperatures for the high stringency cycles should be 1–5°C higher than the annealing temperatures of the specific primers being used. Furthermore, the difference between the melting temperatures of the degenerate and nondegenerate primers should be at least 10°C. These temperatures should

be viewed as guidelines only, and you will need to optimize the procedure for your primer sets. However, providing primers are designed in a similar fashion, there should be no problem with using the same temperature parameters. More than one insert-specific amplification product is not uncommon due to concatemeric T-DNAs and nonspecific AD primers priming in nested positions. TAIL-PCR products can be sequenced directly *(11,12)*. This should help in assessing the origins of multiple insert-specific PCR products.

4. If the phenotype is a seedling-lethal that is difficult to distinguish from kanamycin-sensitive individuals, it may be best to score both kanamycin-sensitive and seedling-lethal individuals as one group.

5. It is important that the alcohol evaporate rapidly. If seeds are left in alcohol for any period of time, the germination rate will be substantially reduced. As the seeds have been sterilized with bleach, they will stick to the filter paper. As such, drying can be hastened by leaning the filter paper supporting the seeds against a beaker, into the flow of air.

6. The kanamycin-sensitive phenotype is characterized by smaller bleached cotyledons and a lack of root growth. However, hypocotyls do elongate. This phenotype can be easily scored within 10 d. Kanamycin-resistance can be scored by the presence of root growth, and a lack of chlorosis. Plants growing at the plate edge can mimic the resistant phenotype, so it is best to plate the seeds carefully. In our hands, *Arabidopsis* plants grown on plates, especially at high densities become less healthy over time. If the mutant phenotype is not obvious early in development, you will need to transplant Kan^R seedlings and only use seed from phenotypic wild-types for further analysis. In this case, lines should be planted on Kan media to score for resistance, MS media as a control, and soil to observe the segregation of your phenotype.

7. Linkage confidence levels can be calculated using the formula $n = \log P/\log f$ where n is the number of individuals tested, P is the alpha value, and f is the probability the event does not occur. For >95% confidence level of linkage within 1 centimorgan the equation would be as follows: $299 > \log(0.05)/\log(0.99)$; for >99% confidence $459 > \log(0.01)/\log(0.99)$; and for a >99% confidence of linkage within 10 cM $44 > \log(0.01)/\log(0.90)$. This will work for the majority of cases. However, once linkage to any degree is established, you may prefer to start the molecular cloning experiments. The low spontaneous mutation rate, and infrequency of linked T-DNA insertions in seed infection generated mutant lines, has been used as justification for only discriminating between linked and unlinked inserts before cloning begins.

8. For the rapid analysis of few progeny from many lines, multiple lines can be placed on the same Petri dish by placing each line in a given sector of the plate, much the way *E. coli* or yeast strains are streaked.

9. Plasmid rescue vectors may differ in restriction endonuclease sites. The 3850:1003 plasmid T-DNA used to make the transformed lines of Forsthoefel et al. *(3)* available at the ABRC has *Eco*RI sites delimiting the right side pBR322 and right border, and a *Sal*I site delimiting the left border and left pBR322.

Furthermore, an intact right border should have a *DraI* site just before the integration site and the left border should have an analogous *ClaI* site.
10. The DNA concentration at this point is critical to balance the rate of ligation, and the rate of unwanted intermolecular ligations. For an in-depth analysis see Sambrook et al. *(21).*
11. Cloning of the proper plant flanking DNA is often complicated by the complexity of the insert. In our experience, as well as that of other laboratories *(6,22,23),* the T-DNAs from the seed transformation protocol are often in complex concatemers.
12. AD Primers should be approximately 16–18 bp in length and designed so that the melting temperature is 47–48°C. The %GC content of *Arabidopsis* is approx 41% *(25)*; primers should be near this value. Liu et al. *(11)* used AD primers ranging from 64–256-fold degeneracy with the success rate correlated with increasing degeneracy. Increasing degeneracy was also positively correlated with increasing optimal primer concentration.

References

1. Feldmann, K. A. (1991) T-DNA insertion mutagenesis in Arabidopsis: mutational spectrum. *Plant J.* **1,** 71–82.
2. Koncz, C., Nemeth, K., Redei, G. P., and Schell, J. (1992) T-DNA insertional mutagenesis in *Arabidopsis. Plant Mol. Biol.* **20,** 963–976.
3. Fortsthoefel, N. R., Wu, Y., Schulz, B., Bennett, M. J., and Feldmann, K. A. (1992) T-DNA insertion mutagenesis in *Arabidopsis*: prospects and perspectives. *Aust. J. Plant Physiol.* **19,** 353–366.
4. Bechtold, N., Ellis, J., and Pelletier, G. (1993) In planta *Agrobacterium* mediated gene transfer by infiltration of adult *Arabidopsis thaliana* plants. *C.R. Acad. Sci. Paris* **316,** 1188–1193.
5. Feldmann, K. A. (1992) T-DNA insertion mutagenesis in *Arabidopsis*: seed infection/transformation; in *Methods in* Arabidopsis *Research.* (Koncz, C., Chua, N.-H., and Schell, J., eds.), World Scientific, London, pp. 274–289.
6. Castle, L., Errampalli, D., Atherton, T. L., Franzmann, L. H., Yoon, E. S., and Meinke, D. W. (1993) Genetic and molecular characterization of embryonic mutants identified following seed transformation of *Arabidopsis. Mol. Gen. Genet.* **241,** 504–541.
7. McNevin, J. P., Woodward, W., Hannoufa, A., Feldmann, K. A., and Lemieux, B. (1993) Isolation and characterization of *Eceriferum (cer)* mutants induced by T-DNA insertions in *Arabidopsis thaliana. Genome* **36,** 610–618.
8. Deng, X.-W., Matsui, M., Wei, N., Wagner, D., Chu, A. M., Feldmann, K. A., and Quail, P. H. (1992) COP1, an Arabidopsis photomorphogenic regulatory gene, encodes a novel protein with both a Zn-binding motif and a domain homologous to the β-subunit of trimeric G-proteins. *Cell* **71,** 791–801.
9. Callos, J. D., DiRado, M., Xu, B., Behringer, F. J., Link, B. M . and Medford, J. I. (1994) The *forever young* gene encodes an oxidoreductase required for proper development of the *Arabidopsis* shoot apex. *Plant J.* **6,** 835–847.

10. Schulz, B., Bennett, M. J., Dilkes, B. P., and Feldmann, K .A. (1995) T-DNA tagging in *Arabidopsis thaliana*: cloning by gene disruption in *Plant Molecular Biology Manual* (Gelvin, S., ed.), Kluwer Academic Publishers, The Netherlands, pp. K3:1–17.
11. Liu, Y. G., Mitsukawa, N., Oosumi, T., and Whittier, R. F. (1995) Efficient isolation and mapping of *Arabidopsis thaliana* T-DNA insert junctions by thermal asymmetric interlaced PCR. *Plant J.* **8,** 457–463.
12. Liu, Y. G. and Whittier, R. F. (1995) Thermal asymmetric interlaced PCR: automatable amplification and sequencing of insert end fragments from P1 and YAC clones for chromosome walkings. *Genomics* **25,** 674–681.
13. Koncz, C., Mayerhofer, R., Koncz-Kalman, Z., Nawrath, C., Reiss, B., Redei, G. P., and Schell, J. (1990) Isolation of a gene encoding a novel chloroplast protein by T-DNA tagging in *Arabidopsis thaliana*. *EMBO J.* **9,** 1337–1346.
14. Yanofsky, M. F., Ma, H., Bowman, J. L., Drews, G. N., Feldmann, K. A., and Meyerowitz, E. M. (1990) The protein encoded by the *Arabidopsis* homeotic gene agamous resembles transcription factors. *Nature* **346,** 35–39.
15. Kieber, J. J., Rothenberg, M., Roman, G., Feldmann, K. A., and Ecker, J. R. (1993) The ethylene response pathway in *Arabidopsis thaliana* is negatively regulated by *CTR1*, a predicted member of the *Raf* family of protein kinases. *Cell* **72,** 427–441.
16. Feldmann, K. A. and Marks, M. D. (1987) *Agrobacterium*-mediated transformation of germinating seeds of *Arabidopsis thaliana*: a non-tissue culture approach. *Mol. Gen. Genet.* **208,** 1–9.
17. Velten, J. and Schell, J. (1985) Selection expression plasmid vectors for use in genetic transformation of higher plants. *Nuc. Acids Res.* **13,** 6981–6998.
18. Koncz, C., Martini, N., Mayerhofer, R., Koncz-Kalman, Z., Korber, H., Redei, G. P., and Schell, J. (1989) High-frequency T-DNA-mediated gene tagging in plants. *Proc. Natl. Acad. Sci. USA* **86,** 8467–8471.
19. Bouchez, D., Camilleri, C., and Caboche, M. (1993) A binary vector based on Basta resistance for in planta transformation of *Arabidopsis thaliana*. *C.R. Acad. Sci. Paris* **316,** 1188–1193.
20. Behringer, F. J. and Medford, J. I. (1992) A plasmid rescue technique for the recovery of plant DNA disrupted by T-DNA insertion. *Plant Mol. Bio. Rep.* **10,** 190–198.
21. Sambrook, J., Fritsch, E. F., and Maniatis, T. (1989) *Molecular Cloning: A Laboratory Manual,* 2nd ed. Cold Spring Harbor Laboratory Press, Cold Spring Harbor, NY.
22. Castle, L. and Meinke, D. W. (1994) A FUSCA gene of *Arabidopsis* encodes a novel protein essential for plant development. *Plant Cell* **6,** 25–41.
23. Feldmann, K. A. Marks, M. D., Christianson, M. L., and Quatrano, R. S. (1989) A dwarf mutant of *Arabidopsis* generated by T-DNA insertion mutagenesis. *Science* **243,** 1351–1354.
24. Liu, Y. G., Mitsukawa, N., and Whittier, R. F. (1993) Rapid sequencing of unpurified PCR products by thermal assymetric PCR cycle sequencing using unlabeled sequencing primers. *Nucleic acids Res.* **21,** 3333, 3334.
25. Leutwiler, L. S., Hough-Evans, B. R., and Meyerowitz, E. M. (1984) The DNA of *Arabidopsis thaliana*. *Mol. Gen. Genet.* **194,** 15–23
26. Mazars, G. R., Moyret, C., Jeanteur, P., and Theillet, C. G. (1991) Direct sequencing by thermal asymmetric PCR. *Nucleic Acids Res.* **19,** 4783.

VII

GENE EXPRESSION ANALYSES

35

In Situ Hybridization

Gary N. Drews

1. Introduction

In situ hybridization has proven to be a valuable technique for the analysis of gene expression in plants. It allows the investigator to determine the specific cell and tissue types expressing a particular gene. The basic procedure is to hybridize a labeled probe with tissue sections *in situ*. The nonspecifically bound probe is then washed off and the tissue sections are exposed to photographic film for autoradioagraphy.

In this procedure, the tissue is fixed with formaldehyde and embedded in paraffin. Single-stranded RNA probes are used because they have excellent hybridization efficiency. The probes are labeled with ^{35}S, which provides a good balance between sensitivity and cellular resolution. Just before hybridization, the tissue sections are dewaxed and rehydrated. The tissue sections are then incubated with proteinase K to enhance probe penetration. Hybridization solution is then applied directly to the tissue sections. Following hybridization, the sections are subjected to several wash steps including a RNase treatment to remove nonspecifically bound probe. The tissue sections are then dipped in liquid emulsion for autoradiography. After the appropriate amount of exposure, the slides are developed and stained. Autoradiography results in the production of silver grains in the emulsion over the tissue. A high concentration of silver grains will be present over the cells containing hybridizing mRNA. The silver grains appear as small black spots against a bright background using bright-field optics and as bright white spots against a dark background using dark-field optics. The procedure described here is adapted from **refs. *1–3*.**

The procedure described in this chapter works well for a wide variety of tissues and probes. In some cases, however, paraffin-embedded tissue sections may not be adequate. For example, the analysis of small structures (e.g., ovules, embryo sacs, young embryos) requires the preparation of thinner sections (1–5 µm) than can be achieved with paraffin-embedded material. Under these circumstances, the tissue must be embedded in a plastic resin. The analysis of small structures may also require better cellular resolution than can be achieved with ^{35}S-labeled probes. A dramatic improvement in cellular resolution can be achieved with nonradioactive probes. Procedures for *in situ* hybridization with plastic-embedded plant tissue and/or with nonradioactive probes have been published *(4–8)*.

2. Materials
2.1. Fixation, Embedding, and Sectioning

1. Fixative solution: 50% ethanol, 5% acetic acid, 3.7% formaldehyde (3.7% formaldehyde is a 1:10 dilution of the manufacturer's solution).
2. Ethanol solutions: 100, 95, 85, 70, 60, and 50%.
3. 0.1% Eosin Y in 95% ethanol.
4. Xylene.
5. Xylene solutions in ethanol: 75, 50, and 25%. These solutions must be made with 100% ethanol. If 95% ethanol is used, a white emulsion will form.
6. Paraplast Plus tissue embedding medium (Sigma, St. Louis, MO).
7. Molten Paraplast Plus: fill a beaker with Paraplast Plus chips and place in a 60°C oven for at least 8 h.
8. 20-mL Glass scintillation vials.
9. Aluminum weighing dishes.
10. Dissecting needle.
11. Alcohol lamp.
12. Superfrost Plus microscope slides (Fisher, Pittsburg, PA).
13. Vacuum desiccator.
14. Micro slide warming table (Eberbach, Ann Arbor, MI).
15. Microtome suitable for paraffin sectioning.
16. Slide warmer set to 45°C.
17. Oven set to 60°C.
18. Incubator set to 42°C.

2.2. Probe Synthesis

1. RNA transcription kit.
2. 650 Ci/mmol uridine 5'[^{35}S]Triphosphate ([^{35}S]-UTP) or equivalent.
3. Whatman DE-81 ion exchange paper.
4. RNase-free DNase (may be included in the transcription kit).
5. 10 mg/mL RNase-free tRNA.
6. RNase-free H_2O (*see* **Note 1**).
7. RNase-free TE: 10 mM Tris-HCl, 1 mM ethylenediaminetetra-acetic acid (EDTA), pH 7.5 (*see* **Note 1**).
8. 70% Ethanol made with RNase-free H_2O.

9. 0.5M Phosphate buffer. Mix 46.3 mL of 0.5M sodium phosphate dibasic and 53.7 mL of 0.5M sodium phosphate monobasic.
10. 200 mM Sodium carbonate.
11. 200 mM Sodium bicarbonate.
12. RNase-free TED solution: 10 mM Tris-HCl, pH 7.5, 1 mM EDTA, and 10 mM dithiothreitol (DTT) (*see* **Note 1**).
13. Heat lamp.
14. Scintillation counter.

2.3. Hybridization

1. Xylene.
2. RNase-free H$_2$O (*see* **Note 1**).
3. Ethanol solutions made with RNase-free H$_2$O: 100, 95, 85, 70, 50, 30, and 15%.
4. Proteinase K solution: 100 mM Tris-HCl, 50 mM EDTA, pH 7.5.
5. RNase-free 10X Hybridization salts: 3M sodium chloride, 100 mM Tris-HCl, 10 mM EDTA, pH 7.5 (*see* **Note 1**).
6. Denhardt's reagent made with RNase-free H$_2$O: 1% Ficoll 400 (type 400, Pharmacia, Uppsala, Sweden), 1% polyvinylpyrrolidone, 1% bovine serum albumin (Fraction V, Sigma).
7. 10 mg/mL RNase-free tRNA.
8. 50% Dextran sulfate made with RNase-free H$_2$O.
9. 1M DTT made with RNase-free H$_2$O.
10. Hybridization solution: 50% formamide, 1X hybridization salts, 1X Denhardt's reagent, 10% dextran sulfate, 70 mM DTT, 150 µg/mL tRNA. Make up using RNase-free stock solutions. This solution can be made up in large quantities and stored in the freezer.
11. 10 mg/mL Proteinase K (Boehringer Mannheim, Mannheim, Germany).
12. Magnetic stirrer.
13. 500 mL Staining dishes with covers and slide racks. Purchase staining dishes with stainless steel slide racks (Mercer Glass Works) because they are cheaper, are more durable, and hold more (30) slides. You will need 15 staining dishes with covers, at least one slide rack, and at least one handle.
14. Vacuum desiccator.
15. Covered plastic container for hybridization. The container should be wide enough to hold 5-mL pipets (≥30 cm) and not very tall (ideally ≤5 cm) so that the humidity can be kept high during the hybridization. A 22 × 33 × 6 cm container can be purchased from Life Science Products (Denver, CO).
16. Plastic disposable pipets, 5 or 10 mL.
17. Incubator set to 42°C.

2.4. Wash

1. 20X SSPE: 3M sodium chloride, 200 mM sodium phosphate monobasic, 20 mM EDTA, pH 7.4.
2. First wash buffer: 4X SSPE, 5 mM DTT.

3. RNase buffer: 500 mM sodium chloride, 10 mM Tris-HCl, 1 mM EDTA, pH 7.5.
4. Low stringency wash buffer: 2X SSPE, 5 mM DTT.
5. High stringency wash buffer: 0.1X SSPE, 1 mM DTT. Note: The DTT concentration is lowered in this buffer only because the odor of 5 mM DTT at 55°C is too strong.
6. 25 mg/mL RNase A (Sigma).
7. Ethanol solutions: 100, 95, 85, 70, 50, 30, and 15%.
8. RNase glassware: Four staining dishes are used for the RNase incubation and the first three washes.
9. One plastic container for the low stringency wash. It should be more wide than tall so that the magnetic stirrer can be used. Rubbermaid makes one that is 7.8 L (33 cups) and measures 35 × 24 × 12 cm.
10. Two, 4-L beakers for the high stringency wash.
11. Magnetic stirrer and stir bars.
12. Vacuum desiccator.
13. Shaking water bath set to 55°C.

2.5. Autoradiography

1. Kodak NTB-2 emulsion.
2. Slide mailers (Baxter, McGaw Park, IL).
3. Aluminum foil.
4. Test tube rack with 2-cm holes (e.g., Nalgene, Rochester, NY).
5. Black wooden box large enough to hold two racks and a beaker of desiccant (id: of 15 × 9 × 7 in. if Nalgene racks are used).
6. Desiccant.
7. Kodak D-19 developer.
8. Kodak fixer.
9. Four staining dishes.
10. Black plastic microscope slide boxes.
11. Refrigerator free of ^{32}P for storing the emulsion aliquoits.
12. Photographic darkroom containing a dark red (Wratten) safe light.

2.6. Staining

1. 0.1% Toluidine blue O in water. This solution should be filtered.
2. Coplin staining jars. Purchase jars with a screw cap (Wheaton). Ten jars are needed.

2.7. Microscopy

1. Compound microscope with bright-field and dark-field condensers.

3. Methods

3.1. Fixation, Embedding, and Sectioning

3.1.1. Fixation (see **Note 3**)

1. Place 10–15 mL of fixative solution into 20-mL scintillation vials.

2. Cut tissue and place in the fixative solution. Place about 30 pieces of tissue in each vial (*see* **Note 4**).
 3. Place the vials of tissue in a vacuum desiccator and pull a vacuum. Pull the vacuum very slowly. Close the desiccator and leave under a vacuum for 15 min. Release the vacuum very slowly (*see* **Note 5**).
 4. Repeat **step 3**.
 5. If the tissue floats at this point, repeat *step 3* until it sinks (*see* **Note 6**).
 6. Incubate at room temperature until the tissue has been exposed to fixative solution for 2 h.
 7. Pull a vacuum again for 15 min, and then release.
 8. Incubate at room temperature for another 2 h without vacuum. At the end, the tissue should be exposed to the fixative for about 4.5 h.
 9. Remove fixative and add 15 mL of 50% ethanol. Incubate at room temperature for 30 min.
10. Repeat **step 9**.

3.1.2. Dehydration

1. Remove the 50% ethanol and replace with 15 mL of 60% ethanol. Incubate for 30 min at room temperature.
2. Remove the 60% ethanol and replace with 15 mL of 70% ethanol. Incubate for 30 min at room temperature.
3. Remove the 70% ethanol and replace with 15 mL of 85% ethanol. Incubate for 30 min at room temperature.
4. Remove the 85% ethanol and replace with 15 mL of 0.1% eosin Y in 95% ethanol. Leave in this solution several hours or overnight (*see* **Note 7**).
5. Remove as much as possible of the eosin-ethanol solution and replace with 15 mL of 100% ethanol. Incubate for 1 h at room temperature (*see* **Note 8**).
6. Remove as much as possible of the 100% ethanol and replace with 15 mL of fresh 100% ethanol. Incubate for 30 min at room temperature.

3.1.3. Clearing

1. Remove the 100% ethanol and replace with 15 mL of 25% xylene:75% ethanol. Incubate at room temperature for 30 min (*see* **Notes 9** and **10**).
2. Remove the 25% xylene solution and replace with 15 mL of 50% xylene:50% ethanol. Incubate at room temperature for 30 min.
3. Remove the 50% xylene solution and replace with 15 mL of 75% xylene:25% ethanol. Incubate at room temperature for 30 min.
4. Remove the 75% xylene solution and replace with 15 mL of 100% xylene. Incubate at room temperature for 30 min.
5. Remove the 100% xylene and replace with 15 mL of fresh 100% xylene. Incubate at room temperature for 30 min.
6. Remove the 100% xylene and replace with 10 mL of fresh 100% xylene.

3.1.4. Infiltration

1. Add 20 chips of Paraplast Plus to each vial. Incubate overnight at room temperature.

2. After the overnight incubation, the Paraplast will be only partially into solution. Place vials in a 42°C incubator. After approx 30 min, the Paraplast chips will be in solution.
3. Add 20 more Paraplast Plus chips, and incubate at 42°C until the chips are in solution. It will require approx 30 min. Swirl occasionally.
4. Repeat **step 3** until the vial is full (four to five times; total of approx 100 chips).
5. Pour off the xylene/Paraplast solution. Add molten Paraplast Plus. Swirl to mix. Incubate at 57–62°C for at least 4 h.
6. Repeat **step 5** three to six times (*see* **Note 11**).

3.1.5. Pouring Boats

1. Prewarm the micro slide warming table for approx 30 min.
2. Place an aluminum weighing dish on the hottest part of the micro slide warming table.
3. Pour the infiltrated tissue into the aluminum weighing dish and top off with molten Paraplast Plus.
4. Arrange the tissue into a regular array using a hot dissecting needle. Use the flame of an alcohol lamp to heat the needle before placing into the hot Paraplast. Tissue pieces must be at least 5 mm apart.
5. Gradually move the boat to the cooler part of the hot plate. It will probably be necessary to arrange the tissue again.
6. Label each boat by inserting a flagged string.
7. Let harden and then float the boat for 30 min (15 min per side) in water at room temperature.

3.1.6. Sectioning (see **Note 12**)

1. Cut out blocks of embedded tissue and mount onto microtome blocks.
2. Section tissue at 8 µm.
3. Cut ribbons into 1.5-cm pieces. Float ribbon pieces on 42°C water for >1 min or until the tissue is no longer compressed.
4. Put a Superfrost Plus microscope slide in the water just under the floating ribbon.
5. Bring the slide up so as to catch the ribbon. Use a spatula to position the ribbon.
6. Repeat for as many different ribbon pieces as you want to place on a given slide.
7. Incubate the microscope slide on a slide warmer overnight to "bake" the ribbon piece onto the slide. This is a very important step. This must be done immediately, and the slide warmer must be set to 45–50°C, or the sections may fall off during the hybridization and wash steps.

3.2. Probe Synthesis

The procedure given here is for the synthesis of single-stranded RNA probes (*see* **Note 13**). Generally two probes are synthesized. The first probe hybridizes with the RNA of interest, and is called antisense, anti-mRNA, or (+) strand probes. You must also synthesize a control probe that will not hybridize with the RNA of interest. The most convenient thing to do is to synthesize an RNA probe in the opposite orientation as the anti-RNA probe. Probes of this type are called sense, mRNA, or (-) strand probes.

3.2.1. General Comments

1. Gloves should be worn at all times to avoid the possibility of RNase contamination.
2. All solutions must be RNase-free (see **Note 1**).
3. All glassware and plasticware must be RNase-free (see **Note 2**).

3.2.2. Template Preparation

1. Digest with the appropriate restriction enzyme to linearize the template so that a "run-off" transcript can be generated (see **Note 14**).
2. Extract once with phenol/chloroform.
3. Extract twice with chloroform.
4. Precipitate the DNA with ethanol.
5. Spin down the DNA, wash with 70% ethanol, and dry.
6. Resuspend the DNA in RNase-free TE at a concentration of about 0.5 mg/mL.

3.2.3. Probe Synthesis (see **Note 15**)

1. Dry down 7.5×10^{-10} moles of [^{35}S]-UTP (see **Note 16**).
2. Resuspend in RNase-free H_2O. Resuspend in the amount of H_2O needed to reach a final volume of 50 µL in the reaction below.
3. Set up a 50 µL reaction. Mix the following reagents to achieve the appropriate final concentrations.

Reagent	Final concentration
[^{35}S]-UTP	15 µM
Transcription buffer	1X
DTT	10 mM
ATP	500 µM
GTP	500 µM
CTP	500 µM
Linerarized plasmid	50 ng/µL
RNasin	50 U

4. Remove a 1µL sample from the reaction to be used to assay the reaction in **Subheading 3.2.4.** Add this sample to 100 µL of H_2O. Label this the prereaction sample.
5. Add 2 µL of RNA polymerase (T3, T7, or SP6).
6. Incubate at 40°C for 1 h.
7. Remove a 1 µL sample from the reaction to be used to assay the reaction in **Subheading 3.2.4.** Add this sample to 100 µL of H_2O. Label this the postreaction sample.
8. Place the reaction tube on ice until you know that the reaction has worked well.

3.2.4. Determine Amount of RNA Synthesized

1. Label four Whatman DE-81 filters as follows: Pre/washed, pre/unwashed, post/washed, and post/unwashed (see **Note 17**).

2. Spot 10 µL of the prereaction sample onto the pre/washed filter and another 10 µL onto the pre/unwashed filter.
3. Spot 10 µL of the postreaction sample onto the post/washed filter and another 10 µL onto the post/unwashed filter.
4. Dry the filters under a heat lamp (approx 10 min).
5. Place the pre/washed and post/washed filters in at least 100 mL of 0.5M phosphate buffer. Stir the solution for 5 min at room temperature, and then pour off the phosphate buffer.
6. Repeat **step 5** at least three times.
7. Add at least 100 mL of H_2O to the filters, stir for 5 min, and then pour off the H_2O.
8. Add 100 mL of ethanol to the filters, stir for 1 min, and then pour off the ethanol.
9. Dry the filters under a heat lamp (approx 5 min).
10. Determine the amount of radioactivity on each of the four filters using a scintillation counter.
11. Calculate the amount of RNA synthesized. The unwashed filters represent total counts per minute (cpm) and the washed filters represent incorporated cpm. The ratio of the two numbers is the percent incorporation. The reaction has 255 ng of UTP. Therefore, the maximum theoretical yield of RNA is 1020 ng. Calculate the amount of RNA synthesized by multiplying the percent incorporation by 1020 ng (*see* **Note 18**).

3.2.5. Removal of DNA Template

1. If the incorporation is good, add 1 µL of RNAse-free DNase per µg of template, and incubate at 37°C for 15 min.
2. Add 150 µL of TED buffer and 2 µL of 10 mg/mL tRNA.
3. Extract once with phenol/chloroform.
4. Extract once with chloroform.
5. Ethanol precipitate.
6. Spin down the probe and wash the pellet with 70% ethanol.
7. Dry the pellet and resuspend in 50 µL of RNase free H_2O.
8. Remove a 1-µL sample and determine the number of cpm using a scintillation counter.
9. Take off a sample of about 10^7 cpm to run on a gel (prehydrolysis sample).

3.2.6. Probe Hydrolysis (see **Note 19**)

1. Calculate the amount of time to hydrolyze the probe (*see* **Note 20**). The required hydrolysis time is calculated using the following formula: $t = (L_o - L_f) / (k)(L_o)(L_f)$, where t is hydrolysis time in minutes, L_o is starting length in kilobasepairs, L_f is final length in kilobasepairs (=0.1 kb), and k is rate constant for hydrolysis (= 0.11 kb/min).
2. Mix: 50 µL RNA, 30 µL 200 mM sodium carbonate, and 20 µL 200 mM sodium bicarbonate.
3. Incubate at 60°C for the calculated time.
4. Stop the reaction by putting on ice and adding (*see* **Note 20**): 3.5 µL 3M sodium acetate (pH 6.0) and 5.0 µL 10% gracial acetic acid.

In Situ Hybridization

5. Take off a sample of approx 10^7 cpm to run on a gel (posthydrolysis sample).
6. Precipitate the probes: 1 μL 10 mg/mL tRNA, 8 μL 3M sodium acetate, and 250 μL ethanol.
7. To determine probe size, run the pre- and posthydrolysis samples along with a size standard on a denaturing gel (*see* **Note 21**). Dry down the gel and expose to X-ray film.
8. Store the probe as an ethanol precipitate until needed. After spinning down the probe, resuspend it in 25–50 uL of TED buffer.
9. Count 1 μL. Using the probe-specific activity, calculate the RNA concentration. You will need this number to know the amount of probe to add to the hybridization (*see* **Note 22**).

3.3. Hybridization

3.3.1. General Comments

1. Gloves should be worn at all times to avoid the possibility of RNase contamination.
2. All solutions must be RNase-free (*see* **Note 1**).
3. All glassware and plasticware must be RNase-free (*see* **Note 2**).
4. For all steps involving xylene, wear gloves and perform in a fume hood.

3.3.2. Set-Up For Hybridization Steps

1. Treat all staining dishes with DEPC (*see* **Note 2**).
2. Remove the hybridization solution from the freezer and place at room temperature.
3. Remove the 10 mg/mL proteinase K stock solution from the freezer and thaw.
4. Place 400 mL of proteinase K buffer in a staining dish. Place this dish in a 37°C water bath to preheat.
5. Set up 15 staining dishes containing the following solutions:
    a. 100% Xylene (drop a small stir bar into this dish).
    b. 100% Xylene (drop a small stir bar into this dish).
    c. 100% Ethanol.
    d. 100% Ethanol.
    e. 95% Ethanol (made with RNase-free H_2O).
    f. 85% Ethanol (made with RNase-free H_2O).
    g. 70% Ethanol (made with RNase-free H_2O).
    h. 50% Ethanol (made with RNase-free H_2O).
    i. 30% Ethanol (made with RNase-free H_2O).
    j. 15% Ethanol (made with RNase-free H_2O).
    k. RNase-free H_2O.
    l. RNase-free H_2O.
    m. Proteinase K solution (preheat to 37°C).
    n. RNase-free H_2O.
    o. RNase-free H_2O.

3.3.3 Dewaxing and Hydration

1. Place 20 slides (*see* **Note 12**) in a staining dish slide rack.
2. Place the slides into the first xylene. Stir for 10 min using a magnetic stirrer. Stir very slowly—just enough to move the solution around a bit.
3. Remove the slides from the first xylene, place into the second xylene, and stir for 10 min.
4. Remove the slides from the xylene and place into the first 100% ethanol. Dip up and down about 10 times or until the "streaks" go away.
5. Repeat **step 4** with the second 100% ethanol.
6. Process the slides sequentially through the following solutions: 95, 85, 70, 50, 30, and 15% ethanol, H_2O, and H_2O. In each solution, dip up and down 10 times or until the "streaks" go away. Begin with 95% ethanol and end with H_2O so as to hydrate gradually.

3.3.4. Proteinase K Digestion and Dehydration (see **Notes 23** and **24**).

1. Add proteinase K to the preheated solution to a final concentration of 1 µg/mL.
2. Incubate for 30 min at 37°C.
3. After the incubation, remove from the proteinase K solution and place into a staining dish filled with H_2O. Dip up and down a few time to rinse off the proteinase K.
4. Remove slides and place them into a second staining dish filled with H_2O.
5. Process the slides sequentially through the following ethanol solutions: 15, 30, 50, 70, 85, and 95%. Use the same solutions as used in **Subheading 3.3.3.** In each solution, dip up and down 10 times or until the "streaks" go away. Begin with 15% ethanol and end with 95% ethanol so as to dehydrate gradually.
6. Remove the slides from the 95% ethanol and place into a staining dish containing fresh 100% ethanol. Dip up and down 10 times or until the "streaks" go away.
7. Remove the slides from the 100% ethanol. Remove excess ethanol by blotting with paper towels.
8. Place slides with slide rack in a vacuum desiccator and dry the slides under a vacuum. This will take approx 1 h.

3.3.5. Hybridization

1. Before applying the hybridization solution, look at the slides and rank them according to quality. Hybridize 15 slides with the antisense probe and 5 slides with the control (sense) probe. Use the worst 2–3 slides as tester slides to gage the exposure length. Hybridize the tester slides with the antisense probe. Be sure to hybridize some good-quality slides with the control (sense) probe. Hybridize the best 3–4 slides with the antisense probe and do not develop these slide until the best exposure length is known.
2. Make up antisense hybridization solution for 15 slides. If the entire slide is filled with sections, you will need approx 150 µL of hybridization solution per slide or

a total of approx 2.25 mL. Use a probe concentration of 200 ng/mL/kb of probe complexity (*see* **Note 25**). Thus, for a probe of 1 kb complexity, mix 450 ng of probe with 2.25 mL of hybridization solution.
3. Make up sense hybridization solution for 5 slides. You will need 750 µL of hybridization solution. The probe concentrations in the sense and antisense hybridization solutions must be exactly the same.
4. Prewarm the hybridization solution to 42°C. This makes it easier to spread.
5. Process one slide at a time. Apply the hybridization solution and then immediately cover with a coverslip. To apply the hybridization solution, elevate one side of the slide approx 1 cm. Apply the hybridization solution in a pool to the elevated side and right onto the tissue. Spread the hybridization solution down the slide using a yellow pipet tip as a spatula. Make sure the tissue is wet—sometimes the hybridization solution has a tendency to surround but not penetrate the tissue. This step is made easier by using a lot (150 µL) of hybridization solution. Keep the slide slanted until a small pool of hybridization solution accumulates toward the bottom end. Place the slide on the benchtop so that it is now in a horizontal position. Hold the cover slip at one end with a jeweler's forceps and place the other cover slip end on the slide end that has the small pool of hybridization solution. Slowly lower the cover slip to completely cover the slide. Avoid bubbles—If you get a lot of bubbles, start over.
6. Place the slides in a humidified box and incubate at 42°C overnight. Place several layers of paper towels that are soaked with 50% formamide at the bottom of a Tupperware box. Place the slides onto a pair of 5-mL plastic pipets to elevate them so that the probe does not get wicked away.

3.4. Wash

3.4.1. Set-Up for Wash Steps

1. Place 400 mL of RNase buffer into an RNase staining dish. Place this dish into a water bath set to 37°C to preheat.
2. Remove the 25 mg/mL RNase A stock solution from the freezer and thaw.
3. Place 2 L of high stringency wash buffer into each of two, 4 L beakers. Place these beakers in a water bath set to 57°C to preheat.

3.4.2. Removal of Cover slips

1. Fill two 150–250-mL beakers and a staining dish with first wash buffer. Place a staining dish slide rack into the staining dish.
2. Process one slide at a time. Process the five control (sense) slides first, change the first wash buffer in the two beakers, and then process the 15 antisense slides.
3. Dip the first slide up and down in the first beaker until its cover slip comes off.
4. Dip up and down 5–10 times in the second beaker of first wash buffer to completely rinse off the hybridization solution.
5. Place the slide into the staining dish.
6. Repeat for the other 19 slides. When all slides have been processed, proceed to the next step.

3.4.3. RNase Treatment (see **Note 26**)

1. Transfer the slides to a staining dish containing RNase Buffer at room temperature. Dip up and down 10 times and then soak the slides in this solution for 5 min to ensure that all of the hybridization solution has diffused out of the tissue.
2. Add RNase to the preheated RNase Buffer to a final concentration of 25 µg/mL. Mix by stirring with a plastic pipet. RNase precaution. Do not use a pipetman to measure out the RNase stock, use a plastic 1-mL pipet.
3. Place the slides into the RNase solution, and incubate at 37°C for 30 min. Dip up and down a few times during the incubation.
4. After the 30-min incubation, remove slides from the RNase solution and place into a staining dish containing low stringency wash buffer. Dip up and down 15 times to rinse off the RNase. Use the RNase staining dishes.
5. Repeat **step 4** twice for a total of three rinses.

3.4.4. Low- and High-Stringency Washes

1. Place 4 L of low stringency wash buffer into a large plastic container (8-L container). Place the plastic container on top of a magnetic stirrer.
2. Place the slides (still in slide rack) in the buffer, and wash with stirring for 30 min.
3. Place the slides (still in slide rack) in a 4-L beaker containing 2 L of high stringency wash buffer at 57°C. Place this beaker in a shaking water bath set to 57°C. Shake slowly to agitate the buffer a bit. Dip slides up and down occasionally to dislodge the bubbles that accumulate on the slides.
4. Repeat **step 3** with fresh high stringency wash buffer.

3.4.5. Dehydration

1. Process the slides sequentially through the following ethanol solutions: 15, 30, 50, 70, 85, 95, and 100%. In each solution, dip up and down 10 times or until the "streaks" go away. Begin with 15% ethanol and end with 95% ethanol so as to dehydrate gradually.
2. Remove the slides from the 100% ethanol. Remove excess ethanol by blotting with paper towels.
3. Place slides with slide rack in a vacuum desiccator and dry under a vacuum. This will take approx 1 h.

3.5. Autoradiography

3.5.1. Preparing Emulsion

1. This step must be done in a photographic darkroom with a safelight.
2. Preheat the emulsion to 42°C in a water bath for approx 30 min.
3. Place an equal volume of H_2O in a 500 mL flask and preheat in a water bath to 42°C.
4. Pour the emulsion into the flask, swirl to mix, return to the 42°C water bath.
5. Aliquot approx 20 mL into slide mailers.
6. Wrap the slide mailers in two layers of aluminum foil and store at 4°C for up to 6 mo.

3.5.2. Dipping Slides

1. This step must be done in a photographic darkroom with a safelight.
2. Remove an aliquot of emulsion, unwrap the aluminum foil, and place in a 42°C water bath for about 1 h to melt.
3. Dip each slide into liquid emulsion. Dip slowly, and one time only to avoid streaks.
4. Dry slides completely—place in a rack, inside a black wooden box with desiccant. This will take 30–60 min.
5. Place slides into black plastic slide boxes. Wrap the slide boxes with two layers of aluminum foil.
6. Expose at 4°C for the appropriate time (*see* **Note 27**).

3.5.3. Develop Slides

1. Fill three staining dishes with developer, water, and fixer. Use fresh solutions each time. Cool solutions to 15°C (*see* **Note 28**).
2. In a photographic darkroom with safelight, remove the appropriate slides from the black plastic slide box and place into a staining dish slide rack.
3. Place slides into developer, dip up and down five times, and let sit in developer for a total of 2.5 min.
4. Remove slides from the developer, drain briefly, and place into the water. Dip up and down continuously for 30 s.
5. Remove slides from the water, drain briefly, and place in fixer. Dip up and down five times, and let sit in fixer for at least 5 min.
6. Remove slides from fixer, drain briefly, and place into water until ready to stain. Rinse in running water for 30 min to remove fixer.

3.6. Staining and Mounting Slides

3.6.1. Staining

1. Fill a slide mailer with 0.1% Toluidine blue. Fill nine coplin jars with the following solutions: H_2O, H_2O, 25, 50, 75, 100, and 100% ethanol, 100, and 100% xylene.
2. Soak slides in 0.1% Toluidine blue until tissue is deeply stained (15–30 s). The emulsion and tissue will be stained very deeply.
3. Rinse off briefly in the first H_2O to get rid of the excess stain.
4. Soak slides in the second water to destain. Soak until the emulsion is only faintly stained but the tissue is still darkly stained (*see* **Note 29**). This step can take >1 h.
5. Process the slides progressively through the following ethanol solutions: 25, 50, 75, 100, and 100%. Dip up and down 5–10 times in each of the following ethanol solutions: Move quickly through the ethanols because the tissue becomes destained in these solutions. It is best to process the slides one at a time through the ethanols.
5. In a chemical fume hood, dip the slides in the first xylene 15 times or until the streaks go away. No need to move quickly because the tissue does not destain in xylene.
6. Place the slides into the second xylene (in fume hood). Let the slides sit in xylene until all slides are processed.

3.6.2. Mounting Cover Slips

1. Process the slides one at a time. Do this step in a chemical fume hood.
2. Remove a slide from xylene and place onto a paper towel.
3. Quickly add two drops of permount to the surface of the slide.
4. Quickly place cover slip on slide.
5. Squeeze out the excess xylene/permount—wipe surface of cover slip while holding down.
6. Let dry for at least 1 h.
7. When the permount is hardened, clean the surface of the cover slip by wiping with xylene (*see* **Note 30**).
8. Scrape off the emulsion on the backside of the slide using a razor blade.

3.7. Microscopy

1. Before observing the slides with a compound microscope, clean the surfaces of the slide with xylene to remove the excess permount.
2. Place the cleaned slide on a microscope fitted with a bright-field condenser.
3. Using a low-power objective (×5), search for a good, quality tissue section.
4. Once a suitable section is found, switch to a higher power. Under bright-field optics, the silver grains appear as small black spots against a bright background (*see* **Fig. 1A**). If the signal is extremely strong, black areas can be seen. In general, however, it is very difficult to observe silver grains using bright-field microscopy.
5. Without moving the slide, remove the bright-field condenser and replace it with a dark-field condenser. Under dark-field optics, the silver grains appear as white spots against a dark background. Because of the dramatic contrast, every silver grain can be observed (*see* **Fig. 1B**).
6. Observe the signal. The emulsion over the cells that have hybridized with the probe will contain a high density of silver grains. To be a convincing signal, this grain density should be at least five times the background grain density.
7. Assess the level of emulsion background. Emulsion background is an even distribution of silver grains throughout the emulsion (*see* **Note 31**). The emulsion background level can be determined by counting the grain density in the region between tissue sections or on a blank slide dipped in emulsion. It should be >10 times lower than the grain density caused by signal.
8. Assess the level of tissue background. Tissue background is caused by probe sticking nonspecifically to the tissue. Tissue background will be slightly higher than emulsion background and can be assessed by counting the grain density over the nonexpressing cells, as well as in the tissue hybridized with the control (sense) probe. Tissue background is generally about one grain per cell (*see* **Note 32**).
9. For publication, you will need both bright-field and dark-field photographs. For the bright-field photograph, deeply stained tissue works best. For the dark-field photograph, lightly stained or unstained tissue works best. Thus, you will have to stain deeply for the bright-field photographs and then destain the tissue for the dark-field photographs. As an alternative to making separate bright-field and dark-field photographs, a bright-field/dark-field double exposure can be prepared.

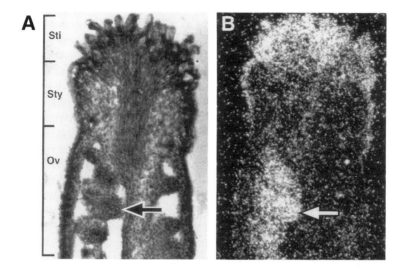

Fig. 1. *In situ* hybridization of an *AGAMOUS* probe with a tissue section of an *Arabidopsis* pistil. **(A)** Bright-field micrograph showing the stigma (Sti), style (Sty), and ovary (Ov) of the pistil. Ovules (arrow) are found within the ovary. **(B)** Dark-field micrograph showing a high density of silver grains over the the stigma and ovules.

To enhance the appearance of the silver grains against the blue-stained tissue, use a red filter for the dark-field exposure.

4. Notes

1. To make solutions RNase-free, add DEPC to 0.05%, shake hard, incubate at room temperature for ≥2 h, and autoclave for 20 min. If a solution cannot be autoclaved (e.g., alcohol solutions, Denhardt's reagent, 50% dextran sulfate), make it up in RNase-free H_2O using RNase-free glassware (*see* **Note 2**). DEPC reacts with amines and, thus, cannot be used to treat Tris solutions. For making RNase-free Tris solutions, set aside a special bottle of Tris that is used only for making RNA solutions. Dissolve the Tris in RNase-free H_2O using RNase-free glassware (*see* **Note 2**).
2. To make glassware and plasticware RNase-free, soak in 0.1% DEPC for a few minutes, wrap in aluminum foil, and autoclave for 20 min. Sterile and disposable plasticware can be assumed to be RNase-free unless contaminated by the investigator—wear gloves at all times.
3. Fixation is one of the most critical steps of *in situ* hybridization. It preserves the morphology of the tissue. Fixation also crosslinks the RNA to proteins and other macromolecules and, thus, is critical for RNA retention during the hybridization and wash steps. Signal can be reduced or eliminated if the tissue is either over- or underfixed. If the tissue is overfixed, the probe cannot penetrate the tissue. If the tissue is underfixed, the RNA gets washed out of the tissue. Thus, to maximize signal, the fixation procedure must be optimized. Unfortunately, each tissue type

is likely to have a different fixation procedure that is optimal. The fixation procedure given in this protocol works well with *Arabidopsis* floral tissue. *Arabidopsis* floral tissue fixed in paraformaldehyde is almost as good, but floral tissue fixed in glutaraldehyde gives no signal.
4. Getting the fixative into the tissue is the most important step of the fixation procedure. To allow optimal penetration, the tissue must be cut into small pieces just before placing into the fixative solution. For hard tissue (e.g., stems), the distance between cut edges should be ≤2 mm. For softer tissue (e.g., petals, leaves, roots), cut into 3 × 3 mm pieces. For *Arabidopsis* floral tissue, cut a cluster of flowers at the apex of the floral stem that includes stages 1–12. Cut so that 1–2 mm of floral stem is present—this helps to orient the tissue when sectioning.
5. Pulling a vacuum during fixation pulls the air out of the tissue and allows the fixative to penetrate better.
6. Tissue flotation during fixation is bad because the tissue will not be exposed as intensly to the fixative. Tissue generally floats because it has air in it. Pulling a vacuum during fixation generally causes the tissue to sink. Sometimes the vacuum causes air bubbles to come out of the tissue or fixative solution. These air bubbles can get trapped in the trichomes and cause the tissue to float. Swirling the solution generally dislodges these air bubbles.
7. The tissue is stained with Eosin Y to make it more visible in the wax blocks. It dramatically helps to orient the tissue during sectioning.
8. Eosin destains slowly in 100% ethanol. Thus, do not incubate for long periods (overnight) in 100% ethanol.
9. The tissue must be completely dehydrated when it hits the xylenes; otherwise, the water and xylenes will form a white emulsion.
10. Eosin will not diffuse from the tissue in xylene.
11. The purpose of the Paraplast changes is to dilute out any remaining xylene. If you are able to remove almost all of the molten Paraplast each time, three changes should be adequate. Some tissues will float in the molten paraplast, making it difficult to remove all of the molten paraplast each time. Under these circumstances, six or more changes may be needed.
12. Twenty slides for a one-probe experiment will need to be prepared. Generally, 15 of these will be hybridized with the antisense probe and 5 with the control (sense) probe.
13. Single-stranded RNA probes offer several advantages over double-stranded DNA probes. First, asymmetric probes cannot anneal during hybridization. Second, RNA:RNA hybrids are considerably more stable than RNA:DNA hybrids. Third, posthybridization RNase digestion dramatically reduces background caused by nonspecific binding. Probably for all of these reasons, single-stranded RNA probes hybridize with eight-fold greater efficiency than double-stranded DNA probes *(1)*.
14. When linearizing the template, avoid the use of enzymes that generate 3' overhangs. Both of the 3' overhanging ends act as sites of transcriptional initiation, leading to the production of RNA containing vector sequences, as well as RNA complementary to the desired runoff transcript *(9)*.

15. A typical experiment consists of 15 slides hybridized with the antisense probe and five slides hybridized with the control probe (*see* **Note 12**). Use a probe concentration of 200 ng/mL per kb of probe complexity (*see* **Note 25**). For a surface area of 24 × 50 mm, use approx 150 uL of hybridization solution/slide. Thus, a typical experiment will require 450 ng of antisense probe and 150 ng of sense probe/kb of probe complexity.
16. 7.5×10^{-10} Mol of UTP in a 50 µL reaction is a concentration of 15 µM. In the transcription reaction, good yields of RNA can be achieved with as little as 7.5 µM UTP; however, the reaction is much more reliable if the UTP concentration is 15 µM.
17. DE-81 Filters contain a high density of positive charges. Incorporated nucleotides adsorb strongly to DE-81 filters and are not washed off during the phosphate buffer washes. Unincorporated nucleotides adsorb less strongly and are washed off during the phosphate buffer washes.
18. In the transcription reaction, 50–90% incorporation generally occurs. However, some plasmids consistently result in poor incorporation. For those plasmids that result in poor incorporation, try boosting up the [UTP] by adding either more label or cold UTP.
19. Shorter probe fragments give higher signals for *in situ* hybridization, presumably because they penetrate the tissue better *(10–13)*. The optimal size is unknown; however, it has been shown that for formaldehyde-fixed tissue, probes of mean length 70 nt give higher signals than probes of mean length 140 nt *(13)*.
20. Probe size is reduced by limited alkaline hydrolysis at 60°C in 40 mM sodium bicarbonate, 60 mM sodium carbonate, pH 10.2. The reaction is neutralized by adding sodium acetate, pH 6.0 to 100 mM, and glacial acetic acid to 0.5% *(1)*.
21. The type of denaturing gel that works best is a 10% polyacrylamide gel in 8M urea, basically a short sequencing gel. If you do not have the capability of running polyacrylamide gels, you can run a denaturing agarose gel. The choices here are glyoxal/dimethyl sulfoxide, formaldehyde, and methylmercuric hydroxide. Instructions on how to runs these gels can be found in **ref. 15**.
22. Probe specific activity in dpm/µg is calculated using the formula: ([Specific Activity of $\{^{35}S\}$-UTP in mCi/µmol] [2.2×10^9 dpm/mCi] [1 µmol/340 ug] [1/4]). For the 650 Ci/mmol [^{35}S]-UTP, the probe specific activity is 1.05×10^9 dpm/µg. Note that dpm is not equal to cpm with [^{35}S]-UTP. Consult the instruction manual for your scintillation counter to make the conversion.
23. Proteinase K is used to partially digest the tissue to allow better probe penetration. This step increases the signal. Proteinase K-treated tissue gives at least five times higher signal than untreated tissue. For *Arabidopsis* floral tissue, the best incubation time is 30–45 min. Less time gives a weaker signal, and greater time tends to destroy the morphology of the tissue. For different species or tissues, a different incubation time may be optimal.
24. The purpose of dehydrating the tissue just before applying the hybridization solution is to completely dry the sections so that when the hybridization solution is applied, it is sucked into the tissue.

25. As probe concentration is increased, signal increases linearly but then plateaus when the saturating probe concentration is achieved. For the hybridization of both a histone probe with sea urchin embryos and an actin probe hybridized with sea urchin plutei, the saturating probe concentration is about 300 ng/mL *(1,14)*. The target RNAs for these two probes differ in concentration by 40–80-fold, suggesting that saturating probe concentration is independent of target density *(14)*. To obtain the best signal to noise ratio it is important to use a probe concentration at or below saturation. Thus, a concentration of 200 ng/mL is likely to work well for most probes.
26. The RNase step reduces background significantly (two- to five-fold). Raising the RNase concentration above 25 μg/mL does not help to reduce background further. Very high RNase concentrations (100–200 μg/mL) will reduce or eliminate signal.
27. The amount of time required for exposure is highly variable. For most floral mRNAs, 2–3 wk of exposure are required when the 650 Ci/mmol [^{35}S]-UTP is used. If you suspect that the hybridizing mRNA is highly prevalent in the expressed cells, develop the first tester slide at about one day of exposure. Generally, the first tester slide is developed at 3.5 d of exposure. If no signal is seen, develop a second tester slide at one week of exposure. Continue this process, each time doubling the exposure, until the optimal exposure is determined. If you do not see a signal within 3–4 wk of exposure, try a higher specific activity [^{35}S]-UTP (e.g., 1250 Ci/mmol).
28. Developing the slides at reduced temperature (15°C) and for a short time (2.5 min) reduces grain density in the emulsion background.
29. If the tissue is too deeply stained, the hybridization grains will be obscured.
30. The coverslips can be removed at any time by soaking in xylene for 1–5 d.
31. All emulsion has a background of silver grains. To determine the emulsion background level, dip a blank slide and develop it. The emulsion background is somewhat variable in that some emulsion batches have more background than others. Furthermore, different emulsion aliquots from the same batch have different background emulsion levels. At least three things contribute to emulsion background. The first is emulsion age. Older emulsion has a higher emulsion background. For this reason, try to purchase the freshest possible batch of emulsion. The second factor contributing to emulsion background is handling. Rough handling (e.g., dropping the solid emulsion) increases emulsion background. Finally, exposure to a source of radiation will dramatically increase emulsion background. It is therefore important to store the emulsion in a refrigerator that is ^{32}P-free.
32. In general, the tissue background will be evenly distributed over the tissue. Some cell types, however, consistently produce a high background, presumably because they contain large amounts of positively charged molecules. For example, the abaxial epidermis of *Arabidopsis* sepals consistently produces an extremely high background *(16)*.

References

1. Cox, K. H., DeLeon, D. V., Angerer, L. M., and Angerer, R. C. (1984) Detection of mRNAs in sea urchin embryos by *in situ* hybridization using asymmetric RNA probes. *Develop. Biol.* **101,** 485–502.

2. Cox, K. H. and Goldberg, R. B. (1988) Analysis of plant gene expression. In *Plant Molecular Biology, A Practical Approach* (Shaw, C. H., ed.), IRL Press, Oxford, UK, pp. 1–34.
3. Meyerowitz, E. M. (1987) *In situ* hybridization to RNA in plant tissue. *Plant Molec. Biol. Rep.* **5,** 242–250
4. McFadden, G. I., Bonig, I., Cornish, E. C., and Clarke, A. E. (1988) A simple fixation and embedding method for use in hybridization histochemistry on plant tissues. *Histochem J.* **20,** 575–586.
5. McFadden, G. I. (1989) *In situ* hybridization in plants: from macroscopic to ultrastructural resolution. *Cell Biol. Int. Rep.* **13,** 3–21.
6. Kronenberger, J., Desprez, T., Hofte, H., Caboche, M., and Traas, J. (1993) A methacrylate embedding procedure developed for immunolocalization on plant tissues is also compatible with *in situ* hybridization. *Cell Biol. Int.* **17,** 1013–1021.
7. DeBlock, M. and Debrouwer, D. (1993) RNA-RNA *in situ* hybridization using digoxigenin-labeled probes: The use of high-molecular-weight polyvinyl alcohol in the alkaline phosphatase indoxyl-nitroblue tetrazolium reation. *Analyt. Biochem.* **215,** 86–89.
8. Coen, E. S., Romero, J. M., Doyle, S., Elliot, R., Murphy, G., and Carpenter, R. (1990) Floircaula: A homeotic gene required for flower development in *Anthirrhinum majus*. *Cell* **63,** 1311–1322.
9. Schenborn, E. T. and Mierendorf, R. C. (1985) A novel transcription property of SP6 and T7 polymerases: Dependence on template structure. *Nucl. Acids. Res.* **13,** 6223–6236.
10. Angerer, L. M. and Angerer, R. C. (1981) Detection of poly A^+ RNA in sea urchin eggs and embryos by quantitative in situ hybridization. *Nucleic Acids Res.* **9,** 2819–2840.
11. Brahic, M. and Haase, A. T. (1978) Detection of viral sequences of low reiteration frequency by in situ hybridization. *Proc. Natl. Acad. Sci. USA* **75,** 6125–6129.
12. Gee, C. E. and Roberts, J. L. (1983) *In situ* hybridization histochemistry: a technique for the study of gene expression in single cells. *DNA* **2,** 157–163.
13. Moench, T. R., Gendelman, H. E., Clements, J. E., Narayan, O., and Griffin, D. E. (1985) Efficiency of *in situ* hybridization as a function of probe size and fixation technique. *J Virol. Meth.* **11,** 119–130.
14. Cox, K. H., Angerer, L. M., Lee, J .J., Davidson, E. H., and Angerer, R. C. (1986) Cell lineage-specific programs of expression of multiple actin genes during sea urchin embyrogenesis. *J. Molec. Biol.* **188,** 159–172.
15. Sambrook, J., Fritsch, E. F., and Maniatis, T. (1989) *Molecular Cloning, A Laboratory Manual,* 2nd ed. Cold Spring Harbor Laboratory, Cold Spring Harbor, NY.
16. Drews, G. N., Bowman, J. L., and Meyerowitz, E. M. (1991) Negative regulation of the *Arabidopsis* homeotic gene *AGAMOUS* by the *APETALA2* product. *Cell* **65,** 991–1002.

36

Whole-Mount *In Situ* Hybridization in Plants

Janice de Almeida Engler, Marc Van Montagu, and Gilbert Engler

1. Introduction

Messenger RNA *in situ* hybridization (ISH) can be defined as a technique for locating specific transcripts. This powerful technique opened the way to study the temporal and spatial patterns of expression of animal and plant genes in a morphological context and on a cell-to-cell basis. Most *in situ* protocols are based on the hybridization of a labeled DNA or RNA probe to tissue sections in which mRNA is preserved. Radioactive as well as nonradioactive *in situ* protocols have been applied with success for various tissues originating from plants and animals *(1-5)*.

Ideally, an ISH procedure should fulfill a number of criteria such as a good tissue preservation, a high resolution, a high sensitivity, and a high selectivity. Moreover, such a method should be quick, inexpensive, and nonhazardous. Since none of the existing protocols meet all these standards, the researcher has to choose a protocol according to his or her needs. Whereas the radioactive mRNA ISH method combines good tissue preservation with high sensitivity, the nonradioactive method offers a superior resolution and is rapid.

Some years ago, a whole-mount *in situ* hybridization (WISH) procedure using nonradioactive probes was developed for locating mRNAs in *Drosophila melanogaster* embryos *(6)*. Such a method offers the unique possibility to visualize transcripts in individual cells of an intact organism allowing a very rapid analysis when compared with the use of tissue sections. Based on the WISH procedures described by Tautz and Pfeifle *(6)* and Ludevid et al. *(7)*, we developed a modified version of this technique, using digoxigenin (DIG)-labeled RNA probes *(8)*. This method allowed us to detect reproducibly various mRNA species in different tissues of intact *Arabidopsis thaliana* seedlings with virtually no background. It appears that probe as well as antibody penetration, at least in young *Arabidopsis* seedlings is not a problem. The method includes tissue preservation with formaldehyde, permeabilization with polyoxyethylenesorbitan (Tween-20), dimethyl sulfoxide (DMSO), and heptane. Hybridization signal is detected using colloidal gold or alkaline phosphatase conjugated anti-DIG antibodies.

Despite the fact that the WISH method is faster and more sensitive than conventional protocols using tissue sections, it is only applicable on plantlets or dissected organs. Certain differentiated plant cells are surrounded by a cell wall resistant to our method of permeabilization. This problem, however, can be circumvented by the use of fresh vibroslices. The WISH method could be applied to detect mRNAs in thick vibroslices of mature organs derived from other plant species than *Arabidopsis*. Typical examples of WISH are shown in **Fig. 1**.

2. Materials

Gloves should be used during the entire procedure to prevent RNase contamination. All solutions are treated with diethyl pyrocarbonate (DEPC). Material which is not sterile can be cleaned with a mixture of ethanol containing 5% DEPC.

2.1. Sample Preparation and Fixation

1. Glass vials (Laborimpex, Brussels, Belgium).
2. Scalpel and forceps.
3. Phosphate-buffered saline (PBS): 120 mM NaCl, 7 mM Na$_2$HPO$_4$, 3 mM NaH$_2$PO$_4$, 2.7 mM KCl, pH 7.4.

Fig. 1. *(opposite page)* **(A)** WISH on an *Arabidopsis thaliana* seedling hybridized with an antisense *cdc2a* probe detected by gold-labeled antibodies. *Cdc2a* DIG RNA preferentially hybridizes in the apical meristem of the shoot and in the basal part of the two-leaf primordia of 4-d-old *Arabidopsis* seedlings *(9)*. The signal is visible as a black precipitate resulting from the silver amplification reaction. **(B)** *In situ* localization of *cdc2a* mRNA on a vibroslice of an *Arabidopsis* root infected with a root-knot nematode (de Almeida et al., in preparation). Fresh vibrosections were processed according to the WISH protocol and detected with AP as an enzymatic marker. The blue precipitate resulted from the histochemical reaction with the substrates X-phosphate and NBT. **(C)** Cotyledon hybridized with an antisense *rha1* probe. Hybrids were visualized by silver amplification of specifically bound gold-labeled antibodies seen as a dark precipitate. The *rha1* gene encodes a small GTP-binding protein *(10)* and its mRNA is predominantly present in particular stomata of young cotyledons. **(D)** Chicory root hybridized with an antisense nitrate reductase probe detected by AP-labeled antibodies *(11)*. A strong expression is visible in the vascular cylinder and in the root meristem. The sample was cleared in CLP for better visualization of the signal. (*See* color insert following p. 230.)

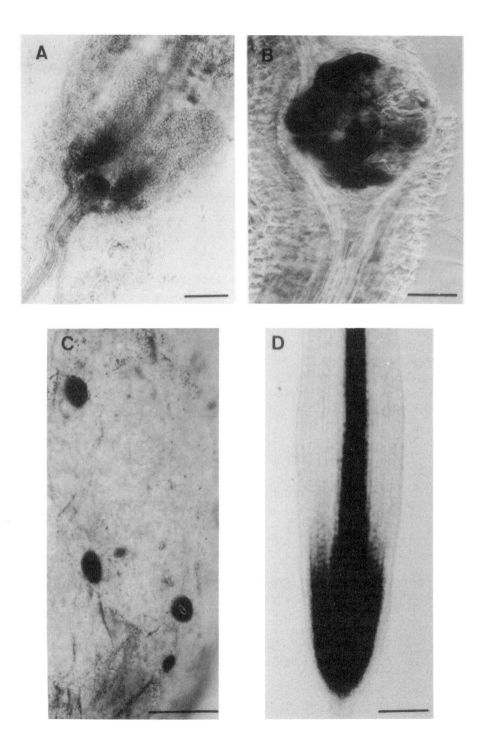

Fig. 1

4. PBT: PBS containing 0.1% Tween-20 (Pierce, Rockford, IL)
5. Fixation buffer (FB): PBT containing $0.08M$ ethylene glycol-bis(β-aminoethyl ether) N,N,N',N'-tetra-acetic acid, 5% formaldehyde (40%) (Merck, Darmstadt, Germany), and 10% DMSO (UCB, Leuven, Belgium), pH 7.4.
6. Heptane (Sigma, St. Louis, MO).
7. Absolute ethanol.
8. Methanol.
9. Horizontal shaker.

2.2. Vibrosections

1. 5% Agarose (Gibco-BRL, Gaithersburg, MD).
2. Plastic rings.
3. Superglue.
4. Vibroslicer.

2.3. Probe Synthesis

1. Phenol (Carlo Erba, Milano, Italy) equilibrated with tris(hydroxymethyl)-aminomethane (Tris).
2. Chloroform (Merck).
3. Ether (UCB).
4. $3M$ Sodium acetate, pH 6.0 (Merck).
5. Agarose, ethidium bromide, gel electrophoresis apparatus, and UV transilluminator.
6. Molecular weight marker (λ, PstI).
7. DIG RNA labeling kit (Sp6/T7) (Boehringer, Mannheim, Germany).
8. Incubator at 37°C.
9. tRNA (10 mg/mL stock) (Boeringer).
10. RNase inhibitor (Pharmacia, Uppsala, Sweden).
11. Bio spin columns (BioRad, Hercules, CA).
12. Speed vacuum.
13. Centrifuge at 13,000g.

2.4. Prehybridization and Hybridization Treatments and Posthybridization Washes

1. Six-well tissue-culture plate (Falcon® 3046; Becton Dickinson, Lincoln Park, NJ).
2. Cell strainers (Falcon V5, Becton Dickinson).
3. 1:1 Ethanol:xylene mixture.
4. 1:1 Methanol:PBT mixture.
5. 5% Formaldehyde in PBT.
6. 40 mg/mL Proteinase K in PBT (Merck).
7. 0.2% Glycine in PBT.
8. 1:1 PBT:hybridization solution (HS) solution.
9. Shaker water bath at 55°C.
10. 2-mL Eppendorf tubes (Eppendorf, Hamburg, Germany).

11. HS: 50% formamide (UCB), 5X SSC (1X SSC: 150 mM NaCl, 15 mM Na$_3$–citrate, pH 7.0), 50 μg/mL heparin.
12. Salmon sperm DNA (Sigma).
13. Poly(A) (Boehringer).
14. RNase A (Boehringer).
15. NTE: 500 mM NaCl, 10 mM Tris-HCl, pH 7.5, 1 mM ethylenediaminetetra-acetic acid (EDTA).

2.5. Plant Powder

1. Mortar and pestle.
2. Ice-cold 90% acetone.
3. Vortex.
4. Filter paper (Whatman, Maidstone, UK).
5. Desiccator.

2.6. Signal Detection: Chromogenic Alkaline Phosphatase (AP) Reaction and Silver Amplification

1. Blocking solution (BS): 2% bovine serum albumin Fraction V (Acros, Beerse, Belgium), 0.1% Tween-20 in PBT.
2. Anti-DIG AP conjugate F$_{ab}$ fragments (Boehringer).
3. Anti-DIG-gold ≤ 0.8 mm (Boehringer).
4. Levamisole (Sigma).
5. Nitro blue tetrazolium (NBT; Pierce).
6. 5-Bromo-4-chloro-3-indolylphosphate (BCIP; Pierce).
7. Staining buffer (SB): 100 mM NaCl, 50 mM MgCl$_2$, 100 mM Tris-HCl, pH 9.5, and 0.1% Tween-20.
8. Stain stop buffer (SSB): PBT containing 20 mM EDTA.
9. Deionized water.
10. Intense™ silver enhancement kit (Amersham, Aylesbury, UK).
11. Glycerol.
12. Glass slides (Vel, Leuven, Belgium) and coverslips.
13. Stereo microscope.
14. Light microscope.
15. Ethanol series: 10, 50, and 70.
16. Glutaraldehyde (Agar, Stansted, UK).
17. Clearing agent, chlorallactophenol (CLP): 2:1:1 chloral hydrate (Acros), lactic acid (Acros), and phenol (Merck).

3. Methods

3.1. Plant Tissue Preparation

1. Soil- or in vitro-grown intact seedlings (*see* **Note 1**) or organs, such as roots and flowers (*see* **Note 2**) are collected and transferred to glass vials containing FB (*see* **Note 3**).

3.2. Vibrosections

1. Cut plant material that was fixed for 30 min (*see* **Subheading 3.3.1.**) in tissue blocks and transfer them immediately into small plastic rings placed on a Petri dish. Pour melted 5% agarose over the tissue. Orient material in the agarose and allow to solidify.
2. Remove the plastic ring from the solidified agar block and glue the sample onto the specimen holder of the vibroslicer. Submerge sample in PBS.
3. Cut vibrosections of 50–150 μm or thicker and transfer them back to FB (*see* **Note 4**).

3.3. Tissue fixation

1. Fix plant material (whole seedlings, organs or vibrosections) for 30 min in 5 mL of fixation cocktail (1:1 ratio of heptane:FB; *see* **Note 5**). Vial should be shaken vigorously for 15 min to mix the organic and water phases.
2. Dehydrate twice for 5 min with absolute methanol and three times for 5 min with absolute ethanol.
3. Samples are stored in ethanol for 1–2 d at –20°C until the prehybridization treatment (*see* **Note 6**).

3.4. Digoxigenin Probe Synthesis

1. RNA probes are prepared in vitro using a DIG RNA labeling kit (Sp6/T7) (Boehringer Mannheim). The reaction is carried out according to the manufacturer's protocol.
2. Mix 1 μg linearized plasmid containing the insert, 2 μL of 10X transcription buffer, 2 μL of nucleotide triphosphate (NTP)-labeling mixture, 1 μL RNA inhibitor, and 2 μL of RNA polymerase (SP6, T7, or T3); add DEPC-treated deionized H_2O to a final volume of 20 μL and incubate for 2 h at 37°C.
3. Check transcript in an 0.9% agarose gel containing a molecular weight marker (λ, *Pst*I) and a known concentration of DIG-labeled RNA provided by the kit. Compare transcript size and amount from the template with those of the control DIG RNA to estimate synthesis (*see* **Note 7**).
4. Add 40 U DNAse and incubate for 15 min at 37°C to remove plasmid DNA.
5. Complete the volume of the probe to 50 μL with DEPC-treated water and hydrolyze at 60°C by adding 30 μL of $0.2M$ Na_2CO_3 and 20 μL $0.2M$ $NaHCO_3$. Riboprobes are digested to an average length of 200 nucleotides by controlled alkaline hydrolysis *(11)* (*see* **Note 8**).
6. Add 5 μL of 10% acetic acid to stop the alkaline hydrolysis.
7. Precipitate probe by adding 11 μL of $3M$ Na-acetate, 2.5 vol of cold ethanol, 1 μg of tRNA, and incubate for 4 h at –20°C.
8. Collect the pellet by centrifugation and redissolve in 50 μL of RNAse-free water containing 40 U RNase inhibitor.
9. Store probe in aliquots at –20°C.

3.5. Prehybridization Treatment

1. Rinse fixed samples once in absolute ethanol and incubate once for 30 min in a 1:1 ethanol/xylene mixture (*see* **Note 9**).
2. Transfer samples to cell strainers, wash twice for 5 min in absolute ethanol and twice for 5 min in absolute methanol.
3. Wash samples for 5 min in a 1:1 methanol/PBT mixture.
4. Postfix samples in PBT containing 5% formaldehyde for 30 min.
5. Rinse twice in PBT.
6. Incubate samples with 40 µg/mL of proteinase K in PBT for 10–15 min at room temperature (*see* **Note 10**).
7. Stop the proteinase K digestion by washing the samples in PBT containing 0.2% glycine for 5 min.
8. Rinse twice in PBT.
9. Postfix samples for 30 min by adding PBT containing 5% formaldehyde.
10. Rinse three times with PBT.
11. Equilibrate samples for 5 min in a 1:1 PBT:HS solution.
12. Rinse twice in HS. Prehybridize for 1–2 h at 55°C.

3.6. Hybridization

1. Transfer plant material to Eppendorf tubes and add 100–500 µL of fresh HS containing DIG-labeled RNA probe (final concentration 0.5–3 µg/mL) and salmon testes DNA (final concentration 100 µg/mL), which were both denatured for 5 min at 85°C (*see* **Note 11**).
2. The hybridization is done at 55°C in a humid environment for 20 h under continuous shaking.

3.7. Posthybridization Washes

1. Remove the hybridization mix and wash the samples twice for 30 min at 55°C in fresh HS (*see* **Note 12**).
2. Wash samples in a 1:1 HS:NTE solution for 15 min at room temperature.
3. Rinse samples twice with NTE.
4. Replace by fresh NTE buffer containing 40 µg/mL of RNase A and incubate for 45 min at 37°C.
5. Wash with NTE at 37°C for 15 min.
6. Transfer samples from Eppendorf tubes to glass vials or to cell strainers. Wash at least five more times for 15 min with NTE at room temperature (*see* **Note 13**).
7. Equilibrate samples in 1:1 NTE:PBT for 5 min.
8. Rinse with PBT and incubate plant material with BS for at least 30 min (*see* **Note 14**).

3.8. Plant Powder Preparation

1. Homogenize an equal amount of fixed and fresh plant material in a small volume of ice-cold 90% acetone by grinding in liquid nitrogen.
2. Add more 90% acetone and vortex vigorously. Store overnight at 4°C.

3. Centrifuge at 13,000g for 5 min. Remove the supernatant, resuspend the pellet in a small volume of 90% acetone, and distribute the suspension over a filter paper. Air dry (using a desiccator).
4. Remove the dried powder from the filter paper and store at 4°C.

3.9. Incubation of Anti-DIG Antibody Conjugate with Plant Tissue

1. Add approx 30 mg of the powder in 400 µL of PBT containing 2% bovine serum albumin (fraction V; see **Note 15**).
2. Add 20 µL of anti-DIG F_{ab} fragment AP or 0.8 nm gold antibody conjugate (Boehringer Mannheim). Incubate in the dark at room temperature for 24 h.
3. Clear the preabsorbed mix by centrifugation at 13,000g for 3 min.
4. Replace BS by 6 mL of freshly made solution. Add 60 µL of the preabsorbed antibody. Incubate overnight at 4°C. The anti-DIG AP conjugate is diluted 1:2000 and the anti-DIG-gold to 1:500.
5. Replace antibody solution by fresh BS and incubate for 10 min.
6. Wash samples four times with PBT for 30 min (see **Note 16**).

3.10. Chromogenic Reaction for AP Antibody Conjugates

1. Equilibrate a fraction of the samples in SB twice for 5 min.
2. Transfer samples to glass vials and add 1 mL of fresh SB containing 1 mM of levamisole (see **Note 17**).
3. Add 4.5 µL of NBT (75 mg/mL NBT in 70% [v/v] H_2O/dimethyl formamide), 3.5 µL of BCIP (50 mg/mL in 100% dimethyl formamide), and incubate in the dark (see **Note 18**).
4. Check the samples under the stereo microscope to assess the degree of staining. If a strong background staining in control samples is observed, the remaining fraction of the samples are further washed with PBT and the detection procedure is repeated. When a clear difference is observed in stain intensity between control and antisense probes, the chromogenic reaction is stopped by adding 1 mL of SSB. In general chromogenic reactions are completed in 5–60 min (longer incubations usually give background staining).
5. Rinse twice in SSB.

3.11. Silver Amplification for Gold Antibody Conjugates

1. Transfer the samples from PBT to deionized water and rinse three times for 5 min to remove remaining salts.
2. Prepare the silver enhancement solution according to the Intense™ kit (Amersham) protocol.
3. Apply the silver enhancement mixture on the samples and monitor the silver amplification under the microscope until a signal is observed (see **Note 19**).
4. Wash the samples in an excess of distilled water for 60 min. Mount them on a microscope slide in a mixture of 1:1 water:glycerol, and put the cover slips. Samples can be directly observed using bright-field or differential interference contrast (DIC) microscopy.

3.12. Microscopy and Photography

1. For a more permanent mount, whole-mount tissue or vibroslice samples are fixed in 2.5% glutaraldehyde in phosphate buffer (PB) (pH 7.2) for 2 h at room temperature or at 4°C overnight.
2. Wash in PB for 30 min and dehydrate gradually until 70% ethanol.
3. For a better visualization of signal and tissues, samples can be transferred to a small volume of clearing agent for 60 min to overnight and observed under the microscope (*see* **Note 20**).

4. Notes

1. The WISH protocol allows the localization of mRNA in most tissues of young intact *Arabidopsis* plants. For tissues consisting of cells with a less permeable wall, probe, but mainly antibody penetration may be a limiting factor. Attempts to circumvent these problems are mentioned in the following notes.
2. A common problem is the presence of unspecific staining in enclosed cavities between tightly packed organs owing to trapping of reagents. Maximal dissection of plant organs (e.g., flowers) is required.
3. Plant tissue, which tends to stick on Pipetman tips or which are very fragile, can be manipulated easily in cell strainers placed in a six-well tissue-culture plate. This simple system allows the easy transport of samples from one solution to another and induces minimal damage; it can be used in all steps of the protocol.
4. Remove the agarose surrounding the plant tissue vibroslice, because it tends to give background staining. The use of vibroslices allows the WISH procedure to be applied on large seedlings and mature plant organs. The resolution obtained with vibrosliced plant tissues is often sufficient to localize mRNA at the cellular level so that additional experiments using, for instance, paraffin sections is superfluous. Large numbers of samples can be hybridized simultaneously allowing a rapid analysis *(12,13)*.
5. Tissue fixation has been done in the presence of 5% paraformaldehyde. This condition offers a good morphology without too much crosslinking, leaving the RNA more accessible. The fixation time varies according to the size of the sample. Often permeabilization can be improved by using higher concentrations of DMSO and heptane. Cell wall loosening enzymes, such as cellulase, and higher detergent concentrations (e.g., Tween-20; Triton-100, or sodium dodecyl sulfate) can both be used to increase permeability. Care should be taken that enzymes are RNase free.
6. Soft tissues such as *Arabidopsis* roots have the tendency to become brittle when kept too long in absolute ethanol; less delicate tissues, such as mature leaves, can be stored in ethanol for several days at −20°C. For long-time storage, use 70% ethanol to prevent excessive dehydration.
7. When too few transcripts are observed, add 2 µL more of enzyme and incubate longer. If transcription remains unsatisfactory, purify the linearized DNA with phenol-chloroform from possibly contaminating proteins or residual phenol from a previous extraction. Subsequently, ether extract twice (freezing sample and discarding the supernatant ether without loss of DNA in the interphase) to remove phenol.

8. The formula used to calculate the time needed to hydrolyze an RNA probe is: $t = (L_o - L_f)/(K \times L_o \times L_f)$, with L_o = starting length (kb), L_f = final length (kb), K = rate constant (0.11 kb/min in the used conditions), and t = in minutes *(14)*.
9. Permeabilization of tissues prior to ISH is critical and for better results the use of pure xylene may be recommended.
10. A too-short digestion prevents efficient probe and antibody penetration. A too-long digestion may affect tissue integrity, so that proteins and nucleic acids are lost. As a rule, digestion should be carried out as long as no visible damage occurs.
11. Use an as low concentration of probe as possible (0.5 μg/mL) to avoid background due to unspecific sticking or trapping of probe on plant tissue. In the hybridization mix, cold single-stranded DNA can be replaced by 150 μg/mL tRNA and 500 μg/mL poly(A) which usually gives a similar background reduction.
12. Posthybridization washing conditions can be slightly modified without loss of signal. After the hybridization, samples can be washed twice for 60 min in 2X SSC at room temperature and once for 15 min in 0.2X SSC at 55°C containing either 0.2% Tween-20 or 0.1% sodium dodecyl sulfate. Washing time can be increased if needed.
13. Probe RNA that is not annealed to target RNA is removed by ribonuclease treatment and high-stringency washes. At moderate temperature (37°C) and under high salt conditions (0.5M), the RNA/RNA duplex is resistant to RNase A, whereas single-stranded RNA molecules are hydrolyzed. It is important to remove all RNase by extensive washes to avoid the introduction of nicks into the hybrids during the more stringent washes later in the procedure *(1)*.
14. To avoid background staining caused by unspecific binding of antibody conjugates to plant tissue, 5% normal sheep serum can be added in the BS; alternatively, up to 5% bovine serum albumin fraction V may be included. Longer incubation schedules with BS may be required for large and rigid samples.
15. It is crucial that the diluted antibody complex is preabsorbed against plant powder to prevent posterior nonspecific binding on plant tissue. An alternative for the use of plant powder is the blocking of antibodies with fixed whole mounts or vibroslices from plant material. In our opinion, much better results are obtained when using plant powder.
16. Postwashes after antibody incubation can be done in an excess of PBT buffer and with more frequent changes than mentioned in the protocol to minimize unspecific binding on plant tissue.
17. Levamisole acts as an inhibitor for endogenous APs. Samples can also be incubated at 70°C to inactivate endogenous phosphatases causing nonspecific background. The histochemical reaction should be carried out in glass containers because a precipitate can form if a plastic container is used *(13)*. AP activity is best detected when reacted with the color substrate NBT plus BCIP producing a dark blue color *(13)*. This substrate combination is considered to be more sensitive than any other commercially available substrate combinations such as naphthol-phosphate/Fast Red TR *(13)*.
18. Lower concentrations (up to 1/3) of NBT and BCIP can be used but inevitably increase the detection time. It is a good strategy for abundantly expressed genes

in which a signal is often observed in a few minutes to reduce the color substrate concentration.
19. The silver amplification detection system seems to be less sensitive than the AP-based method. Its limited sensitivity may be explained by the fact that intact antibodies are bound to 1-nm gold particles instead of the smaller F_{ab} fragments that are bound to AP. F_{ab} fragments will penetrate more easily through the cell wall. The charged surface of gold particles may also hamper penetration. Another limiting factor of the silver method is that the time available to perform a silver detection is restricted to approx 40 min. Longer incubation leads to self-nucleation of silver, causing too much background and, thus, reducing its applicability.
20. The NBT/BCIP precipitate looses some intensity and changes color from brown to blue when dehydrated or cleared in CLP.

Acknowledgments

The authors thank Martine De Cock for help preparing the manuscript. This work was supported by grants from the Belgian Programme on Interuniversity Poles of Attraction (Prime Minister's Office, Science Policy Programming, No. 38) and the Vlaams Actieprogramma Biotechnologie (ETC 002). GE is a Research Engineer of the Institut National de la Recherche Agronomique (France).

References

1. Angerer, L. M. and Angerer, R. C. (1981) Detection of polyA$^+$ RNA in sea urchin eggs and embryos by quantitative in situ hybridization. *Nucleic Acids. Res.* **9**, 2819–2840.
2. Meyerowitz, E. M. (1987) *In situ* hybridization to RNA in plant tissue. *Plant Mol. Biol. Rep.* **5**, 242–250.
3. Cox, K. H. and Goldberg, R. B. (1988) Analysis of plant gene expression; in *Plant Molecular Biology, A Practical Approach* (Shaw, C. H., ed.), IRL, Oxford, UK, pp. 1–35.
4. Coen, E. S., Romero, J. M., Doyle, S., Elliott, R., Murphy, G., and Carpenter, R. (1990) *floricaula*: a homeotic gene required for flower development in *Antirrhinum majus*. *Cell* **63**, 1311–1322.
5. Jackson, D., Veit, B., and Hake, S. (1994) Expression of maize KNOTTED1 related homeobox genes in the shoot apical meristem predicts patterns of morphogenesis in the vegetative shoot. *Development* **120**, 405–413.
6. Tautz, D. and Pfeifle, C. (1989) A non-radioactive in situ hybridization method for the localization of specific RNAs in *Drosophila* embryos reveals translational control of the segmentation gene *hunchback*. *Chromosoma* **98**, 81–85.
7. Ludevid, D., Höfte, H., Himelblau, E., and Chrispeels, M. J. (1992) The expression pattern of the tonoplast intrinsic protein γ-TIP in *Arabidopsis thaliana* is correlated with cell enlargement. *Plant Physiol.* **100**, 1633–1639.
8. de Almeida Engler, J., Van Montagu, M., and Engler, G. (1994) Whole-mount messenger RNA *in situ* hybridization in plants. *Plant Mol. Biol. Rep.* **12**, 319–329.

9. Hemerly, A. S., Ferreira, P. C. G., de Almeida Engler, J., Van Montagu, M., Engler, G., and Inzé, D. (1993) *cdc2a* expression in *Arabidopsis thaliana* is linked with competence for cell division. *Plant Cell* **5,** 1711–1723.
10. Terryn, N., Brito Arias, M., Engler, G., Tiré, C., Villarroel, R., Van Montagu, M., and Inzé, D. (1993) *rha1*, a gene encoding a small GTP binding protein from *Arabidopsis* primarily expressed in developing guard cells. *Plant Cell* **5,** 1761–1769.
11. Palms, B., Goupil, P., de Almeida Engler, J., Van Der Sraeten, D., Van Montagu, M., and Rambour, S. (1996) Evidence for the nitrate-dependent spatial regualtion of the nitrate reductase gene in chicory roots. *Planta* **200,** 20–27.
12. Dickson, M. C., Slager, H. G., Duffie, E., Mummery, C. L., and Akburst, R. J. (1993) RNA and protein localisations of TGFβ2 in the early mouse embryo suggest an involvement in cardiac development. *Development* **117,** 625–639.
13. Wilkinson, D.G. (1993) Whole mount *in situ* hybridization of vertebrate embryos, in In situ *Hybridization: A Practical Approach* (Wilkinson, D. G., ed.), IRL, Oxford, UK, pp. 75-83.
14. Cox, K., DeLeon, D. V., Angerer, L. M., and Angerer, R. C. (1984) Detection of mRNAs in sea urchin embryos by *in situ* hybridization using asymmetric RNA probes. *Dev. Biol.* **101,** 485–502.

37

Bacterial and Coelenterate Luciferases as Reporter Genes in Plant Cells

William H. R. Langridge and Aladar A. Szalay

1. Introduction

The study of gene regulation has been greatly enhanced by the use of reporter gene systems such as beta galactosidase (β-gal), neomycin phosphotransferase (APH[3]II), chloramphenicol acetyl transferase (CAT), beta glucuronidase (GUS), and dihydrofolate reductase (DHFR). In the past several years, development of marker gene systems based on bioluminescence have extended the power of marker gene technology from enzymatic or color based in vitro assays to more sensitive single photon counting methods in vitro and in vivo. A variety of proteins that catalyze bioluminescent reactions have been isolated and characterized. For several of these proteins, the genes are available and are currently being used as reporters for gene expression studies. Concomitant advances in single photon detection technology have recently made it possible to measure gene expression noninvasively in real time. We describe here the application of bacterial luciferase from *Vibrio harveyi* and eukaryotic luciferase from *Renilla reniformis* as markers for transformation and reporters of gene expression in transgenic plants. The major advantages of the luciferase gene expression system are its simplicity, sensitivity, safety for the investigator and available nondestructive assay conditions, which permit real-time measurements of gene expression continuously throughout development of transgenic plants.

1.1 Bacterial Luciferase

Bacterial luciferases were originally isolated from deep sea fish associated *Vibrio harveyi* or *fischeri*, *(1,2)*. Both *V. harveyi luxA* and *luxB* genes have been cloned and sequenced *(3–5)*. The *luxA* and *luxB* genes encode a heterodimeric, mixed function oxidase composed of one 40-kDa subunit and one 36-kDa subunit protein *(4–12)*. The mixed function oxidase, catalyzes the oxidation of reduced flavin mononucleotide (FMNH$_2$) and long chain aldehydes with molecular oxygen to yield FMN, the corresponding carboxylic acid, H$_2$O, and a photon of blue-green light according to the reaction: RCHO + O$_2$ + FMNH$_2$ → RCOOH + FMN + H$_2$O + hv (490 nm). The products of the enzymatic reaction are a long-lived flavin peroxide intermediate (an enzyme bound hydroperoxide, FMN-OOH), which reacts with a 10-carbon aliphatic aldehyde (decanal) leading to formation of an unstable hydroxy-flavin emitter in its singlet excited state which decays to ground state giving off a photon of light in approx 40% of the decay transitions *(13–16)*. In the presence of excess FMNH$_2$ and aldehyde substrates, light production is proportional to the number of enzyme molecules as the rapid oxidation of free FMNH$_2$ and the long relaxation time of the enzyme from its altered conformational state permits only one catalytic cycle in vitro *(17)*. The light emitting reporter gene assay system is easy to use, inexpensive, and eliminates the need for radioactivity. Rapid quantitative analysis of marker gene expression can be obtained by in vitro measurement of luciferase activity in homogenates of transgenic plant cells. Promoter activation and the precise temporal and spacial localization of gene expression in cells and tissues of transgenic plants can be detected continuously in vivo throughout development by intensified CCD camera based image analysis methods. Expression vectors have been assembled in which the *luxA* and *luxB* coding sequences isolated from the *V. harveyi lux* operon are fused separately to the TR 1' and 2' bidirectional promoters of the *Agrohacterium tumefaciens* mannopine synthase genes. The *luxA* and *luxB* gene containing vectors have been used to transform tobacco and *Arabidopsis thaliana* cells. Analysis of light emission and immunological detection of luciferase subunits in regenerated transgenic plants and transgenic plant extracts have indicated that assembly of a functional heterodimeric luciferase occurs in the cytoplasm of the transformed plant cells *(18)*. Further, a *luxA* and *luxB* structural gene fusion has been constructed resulting in a single gene *(luxF)*, which encodes a monomeric bacterial luciferase alpha/beta fusion protein *(19,20)*. Both the two genie (*luxA* and *luxB* genes) and the single *luxF* fusion gene have been used as reported in transformation experiments with plant cells *(21,22)*. Constitutive and organ-specific activation of plant genes can be followed indirectly throughout plant development by visualization in vivo of the expression of light emitting gene fusions.

The low levels of bioluminescence emitted from plant cells in which substrate penetration may be retarded by a waxy cuticle, thick cellulose cell wall, limiting endogenous FMNH$_2$ concentrations, or the quenching of light emission from cells in underlying tissues, requires development of a sensitive method for detection of in vivo luciferase gene expression. Computer-aided

Plant Cell Reporter Genes

intensified CCD camera-based image analysis methods can easily monitor the low levels of light emission from cell aggregates, organs, tissues, or even entire plant. Qualitative measurement of luciferase gene expression *in situ* can be accomplished by placing the transgenic plantlet or plant organ in a closed transparent chamber (plastic culture dish) in the presence of decanal vapors which rapidly penetrate the tissues. The plate containing the plant tissue is placed in a dark chamber and the photons emitted by the tissue are detected by an image intensifier coupled to an 8 bit VIM CCD camera. Alternatively, marker gene expression in plant tissue homogenates can be quantitatively assayed by luminometer based methods. Quantitative measurement of luciferase activity in homogenates of transgenic plant cells is performed by simultaneous addition of $FMNH_2$ and decanal directly into the cell homogenate followed by photometric detection of the emitted photons. A variety of methods for detection and measurement of bioluminescence have been described *(21,23)*.

1.2. Eukaryotic Luciferase

With eukaryotic luciferins, peroxide formation occurs in all luciferase reactions which produce CO_2 and H_2O. The excited oxyluciferin is the primary emitter, the energy is transmitted to a secondary light emitter, e.g., a protein-bound flavin chromophore. Light emission from the soft coral coelenterate *Renilla reniformis* (phylum *Coelenterata*, order *Cnidaria*) is catalyzed by a luciferase *(24,25)*. Oxidation of the luciferin substrate, coelenterazine, by this luciferase leads to an excited state product (oxyluciferin) which yields blue light (wavelength = 480 nm), *(25)*. *R. reniformis* luciferase cDNA has been cloned and expressed in plants *(26,27)* substrate coelenterazine or 2-benzyl luciferin, is nontoxic and is rapidly taken up by the cells. Light emission from homogenates of alfalfa protoplasts expressing *R. reniformis* luciferase was 16-fold higher than that of protoplasts expressing the bacterial luciferase genes. Under the control of the same promoter, application of a 3 µ*M* aqueous solution of 2-benzyl luciferin onto calli, leaves, roots, and slices of tomato fruits and potato tubers from transformed plants constitutively expressed the *R. reniformis* luciferase cDNA resulting in strong light emission which can be visualized within seconds by an intensified photon counting camera.

Bacterial and eukaryotic luciferases possess unique attributes which make each of them well suited for the measurement of gene expression in plants under different sets of conditions. Bacterial luciferase excels in ease of activity measurement due to its inexpensive aldehyde substrate which efficiently penetrates plant cell walls and membranes. The volatile decanal substrate permits monitoring of temporal and special gene expression in real time without destruction of the transgenic organism. Therefore the *lux* system is suitable for identifying the location of gene expression in vivo in *Arabidopsis* and other transgenic plant tissues as well as the determination of changes in gene expression that occur in tissues and organs of the plant throughout development. A potential drawback to the use of bacterial luciferase is that prolonged exposure to decanal can increase the rate of morbidity of affected plant tissues. On the other hand, because of the higher quantum efficiency of *Renilla* luciferase, light emission obtainable from tissue homogenates of *Arabidopsis* and tobacco plants containing a single copy of the *Renilla* luciferase gene is about 360-fold higher than that obtained from plants transformed with a single copy of the bacterial *luxF* gene. In addition, the low toxicity of its luciferin makes *Renilla* luciferase a better marker for detection of early gene activation, or the detection of low levels of constitutive promoter activity in vivo. When transferred into plant cells, both *Renilla* and bacterial *lux* gene fusion systems provide simple assay methods for the detection of gene expression in whole plants, organs, tissues, cell layers, and single cells, reaching levels of sensitivity equivalent to radioactivity-based assay methods. Improvements in the efficiency of photon emitting substrates and the sensitivity of photon detection equipment DNA or protein luciferase ligands will increasingly favor the replacement of radioactive based nucleic acid and protein blotting methods, reducing the biohazard and the accumulation of radioactive waste in the environment. Here we describe two protocols which can be used for the localization and quantitative measurement of light emitting marker gene systems in transgenic plant tissues. In **Subheading 3.1.**, we describe the quantitative luminometric measurement of bacterial luciferase activity in homogenates of transgenic plant tissues and organs by low light image analysis methods. In **Subheading 3.2.**, we describe the protocols for measurement of a eukaryotic luciferase from *R. reniformis* in transgenic plant tissues and cell homogenates.

2. Materials

2.1 Bacterial Luciferase

1. Luciferase assay buffer: 50 mM Na$_2$HPO$_4$, pH 7.0, 50 mM β-mercaptoethanol, 0.4M sucrose.
2. FMNH$_2$ substrate: 100 μM FMN in 200 mM tricine buffer, pH 7.0.
3. Decanal substrate (10 mL): 50 mM Na$_2$HPO$_4$, pH 7.4 and 10 μL decanal (sonicate the mixture 10 s, or until a stable turbid emulsion is obtained).
4. Diaphorase (NADH-FMN oxidoreductase), Boehringer Manaheim (Mannheim, Germany), 1 U enzyme/100 μL luciferase assay buffer containing 50 mM Na$_2$HPO$_4$; 50 mM β-mercapto-ethanol and 0.1% bovine serum albumin.
5. NADH (500 μM): 4.0 mg NADH/10 mL 50 mM Tris-HCl, pH 8.0, aliquot and store at –20°C.
6. FMN (100 uM) in 200 mM tricine buffer, pH 7.0.
7. *V. harveyi* luciferase standard; 1 μg luciferase (Sigma, St. Louis, MO)/μL luciferase assay buffer + 0.1% BSA.

Plant Cell Reporter Genes

8. Decanal (1:1000 dilution) in distilled water. Sonicate the mixture for 10 s at one-half power to create an emulsion.
9. Microfuge (Marathon 16 Km, Fisher, Pittsburg, PA).
10. Plastic pestles which fit 1.5-mL Eppendorf tubes (Kontes Glass, Vineland, NJ).
11. Glass scintillation vials.
12. Sonicator (Fisher).
13. Luminometer (Turner model TD-20e, Sunny Vale, CA).
14. Photon counting camera system (Argus-100, Hamamatsu Phontonics, Hamamatsu, Japan).
15. Filter paper (Whatman #1).
16. 16-cm Plastic culture dishes (Falcon, Los Angeles, CA.).

2.2. Renilla *Luciferase*

1. Assay buffer: $0.5 M$ NaCl, $0.1 M$ potassium phosphate, pH 7.4, and 1 mM EDTA.
2. Luciferin substrate: 10.0 µM solution of luciferin 2-benzyl coelenterazine (Molecular Probes, Eugene, OR), in distilled water.
3. 0.02% BSA.
4. Glass scintillation vials.
5. Sonicator.
6. Luminometer.

3. Methods
3.1. Bacterial Luciferase
3.1.1. Homogenates

1. Using a tapered plastic pestle, grind 30–50 mg fresh weight of transgenic *Arabidopsis* or other plant callus, leaf, stem, flower, or root tissue in 1.0-mL of ice cold luciferase assay buffer in a 1.5-mLvolume polypropylene microcentrifuge tube (*see* **Note 1**).
2. Centrifuge the homogenate for 30 s, at 4°C in a microfuge to remove cell debris.
3. Transfer the supernatant to a clean microfuge tube on ice.
4. Place 0.2 mL of the cleared homogenate in a separate microfuge tube and hold on ice for later protein determination *(28)*, (*see* **Note 2**).
5. Transfer 0.4 mL of the cleared homogenate to a prewashed glass scintillation vial (*see* **Note 3**).
6. Introduce the vial containing the sample into the photon counting chamber of the luminometer, and close the chamber to dark adapt the sample. A considerable amount of fluorescence is emitted from chlorophyll molecules in green samples, allow 30 s incubation of the sample in the dark chamber for the fluorescence to disappear.
7. Fill a 1.0 µL graduated tuberculin syringe fitted with a 25-gauge needle with 0.5 mL of light-reduced $FMNH_2$ substrate (*see* **Note 4**). Continue to fill the syringe with 20 µL of the diluted decanal emulsion. Be careful at this stage to avoid introduction of air bubbles into the syringe which will promptly oxidize the $FMNH_2$ (*see* **Note 5**).

8. The $FMNH_2$/decanal mixture is injected rapidly into the sample through the rubber septum in the top of the chamber. Begin photon measurement immediately and record the light emission in arbitrary light units (LU), from the sample at 490 nm over a 10-s time interval (in our luminometer, 1 LU = 1.0×10^6 photons/s), the exact LU value will depend on individual instrument phototube sensitivity. Luminometer sensitivity should be calibrated once each year with a light emitting (radioactive/phosphor) standard to determine changes in photon measurement efficiency. A gradual decline in phototube sensitivity is expected with time (*see* **Note 6**).
9. The luciferase assay procedure is repeated with the remaining 0.4 mL of homogenate to provide data on assay reproducibility (*see* **Note 7**).
10. The protein concentrations of each sample is determined *(28)*, and the luciferase activity calculated (LU/µg protein).

3.1.2. In Vivo

Reduced penetration of decanal into intact transgenic plant tissues coupled with low intracellular $FMNH_2$ concentrations may require the application of image enhancement methods coupled with single photon-counting camera technology to detect the cellular location of luciferase activity in intact transgenic plant tissues (**Fig. 1**).

1. Leaf, stem, flower, and root tissues from both transgenic and untransformed plants are excised and placed adjacent to each other in a plastic culture dish containing a 1-cm wide ring of filter paper in the base of the dish saturated with water to protect the tissues from dehydration during photon counting and to serve as a reservoir for the volatile decanal.
2. Dispense a total of 10 µL decanal evenly around the water saturated filter paper ring so that aldehyde vapors diffuse throughout the dish.
3. Place the covered dish containing the plant tissue the photon counting chamber, under the lens of the Argus-100 intensified photon-counting camera (*see* **Note 8**).
4. Close the chamber to exclude light and incubate the sample in the dark for 1 min to permit nonspecific fluorescence from chlorophyll molecules to dissipate.
5. The level of bioluminescence emitted by the plant tissues is observed on the video monitor, and photon emissions collected for a time interval from 10 min to 1 h (*see* **Notes 9** and **10**). Longer photon accumulation intervals may result in unfavorable signal to noise levels which obliterate photon emissions obtained from the transgenic plant tissues.
6. Following collection of the emitted photon image, the bioluminescent sample is illuminated and a video image of the plant tissue or organ is displayed on the video monitor and stored in computer memory.
7. The bioluminescent image of the transformed plant tissues stored in memory is overlaid on the video image. Superimposition of these two images reveals the precise location of the bioluminescence (luciferase gene expression) in the transgenic plant tissue.

Plant Cell Reporter Genes

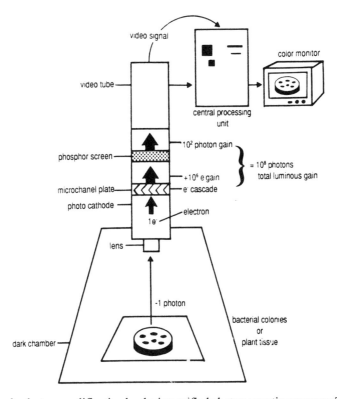

Fig. 1. Single photon amplification by the intensified photon counting camera: This figure describes the amplification of a photon emitted from bioluminescent plant tissues as it passes through the intensified CCD camera. In the presence of decanal substrate vapors the plant cells emit photons according to the bacterial luciferase mediated reaction: $RCHO + O_2 + FMNH_2 \rightarrow RCOOH + FMN + H_2O + h\nu$ 490 nm. The photons pass from the subject through a glass lens focused on a charged photocathode metal film-coated membrane of the image intensifier. One electron is ejected from the photocathode into the image intensifier for every contacting photon. The ejected electron enters a focused electromagnetic field, traverses the photon counting tube, and enters a charged glass microcapillary tube within the microchannel plate at the same x,y coordinates from which the electron was ejected from the photocathode. Repeated collisions of the electron with the wall of the angled charged capillary tube results in the release of a cascade of electrons at the coordinates of the microcapillary tube resulting in a gain of 10^6 electrons for every electron entering the microchannel plate. Electrons released from the microchannel plate bombard a phosphor screen which converts the electrical energy back to photons with a gain of 10^2 photons for every electron which excites a phosphor molecule. Thus, for every photon which enters the photon counting tube, a total of 10^8 photons exit the image intensifier and enter a vidicon video tube capable of detecting single-photons. The amplified video signal is relayed to a central processing unit where it is simultaneously stored in RAM memory and displayed in real time as a pseudo-colored image on a high-resolution color video monitor.

8. Bioluminescence and video images are transferred from memory to a floppy disk for quantitative or graphic analysis and photography. The disk provides a permanent record of reporter gene expression for comparison of light emission during further developmental stages.
9. After measurement of bioluminescence, transgenic plant tissues or intact plantlets can be transferred to appropriate media for further growth and development.

3.2. R. reniformis Luciferase

For quantitative measurement of *Renilla* luciferase activity in small (2 cm) *A. thaliana* plants, the leaves, stems, flowers or roots of 10 plants are pooled and analyzed together. However, tissue samples from larger plant species e.g., tobacco, potato, and tomato are analyzed independently.

3.2.1. Homogenates

1. To measure *R. reniformis* luciferase activity, transgenic *Arabidopsis*, other plant tissue samples, or sedimented transformed protoplasts (1×10^6 protoplasts/mL), are homogenized in 1–4 mL assay buffer. The plant cell homogenates are cleared by centrifugation in a microcentrifuge for 5 min at 4°C and a 0.1-mL aliquot of the supernatant is held on ice for protein assay *(28)*.
2. A 1:100 dilution of the cleared homogenate in assay buffer, supplemented with 0.02% BSA is used for luminometric analysis (*see* **Note 11**). The 2-benzyl derivative of the coelenterazine substrate (luciferin), is used in all assays. The substrate 2-Bengal coelenterazine is about 60% as active as coelenterazine in vitro *(29)* (*see* **Notes 12** and **13**).
3. The bioluminescence reaction is initiated by injecting 400 µL of 1 µM substrate into 400 µL of plant extract in a scintillation vial, in a Turner TD 20e luminometer (Turner). The integral of the emission peak produced during the first 10 s of the reaction is taken as the luciferase activity (LU).
4. To ensure a linear relationship between light yields and luciferase concentration, a standard calibration curve is prepared with dilutions of purified recombinant *Renilla* luciferase protein (obtained from Dennis O'Kane, Mayo Clinic, Rochester MN), in assay buffer, supplemented with 0.02% BSA to stabilize the enzyme.

3.2.2. In vivo

1. Bioluminescence is visualized in plant tissues in vivo with the Argus-100 intensified VIM 3 camera based imaging system exactly as described for bacterial luciferase earlier in this work.
2. Small leaves and calli from transgenic *Arabidopsis* plants, slices of potato tubers or tomato fruit or whole plantlets are sprayed with an aqueous solution of luciferin. Photon counting is initiated after incubation of the plant material for 2 min in the dark to quench fluorescence when chlorophyll is present.
3. Because of the high yield of photon emission of *Renilla* luciferase, light emission from callus), tubers and fruit slices is collected for 10 s and from leaves and

whole plants up to 10 min. To study substrate uptake via the vascular system, roots of whole plantlets are immersed in 3.0 μ*M* of the luciferin substrate and incubated for selected time periods prior to low light image analysis (*see* **Note 14**).

4. Notes

1. To measure bacterial luciferase activity in transformed plant protoplasts the protoplasts are resuspended in 0.5–1.0 mL of ice-cold luciferase assay buffer and disrupted for 3–5 s at the lowest energy setting of a probe sonication device. The precise sonication time required to obtain 95–100% cell disruption is determined empirically by light microscopy. A volume of 0.1 mL of the homogenate is transferred to a clean microfuge tube at 4°C for protein assay, and a volume of 0.4 mL of the homogenate is transferred into a clean glass scintillation vial to perform the luminometric assay. Luciferase activity is detected in transformed tobacco cv. BY-2 suspension culture protoplasts 12–16 h following gene transfer and bioluminescence in the protoplasts reaches a maximum from 20–48 h after transformation with 10 μg DNA/mL of protoplast suspension.
2. Protein assays of the plant homogenates should be performed on the same day as the luminometer assays to avoid protein precipitation during freezing storage which results in decreased protein concentration values.
3. Prewashing the scintillation vials removes contaminants which inhibit luciferase activity and create a source of assay variability.
4. Reduction of FMN is accomplished by suspending a 1.0 mL volume tuberculin syringe containing 0.5 mL FMN-tricine buffer solution by the needle (23-gauge) in a Styrofoam block placed adjacent to a cool white fluorescent light for approx 3 min or until the oxidized (yellow) FMN solution becomes reduced (colorless).
5. The force of substrate injection from the syringe must be held as constant as possible between samples to obtain reproducible bioluminescence (LU) values. To reduce this source of variability, an automated mechanical sample injection device (Turner) is recommended.
6. When measuring pure luciferase standards, include BSA (0.1%) in the assay buffer to protect the enzyme from denaturation due to dilution.
7. To obtain detectable levels of luciferase activity from tissue homogenates over time intervals longer than 10 s, the bioluminescent luciferase reaction may be coupled to an NADH-FMN oxidoreductase (diaphorase), capable of specific continuous reduction of FMN in the presence of an exogenous source of reducing power (NADH). For this purpose, to the scintillation vial containing the plant tissue extract, add 10 μL of diaphorase stock solution and 500 μL of 500 μ*M* NADH and mix. Insert the vial into the counting chamber of the luminometer and initiate the reaction by injecting 500 μL of light-reduced $FMNH_2$ + 20 μL decanal emulsion (**Subheading 3.1., step 8**). Measure light emission from 1–10 min and calculate total light emission (LU) per unit time per μg protein. When performing the coupled luciferase-diaphorase reaction using the photon counting camera to accumulate photons, 1–10 picograms of luciferase can be detected using a 10 min photon collecting interval *(21)*.

8. The photon counting camera can be coupled to a stereo microscope to obtain magnification levels up to 64 times. This level of magnification permits accurate determination of the cellular location of luciferase gene expression in transgenic plant tissue sections *(21)*.
9. In our experience, a 10–30 min exposure to decanal vapors will not inhibit further growth and development of *Arabidopsis* or tobacco calli, or plantlets 10–15 cm tall. To prevent losses of valuable transgenic material, the optimal concentration of decanal substrate must be determined for each plant species.
10. Quantitative luciferase assay measurements in homogenates of plant tissues may be obtained by luminometry over a 10 s counting interval. However, the detection of luciferase generated bioluminescence in intact tissues requires 15 min to 2 h of low light image collection.
11. Because of the high yield of photon emission, very diluted plant extracts can be analyzed.
12. *Renilla* luciferin substrate is relatively unstable and should be protected from autooxidation by storage of the dry lyophilized compound under argon gas at –80°C.
13. The luciferin substrate is subject to oxidation in aqueous solution at room temperature and only small aliquots required for each assay should be thawed and used up.
14. The apparent failure to observe transport of luciferin through the xylem may be the result of its rapid degradation in the vascular system. In contrast to the bacterial luciferin decanal, the *Renilla* luciferin does not appear to be toxic to plant cells.

References

1. Engbrecht, J., Simon, M,. and Silberman, M. (1985) Measuring gene expression with light. *Science* **227**, 1345–1347.
2. Hastings, W. J. and Nealson, K. H.(1977) Bacterial bioluminescence. *Ann. Rev. Microbiol.* **31**, 549–595.
3. Baldwin, T. O., Berends, T.,, Bunch, T. A., Holzman, T. T., Rausch, S. K., Shamansky, L., Treat, M. L., and Ziegler, M. M. (1984) Cloning of luciferase structural genes from *Vibrio harveyi* and expression of bioluminescence in *Escherichia coli. Biochemistry* **23**, 3663–3667.
4. Cohn, D. H., Mileham, A. J., Simon, M. I., Nealson, K. H., Rausch, S. K., Bonam, D., and Baldwin, T. O. (1985) Nucleotide sequence of the *luxA* gene of *Vibrio harveyi* and the complete amino acid sequence of the a subunit of bacterial luciferase. *J. Biol. Chem.* **260**, 6139–6146.
5. Johnston, T. C., Thompson, R. B., and Baldwin, T. O. (1986) Nucleotide sequence of the *luxB* gene of *Vibrio harveyi* and the complete amino acid sequence of the B subunit of bacterial luciferase. *J. Biol. Chem.* **261**, 4805–4811.
6. Haygood, M. G. and Cohn, D. M. (1986) Luciferase genes cloned from the unculturable luminous bacteroid symbiont of the Caribbean flash-light fish, *Kryptophanaron alfredi. Gene* **45**, 203–209.

7. Miyamoto, C., Byers, D., Graham, A. F., and Meighen, E. A. (1987) Expression of bioluminescence by *Eschrichia coli* containing recombinant *Vibrio harveyi* DNA. *J. Bacteriol.* **169,** 247–253.
8. Miyamoto, C., Graham, A. D., Boylan, M., Evans, J. R., Hasel, K. W., Meighen, E. A., and Graham, A. F. (1985) Polycistronic mRNA codes for polypeptides of the *Vibrio harveyi* luminescence system. *J. Bacteriol.* **161,** 995–1001.
9. Miyamoto, C., Graham, A. F., and Meighen, E. A. (1988) Nucleotide sequence of the *luxC* gene and the upstream DNA from the bioluminescent system of *Vibrio harveyi*. *Nucleic Acids Res.* **16,** 1551–1562.
10. Foran, D. R. and Brown, W. M. (1988) Nucleotide sequence of the *luxA* and *luxB* genes of bioluminescent marine bacterium *Vibrio fscheri*. *Nucleic Acids Res.* **16,** 777.
11. Illarionov, B. A., Protopopova, M. V., Karginov, V. A., Martvetsov, N. P., and Gitelson, J. I. (1988) Nucleotide sequence of part of *Photobacterium leiognathi lux* region. *Nucleic Acids Res.* **16,** 9855.
12. Mancini, J. A., Boylan, M., Soly, R. R., Graham, A. F., and Meighen, E. A. (1988) Cloning and expression of the *Photobacterium phosphoreum* luminescence system demonstrates a unique *lux* genc organization. *J. Biol. Chem.* **263,** 14,308–14,314.
13. Ziegler, M. M. and Baldwin, T. O. (1981) Biochemistry of bacterial bioluminescence. *Curr. Top. Bioeng.* **12,** 65–113.
14. Meighen, E. A., Riendeau, D., and Bognar, A. (1981) Luciferase gene from *Vibrio harveyi*, in: *Bioluminescence and Chemiluminescence* (DeLuca, M. A. and McElroy, W. D. eds.), Academic, NY, pp. 129–137.
15. Engebrecht, J., Nealson, K., and Silverman, M. (1983) Bacterial bioluminescence: isolation and genetic analysis of functions from *Vibrio fischeri*. *Cell* **32,** 773–781.
16. Engebrecht, J. and Silverman, M. (1984) Identification of genes and gene products necessary for bacterial bioluminescence. *Proc. Natl. Acad. Sci. USA* **81,** 4154–4158.
17. Hastings, J. W. and Gibson, Q. H. (1963) Intermediates in the bioluminescent oxidation of reduced flavin mononucleotide. *J. Biol. Chem.* **238,** 2537–2554.
18. Koncz, C., Olsson, O., Langridge, W. H. R., Schell, J., and Szalay, A. A. (1987) Expression and assembly of functional bacterial luciferase in plants. *Proc. Natl. Acad. Sci. USA* **84,** 131–135.
19. Escher, A. P., O'Kane, D., Lee, J., Langridge, W. H. R., and Szalay, A. A. (1989) Construction of a novel functional bacterial luciferase by gene fusion and its use as a gene marker in Low Light Video Image Analysis, in: *New Methods in Microscopy and Low Light Imaging*, (Wampler, J. ed.), *SPIE* **1161,** 230–235.
20. Olsson, O., Escher, A., Sandberg, G., Schell, J., Koncz, C., and Szalay, A. A. (1989) Engineering of monomeric bacterial luciferases by fusion of *luxA* and *luxB* genes in *Vibrio harveyi*. *Gene* **81,** 335–347.
21. Langridge, W. H. R., Escher, A., Baga, M., O'Kane, D., Wampler, J., Koncz, C., Schell, J., and Szalay, A. A. (1989) Use of low light image microscopy to monitor genetically engineered bacterial luciferase gene expression in living cells and gene activation throughout the development of a transgenic organism, in *New Methods in Microscopy and Low Light Imaging* (Wampler, J. E., ed.), *SPIE* **1161,** 216–229.

22. Koncz, C., Martini, N., Mayerhofer, R., Koncz-Kalman, Z., Korber, H., Redei, G., and Schell, J., (1989) High-frequency T-DNA-mediated gene tagging in plants. *Proc. Natl. Acad. Sci. USA* **86**, 8467–8471.
23. VanDyke, K. (1985) *Bioluminescence and Chemiluminescence: Instruments and Applications*. CRC, Florida.
24. Cormier, M. J. (1978) Applications of *Renilla* bioluminescence: an introduction, in *Bioluminescence and Chemiluminescence*, Methods in Enzymology, (DeLuca, M., ed.) **57**, 237–244.
25. Matthews, J. C., Hori, K., and Cormier, M. J. (1977) Purification and propterties of *Renilla reniformis* luciferase. *Biochemistry* **16**, 85–91.
26. Lorenz, W. W., McCann, R. O., Longiaru, M., and Cormier, M. J. (1991) Isolation and expression of cDNA encoding *Renilla reniformis* luciferase. *Proc. Natl. Acad. Sci. USA* **88**, 4438–4442.
27. Mayerhofer, R., Langridge, W. H. R., Cormier, M. H., and Szalay, A. A. (1995) Expression of recombinant *Renilla* luciferase in transgenic plants results in high levels of light emission. *Plant Cell* **7**, 101–108.
28. Hart, R. C., Matthews, J. C., Hori, K., and Cormier, M. J. (1979) *Renilla reniformis* bioluminescence: luciferase-catalysed production of nonradiating excited states from luciferin analogues and elucidation of the excited state species involved in energy transfer to *Renilla* green fluorescent protein. *Biochemistry* **18**, 2204–2210.

38

β-Glucuronidase Enzyme Histochemistry on Semithin Sections of Plastic-Embedded *Arabidopsis* Explants

Marc De Block and Mieke Van Lijsebettens

1. Introduction

Reporter genes are commonly used in prokaryotes and eukaryotes to measure promoter activity of a gene of interest. Transgenic organisms have to be generated by transformation with a chimeric gene consisting of the respective promoter and the coding sequence of the reporter gene. The activity of promoters can be studied by gene fusions without interference of the regulatory 3' or intron sequences that might be present in the endogenous gene. Therefore, gene fusions are an important tool in transgenic research for promoter evaluation. The β-glucuronidase gene (*gusA* or *uidA*), isolated from *Escherichia coli*, has been exploited over many years to monitor plant promoter activity. *In situ* enzyme histochemical methods are used to localize the cells expressing the reporter gene within the cellular context of the explant or tissue. A qualitative histochemical assay to measure GUS activity is based on the substrate 5-bromo-4-chloro-3-indolyl-β-D-glucuronide (X-Gluc) *(1)*. The GUS enzyme hydrolyzes X-Gluc to a water-soluble indoxyl intermediate that is further dimerized into a dichloro-dibromo-indigo blue precipitate by an oxidation reaction (**Fig. 1**). However, this reaction has a serious drawback. Halogen-substituted indoxyls liberated at the site of high enzyme activity can diffuse widely before complete oxidation to insoluble indigo blue can take place. This is especially true at lower pH (< 9.0) *(2–4)*. The indoxyl dimerization can be enhanced by the presence of the oxidant potassium ferri(III)cyanide *(5)*. In optimal reaction conditions, the water-insoluble indigo blue is located at the place of enzyme activity *(6)*. The X-Gluc-based assays that are usually done on intact seedlings, explants, or vibro-sliced tissues suffer from a low histological and cytological

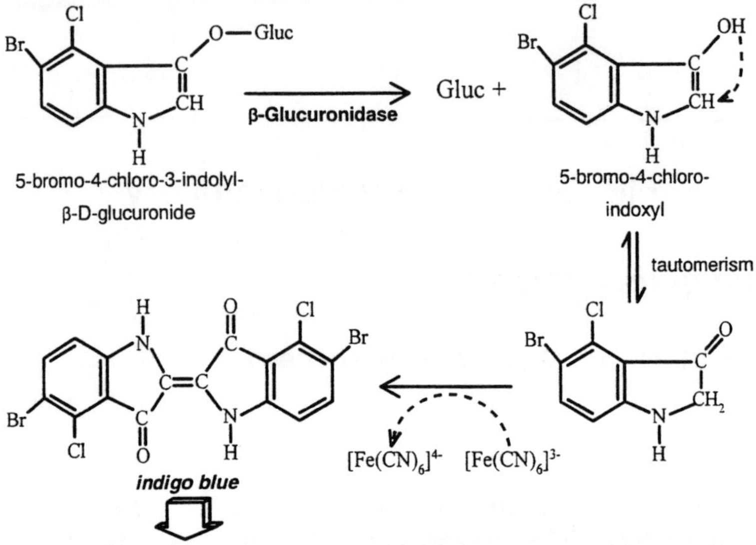

Fig. 1. Chemistry of X-Gluc reaction. Hydrolyzation of X-Gluc by the β-glucuronidase enzyme results in a reactive indoxyl molecule. Two indoxyl molecules are oxidized to indigo blue; ferri(III)cyanide enhances the dimerization.

resolution, unequal permeability of different cell types for the reaction substrates, diffusion of the reaction intermediates from the enzyme location, and uncontrolled reaction conditions in the intact cells, such as the cytoplasmic pH value and the presence of oxidizing or reducing compounds Moreover, localization artifacts might be observed owing to enhanced dimerization of the indoxyl intermediates at sites in the tissue with high peroxidase activity *(6,7)*. Also, artifacts might occur at the level of the *gus* gene expression. Promoter activity might be induced by mechanical, anaerobic, salt, or osmotic stresses generated during manipulation of the explants or during incubation in the enzyme reaction mix.

Water-miscible glycol methacrylate resin has been used for histochemical localization of a variety of enzymes *(6,8–13)*. We developed an X-Gluc-based histochemical assay for GUS enzyme localization on plastic-embedded plant tissue with a high resolution at the singe-cell level and artifact free. This method improves the preservation of tissue morphology, the permeability of the tissue and cells for the substrate, the maintenance of enzyme activity during procession of the material, embedding and sectioning, and the controlled reaction conditions at the site of the enzyme activity. The method is based on the following principles: processing of the tissue on crushed ice, pretreatment with spermidine instead of fixation, partial dehydration with acetone, embedding in water-miscible glycol methacrylate at 5°C, and finally performing the X-Gluc reaction on semi-thin sections (20 μm) of the plastic-embedded material. The assay has been developed for Brassica *(6,14)*, modified for *Arabidopsis thaliana (15)*, and has further been optimized in this species to obtain higher resolution for all tissue types. In a scheme, practical tips are discussed for orientation of the fragile *Arabidopsis* explants during embedding and for trimming of the capsules in order to simplify further sectioning. The interpretation of the cell specificity of the *gus* gene expression has to be correlated with a test for the metabolic activity of each cell type present in the tissue. The activity of succinate dehydrogenase, an enzyme from the Krebs cycle which is present in all living cells, is measured in these semithin sections as a standard for metabolic activity. Finally, it has to be realized that, because of the stability of the GUS enzyme, the histochemical assay allows the monitoring of the induction of promoter activity at high resolution at the single-cell level, but is less accurate in determining the time of switching off.

2. Materials

2.1. Processing of Plant Material and Pretreatment

1. Razor blades.
2. 15-mL Disposable glass vials (30 mm diameter × 40 mm height).
3. Cold room (4–5°C).
4. Crushed ice.
5. Water aspirator.
6. Orbital shaker.
7. Pretreatment solution: Murashige and Skoog medium *(16)* (Duchefa, Haarlem, The Netherlands) containing 1 mM spermidine.

2.2. Dehydration and Embedding

1. Acetone (50, 70, 80, and 90%) dilutions made with 0.85% NaCl.
2. Three-component Historesin-embedding kit (70-2218-500; Reichert-Jung, Heidelberg, Germany): component 1, basic resin (glycol methacrylate monomer, polyethylene glycol 400, hydroquinone); component 2, activator (benzoyl peroxide with 50% plasticizer); component 3, hardener (derivative of barbituric acid in dimethyl sulfoxide). Remove hydroquinone from the basic resin by shaking 100 mL of basic resin with 4 g of activated charcoal (100–400 mesh) for 1 h (*see* **Note 1**). Separate the treated basic resin from the activated charcoal by filtering through a fine mesh filter paper (602H; Schleicher and Schuell, Dassel, Germany). Store the hydroquinone-free basic resin at 4°C in the dark (*see* **Note 2**).
3. Make 50 and 70% basic resin (activated charcoal treated) dilutions with a 90% acetone dilution (made with 0.85% NaCl).
4. Embedding medium: basic resin (activated charcoal treated) + activator (0.6%) + hardener (1 mL for 15 mL basic resin).

5. Polyethylene embedding capsules: TAAB C094 capsules (TAAB, Reading, UK) with a flat bottom, approx 1000 µL volume; BEEM3 capsules (Agar, Cambridge, UK) with a pyramidal truncated cone, approx 300 µL volume.

2.3. Sectioning

1. Scalpel.
2. 1 cm^3 plexiglass blocks.
3. File.
4. Super glue.
5. Small bench vice.
6. Microtome suited for Ralph glass knives.
7. Glass knife maker for Ralph knives (Reichert-Jung)
8. Six-mm thick glass strips for Ralph knives (Agar).
9. Straight tweezers with very fine tips.

2.4. Enzyme Reactions

1. Small glass Petri dishes (diameter 35 mm).
2. Incubators at 26 and 37°C.
3. Humidified boxes: Put a small amount of water in closed plastic boxes (suited for two 35-mm Petri dishes) containing a plastic plate separated from the bottom by two layers of 1-cm thick plexiglass rods.
4. Enzyme reaction buffers:
   a. β-Glucuronidase: 100 mM Tris-HCl, pH 7.0, 50 mM NaCl, 2 mM X-Gluc cyclohexylammonium salt (Biosynth, Staad, Switzerland) (50 mg/mL stock in dimethyl formamide), 1 mM potassium ferri(III)cyanide (*see* **Note 3**).
   b. Succinate dehydrogenase: 100 mM Tris-HCl, pH 7.8, 50 mM NaCl, 100 mM Na-succinate, 1.3% dimethyl formamide, 1 mM nitro-blue tetrazolium (NBT) (BioRad, Hercules, CA) (80 mg/mL stock in 70% dimethyl formamide), 1.5 mM β-nicotinamide adenine dinucleotide.
5. Pyrex® baking dish.
6. Demineralized water.
7. Vectabond-treated slides (Vector, Burlingame, CA).
8. Horizontal staining dishes suited for 10 slides.
9. Mounting medium: Eukitt (O. Kindler GmbH, Freiburg, Germany).
10. Cover glasses.

3. Methods

3.1. General

The whole process is done in the cold room (5°C) on crushed ice and under continuous shaking, unless otherwise indicated. Vacuum infiltration is done by means of a water aspirator filled with crushed ice, unless otherwise indicated.

3.2. Processing of Plant Material and Pretreatment

1. Trim the tissue with a razor blade in such a way that the solutions used during the processing can reach all parts of the tissue (but do not wound the tissue too much);
   a. Roots: pieces of 7–10 mm (so that you can manipulate them easily afterwards).
   b. Leaves: very small leaves are not cut, leaves of approx 5 mm in length are cut longitudinally in two pieces, and fully expanded leaves in strips of approx 2 mm width and approx 5 mm length.
   c. Stem: approx 3 mm in length.
   d. Flowers: very small buds are not cut, whereas the tip is removed from buds larger than 0.5 mm in length.
   e. Siliques: approx 2 mm in length.
   The trimming of the tissue can be done at room temperature or in the cold room.
2. Put the trimmed material immediately in 15-mL disposable glass vials containing 5 mL pretreatment solution cooled on crushed ice (*see* **Note 4**).
3. Incubate for 15 min.
4. Vacuum infiltrate for 10 min.

3.3. Dehydration and Embedding

1. Replace the pretreatment solution with 50% acetone and incubate for 1 h.
2. Continue dehydration scheme as follows: 70% acetone for 1 h, refresh 70% acetone and incubate overnight, 1 h in 80% acetone, twice 1 h in 90% acetone (*see* **Note 5**).
3. Incubate for 4 h in 50% basic resin.
4. Incubate overnight in 70% basic resin.
5. Incubate 4 h in 100% basic resin.
6. Renew the 100% basic resin and vacuum infiltrate for 20 min (*see* **Note 6**).
7. Incubate for approx 24 h.
8. Wash explants for 5–10 min with embedding medium.
9. Place explants in embedding capsules partly filled with polymerized embedding medium: BEEM3 capsules with 50 µL embedding medium polymerized in the conical end of the capsule; TAAB C094 capsules with 500 µL embedding medium (*see* **Note 7**) polymerized in the length of the capsule (**Fig. 2A**). The polymerization is done for approx 40 min at 55°C.
10. Orient the specimen on the sticky surface of the polymerized embedding medium (**Fig. 2A**). **Table 1** summarizes the type of capsule that is convenient to use for each type of explant, depending on the sectioning plane (*see* **Note 8**)
11. Fill the capsules to the top with fresh non-polymerized embedding medium. Close the capsules tightly and put them in racks that allow air flow around the capsules (*see* **Note 9**).
12. Allow polymerization to proceed for 5 d at 5°C in the dark. Store the polymerized Historesin blocks still enclosed by the polythene capsules in a well-closed container (e.g., 50-mL Falcon tubes; Becton Dickinson, Lincoln Park, NJ) at 5 or –20°C. The embedded tissue can be stored in this way for several months during which no significant loss of enzyme activity occurs.

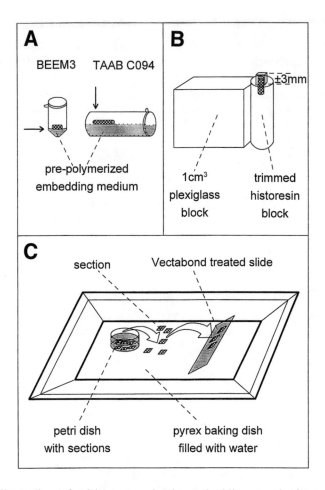

Fig. 2. Illustration of critical steps in the embedding, sectioning, and X-Gluc staining. **(A)** Prepolymerization of Historesin in the capsules and orientation of the explants depending on the plane for sectioning. **(B)** Mounting of the plastic-embedded explant onto a plexiglass block; filing of the Historesin in a cube-like shape. **(C)** After the X-Gluc reactions, the sections are transferred from the Petri dish to the washing solution in a glass container and subsequently transferred to a slide.

3.4. Sectioning

1. Bring the polythene capsule with the embedded tissue to room temperature in a vacuum desiccator.
2. Remove the polythene capsule from the polymerized Historesin block by means of a scalpel. Be sure that the material is at room temperature, otherwise the hydrophillic Historesin will take up condensation water.

Table 1
Type of Capsule Used per Explant for a Given Sectioning Plane

Explant	Cutting plane[a]	Embedding capsule BEEM3	Embedding capsule TAAB C094
Root	T+L[b]		x
Stem	T		x
	L	x	
Leaf	T		x
	L		x
Small flower buds[c]	T		x
	L	x	
Bigger flower buds[d]	T		x
	L	x	
Siliques	T		x
	L	x	
Callus		x	

[a]T, transversal; L, longitudinal.
[b]Several root explants, 3-6, are embedded together in one capsule. Upon cutting, both transversal and longitudinal sections are produced.
[c]Small flower buds are less than 0.5 mm in length. Several small flower buds are taken together for embedding in one capsule.
[d]Bigger flower buds are more than 0.5 mm in length.

3. File the Historesin block to obtain one flat side. Glue the Historesin block with the flat side on a 1-cm^3 plexiglass block (**Fig. 2B**).
4. Pinch the plexiglass block in a small bench vice (*see* **Note 10**).
5. Trim the Historesin block with a file squarely around the specimen (*see* **Note 11**; **Fig. 2B**).
6. Mount on the microtome.
7. Cut 20-µm sections (*see* **Note 12**).

3.5. Enzyme Reactions

1. Perform the reactions in 35-mm glass Petri dishes containing 3 mL of reaction buffer. The GUS reactions are done at 37°C during 1–24 h, depending on the promoter strength. The succinate dehydrogenase reactions are done overnight at 26°C (*see* **Note 13**).
2. Put approx 6–10 sections in the reaction buffer, on the bottom of the Petri dish (**Fig. 2C**; *see* **Note 14**).
3. Stop the reactions by washing the sections in demineralized water, as illustrated in **Fig. 2C**:
   a. Fill the Petri dish carefully to the top with demineralized water.
   b. Bring the Petri dish in a Pyrex® baking dish filled with demineralized water.

c. Pick up the sections with tweezers with very fine tips and place the sections on a Vectabond-treated slide. This is done in the water. Never take the sections with the tweezers out of the water;
   d. Take the slide with the attached sections out of the water (*see* **Note 15**).
4. Wash the slides a few times with demineralized water and incubate them overnight at 4°C in the dark in a horizontal staining dish filled with demineralized water.
5. Dry the slides at room temperature in the dark.
7. Mount the slides (*see* **Note 16**).
8. Visualize under the microscope. The X-Gluc reaction product is blue under bright light illumination or Normasky interference; it is red when in dark field. The NBT reaction product, formazan, is purple under bright light or Normasky interference (*see* **Note 17**).

4. Notes

1. The hydroquinone present in the basic resin has to be removed because it is an antioxidant which inhibits the enzyme reactions.
2. Never open cold Historesin at room temperature because it is very hygroscopic. Hydrated Historesin causes bad polymerization which results in a too-soft plastic that cannot be cut properly. Moreover, a high water content causes a decrease in enzyme activity probably due to a higher protease activity.
3. It is important to add the oxidant potassium ferri(III)cyanide to enhance the dimerization of the indoxyl intermediates at the site of GUS enzyme activity.
4. The trimmed explants have to be transferred immediately to the pretreatment solution to avoid induction of the gene fusion by stress. No fixative is used because it causes loss of enzyme activity. Instead, the tissue is infiltrated with a spermidine solution in plant tissue-culture medium. Spermidine stabilizes membranes, and inhibits protease and RNAse activity, which results in an excellent preservation of tissue morphology and enzyme activity. Plant tissue culture medium is used because the previously used sodium-phosphate buffer *(6)* was shown to activate stress-inducible promoters, resulting in an artifactual induction of GUS enzyme activity.
5. Dehydrate with well-kept acetone (acetone is hygroscopic) until 90%. Do not dehydrate completely because this would destroy the enzymes. A further dehydration to 94% is obtained when transferring the explants to basic resin (contains still 6% water).
6. Release vacuum in cold room, not at room temperature because Historesin is hygroscopic (*see* **Note 2**).
7. The capsules are partly filled with polymerized Historesin to allow the explant to be surrounded by embedding medium at all sides, because polymerization usually is incomplete at the edges. Partly filling of the capsules with Historesin is also done to keep the volume for final embedding of the explant low, hence to prevent that temperature rises too much during polymerization.
8. The sticky surface of the prepolymerized Historesin facilitates the orientation of the explant because it prevents floating of the tissue when subsequently the capsule is filled with embedding medium.

β-Glucuronidase Histochemistry

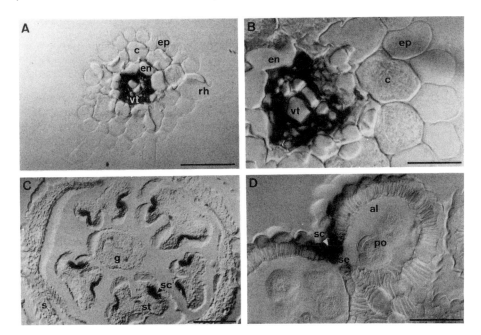

Fig. 3. X-Gluc reactions on *Arabidopsis* tissue. **(A)** Transverse section through a 2-wk-old root of a *pRPS18A-gus* transformed line *(15)*. **(B)** Detail of the vascular tissue; main GUS activity in vascular tissue. **(C)** Transverse section through a young (stage 9) flower of a line transformed with a *gus* gene fused to a stomium-specific tobacco promoter (provided by T. Beals and P. Sanders, Plant Molecular Biology Laboratory, University of California Los Angeles, CA). **(D)** Detail of mature anther. Main GUS activity located at the stomium, the site of anther dehiscence. Abbreviations: al, anther locule; c, cortex; en, endodermis; ep, epidermis; g: gynoecium; po, pollen; rh, root hair; s, sepal; sc, stomium cells; se: septum; st, stamen; vt vascular tissue. Visualization with Normasky interference microscopy. Bar = 50 μm (A), 20 μm (B,D), and 100 μm (C). (*See* color insert following p. 230.)

9. Close the capsule because polymerization of Historesin at low temperature is oxygen sensitive. Polymerization of Historesin is an exothermic process, therefore temperature is kept at 5°C and capsules are in open racks.
10. Do not pinch the Historesin block, because polymerized Historesin is a soft plastic and deforms easily. It is impossible to obtain good sections from a deformed Historesin block.
11. The Historesin block is filed in a cube instead of a cylinder. A square section has more capacity to expand during hydration (low surface/perimeter ratio) than a circular one (high surface/perimeter ratio), preventing the section from deformation during swelling.
12. When the *gus* gene is driven by a strong promoter, 10-μm sections are used (e.g., p*35S-gus* expression in roots). When the *gus* gene is driven by a weak promoter

or when very complex structures are analyzed, such as inflorescences, 30-µm sections are used.
13. Take successive sections, one to determine succinate dehydrogenase activity, the other one to test for GUS enzyme activity. Compare the patterns to decide on the cell specificity of the promoter examined.
14. The penetration of the substrate is much better when sections are put into the reaction buffer compared to the former protocol in which reaction buffer was added on top of the sections attached to a slide *(6)*. This change in the protocol improves the resolution of the method a lot because of a better uptake of the substrates by the sections and a better preservation of enzyme activity, because no attachment step of the sections onto the slide is needed.
15. Take care that the nonattached sections stay submerged during the washing process, otherwise they collapse.
16. When 30-µm sections are dried on a slide, they often detach partly from the slide. However, when they are covered with enough mounting medium, and a light pressure is applied on the cover glass after the mounting medium has spread over the sections, most of the sections are in one plane.
17. X-Gluc staining on semi-thin sections of a plastic-embedded *Arabidopsis* root and flower is shown in **Fig. 3**.

Acknowledgments

The authors thank Martine De Cock for help preparing the manuscript, Rebecca Verbanck for artwork, and Karel Spruyt for photographs.

References

1. Jefferson, R. A., Kavanagh, T. A., and Bevan, M. W. (1987) GUS fusions: ß-glucuronidase as a sensitive and versatile gene fusion marker in higher plants. *EMBO J.* **6,** 3901–3907.
2. Holt, S. J. and Withers, R. F. J. (1952) Cytochemical localization of esterases using indoxyl derivates. *Nature* (Lond.) **170,** 1012–1014.
3. Holt, S. J. and Withers, R. F. J. (1958) Studies in enzyme cytochemistry. V. An appraisal of indigogenic reactions for esterase localization. *Proc. R. Soc. London B* **148,** 520–534.
4. Cotson, S. and Holt, S. J. (1958) Studies in enzyme cytochemistry. IV. Kinetics of aerial oxidation of indoxyl and some of its indoxyl derivatives. *Proc. R. Soc. London B* **148,** 506–519.
5. Jefferson, R. A. (1987) Assaying chimeric genes in plants: the GUS gene fusion system. *Plant Mol. Biol. Rep.* **5,** 387–405.
6. De Block, M. and Debrouwer, D. (1992) *In-situ* enzyme histochemistry on plastic-embedded plant material. The development of an artefact-free ß-glucuronidase assay. *Plant J.* **2,** 261–266.
7. Mascarenhas, J. P. and Hamilton, D. A. (1992) Artifacts in the localization of GUS activity in anthers of petunia transformed with a CaMV 35S-GUS construct. *Plant J.* **2,** 405–408.

8. De Block, M. and Debrouwer, D. (1993) Engineered fertility control in transgenic *Brassica napus* L.: histochemical analysis of anther development. *Planta* **189**, 218–225.
9. Murray, G. I., Burke, M. D., and Ewen, S. W. B. (1988) Enzyme histochemical demonstration of NADH dehydrogenase on resin-embedded tissue. *J. Histochem. Cytochem.* **36**, 815–819.
10. Pretlow, T. P., Lapinsky, A. S., Flowers, L. C., Grane, R. W., and Pretlow, T. G. (1987) Enzyme histochemistry of mouse kidney in plastic. *J. Histochem. Cytochem.* **35**, 483–487.
11. Ashford, A. E., Allaway, W. G., Gubler, F., Lennon, A., and Sleegers, J. (1986) Temperature control in Lowicryl K4M and glycol methacrylate during polymerization: is there a low-temperature embedding method? *J. Microsc.* **144**, 107–126.
12. Soufleris, A. J., Pretlow, T. P., Bartolucci, A. A., Pitts, A. M., MacFadyen, A. J., Boohaker, E. A., and Pretlow, T. G. (1983) Cytological characterisation of pulmonary alveolar macrophages by enzyme histochemistry in plastic. *J. Histochem. Cytochem.* **31**, 1412–1418.
13. Namba, M., Dannenberg, A. M., and Tanaka, F. (1983) Improvement in the histochemical demonstration of acid phosphatase, β-galactosidase and non-specific esterase in glycol methacrylate tissue sections by cold temperatures embedding. *Stain Technol.* **58**, 207–213.
14. De Block, M. (1995) *In situ* enzyme histochemistry on plastic-embedded plant material, in *Methods in Cell Biology*, vol. 49 (Galbraith, D. W., Bohnert, H. J., and Bourque, D. P., eds.) Academic, London, pp. 153-163.
15. Van Lijsebettens, M., Vanderhaeghen, R., De Block, M., Bauw, G., Villarroel, R., and Van Montagu, M. (1994) An S18 ribosomal protein gene copy, encoded at the *Arabidopsis* PFL locus, affects plant development by its specific expression in meristems. *EMBO J.* **13**, 3378–3388.
16. Murashige, T. and Skoog, F. (1962) A revised medium for rapid growth and bio assays with tobacco tissue cultures. *Physiol. Plant* **15**, 473–497.

39

In Situ Hybridization to RNA in Whole *Arabidopsis* Plants

Qingzhong Kong and Anne E. Simon

1. Introduction

The importance of determining spatial distribution and expression of biomolecules in living organisms has been increasingly recognized because of the relevance to biological function. High resolution spatial information at the level of the cell or organism can be obtained through *in situ* detection techniques, including *in situ* hybridization, immunohistochemistry *(1)*, and expression of reporter genes such as the green fluorescent protein *(2,3)*. For plant materials, tissue printing is also widely used *(4)*. This chapter focuses on *in situ* detection of RNA in *Arabidopsis* at the level of the whole plant. Other in situ detection protocols for *Arabidopsis* plants are described in Chapters 35 and 36.

Whole plant *in situ* hybridization can reveal the spatial distribution of a DNA or RNA species in whole plants. Melcher et al. *(5)* first developed an *in situ* hybridization protocol to detect cauliflower mosaic virus (CaMV) DNA in individual turnip leaves, and Leisner et al. *(6)* modified the technique to detect CaMV DNA in whole *Arabidopsis* plants. The following protocol was developed for *in situ* detection of viral RNA in *Arabidopsis*, and has been used to study long-distance movement of RNA viruses *(7–9)*. An example is shown in **Fig. 1**. The protocol should also be amenable for *in situ* detection of abundant mRNAs in *Arabidopsis* plants after minor modifications. In brief, the plant is placed in a Petri dish, fixed with 95% ethanol, digested with pronase to remove barriers for probe access, treated with $0.2N$ HCl to denature the RNA, neutralized with 2X SSC, dried in air, baked for 2 h at 80°C under vacuum, prehybridized, and hybridized with a specific radioactive DNA probe. The plant is then washed, transferred to a piece of acetate sheet, dried in air, wrapped in plastic wrap, and subjected to autoradiography. The steps before

Fig. 1. Time course of turnip crinkle virus (TCV) movement in *Arabidopsis*. One of the two oldest leaves of 2-wk-old *Arabidopsis* ecotype Col-0 (TCV-susceptible) or Di-0 (TCVresistant) seedlings were inoculated by TCV or by inoculation buffer alone (Mock). Numbers on top of the panel denote number of days postinoculation when the plants were collected and frozen at –80° C. The plants were subjected to the whole plant *in situ* hybridization analysis using an oligonucleotide probe specific for TCV genomic RNA, as described in the text. Arrowheads point to the inoculated leaf. Outlines of leaves were drawn on a clear plastic sheet and photographed with the autoradiogram.

autoradiography require 3–4 d. The exposure time for detection of turnip crinkle virus (TCV) RNA with a ^{32}P-end-labeled oligonucleotide probe in infected *Arabidopsis* is as short as 30 min without an intensifying screen. This is a delicate technique, and some practice is needed for best results.

2. Materials

2.1. Pretreatment of Plant Materials

1. Gloves.
2. Freezers (–80 and –20°C).
3. Vaccum oven.
4. Incubator with shaker.
5. 95% Ethanol.
6. Diethyl pyrocarbonate (DEPC; Sigma, St. Louis, MO).
7. 10% Sodium dodecyl sulfate (SDS).
8. 20X SSC (1X SSC is $0.15M$ NaCl plus $0.015M$ sodium citrate).
9. $0.2N$ HCl.
10. DEPC-treated water: Add 0.1% (v/v) of DEPC to distilled water in a capped bottle, shake vigorously for several minutes or stir with a magnetic bar overnight,

and then autoclave for 20 min at 121°C. Solutions can be treated in the same way, but some reagents, such as Tris, cannot be treated with DEPC.
11. Busting solution: 0.1 mM NaN$_3$ (stock 10 mM), 0. 1% SDS (stock 10%), 10 mM EDTA (pH 8.0; stock 0.5M), 0.5 mg/mL pronase E (Sigma; stock 20 mg/mL; store at –20°C). Pronase E needs to be self-digested and it is prudent to check for RNase activity in all the stock solutions before application (*see* **Note 1**). Prepare busting solution from stock solutions just before use.
12. Plastic Petri dishes (60 × 15 mm or 100 × 15 mm; Fisher, Pittsburg, PA).

2.2. Hybridization and Autoradiography

1. 2X SSC made from stock 20X SSC.
2. 100X Denhardt's solution: 2% Ficoll 400 (Sigma), 2% polyvinylpyrrolidone (Sigma), 2% bovine serum albumin (Fraction V; Life Technologies, Gibco-BRL, Gaithersburg, MD). Sterilize by filtering through a filter (pore size 0.2 μm). Store in aliquots at –20°C.
3. 10 mg/mL fish sperm DNA (U.S. Biochemicals, Cleveland, OH): Suspend 500 mg of crude fish sperm DNA in 25 mL of water in a 50 mL-polypropylene centrifuge tube (Corning, Corning, NY) and then microwave for 1–2 min to fragmentize the DNA. Cool to room temperature and extract with phenol: chloroform (1:1) three times. Add 1/10 volume of 3M sodium acetate (pH 5.3), and precipitate DNA with ethanol. Dissolve the pellet in 30 mL of sterile water, measure OD$_{260}$ (an OD$_{260}$ reading of 1.0 equals 50 μg/mL for double-stranded DNA). Adjust the concentration to 10 mg/mL with sterile water. Store in aliquots at –20°C. Fish sperm DNA should be boiled for 5 min and chilled on ice before use.
4. Prehybridization solution: 2X SSC (stock 20X), 5X Denhardt's solution (stock 100X), 0.2% SDS (stock 10%), 1.0 mg/mL fish sperm DNA (stock 10 mg/mL; freshly denatured), and appropriate amount of formamide (*see* **Note 2**).
5. Hybridization solution: Prehybridization solution plus probe.
6. [γ-^{32}P]ATP (150 mCi/mL) or [α-^{32}P] dCTP (10 mCi/mL).
7. T4 polynucleotide kinase (New England Biolabs, supplied with 10X buffer) or random primer DNA labeling system (Life Technologies, Gibco-BRL).
8. Washing solution: 2X SSC, 0.1% SDS.
9. Folacel™ matte acetate sheet (available in art supply stores).
10. X-ray film and cassette.
11. Plastic wrap.
12. Tweezers.

3. Methods
3.1. Pretreatment of Plants

1. Remove *Arabidopsis* plants from the soil by cutting off the roots with a scalpel. Put each plant in one plastic Petri dish of suitable size. Store immediately at –80°C to avoid RNA degradation (*see* **Note 3**).
2. Immerse plants in cold 95% ethanol. Incubate at –80°C overnight, then at –20°C for 6 h to overnight, followed by shaking slowly at room temperature overnight (*see* **Note 4**).

3. Discard ethanol. Rinse with 95% ethanol (*see* **Note 5**).
4. Add 10 mL freshly prepared busting solution, shake gently at 38–40°C for 4–8 h. The optimal duration for busting solution treatment depends on the developmental stage of the plant material (*see* **Note 6**).
5. Remove busting solution and rinse plants four times in sterile DEPC-treated water. Add 10 mL of 0.2*N* HCl and incubate at room temperature for 20–25 min (*see* **Note 7**).
6. Carefully remove HCl and rinse plants two times in 2X SSC. Arrange leaves of each plant in what will be their permanent position. Dry plants in the Petri dish at room temperature for at least 3 h. Bake plant in the Petri dish at 80°C under vacuum for 2 h (*see* **Note 8**). The plants can now be stored dry for at least 6 mo at room temperature.

3.2. Hybridization and Autoradiography

1. Wet plants from below with 2X SSC for 15 min (*see* **Note 9**). Remove solution, then add 5 mL (for 60 × 15 mm dishes) or 10 mL (for 100 × 15 mm dishes) of freshly made prehybridization solution. Shake gently at 38–40°C overnight. Adjust the incubation temperature and formamide concentration in the prehybridization solution based on the probe used (*see* **Note 2**).
2. Oligonucleotide probes work best for this protocol. Label the oligonucleotide using T4 polynucleotide kinase and [γ-^{32}P]ATP. In a 20-µL reaction, use 20 pmol of the oligonucleotide, 2 µL of 10X T4 polynucleotide kinase buffer (supplied with the enzyme), 1–2 µL of [γ-^{32}P]ATP (150 mCi/mL), and 2 µL of T4 polynucleotide kinase (1 U/µL). Incubate at 37°C for 45 min. Inactivate the kinase by incubation at 68°C for 20 min. Store the probe at –20°C. Probes prepared using random primer labeling can also be used with compromised results (*see* **Note 10**).
3. Add 0.2–0.5 pmol of the oligonucleotide probe per plant. Hybridize overnight at 38–40°C (*see* **Note 2** for optimal hybridization temperature determination) with gentle shaking.
4. Pour off the hybridization solution. Rinse twice with 2X SSC. Incubate in 10 mL washing solution (2X SSC, 0.1% SDS) at 50°C for 30 min with gentle shaking. Carefully remove the washing solution and replace with 5–10 mL of fresh washing solution.
5. Cut Folacel™ matte acetate sheet to small circles the size of the Petri dish lid. Place a circle in the lid, cover the dish, and invert the dish containing the plant and washing solution such that the plant falls onto the acetate circle. Arrange leaves on the circle (*see* **Note 8**), then slowly remove the liquid. Remove the acetate circle containing the plant with a pair of tweezers, and let the plant air dry onto the sheet (> 3 h). Cut the unnecessary part of the circle to make it square or rectangular if you wish.
6. Mark the sheet for later identification. Position them onto plastic wrap, cover with a second piece of plastic wrap such that each square is fixed in place. Expose to X-ray film at –80°C for an appropriate time (*see* **Note 11**). Slight background in the stem and stronger background in the flower buds are normal.

4. Notes

1. The active component in the busting solution is pronase E, which needs to be self-digested before use. Dissolve 1 g of pronase E in 50 mL of 10 mM Tris-HCl, pH 7.4, 10 mM NaCl, and incubate at 37°C for 60–90 min. Divide into aliquots and store at –20°C. To check this or other reagents for contaminating RNase activity, combine 5 µL (1 µg/µL) of good quality total RNA with 5 µL of 2X busting solution, incubate at 37°C for 2 h, and then examine RNA for degradation by electrophoresis through a 1.5% agarose gel containing ethidium bromide.
2. The temperature for prehybridization and hybridization depends on the melting point (T_m) of the oligonucleotide probe used. The T_m (in °C) of an oligonucleotide can be estimated by using the equation T_m = (number of G or C bases × 4) + (number of A or T bases × 2). The hybridization temperature should be 5–10°C below the estimated T_m of the oligonucleotide. Formamide is added to the solution for prehybridization and hybridization to reduce the temperature needed for stringent hybridization. For every 1% of formamide added, the hybridization temperature drops by 0.7°C. We recommend using a 20-mer oligonucleotide with a Tm of 55–70°C as the probe, and maintaining the hybridization temperature at 37–40°C by adjusting the concentration of formamide.
3. RNA is vulnerable to degradation by RNases. Precautions include wearing gloves and preparing solutions with DEPC-treated water. Since RNA degradation within plant tissues occurs rapidly following root excision, each plant should be placed in a Petri dish on dry ice immediately after the root is cut off. Plants can then be stored at –80°C for at least several years. Frozen plants are easily broken, and care should be taken to avoid excessive movement. Mark each dish with a water- and alcohol-resistant marker pen for later identification.
4. The ethanol treatment that is used to fix the plant tissue also removes pigments. The gradual increase in temperature is needed because direct incubation of frozen plants in ethanol at room temperature causes degradation of RNA. To avoid breaking plants during ethanol addition, do not pour ethanol directly onto the plant. A pipet is recommended for addition and removal of liquid.
5. Plants are brittle following ethanol treatment. When touching the plant is necessary, use the side (not the tip!) of a pipet to hold the remaining root or edge of the base of a thick petiole.
6. The busting solution treatment serves to remove proteins so that RNA molecules in the plant tissue become accessible to probes. Overdigestion results in disintegration of the plant during this or subsequent steps. Inadequate digestion causes low signal-to-noise ratios. The optimal duration of digestion depends on the developmental stage of the plants. For young plants (up to five expanded leaves), 4–5 h treatment in the busting solution is sufficient; for older plants (plants with at least six fully expanded leaves but no bolts), 6–8 h is suggested; for plants that have bolted, overnight digestion may be necessary. Plants should become soft, flexible, and more transparent after busting solution treatment. It is normal to see some small particles in the solution after treatment. The busting solution should be made just before use to avoid loss of proteolytic activity.

7. Plants are often relatively weak at this point, so care must be taken not to touch areas other than the edge of the base of a petiole or the remaining root when changing solutions. HCl treatment is required to denature the RNA. The optimum duration of HCl treatment depends on the developmental stage of the plant. For young plants, 20–25 min is sufficient; plants that have bolted may require 30–35 min. Over-treatment with HCl results in fragile plant material, and under-treatment with HCl causes a low signal-to-noise ratio.
8. The plant should be arranged before drying since the position of the leaves are mainly fixed after drying. Care should be used since plants are very fragile. Plants should be moved only in the 2X SSC solution. Use a controlled flow of 2X SSC from a pipet to unfold a leaf or separate two overlapping leaves. Leaves should be arranged so that overlapping is minimal. When handling many samples, remove the HCl and add the 2X SSC for all the samples before rinsing with 2X SSC so that no sample receives excessive HCl treatment. Baking the plants is a critical step for low background counts, but over-baking will break the plants and should be avoided.
9. After being baked, plants are fragile and often stick to the dish. Wet plants by slowly adding 2X SSC with a pipet until just sufficient to wet the whole plant from beneath. Do not add 2X SSC directly onto the plant. After 10 min, use a controlled flow of 2X SSC from a pipet to separate the plant from the dish, and sometimes a little push on the edge of the sticking leaf blade or stem may help.
10. Oligonucleotide probes are recommended for this protocol because their small size allows better tissue penetration resulting in better signal-to-noise ratios. When an oligonucleotide is not an option, random-primed DNA probes can be used. However, only probes with high specific activity ($> 1 \times 10^8$ dpm/µg) will produce satisfactory results, and at least 5×10^6 dpm of denatured probe is needed per plant. Use a hybridization solution containing 50% formamide and hybridize at 38–40°C overnight. After hybridization, plants should be rinsed with 2X SSC, and then washed in 2X SSC, 0.1% SDS at 50°C with gentle shaking for 20 min.
11. The optimum exposure time to X-ray film should be determined empirically. A 24-h exposure is required for a Geiger counter reading of 500 cpm.

References

1. Ausubel, F. M., Brent, R., Kingston, R. E., Moore, D. D., Seidman, J. G., Smith, J. A., and Struhl, K. (1994) *Current Protocols in Molecular Biology,* vol. 2. Wiley, New York.
2. Haseloff, J. and Amos, B. (1995) GFP in plants. *Trends Genet.* **11(8),** 328,329.
3. Prasher, D. C. (1995) Using GFP to see the light. *Trends Genet.* **11(8),** 320–323.
4. Varner, J. E. and Ye, Z. (1994) Tissue printing. *FASEB J.* **8(6),** 378–384.
5. Melcher, U., Gardner, C. O., Jr., and Essenberg, R. C. (1981) Clones of cauliflower mosaic virus identified by molecular hybridization in turnip leaves. *Plant Mol. Biol.* **1,** 63–74.
6. Leisner, S. M, Turgeon, R., and Howell, S. H. (1993) Effects of host plant development and genetic determinants on the long-distance movement of cauliflower mosaic virus in *Arabidopsis. Plant Cell* **5(2),** 191–202.

7. Kong, Q., Oh, J.-W., and Simon, A. E. (1995) Symptom attenuation by a normally virulent satellite RNA of turnip crinkle virus is associated with the coat protein open reading frame. *Plant Cell* **7(10),** 1625–1634.
8. Oh, J.-W., Kong, Q., Song, C., Carpenter, C. D., and Simon, A. E. (1995) Open reading frames of turnip crinkle virus involved in satellite symptom expression and incompatibility with *Arabidopsis thaliana* ecotype Dijon. *Mol. Plant–Microbe Interact.* **8(6),** 979–987.
9. Simon, A. E., Li, X. H., Lew, J. E., Stange, R., Zhang, C., Polacco, M., and Carpenter, C. D. (1992) Susceptibility and resistance of *Arabidopsis thaliana* to turnip crinkle virus. *Mol. Plant–Microbe Interact.* **5(6),** 496–503.

40

In Vivo Footprinting in *Arabidopsis*

Anna-Lisa Paul and Robert J. Ferl

1. Introduction

In vivo footprinting is a method of observing DNA/protein interactions of a gene within the context of the living cell. In other words, in vivo footprinting provides an opportunity to observe conformational changes in chromatin associated with the transcriptional machinery of a cell as it responds to environmental and developmental signals in vivo. The DNA/protein interactions targeted by in vivo footprinting can be visualized through the use of genomic sequencing, a technique that can provide strand-specific resolution of single bases *(1)*.

There are basically three requirements for characterizing a gene with in vivo footprinting. First, the sequence of the gene must be known (at least through the region of interest). Second, the tissue source for in vivo analysis must be relatively uniform with regard to cell type. And third, if the in vivo interactions are to be visualized by direct hybridization, a strand-specific radioactive probe of very high specific activity must be available. An alternative method of visualizing the in vivo interactions is through ligation-mediated polymerase chain reaction (LMPCR), but our method of choice for *Arabidopsis* is Direct Hybridization (DH). The small genome of *Arabidopsis* makes it relatively easy to generate a strong hybridization signal, even in single copy genes *(2,3)*. However, we are also successful with DH in maize, which has a genome two orders of magnitude larger *(4,5)*. In addition, many researches using LMPCR to amplify gene sequences from large genomes still use hybridization to visualize the results of the in vivo modifications *(6)*.

The basic steps in the process can be summarized as:

1. Briefly incubate plant tissue with dilute dimethyl sulfate (DMS).
2. Isolate the genomic DNA from the DMS treated tissue.

3. Cleave the DNA at the site of the DMS introduced modifications with piperidine.
4. Cut the genomic DNA at a suitable restriction site to facilitate indirect end labeling.
5. Resolve the DNA on a sequencing gel.
6. Transfer the DNA in the gel to a nylon membrane.
7. Hybridize the membrane with a strand-specific probe and visualize with autoradiography or phosphorimaging.

The theory behind the process is very simple. The technique applies chemical sequencing reagents and procedures to genomic DNA (hence the name genomic sequencing). DMS rapidly diffuses into intact cells and preferentially interacts with the genome by methylating the N7 of guanine residues in the major groove of DNA *(7)*. The DNA that has been modified in vivo can be purified from the cells without any alteration of the DMS reactions, and the "footprinting" is enabled by the fact that when a protein is bound to the DNA in vivo, the ability of DMS to modify a G residue will be altered. Both enhancements and protections of DMS reactivity can occur. The DMS reactions are conducted at concentrations that facilitate single-hit kinetics (i.e., statistically the DMS will modify only one G residue per DNA molecule), thus the segments of DNA that are either overly abundant (enhanced reactivity) or under represented (protections) generate a footprint of the region occupied by the protein when compared to the DMS reaction conducted in protein-free ("naked") DNA in vitro (**Fig. 1**).

2. Materials
2.1. Solutions and Materials for the DMS Treatments
1. DMS: The reagent is purchased from Aldrich (Milwaukee, WI) as a 99+% liquid and stored at 4°C under argon. The liquid is colorless; any hint of color indicates that the reagent is beginning to breakdown and should be discarded (*see* **Note 1** for disposal and inactivation guidelines). DMS is a moderately volatile, potent carcinogen. Treatments of cells and naked DNA with DMS should be conducted in a chemical fume hood while wearing gloves and protective clothing.
2. 10N NaOH for inactivating DMS in waste containers.
3. Liquid nitrogen (LN_2) for freezing tissue, and for freezing samples prior to lyophilization (*see* **Subheading 3.3.**).
4. Vacuum filtration flask and Buchner funnel for collecting and washing cells and tissue after DMS treatments.

2.2. Solutions and Materials for Genomic DNA Isolation
1. Cell lysis buffer: 50 mM Tris-HCl pH 8.0, 50 mM NaCl, 400 µg/mL ethidium bromide, 2% *N*-lauroyl sarcosine. This buffer requires approx 30 min of stirring for the ethidium bromide to fully solubilize. It is then stored in a brown bottle at 4°C.
2. 10 mM Ethidium bromide: Ethidium bromide powder is added to water and allowed to stir several hours. The solution is kept in a brown bottle at 4°C.

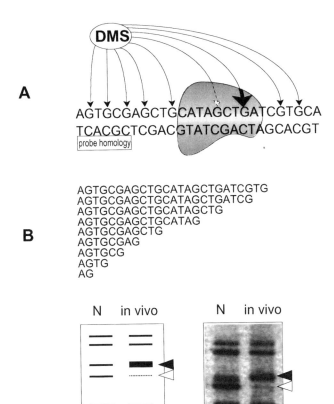

Fig. 1. A schematic of in vivo footprinting. **(A)** DMS interacts with genomic DNA in vivo. A protein bound to the DNA (shaded shape) will either protect guanines from the action of DMS (small open arrow) or enhance the ability of DMS to react with the base (large solid arrow). In direct hybridization, a section of sequence outside of the region of interest is selected for probe homology (This drawing does not represent true scale, whereas **Fig. 2** depicts the regions of probe homology and footprinting in realistic proportions). **(B)** When the DMS reactions are conducted to achieve single hit kinetics, a population of DNA fragment are generated that differ in length by the distance from one G residue to the next. The fragments containing the G residues that were protected from DMS will be underrepresented and the fragments containing G residues that had enhanced reactivity with DMS will be over-represented compared to the fragments containing the surrounding bases. **(C)** When these fragments are resolved on a sequencing gel and visualized with hybridization, they are seen as discrete bands. Compared to the naked genomic DNA control (N), the protein mediated protections and enhancements of the in vivo treatments appear as darker (solid triangle) and lighter (open triangle) bands. **(D)** A section of an in vivo footprint in the *Arabidopsis Adh* gene showing the naked genomic control (N) and the DNA from the in vivo treatments (in vivo).

3. Water saturated n-butanol: Fill a glass bottle 1/2 full with n-butanol then add 1/4 vol of distilled water. Shake vigorously, then let settle into two phases (may take more than an hour). The top phase contains the water saturated butanol. It is used to remove ethidium bromide ("decolorize") from DNA recovered from cesium chloride (CsCl) gradients.
4. 1X TE: 10 mM Tris-HCl pH 8.0, 1 mM EDTA. A standard buffer for resuspending precipitated DNA and in dialysis.
5. CsCl: Molecular biology grade from BRL.
6. Small household coffee grinder for grinding frozen tissue.

2.3. Solutions for Preparing the Genomic DNA for Sequencing

1. Phenol/chloroform/isoamyl alcohol: Made in the ratios of 25:24:1, for the organic extraction of DNA. Prepared with buffered phenol (BRL).
2. Chloroform for the organic extraction of DNA.
3. 7.5M ammonium acetate (AmOAc): Used at 1/3X vol to precipitate DNA from solution with ethanol.
4. 95% Ethanol: Used at 2.5X vol for the precipitation of DNA from solution.
5. Piperidine: The 99.5% liquid is purchased from Fisher (Pittsburg, PA). It is stored at 4°C under argon. The liquid is colorless and any change is indicative of breakdown products. We typically make working aliquots (approx 10-mL) to minimize the opening of the stock bottle.

2.4. Solutions and Materials for Sequencing.

1. 10X TBE: 0.89M Tris, 0.89M boric acid, 26 mM EDTA, pH 8.0. 1X TBE is used as the running buffer for the sequencing gel and in the electrotransfer of the acrylamide gel to the nylon membrane.
2. 40% acrylamide stock (250 mL): 95 g acrylamide, 5 g bis-acrylamide, water to make 250 mL total volume. Stir for 30–60 min, filter solution through a 0.2-micron nitrocellulose filter and store at 4°C in a brown bottle.
3. Sequencing gel solution for 6% acrylamide gel (150 mL): 22 mL 40% acrylamide stock solution, 15 mL 10X TBE, 63 g urea, 60 mL distilled water. Add the other components to the water then stir at room temperature until urea dissolves. Filter solution through a 0.2-micron nitrocellulose filter just prior to use.
4. 10% Ammonium persulfate: Use in the polymerization of the acrylamide gel.
5. TEMED: N,N,N',N'-tetramethylethylenediamine, BRL Ultra Pure. Used in the polymerization of the acrylamide gel.
6. Urea.
7. Sequencing dye: 98% deionized formamide, 10 mM EDTA (pH 8.0), 0.025% xylene cyanol FF, 0.025% bromophenol blue.
8. Any commercial sequencing apparatus with 0.75 mm spacers and combs.
9. GeneScreen (DuPont, Boston, MA), or any uncharged nylon membrane.
10. Whatman 3-mm filter paper in large sheets.

2.5. Materials for Electrotransfer

1. 1X TBE (*see* **Subheading 2.4., item 1**).
2. Electrotransfer apparatus: We use a 20-L horizontal electrotransfer tank, but any commercially available apparatus can be adapted for genomic sequencing. Our unit consists of a submersible sandwich that holds the gel and nylon membrane in close contact throughout the transfer. The sandwich is made from two, 50 × 40 cm plastic grids that hold two pieces of Scotch-Brite cut to the same size. The membrane/gel/paper piece described in the **Subheading 3.** is supported between the two Scotch-Brite pads. The whole sandwich is held tightly together with large rubber bands.
3. A power supply capable of delivering up to 2 A. We use the BioRad model 200/2.0.

2.6. Solutions and Materials for Hybridization and Probe Synthesis

1. M13 vectors: M13mp18 and M13mp19 *(8)* are used to generate clones of the probe DNA fragment in either orientation, enabling the production of probes specific for either the top or bottom strand.
2. The standard 17mer sequencing primer: Used for priming the synthesis reaction from the M13 templates.
3. Deoxynucleotides dGAT, dATP, and dTTP at concentrations of 2 mM.
4. *Taq* polymerase and its corresponding 10X buffer (provided with enzyme).
5. Radiolabeled dCTP (Alpha-dCTP-^{32}P).
6. Probe synthesis reaction mixture: 4.5 μg of M13 phage annealed to the 17mer primer (brought to 10 μL), 1 μL each dGAT, dATP, and dTTP (at 2 mM), 4 uL 10X *Taq* polymerase buffer, 1 μL *Taq* polymerase and 25 μL (250 μCi) ^{32}P-CTP.
7. Preparative 6% sequencing gel: We use a small (20 × 15 cm) gel cast with 1.5 mm spacers and a large three-well comb.
8. Polaroid type 57 film and developing cassette.
9. Hybridization solution: 0.5M Na phosphate pH 7.2, 7% sodium dodecyl sulfate (SDS) (Ultra Pure, from BRL), 1% bovine serum albumin (BSA) (Sigma A-4378), 1 mM EDTA. The pH of this buffer must be adjusted with H$_3$PO$_4$. The use of a poorer quality of BSA will result in a viscous solution that will give less than optimal results. It is stored at room temperature, but can be gently warmed to resolubilize any precipitated components if the solution cools.
10. Wash solution: 40 mM Na phosphate, 1 mM EDTA, 33 mM NaCl, 0.1% SDS. We typically make this as a 5X concentrated stock, then dilute as needed. You cannot make a concentration greater than 5X without some precipitation of the components.
11. Hybridization apparatus: We use a rotating water bath that contains Lucite tubes (2.0-cm inner diameter). The membrane is wrapped around a Plexiglass rod (1.2-cm diameter) that is about the same length as the width of the membrane and placed inside a Lucite tube with 5–10 mL of hybridization buffer. Incubator ovens with roller bottles are also suitable.

3. Methods

3.1. Treating the Cells and Tissue with Dimethyl Sulfate

In our hands, *Arabidopsis* leaves and cultured cells lend themselves to footprinting equally well. The leaves should be harvested before the onset of flowering and the cultures should be used one or two days past their last transfer to fresh media.

1. Set up a vacuum filtration flask (1000 mL) with 30 mL of 10N NaOH in the bottom of the flask. The NaOH is necessary to inactivate the DMS as it is filtered away from the treated tissue (*see* **Note 1**).
2. Harvest the leaves and keep on ice until 3–5 g of tissue has been accumulated. Collect two sets, one for the DMS treatment, and one to be used to isolate untreated genomic DNA that will be used for the naked DNA controls. The control tissue can be frozen in LN_2 and stored at –80°C until the DMS treated tissue has reached the DNA purification step.
3. It may be necessary to lightly mince older leaves with a razor blade (creating approx 2 mm sections), but we have used whole leaves when they are young. Drop the leaves into a 250 mL flask containing 100 mL of water at room temperature.
4. Add DMS to a concentration of 0.2% (200 µL) and swirl the flask for 2 min. If cultured cell suspensions are being used, simply bring the volume of the liquid culture medium to 100 mL and treat in the same way.
5. Filter the tissue away from the dilute DMS with vacuum filtration (we use a 4-in. Buchner funnel lined with Miracloth [Calbiochem, La Jolla, CA]) and wash with approx 800 mL of water without breaking the vacuum.
6. Freeze the recovered tissue (or cultured cells) with LN_2 and store at –80°C.

3.2. Isolation of Genomic DNA (see Note 2)

1. Place 4 mL of lysis buffer into each of two screw-top preparative centrifuge tubes.
2. Remove the frozen tissue described in the previous section from the freezer and weigh out 4 g each of the control and DMS treatment.
3. Place the tissue in a weigh boat and cover with LN_2. When most of the LN_2 has evaporated, place the tissue into a small household coffee grinder that has been prechilled with a little liquid nitrogen and grind to a fine powder (*see* **Note 3**).
4. Drop the powdered tissue into the 4 mL of lysis buffer described in **step 1** and stir with a glass rod until the powder has been evenly wet. Place the tubes in a room temperature water bath and stir occasionally until the mixture thaws completely. The mixture should look very viscous.
5. Centrifuge the tubes at 25,000g for 10 min at 5°C.
6. Transfer the supernatant to new screw-top tubes and add 1 g of CsCl for every mL of supernatant (it should be close to 4 mL in volume, but measure accurately before adding CsCl).
7. Rock gently to dissolve the CsCl. When the CsCl is in solution, respin the tubes as above then carefully decant the supernatant away from the pellet and possible

pelicle (*see* **Note 4**). Pour the supernatant into ultracentrifuge tubes. This protocol is ideal for the Beckman 5.1-mL heat-sealed tubes that fit the VTi65 rotor. In this system we centrifuge the samples for 4 h at 65,000 rpm then remove the banded genomic DNA with a syringe (with an 18-gage, or larger, needle).
8. Decolorize the recovered DNA with 3–4 sequential volumes of water-saturated n-butanol then dialyze against 1X TE.
9. Check the concentration of the DNA. A concentration of at least 0.1 μg/uL will enable the subsequent steps to be conducted in a single microcentrifuge tube.

3.3. Preparing the Genomic DNA for Sequencing

The ensuing steps are applied identically to both the genomic DNA treated with DMS in vivo and the control DNA, with one exception; the control DNA is treated with DMS in vitro to establish a pattern of guanine modifications characteristic of genomic DNA that is free of associated proteins. Be careful not to "re-DMS" the samples from the in vivo treatments. The first step is to restrict the genomic DNAs with suitable endonucleases that define the region of interest and provide a homologous end for the subsequent indirect-end labeling of the hybridization process (**Fig. 1** and **2**).

1. Bring 20 μg of both the control genomic DNA and the DNA from the in vivo DMS treatment to a volume of 200 μL with 1X TE.
2. Adjust with appropriate restriction buffers and incubate with the desired restriction endonuclease(s) designed for your particular system. Digest with two applications of approx 20 U of enzyme, with a duration of 2 h for each application (i.e., 4 h total) (*see* **Note 5**).
3. After the restriction digests, temporarily set the in vivo treatments aside. Proceed with the in vitro DMS treatment of the naked genomic control DNA by adding 0.5 μL of DMS to the restricted sample and vortex. Let it incubate for 2 min then add 200 μL of phenol/chloroform/isoamyl and vortex. The organic extraction removes the DMS from the DNA fraction, so there is no harm in letting sit at this stage while the restricted DNAs from the in vivo treatments are also extracted with 200 μL of phenol/chloroform/isoamyl.
4. After organic extraction, centrifuge both sets for 5 min. Remove the aqueous phase to fresh tubes and extract with 200 μL of chloroform. Centrifuge again and transfer the aqueous phase to fresh tubes.
5. Add 75 μL of 7.5M AmOAc and 500 uL of 95% ethanol and let sit on ice 30 min.
6. Centrifuge at 4°C for 10 min, decant supernatant, and wash pellet with 500 μL of 95% ethanol. Let the pellets air dry, upside down.
7. Resuspend the pellets in 50 μL of 10% piperidine (dilute the piperidine to 10% just prior to resuspending the pellets and keep on ice until use).
8. Incubate the pellets 20 min at 95°C in a heat block that can be secured with a metal cover to hold the caps in place (the tubes can also be incubated in a boiling water bath if the tube caps are adequately secured).

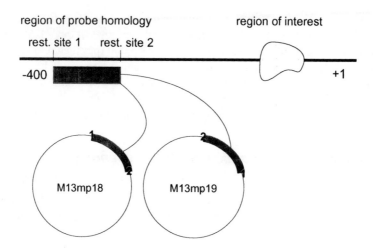

Fig. 2. Designing M13 clones for probe synthesis. Choose two restriction sites (rest. site 1, rest. site 2) outside of the region of interest (but within 400 bp for the best resolution of a particular region) that are between 100 and 200 bp apart. Subclone this segment (gray bar) into the M13 vectors mp18 and 19. The mp18 clone will synthesize a probe homologous to the bottom strand, and thus illuminate top strand modifications. The reverse is characteristic of the mp19 clone.

9. Place the piperidine-treated samples on ice for a few minutes. Spin down the condensation with a 5-s pulse, then add 250 µL of distilled water to each sample.
10. Freeze the samples with LN_2, open the lids, and lyophilize until all the liquid has been evaporated (*see* **Note 6**). Resuspend the dry pellets in 50 µL of distilled water and repeat lyophilization. The final dry pellets are ready to be resuspended in sequencing dye and applied to the sequencing gel.

3.4. Sequencing the DNAs

1. Prepare a standard 6% acrylamide sequencing gel cast 0.75 mm thick (thinner gels have difficulty accommodating 20 µg of genomic DNA and do not withstand the rigors of electrotransfer very well). When preparing the plates for casting, siliconize the back plate only, this will help to ensure that the gel sticks to just the front plate when the gel is removed. Prerun the gel in 1X TBE for at least 30 min.
2. Resuspend the lyophilized samples in 5 µL of sequencing dye, boil 3 min and immediately place on ice.
3. Load the boiled samples onto the prerun gel. Allow the samples to electrophorese (at 10–40 mA, as convenient) until the center of the region of interest (as defined by the restriction digests) has traversed approx 2/3 of the way to the bottom of the gel (*see* **Note 7**).
4. When the desired distance of migration has been reached, turn off the apparatus and allow the plates to cool for a few minutes before separating the plates and

exposing the gel. Attempting to separate the plates while the gel is still hot may result in the gel sticking to both plates and breaking (*see* **Note 8**).
5. While the gel is cooling, cut a piece of GeneScreen to a size that will cover the length of the gel and the width of the sample lanes, then pre-soak in 1X TBE. Also cut a piece of Whatman 3-mm paper slightly larger than the GeneScreen.
6. Cover the exposed gel surface with plastic wrap, then trace the shape of the GeneScreen onto the plastic (we use the membrane's paper backing as a template). Next, cut through the plastic and the underlying gel along the tracing. Use a very sharp scalpel and a smooth motion or the gel may tear.
7. Carefully lift the "window" of plastic wrap off of the defined section of gel. Center the previously cut Whatman 3-mm paper above the exposed section of gel and apply it to the surface in a smooth, rolling motion. The dry paper will adhere to the gel, and the gel/paper can be lifted from the glass as a single unit.

3.5. Electrotransfer of the Samples in the Sequencing Gel

1. Place the gel/paper unit gel side up on to the surface of one of the Scotch-Brite pads and plastic support grid of the electrotransfer cassette described in **Subheading 2.**
2. Wet the surface of the gel with 1X TBE (a squirt bottle makes this easy) and overlay with the prewet piece of GeneScreen in a rolling motion to exclude any bubbles.
3. Cover with the other Scotch-Brite pad and plastic support grid, flip over, and remove the first pad to expose the 3-mm paper on the other side of the sandwich. Wet the surface of the paper with 1X TBE, carefully ease the paper from the gel in a rolling motion, wet the surface of the gel, then replace the paper with a rolling motion to exclude any bubbles. Replace the Scotch-Brite pad and grid, secure with large rubber bands, and place in the transfer tank with the GeneScreen side of the sandwich at the positive pole.
4. Electrotransfer at 1.8 A for 2 h.
5. Disassemble the transfer cassette and peel the GeneScreen from the gel surface. Blot the membrane dry on paper towels, wrap in plastic and UV cross link.

3.6. Hybridization

The only requirement for the probe is that it be single stranded and of a very high specific activity (10^9–10^{10} cpm/µg) since the individual bands to be visualized will be present in minuscule amounts in a reaction from a single copy gene. This kind of specific activity can be achieved with DNA probes produced by synthesis from M13 clone templates *(5)* and RNA probes produced by in vitro transcription systems *(10)*. The following protocol describes the synthesis of a DNA probe generated by priming an M13 clone containing sequence that borders on the restriction site chosen for indirect-end labeling *(1)*.

1. Design M13 clones by choosing a restriction fragment in the gene of interest of approx 200 bp that has one end in common with the restriction site chosen for the

genomic digests (**Fig. 2**). Clone this fragment into both M13mp18 and M13mp19 to produce template orientations which will generate probes specific for either the top or bottom strand.
2. Mix 4.5 µg of the M13 phage with 30 ng of the standard 17mer sequencing primer, boil 3 min and place at 50°C for 45 min. This anneals the primer to the phage.
3. Synthesize a radiolabeled single strand by incubating the annealed phage/primer with a reaction mixture containing: 4.5 µg annealed phage/primer (in 10 µL), 1 µL each dGAT, dATP, and dTTP (at 2 mM), 4 µL 10X *Taq* polymerase buffer, 1 µL *Taq* polymerase, and 25 µL (250 µCi) ^{32}P-CTP for 30 min at 65°C.
4. Sequencing gel. The position of the radiolabeled DNA is visualized by covering the gel with plastic wrap then overlaying with a piece of Polaroid type 57 film for 5 min and aligning the developed film with reference points on the gel (i.e., the well and punctures made through the gel and film as it was being exposed).
5. Cut out the section of the gel corresponding to the bright band on the film and chop into small pieces (approx 2 mm^2). Put the pieces into a disposable screw-top test-tube with 10 mL of hybridization buffer and a small scrap of GeneScreen (the latter seems to "soak up" some of the components that contribute to nonspecific hybridization).
6. Incubate at 65°C for at least 1 h.
7. Prehybridize the electrotransferred GeneScreen for at least an hour at 65°C in the hybridization buffer, discard the prehybridization solution and add the eluted probe from **Subheading 3.6., step 4** (gel pieces can be included).
8. Hybridize for at least 12 h, then transfer the membrane to a tray containing approx 300 mL of wash buffer at 65°C. Wash in a total of four changes of wash buffer at 65°C.
9. Blot the membrane dry on paper towels, wrap in plastic, and expose to film for autoradiography or to a phosphorimaging screen.

4. Notes

1. DMS is a potent carcinogen and extreme care needs to be exercised when using it. Fortunately, it is easily inactivated by NaOH. We keep a 250 mL beaker in the fume hood containing about 50 mL of 10N NaOH, and discard used tips and microcentrifuge tubes directly into it. When it is full, simply dilute with water (to render the NaOH less dangerous) and flush down the sink. In addition, the dilute DMS that is filtered away from the cells is inactivated by collecting the filtrate into a 1 L vacuum flask containing approx 30 mL of 10N NaOH. Always wear gloves and protective clothing.
2. We have tried a number of different methods for the isolation of genomic DNA and found that the CsCl method described in the text results in the highest quality footprints. For instance, rapid lysis methods that include phenol extractions and precipitation of unrestricted DNA seem to contain more in-lane background noise and nonspecific single-stranded nicks than in CsCl prepared DNAs.
3. Grinding the cells: The best results are obtained with the small household coffee grinder mentioned in the text. It is important to precool the unit, and grinding up a chunk of dry ice is the best way to do this. If you use LN$_2$ to precool, make sure

all of the LN$_2$ has evaporated before starting to grind. Failure to do so can result in blowing the lid off of the unit. However, there will always be a small amount of residual LN$_2$ left in the tissue to be ground, thus it is a good idea to drill or melt (with a hot needle) a small vent hole in the side of the lid to allow the rapidly expanding nitrogen to escape. The one drawback of this method is that it requires at least 2 g of tissue to be really effective, and 3–5 g will give the most finely ground product.

4. Preparing the lysates for ultracentrifugation: After the CsCl/cell lysate mixture is spun in the preparative centrifuge, you may notice that most of the ethidium bromide from the lysis buffer has been lost to the pellet and pellicle. This phenomenon is especially common with green tissue. A solution of 10 mM ethidium bromide can be used to adjust the sample back to its previous saturation (added empirically, 50–100 µL will be sufficient).

5. Genomic markers: While the DNAs are being restricted to facilitate indirect end labeling it is convenient to make genomic DNA markers that can be run on the sequencing gel for reference. For example, referring to **Fig. 2**, cutting with the enzyme for site 1 alone gives you the "parent band" that should be visible in each lane of your DMS reactions as well. It represents the fragment created by cutting at site 1 and at the next incidence of that restriction motif, the position of which may or may not be known (it does not really matter as long as it is outside of the region of interest). Cutting with restriction enzyme 1 and an enzyme that cuts anywhere else within the region of interest will generate a band at the position of the second restriction site. Having one or more of these references for position is a great aid in deciphering the sequence of G residues. It is also helpful to treat genomic DNA with other chemical sequencing reagents to generate A + G and C ladders *(9)*. If you use plasmid DNA to generate markers it will need to be diluted approx 10,000x (determine empirically) to match the sequence abundance represented in a single copy gene in *Arabidopsis*.

6. Lyophilization: This process is employed to remove the piperidine. A common alternative used with plasmid DNA is butanol precipitation. This procedure does not seem to be as effective with genomic DNAs, and the result of using butanol to remove the piperidine is a good deal of smearing in the lanes. The lyophilization can be accomplished with a speed-vac microcentrifuge with the heater shut off, or a canister-style freeze drier.

7. Estimating migration in the sequencing gel: In a 6% acrylamide gel the bromophenol blue (BɸB) dye comigrates with approx 25 base fragments and the xylene cyanol FF (XC) comigrates with approx 110 base fragments. These dyes can be used as guides to estimate when your region of interest has been optimally resolved in the gel, but it is an empirical process to fine tune the precise distance. For instance, to resolve a sequence between 300 and 350 bases away from the restricted end, the XC is allowed to migrate 60 cm. Of course, the XC of the original loading will run off the bottom of the gel at 40 cm, so the progress of the dye is monitored through the use of a second loading of dye. When the XC from the first loading reaches 30 cm, a second aliquot of dye is loaded in one of the wells and its

progress is monitored. When the second loading reaches the 30 cm mark, the XC from the initial loading has reached 60 cm.

8. Occasionally the gel will stick to both plates or crumple up as the plates are pried apart. The reason for this problem is usually that either the plates were not scrupulously clean (clean with glass cleaner, then 95% ethanol) or that the plates were still too hot. To ease a partially stuck gel off of the top plate, hold it down at a very shallow angle over the bottom plate and coax the gel into peeling off the plate by gently putting some upward tension on the top plate at the junction. This process will stretch the gel somewhat and may distort the bands a bit, but you will be able to get readable information from your experiment. If the gel just crumples into a heap, or has serious folds and bubbles in it, first inject a little 1X TBE under the folds and creases and then use a combination of "jiggling" of the plate and smoothing with gloved fingers to straighten out the gel. We use a 10 mL syringe fitted with a pipet tip to apply the 1X TBE in these situations. It has been our experience that a little patience and perserverance can recover useful information from even the most hopeless looking mess.

Acknowledgments

Aspects of the work described herein were supported by United States Department of Agriculture Grant NRI 94-37301-0565 to R. J. Ferl and A.-L. Paul, and National Institute of Health Grant R01-GM-40061 to R. J. Ferl. The authors would like to thank Beth Laughner for her development and maintenance of the *Arabidopsis* cultured cell line used in our laboratory. This is journal series number R-04919 from the Florida Agriculture Experiment Station.

References

1. Church, G. M. and Gilbert, W. (1984) Genomic sequencing. *Proc. Natl. Acad. Sci. USA* **81,** 1991–1995.
2. Ferl, R. J. and Laughner, B. H. (1989) *In vivo* detection of regulatory factor binding sites of *Arabidopsis thaliana Adh. Plant Mol. Biol.* **12,** 357–366.
3. Ferl, R. J. and Nick, H. S. (1987) *In vivo* detection of regulatory factor binding sites in the 5' flanking region of maize *Adh1. J. Biol. Chem.* **262,** 7947–7950.
4. Hornstra, I. K. and Yang, T. P. (1993) *In vivo* footprinting and genomic sequencing by ligation mediated PCR. *Anal. Biochem.* **213,** 179–193.
5. Maxam, A. M. and Gilbert, W. (1980) Sequencing end-labelled DNA with base-specific chemical cleavages. *Methods Enzymol.* **65,** 499–560.
6. McKendree, W. L., Paul, A.-L., DeLisle, A. J., and Ferl, R. J. (1990) *In vivo* and *in vitro* characterization of protein interactions with the G-box of the *Arabidopsis Adh* gene. *Plant Cell* **2,** 207–214.
7. Messing, J. (1983) New M13 vectors for cloning. *Methods Enzymol.* **101,** 20.
8. Paul, A.-L. and Ferl, R.J. (1991) *In vivo* footprinting reveals unique cis-elements and different modes of hypoxic induction in maize *Adh1* and *Adh2. Plant Cell* **3,** 159–168.

9. Sambrook, J., Fritsch, E. F., and Maniatis, T. (1989) *Molecular Cloning: A Laboratory Manual*, 2nd ed. Cold Spring Harbor Laboratory, Cold Spring Harbor, NY.
10. Vega-Palas, M. A. V. and Ferl, R. J. (1995) The *Arabidopsis* Adh gene exhibits diverse nucleosome arrangements within a small DNase I sensitive domain. *Plant Cell* **7,** 1923–1932.

Index

A

Ac/Ds,
  biological mutagens, 91
  genetic strategy, 316
  mutation isolation, 316
  transposase genes, 316, 317
  transposon tagging, 315–318
Activator/dissociation (*Ac/Ds*), *see Ac/Ds*
AFLP, DNA fingerprinting technique, 147, 170
AFLP amplification, 151–154
AFLP analysis, 307, 312
  polymorphism screens, 306, 312
AFLP-based strategy,
  analysis, 307, 312
  chromosome landing, 305, 306
  genomic DNA extraction, 306–308
  isolation and cloning, 307, 310, 312
  materials, 306, 307
  methods, 307–312
  recessive mutation marker identification, 306
  target locus identification, 308–310, 312
AFLP fingerprinting,
  materials, 149, 150
  methods, 150–152
  pattern, 160, 161
AFLP image analysis, software package, 170
AFLP markers,
  BAC clones, 311, 312
  bi-allelic, 171
  identification, 309
  isolation and cloning, 307, 310–313
  list, 310
  primer combinations, 162–167
  YAC clones, 311, 312
AFLP-PCR amplification, 310
AFLP protocol, schematic representation, 148
AFLP reaction products, gel analysis, 152
AFLP reactions, 149, 150, 170
AFLP technique,
  genomic DNA, 147
  procedure, 147–149

Agarose gel analysis, PCR reaction products, 205, 206
Agarose plugs, megabase DNA, 62
Agrobacterium binary transformation system, DNA transfer, 245
Agrobacterium-mediated transfer methods, development, 227, 228
Agrobacterium-mediated transformation
  infiltration method, 259, 260
  in planta method, 259, 260
  vacuum infiltration, 259, 260
Agrobacterium T-DNA transformation, problems, 245
Agrobacterium tumefaciens,
  leaf disc transformation, 245–247
  root transformation, 227–229
Algae, soil and plants, 10
Allelism tests, genetic analysis, 109
Anti-DIG antibody conjugate, incubation, 380, 382
Aphids, 20, 21
ARMS, mapping mutations, 183–188
ARMS markers,
  autoradiography, 184
  chromosomal position, 185–187
  cross selection, 190–193
  DNA fragments, 183
  evaluation, 189, 190, 194, 195
  hybridization, 189, 193, 194, 196
  LOD value, 197
  materials, 188–190
  methods, 190–195
Autoclaving, 29

B

BAC clones, AFLP markers, 311, 312
BACs, genomic libraries, 278
Bacterial artificial chromosomes (BACs), *see* BACs
Bacterial luciferases,
  historical development, 385, 386
  homogenates, 389, 390, 393

432

materials, 388, 389
methods, 389–392
reporter genes, 385–387, 388
Biolistic system, transient expression, 219
Biological mutagens, listed, 91
Bombardment-mediated transformation,
   foreign gene delivery, 219, 220
   β-glucuronidase (GUS) gene, 220–224
Botrytis, 24

C

Callus culture, 31
   materials, 32
   methods, 32, 33
   transfer, 33
   tubes, 33
CAPS markers,
   chromosome I, 176
   development, 175–177
   genetic map distances, 180
   limitations, 177
   materials, 177, 178
   methods, 178–180
   PCR detection, 175
   reactions, 178–181
CAPS method, illustrated, 174
cDNA cloning, 85
Cell suspension cultures,
   aseptic conditions, 29
   contamination, 30
   establishment, 27
   growth rate, 29
   hormones, 29
   incubator, 29
   initiation, 28, 29
   maintenance, 27, 29
   materials, 27, 28
   methods, 28, 29
   protoplasts, 39
   seed-derived calli, 29
Chemical mutagens, 100
Chi-square test, gene segregation, 113
Chloramphenicol acetyl transferase (CAT), 209
Chloroplast DNA isolation,
   banding pattern, 74–76
   laboratory procedure, 71, 72
   materials, 72, 73
   methods, 73, 74
Chloroplast isolation,

materials, 45
methods, 46, 47
procedures, 43, 44
Chloroplast preparation, 43, 47
   examination, 48
   leaf-stripping method, 47
   materials, 44, 45
   metabolic activities, 43
   methods, 45–47
Chloroplast protocols, 43
Chlorotic somatic sector, seed mutagenesis, 102
Chromogenic alkaline phosphatase (AP),
   reactions, 380, 382, 383
   signal detection, 377
Chromogenic reaction, AP antibody
   conjugates, 380, 382, 383
Chromosome examination, 122, 123
Chromosome landing,
   AFLP-based strategy, 305, 306
   chromosome walking, 305
   gene isolation, 306
Chromosome preparation,
   materials, 121–123
   methods, 121, 123, 124
Chromosome staining, 122, 124–126
Chromosome walking,
   chromosome landing, 305
   cloning genes, 277–279
   genetic mapping, 277
   genomic libraries, 278
   principles, 277
   RFLP marker, 277
   segregation analysis, 277
   YAC libraries, 278
Clearing, hybridization technique, methods, 357
Cleaved amplified polymorphic sequences
   (CAPS), genetic markers, 173–177
Clonal propagation, plant genes, 227
Cloning genes,
   chromosome walking, 277–279
   T-DNA tagged mutants, 339, 340
Codominant markers, RFLPs, 183
Coelenterate luciferases, reporter genes, 385,
   387, 388
Colony hybridization, yeast, 284, 285, 292, 293
Colorimetric GUS assay, mesophyll
   protoplasts, 211, 213, 215, 216
Cosegregation analysis,
   materials, 340, 341
   methods, 342, 343, 349

Index

Cosmid, subcloning YACs, 287–291, 295–301
Cosmid arms, preparation, 283, 284
Cytogenetic analysis,
    aneuploidy examination, 119
    chromosome/heredity correlation, 119
    chromosome number determination, 120, 121
    materials, 121–123
    methods, 121
    plant ploidy determination, 119

D

Digoxigenin probe synthesis, whole-mount plant hybridization, methods, 378, 381, 382
Direct gene transfer, root transformation, 227–229
Diseases, treatment and types, 20, 24, 25
DMS treatments,
    footprinting,
        materials, 418
        methods, 422, 426
DNA,
    dissociation, 324
    PEG, 215
    PEG-mediated protoplast transformation, 267, 268
    primers, 324
DNA amplification, YAC clones, 204, 205
DNA blot analysis, RFLP markers, 173
DNA extraction, 55
    large molecules, 61
DNA fragments, 61
    materials, 62–65
    methods, 66–68
    nuclei preparation, 64–66
    PFGE, 61, 65, 67, 68
    plant cultures, 62, 63
    plug preparation, 65, 66, 68, 69
    protoplast isolation, 63, 64, 66
    YACs, 61
    YAC clones, 203, 204, 207
DNA fingerprinting, AFLP technique, 147
DNA isolation, 55
    factors, 55, 56
    methods, 59
DNA polymorphisms, gene mapping, 137
DNA preparation,
    contamination, 58
    materials, 56
    methods, 56, 57, 150–152
    PCR analysis, 57, 59
    pellet, 59
    phenol/chloroform extraction, 58
DNA probes, labeling, 189, 192, 193
DNA stains, color print films, 125
DNA and yeast isolation, subcloning YACs, 282–284
Double mutant isolation, genetic analysis, 109, 110

E

Electrotransfer,
    footprinting,
        materials, 421
        methods, 425
EMS,
    mutagenesis, 98, 99, 101
    precautions, 97, 98
EMS mutagenesis, seed mutagenesis, methods, 98, 99, 101, 102
Enhancer-Inhibitor (*En-I*), *see En-I*
En-I system,
    T-DNA, 330
    transposon tagging, 329–331
En-I transposon tagging,
    DNA isolation, 334
    DNA probe isolation, 334, 335
    materials, 331
    methods, 331–335
    random, 331, 332, 335
    tagged mutant analysis, 333, 334, 336
    targeted, 332, 333, 335, 336
En transposase, CaMV 35S, 330, 331
Eukaryotic luciferases, reporter genes, 385, 387, 388
Expressed sequence tags (ESTs), 129

F

Fixation,
    hybridization technique,
        materials, 356
        methods, 356, 357, 367, 368
        sources, 121, 122
Fluorochrome stains,
    exposure, 125, 126
    nucleic acids, 125
Footprinting, 417, 418
    DMS treatments, 418, 426
    dimethyl sulfate treatment, 422, 426

electrotransfer, 421, 425
gene characterization requirements, 417
genomic DNA isolation, 418–420, 422, 423, 426
genomic DNA sequencing, 420, 423–425, 427, 428
hybridization, 421, 422, 425, 426
materials, 418–421
methods, 421–428
probe synthesis, 421, 422
sequencing, 420
schematic, 419
steps, 417, 418
Foreign genes, transient expression, 219, 220
Fragment ligation, plasmid rescue, 344, 350
Fungicide treatments, 26
Fungus fly, 21

G

Gel analysis,
    AFLP reaction products, 152, 154
    PCR reaction products, 205, 206
Gel electrophoresis, 150, 205
Gene cloning,
    chromosome walking, 277–279
    transposon insertion, 319–326
Gene expression analysis, hybridization, 353, 354
Gene fusion markers, list, 209
Gene isolation, chromosome landing, 306
Gene mapping, DNA level polymorphisms, 137
Gene segregation, Chi-square test, 113
Genetic analysis,
    allelism tests, 109
    crosses, 108, 111, 112
    desired results, 105
    epistatic relationships, 109, 110
    gene location, 106
    linkage, 110, 111
    mapping function, 107
    materials, 107, 108
    methods, 108–111
    molecular markers, 106
    quantification, 107
    segregation, 108, 109, 112, 113
Genetic maps,
    CAPS marker distances, 180
    chromosome walking, 277
    high-density, 157, 158
    RILs, 137–139

segregation analysis, 277
software packages, 107
vegetative propagation, 138
Genetic markers, CAPS, 173–177
Genetic strategy, Ac/Ds system, 316
Genetic transformation, plant genes, 227
Genetic variation,
    analysis, 105
    identification, 105, 106
Gene transformation protocol,
    materials, 229–233
    methods, 234, 236, 238
    sterilization and growth, 230, 234, 238
    tissue culture, 229, 234–237
    transformant analysis, 233, 236, 241, 242
    transformation procedure, 232–234, 235, 238, 240
Genomic DNA,
    AFLP technique, 147
    digestion separation and blotting, 188, 190, 191
Genomic DNA digestion, plasmid rescue, 344, 349, 350
Genomic DNA extraction, AFLP-based strategy, 306–308
Genomic DNA isolation,
    footprinting,
    materials, 418–420
    methods, 422, 426
Genomic DNA sequencing,
    footprinting,
    materials, 420
    methods, 423–425, 427, 428
Genomic lambda library isolation, plant flanking DNA isolation technique, 340, 348
Genomic libraries,
    BACs, 278
    chromosome walking, 278
    YACs, 278
Genomic regions, 145
β-glucuronidase (GUS), 209, 270, 273
β-glucuronidase (GUS) enzyme,
    histochemistry, 397–399
β-glucuronidase (GUS) gene,
    bombardment, 222, 224
    bombardment-mediated transformation, 220–224
    DNA coating, 222
    materials, 220, 223

Index

methods, 221–223
plant promoter activity, 397
preparation, 221–223
transient assay, 222–224
Gold antibody conjugates,
  signal detection, 380, 383
  silver amplification, 380, 383
Green fluorescence protein (GFP), 209, 268, 270, 273, 274
GUS, *see* β-glucuronidase
GUS activity, qualitative histochemical assay, 397
GUS assays,
  colorimetric, 211, 213, 215, 216
  histochemical, 210, 213, 215
GUS enzyme, X-gluc-based histochemical assay, 398
GUS enzyme histochemical assay,
  dehydration and embedding, 399–401, 404, 405
  enzyme reaction, 400, 403, 404, 406
  materials, 399, 400, 404
  methods, 400–404
  sectioning, 400, 402, 403, 405

H

Harvest,
  materials, 1, 2
  methods, 2, 3, 5–7
Heat protocol, DNA isolation, 59
Heterologous transposons, biological mutagens, 91
High-density genetic map,
  AFLP technology, 157, 158
  enzyme combination, 158
  linkage groups, 168
  materials, 158–160
  methods, 160, 168, 169
Histochemical GUS assay, mesophyll protoplasts, 210, 213, 215
Hybridization,
  autoradiography, 353
  footprinting,
    materials, 421
    methods, 425, 426
  gene expression analysis, 353, 354
Hybridization technique,
  materials, 354–356
  methods, 356–370

I

Incubator, cell suspension cultures, 29
Infiltration, hybridization technique, methods, 357
Infiltration medium, 260
Infiltration method, *Agrobacterium*-mediated transformation, 259, 260
Infiltration transformation,
  growth conditions, 262, 264
  infiltration, 262–265
  materials, 261, 262
  methods, 262–264
  preparation, 262, 264
  transformant screen, 263–265
Insertion mutagens,
  T-DNAs, 339
  transposons, 339
Inverse polymerase chain reaction (IPCR)
  plant flanking DNA isolation technique, 340, 348
  YAC end probes, 286, 287, 294, 295
Isolation procedure, mitochondria purification, 51

J

J-detector pooling scheme, 200

L

Lambda library, 348
Leaf disc transformation, *Agrobacterium tumefaciens*, 245–247
Leaf mesophyll protoplasts, 37
Leaf-stripping method, chloroplasts, 47
Leaf transformation protocol, 246, 247
  bacterial infection, 249, 251, 254, 255
  bacterial strains, 248, 253, 254
  bacterial strains and growth conditions, 251
  callus induction, 248, 250, 251, 254
  genetic analysis, 249, 252, 253, 256
  growth conditions, 248
  materials, 247–249
  methods, 249–253
  media and chemicals, 247–250, 253
  preparation, 248, 250, 253
  regeneration, 249, 252, 253, 255, 256
  washing, 249, 251–254
Light emitting marker gene system,
  bacterial luciferase, 388–394
  materials, 388, 389
  methods, 389–393
  renilla luciferase, 389, 392–394

Linkage analysis, 110–112
    genetic analysis, 110, 111, 113–116
Linkage map construction, genetic analysis, 110, 111, 113–116
Linkage maps, construction, 110, 111
LOD value,
    ARMS markers, 197
    linkage maps, 113–116
    QTL interval, 144, 145
Luciferase gene expression system, 385–388
Luciferase (LUC), 209, 268

M

Mapping,
    functions, 114–116
    methods, 183
    program types, 116
    recessive mutation strategies, 305, 312
Mapping mutations, ARMS, 183–188, 191, 192
Marker genes, transposition monitor, 316
Marker gene system, development, 385
Marker technology, strategies, 137
Mesophyll protoplasts,
    colorimetric GUS assay, 211, 213, 215, 216
    extraction, 211, 213
    histochemical GUS assay, 210, 213, 215
    materials, 210, 211
    methods, 211–214
    preparation, 210–212, 215
    transfection, 210
    transformation, 212, 213, 215
Messenger RNA, whole-mount plant hybridization, 373
Methanesulfonic acid ethyl ester (EMS), 97
Microscopy,
    hybridization technique,
        materials, 356
        methods, 366, 367, 370
        whole-mount plant hybridization, methods, 381, 383
Mitochondria agarose-embedded, PFGE analysis, 82–84
Mitochondria isolation, 80, 81
Mitochondrial DNA extraction, 81, 82
Mitochondrial DNA purification,
    contamination, 79
    green tissue, 79
    materials, 80
    methods, 80–83
    PFGE analysis, 79
Mitochondria purification,
    applications, 49
    cell debris, 52
    isolation procedure, 51
    materials, 50
    methods, 51, 52
    pellet, 52, 53
    percoll step procedure, 49
    storage, 53
Monogenic markers, 138
Morphogenesis, phytohormones, 245, 246
Mushroom fly, 21
Mutagenesis experiments, 99, 100
Mutation isolation, *Ac/Ds* system, 316

N

NaCl gradients, 290
NAOH extraction method, DNA isolation, 59
Neomycin phosphotransferase (NPTII), 209
Nopaline synthase (NOS), 209

P

PCR analysis, DNA preparation, 57, 59
PCR identification, T-DNA mutants, 129–133
PCR product analysis, 133
PCR reactions, YAC clones, 199
PEG, DNA, 215
PEG-mediated protoplast transformation,
    DNA, 267, 268
    gene assays, 270–273
    GFP reporter, 270, 273, 274
    GUS reporter, 270, 273
    isolation, 269
    materials, 268, 269
    media and solutions, 268, 269
    methods, 269–273
    regeneration, 272, 273, 275
    transformant selection, 272, 274
Percoll density gradient centrifugation, 51, 52
Pest control, prevention, 19
Pesticide treatments, 26
Pest management, infection, 19, 20
Pests, identification and treatment, 20–23
Petri plate screens, categories, 13, 16, 17
PFGE analysis,
    mitochondria, agarose-embedded, 82, 83
    mitochondrial DNA purification, 79

Index

Phenol/chloroform extraction, DNA
    preparation, 58
Phenol extractions, RNA preparation, 89
Photography, whole-mount plant
    hybridization, methods, 381, 383
Photomicroscopy, 124–126
    Photon amplification, procedure, 391
Phytohormones, morphogenesis, 245, 246
Plant flanking DNA,
    isolation techniques,
        genomic lambda library, 340, 348
        IPCR, 340, 348
        plasmid rescue, 340, 341, 343–346, 348–350
        TAIL-PCR, 340–342, 346–350
Plant flanking DNA isolation,
    cosegregation analysis, 340–343, 349
    materials, 340–342
    methods, 342–348
    plasmid rescue, 341, 343–346, 349, 350
    TAIL-PCR, 341, 342, 346–350
Plant growth,
    environmental factors, 1, 11
    environmental settings, 1, 11
    materials, 1, 2
    methods, 2–10
    proper conditions, 3, 5, 11
Plant regeneration, protoplast, derived calli, 39, 40
Plasmid rescue,
    clone analysis, 345, 346
    fragment ligation, 344, 350
    genomic DNA digestion, 344, 349, 350
    materials, 341
    methods, 343, 346, 349, 350
    transformation, 345
Ploidy analysis, pollen morphology, 121
Ploidy determination, 123
Pollen examination, 123
Pollen morphology, ploidy analysis, 121
Polyethylene glycol (PEG), *see* PEG
Polymorphism screens, AFLP analysis, 306, 312
Polytron homogenization, 44
Polytron homogenizer, chloroplast
    preparation, 45, 46
Pooling strategy,
    materials, 132
    methods, 132, 133
    PCR reactions, 135
    T-DNA, 129–131

Powdery mildew, 24
Primers, DNA, 324
Probe synthesis,
    footprinting, materials, 421
    hybridization technique,
        materials, 354, 355, 367
        methods, 358–361, 367–369
    whole-mount plant hybridization,
        materials, 376
        methods, 378, 381, 382
Protoplast cultures, 37, 40
    survival rates, 41
Protoplast-derived calli, plant regeneration, 39, 40
Protoplast isolation, 35, 40
    culture tubes, 41
    depolymerization, 41
    DNA extraction, 63, 64
    materials, 35, 36
    mesophyll tissue, 35
    methods, 37–40
    protocols, 35
Protoplast medium (PM), 36
Protoplasts,
    cell suspension cultures, 39
    transient gene expression, 209, 210
Pulsed-field agarose gel electrophoresis
    (PFGE), DNA analysis, 61, 65, 67, 68
Pulsed-field gel-electrophoresis
    YAC clone analysis, 285, 293, 294
    yeast colony screen, 281

Q

QTL, RILs, 138
QTL analysis, 143
QTL detection methods, 143, 144
QTL interval, LOD value, 144, 145
QTL locations, 145
Qualitative markers, 138
Quantitative trait loci (QTL), *see* QTL

R

RACE polymerase chain reaction (PCR), 85
Recombinant inbred lines (RILs),
    genetic mapping, 137–139
    harvesting, 141
    markers, 141, 142
    materials, 139

methods, 139, 140
recombination rates, 142, 143
software analyzation packages, 142–144
Recombinant screens, T-DNA crosses, 305
Red spider mite, 21
Regeneration, 31
materials, 32
methods, 32, 33
Renilla luciferase,
homogenates, 392, 394
materials, 389
methods, 392, 393, 394
Replica plating, yeast colony screen, 284, 291, 292
Reporter genes,
list, 209
prokaryotes and eukaryotes, 397
spatial information, 409, 410
Reporter gene system, types, 385
RFLP mapping, genetic location, 173
RFLP marker, chromosome walking, 277
RFLP markers,
codominant, 183
DNA blot analysis, 173
hybridization probe, 173
YAC clones, 279
RNA blot hybridizations, 85
RNA hybridization,
autoradiography, 411–414
high resolution spatial information, 409, 410
materials, 410, 411, 413
methods, 411–414
RNA isolation, RNase contamination, 88
RNA preparation,
gene expression, 85
lysis and denaturation, 85
materials, 85, 86, 88
methods, 86–89
phenol extractions, 89
procedure, 85
RNase contamination, RNA isolation, 88
RNA sequencing, 85
Root cultures, initiation, 28–30
Root-derived protoplasts, 38, 40, 41
Root transformation,
Agrobacterium tumefaciens, 227–229
composition, 231
ROSE method, DNA isolation, 59

S

Salt gradients, 290

Sciarid fly, 21
Sectioning,
GUS enzyme histochemical assay
illustrated steps, 402
materials, 400
methods, 402, 403, 405
hybridization technique, methods, 358
Seed cleaning, 7, 11
Seed-derived Calli, initiated cell suspensions, 29
Seed drying, 7
Seed fertilization,
application, 5
preparation, 2
types, 11
Seed moisture, content determination, 7, 8
Seed mutagenesis,
availability, 91, 92
chemical, 91
chlorotic somatic sector, 102
choice, 92, 93
dose, 95, 96
EMS mutagenesis, 98, 99, 101, 102
harvesting strategies, 94, 95
materials, 96, 97
methods, 97–99
population size, 93, 94
rates, 100
safety, 97, 98
Seed packaging, storage, 8
Seed planting, methods, 10
Seed preservation,
materials, 2
methods, 3, 4, 6–10
Seed production,
bulk, 5, 6
commercial, 6
Seed quality, methods, 3, 4, 6–8
Seed storage, 8, 9
Seed testing, procedure, 9, 10
Seed threshing, 7, 11
Seed viability, 9, 10
Segregation analysis,
chromosome walking, 277
genetic mapping, 277
genetic method, 108, 109
Silver amplification,
gold antibody conjugates, 380, 383
signal detection, 377
Spatial information, detection techniques, 409, 410
Spermidine, 398

Spm-dSpm system, transposon tagging, 329, 330
Staining slides,
   hybridization technique,
      materials, 356
      methods, 365, 366, 370
Sterile techniques,
   harvesting, 16
   materials, 14
   methods, 14, 15
   mutant screens, 13, 16
   plating, 14, 15, 17
   storage, 15
   transplantation, 15
Subcloning YACs, DNA and yeast isolation, 282–284
Supressor mutator (SPM), biological mutagens, 91

T

TAIL-PCR,
   amplification, 347, 348
   confirmation, 348
   DNA isolation, 346
   materials, 341, 342, 348, 349
   methods, 346–348, 350
   primary cycles, 347
   primer design, 346, 347
   secondary cycles, 347
   tertiary cycles, 347, 348
T-DNA,
   biological mutagens, 91
   En-I system, 330
   identification, 183
   pooling strategy, 129
T-DNA crosses, recombinant screens, 305
T-DNA mutants, PCR identification, 129–135
T-DNAs,
   advantages, 339
   insertion mutagens, 339
T-DNA structure, 345
T-DNA tagged mutants, cloning genes, 339–340
Template preparation, 150, 151
Thrips, 22, 23
Transient expression,
   biolistic system, 219
   foreign genes, 219, 220
Transposase genes,
   Ac/Ds, 316, 317
   transposon tagging, 316, 317
Transposed Ds identification

gene cloning, 319, 320, 322–325, 326
   materials, 318–320
   methods, 320–325
   mutant phenotype identification, 319
   seedling selection, 318–321, 325, 326
   testing, 319, 321, 322, 326
Transposon-induced mutations, identification, 183
Transposon insertion,
   gene cloning, 319, 320, 321–326
   testing, 319, 321, 322
Transposons,
   advantages, 339
   insertion mutagens, 339
Transposon tagging,
   Ac/Ds, 315–318, 329
   En-I system, 329–331
   particular gene, 318
   plant gene isolation, 315
   Spm-dSpm system, 329
   transposase genes, 316, 317

V

Vacuum infiltration, *Agrobacterium*-mediated transformation, 259, 260
Vibrosections,
   whole-mount plant hybridization,
      materials, 376
      methods, 378, 381

W

Whitefly, 23
Whole-mount plant hybridization, 373, 374
   chromogenic reaction, 380, 382, 383
   digoxigenin (DIG) labeled RNA probes, 373
   digoxigenin probe synthesis, 378, 381, 382
   incubation, 380, 382
   ISH, 373
   materials, 374–377
   messenger RNA, 373
   methods, 377–383
   microscopy and photography, 381, 383
   plant powder, 377
   plant powder preparation, 379, 380
   plant tissue preparation, 377, 378, 381
   probe synthesis, 376
   signal detection, 377
   silver amplification, 380, 383
   tissue fixation, 378, 381
   vibrosections, 376, 378, 381
WISH, 373, 374

X

X-gluc-based histochemical assay,
   GUS enzyme, 398
   reactions, 405

Y

YAC clones,
   AFLP markers, 311, 312
   DNA amplification, 204, 205
   DNA extraction, 203, 204, 207
   growth, 207
   materials, 201, 202
   methods, 203–206
   PCR based maps, 204–206
   PCR reactions, 199
   propagation, 203
   pulsed-field gel electrophoresis, 279
   restriction map, 280
   RFLP markers, 279
YAC end probes, preparation, 286, 287, 294, 295
YAC inserts, yeast colony screen, 281, 282
YAC libraries, 61, 62
   chromosome walking, 278
YACs,
   cloned sequence maps, 199–201
   DNA analysis, 61, 62
Yeast, colony hybridization, 284, 285, 292, 293
Yeast artificial chromosomes (YACs), *see* YACs
Yeast colony screen,
   gel-electrophoresis, 281
   materials, 279–284
   media and reagents, 279, 280
   subcloning YACs, 282–284
YAC insert end probes, 281, 282
   colony hybridization, 284, 285, 292, 293
   end probe preparation, 286, 287, 294, 295
   gel electrophoresis analysis, 285, 293, 294
   methods, 284–291
   replica plating, 284, 291, 292
   subcloning YACs, 287–291, 295–301
Yeast and DNA isolation, subcloning YACs, 282–284